Lecture Notes in Computer Science　　10388

Commenced Publication in 1973
Founding and Former Series Editors:
Gerhard Goos, Juris Hartmanis, and Jan van Leeuwen

FoLLI Publications on Logic, Language and Information

Subline of Lectures Notes in Computer Science

More information about this series at http://www.springer.com/series/7407

Juliette Kennedy · Ruy J.G.B. de Queiroz (Eds.)

Logic, Language, Information, and Computation

24th International Workshop, WoLLIC 2017
London, UK, July 18–21, 2017
Proceedings

 Springer

Editors

Juliette Kennedy
Department of Mathematics and Statistics
University of Helsinki
Helsinki
Finland

Ruy J.G.B. de Queiroz
Centro de Informática
Recife, Pernambuco
Brazil

ISSN 0302-9743   ISSN 1611-3349 (electronic)
Lecture Notes in Computer Science
ISBN 978-3-662-55385-5   ISBN 978-3-662-55386-2 (eBook)
DOI 10.1007/978-3-662-55386-2

Library of Congress Control Number: 2017944357

LNCS Sublibrary: SL1 – Theoretical Computer Science and General Issues

Printed on acid-free paper

This Springer imprint is published by Springer Nature
The registered company is Springer-Verlag GmbH Germany
The registered company address is: Heidelberger Platz 3, 14197 Berlin, Germany

Preface

This volume contains the papers presented at the 24th Workshop on Logic, Language, Information and Computation (WoLLIC 2017) held during July 18–21, 2017, at the Department of Computer Science, University College London, London, UK. The WoLLIC series of workshops started in 1994 with the aim of fostering interdisciplinary research in pure and applied logic. The idea is to have a forum which is large enough in the number of possible interactions between logic and the sciences related to information and computation, and yet is small enough to allow for concrete and useful interaction among participants.

There were 61 submissions. Each submission was reviewed by at least three Program Committee members. The committee decided to accept 28 papers. The program included eight invited lectures by Hazel Brickhill (Bristol) (University of Bristol), Michael Detlefsen (University of Notre Dame), Alexander Kurz (University of Leicester), Frederike Moltmann (New York University), David Pym (University College London), Nicole Schweikardt (Humboldt Universitat), Fan Yang (Delft University), and Boris Zilber (University of Oxford). There were also four tutorials given by Michael Detlefsen, Alexander Kurz, Frederike Moltmann, and Nicole Schweikardt.

As a tribute to a recent project focusing on the cross-cultural connections that are made through mathematics and the impact that Navajo Math Circles can have on a community, there was also a screening of Csicsery's Navajo Math Circles (2016), a one-hour film, documenting the process of a two-year period in which hundreds of Navajo children in recent years have found themselves at the center of a lively collaboration with mathematicians from around the world.

We would very much like to thank all Program Committee members and external reviewers for the work they put into reviewing the submissions. The help provided by the EasyChair system created by Andrei Voronkov is gratefully acknowledged. Finally, we would like to acknowledge the generous financial support by the University College London's Department of Computer Science, and the scientific sponsorship of the following organizations: Interest Group in Pure and Applied Logics (IGPL), The Association for Logic, Language and Information (FoLLI), Association for Symbolic Logic (ASL), European Association for Theoretical Computer Science (EATCS), European Association for Computer Science Logic (EACSL), Sociedade Brasileira de Computação (SBC) and Sociedade Brasileira de Lógica (SBL).

July 2017

Juliette Kennedy
Ruy de Queiroz

Organization

Program Committee

Matthias Baaz	TU Wien, Austria
John Baldwin	University of Illinois at Chicago, USA
Dana Bartozová	Carnegie Mellon University, USA
Agata Ciabattoni	TU Wien, Austria
Ruy De Queiroz	Centro de Informatica, Universidade Federal de Pernambuco, Brazil
Walter Dean	University of Warwick, UK
Erich Grädel	RWTH Aachen University, Germany
Volker Halbach	New College, Oxford University, UK
Juliette Kennedy	Helsinki University, Finland
Dexter Kozen	Cornell University, USA
Johann Makowsky	Technion, Israel Institute of Technology, Haifa
Larry Moss	Indiana University, USA
Alessandra Palmigiano	Technical University of Delft, The Netherlands
Mehrnoosh Sadrzadeh	Queen Mary University of London, UK
Sonja Smets	University of Amsterdam, The Netherlands
Asger Törnquist	University of Copenhagen, Denmark
Andrés Villaveces	Universidad Nacional de Colombia, Colombia
Philip Welch	University of Bristol, UK

Additional Reviewers

Aguilera, Juan	Filmus, Yuval	Leach-Krouse, Graham
Baldi, Paolo	Fiorentini, Camillo	Liang, Fei
Balzer, Stephanie	Frittella, Sabine	Liang, Fey
Bertrand, Nathalie	Ghari, Meghdad	Lolic, Anela
Berwanger, Dietmar	Ghica, Dan	Löding, Christof
Bilkova, Marta	Greco, Giuseppe	Martinez, Maricarmen
Blot, Valentin	Grudzinska, Justyna	Mazza, Damiano
Bollig, Benedikt	Hella, Lauri	Picollo, Lavinia
Carlier, Pierre	Henk, Paula	Pietarinen, Ahti-Veikko
Conradie, Willem	Hoelzel, Matthias	Rafiee Rad, Soroush
Demey, Lorenz	Kishida, Kohei	Ramanayake, Revantha
Eguchi, Naohi	Kuznets, Roman	Rittberg, Colin

Sack, Joshua
Salehi, Saeed
Schindler, Thomas
Sequoiah-Grayson,
    Sebastian

Shi, Chenwei
Tzimoulis, Apostolos
Vaananen, Jouko
Velázquez-Quesada,
    Fernando R.

Visser, Albert
Yang, Fan
Zamansky, Anna

Contents

Graph Turing Machines

Nathanael L. Ackerman[1] and Cameron E. Freer[2]([⊠])

[1] Department of Mathematics, Harvard University, Cambridge, MA 02138, USA
nate@math.harvard.edu
[2] Remine, Falls Church, VA 22043, USA
cameron@remine.com

Abstract. We consider *graph Turing machines*, a model of parallel computation on a graph, which provides a natural generalization of several standard computational models, including ordinary Turing machines and cellular automata. In this extended abstract, we give bounds on the computational strength of functions that graph Turing machines can compute. We also begin the study of the relationship between the computational power of a graph Turing machine and structural properties of its underlying graph.

1 Introduction

When studying large networks, it is important to understand what sorts of computations can be performed in a distributed way on a given network. In particular, it is natural to consider the setting where each node acts independently in parallel, and where the network is specified separately from the computation to be performed. In order to study networks whose size is considerably larger than can be held in memory by the computational unit at any single node, it is often useful to model the network as an infinite graph. (For a discussion of modeling large networks via infinite graphs, see, e.g., [14].)

We define a notion of *graph Turing machine* that is meant to capture this setting. This notion generalizes several other well-known models of computation, including ordinary Turing machines, cellular automata, and parallel graph dynamical systems. Each of these models, in turn, occurs straightforwardly as a special case of a graph Turing machine, suggesting that graph Turing machines capture a natural concept of parallel computation on graphs.

A graph Turing machine (henceforth abbreviated as "graph machine") performs computation on a vertex-labeled edge-colored directed multigraph satisfying certain properties. This notion of computation is designed to capture the idea that in each timestep, every vertex performs a limited amount of computation (in parallel, independently of the other vertices), and can only distinguish vertices connected to it when they are connected by different colors of edges.

In this paper we study the functions that can be computed using graph machines, which we call *graph computable* functions. As we will see, this parallel notion of computation will yield significantly greater computational strength

© Springer-Verlag GmbH Germany 2017
J. Kennedy and R.J.G.B. de Queiroz (Eds.): WoLLIC 2017, LNCS 10388, pp. 1–13, 2017.
DOI: 10.1007/978-3-662-55386-2_1

than ordinary Turing machines. We will see that the computational strength of graph machines is exactly that of $\mathbf{0}^{(\omega)}$, the Turing degree of true arithmetic (thereby providing another natural construction of this degree). We also begin to examine the relationship between various properties of the underlying graph (e.g., finiteness of degree) and the computational strength of the resulting graph machines.

In this extended abstract, we state the main results and provide proofs or proof sketches of several of these results. For detailed proofs and other related results, see the full version at https://arxiv.org/abs/1703.09406.

1.1   Main Results and Overview of the Paper

We begin by introducing the notions of colored graphs, graph machines, and graph computability in Sect. 2.

Our main results fall into two classes: bounds on the computational power of arbitrary computable graph machines, and bounds among machines with an underlying graph every vertex of which has finite degree (in which case we say the graph is of *finite degree*).

Theorem 3.6 states that every graph computable function is Turing reducible to $\mathbf{0}^{(\omega)}$. In the other direction, we show in Theorem 3.10 that this bound is attained by a single graph Turing machine.

Sitting below $\mathbf{0}^{(\omega)}$ are the arithmetical Turing degrees, i.e., those less than $\mathbf{0}^{(n)}$ for some $n \in \mathbb{N}$, where $\mathbf{0}^{(n)}$ denotes the n-fold iterate of the halting problem. We show in Corollary 3.9 that every arithmetical Turing degree contains a function that is graph Turing computable. (It remains open whether every degree below $\mathbf{0}^{(\omega)}$ can be achieved.)

We next show in Corollary 4.9 that functions determined by graph machines with underlying graph of finite degree are reducible to the halting problem, $\mathbf{0}'$. Further, we show in Corollary 4.10 that if we restrict to graph machines where every vertex has the same (finite) degree, then the resulting graph computable function is computable by an ordinary Turing machine.

We also show in Theorem 4.11 that every Turing degree below $\mathbf{0}'$ is the degree of some graph computable function with underlying graph of finite degree.

In Sect. 5, we examine how several other models of computation can be viewed as special cases of graph machines, including ordinary Turing machines, cellular automata, and parallel graph dynamical systems. Note that there have been many other attempts (which we do not discuss at length here) to extend Turing machines to operate on graphs, including [3], [12, pp. 462–463], [13,15]—the first of which calls its different notion a "graph Turing machine" as well.

1.2   Notation

If $f\colon A \to \prod_{i \leq n} B_i$ and $k \leq n \in \mathbb{N}$, then we let $f_{[k]}\colon A \to B_k$ be the composition of f with the projection map onto the k'th coordinate.

Fix an enumeration of computable partial functions, and for $e \in \mathbb{N}$, let $\{e\}$ be the e'th such function in this list. If X and Y are sets with $0 \in Y$, let

$Y^{<X} = \{\eta \colon X \to Y \; : \; |\{a \; : \; \eta(a) \neq 0\}| < \omega\}$, i.e., the collection of functions from X to Y for which all but finitely many inputs yield 0. (Note that by this notation we do *not* mean partial functions from X to Y supported on fewer than $|X|$-many elements.) For a set X, let $\mathfrak{P}(X)$ denote the collection of subsets of X and let $\mathfrak{P}_{<\omega}(X)$ denote the collection of finite subsets of X. Note that the map which takes a subset of X to its characteristic function is a bijection between $\mathfrak{P}_{<\omega}(X)$ and $\{0,1\}^{<X}$.

When working with computable graphs, sometimes the underlying set of the graph will be a finite coproduct of finite sets and $\mathbb{N}^k$ for $k \in \mathbb{N}$. The standard notion of computability for $\mathbb{N}$ transfers naturally to such settings, making implicit use of the computable bijections between $\mathbb{N}^k$ and $\mathbb{N}$, and between $\coprod_{i \leq k} \mathbb{N}$ and $\mathbb{N}$, for $k \in \mathbb{N}$. We will sometimes say *computable set* to refer to some computable subset (with respect to these bijections) of such a finite coproduct X, and *computable function* to refer to a computable function having domain and codomain of that form or of the form $F^{<X}$ for some finite set F.

For sets $X, Y \subseteq \mathbb{N}$, we write $X \leq_{\mathrm{T}} Y$ when X is Turing reducible to Y (and similarly for functions and other computably presented countable objects). We write $X \equiv_{\mathrm{T}} Y$ when $X \leq_{\mathrm{T}} Y$ and $Y \leq_{\mathrm{T}} X$. For more details on results and notation in computability theory, see [16].

2   Graph Computing

We now define graph Turing machines and graph computable functions. Note that these definitions differ from those in the full version of this paper, as here we require the sets of labels and colors to be finite. This simplifies the presentation, while losing little generality (since all of our constructions produce graphs with this property).

Definition 2.1. *A* **colored graph** *is a tuple $\mathcal{G}$ of the form $(G, (L, V), (C, E))$ where*

- *G is a set, called the* **underlying set** *or the* **set of vertices**,
- *L is a finite set, called the* **set of labels**, *and $V \colon G \to L$ is called the* **labeling function**, *and*
- *C is a finite set, called the* **set of colors**, *and $E \colon G \times G \to \mathfrak{P}(C)$ is called the* **edge coloring**.

A **computable colored graph** *is a colored graph such that G is a computable set and V and E are computable functions.*

The intuition is that a colored graph is an edge-colored directed multigraph where each vertex is assigned a label, and such that among the edges from one vertex to another, there is at most one of each color. Eventually, we will allow each vertex to do some fixed finite amount of computation, and we will want vertices with the same label to perform the same computations.

For the rest of the paper by a *graph* we will always mean a colored graph, and will generally write the symbol $\mathcal{G}$ to refer to graphs.

Let $\mathcal{G}$ be a graph with underlying set G, and suppose $A \subseteq G$. Define $\mathcal{G}|_A$ to be the graph with underlying set A having the same set of labels and set of colors as $\mathcal{G}$, such that the labeling function and edge coloring function of $\mathcal{G}|_A$ are the respective restrictions to A.

Definition 2.2. *A* **graph Turing machine,** *or simply* **graph machine,** *is a tuple* $\mathfrak{M} = (\mathcal{G}, (\mathfrak{A}, \{0, 1\}), (S, s), T)$ *where the following hold.*

- $\mathcal{G} = (G, (L, V), (C, E))$ *is a graph, called the* **underlying graph.** *We will speak of the components of the underlying graph as if they were components of the graph machine itself. For example, we will call G the* underlying set *of $\mathfrak{M}$ as well as of $\mathcal{G}$.*
- $\mathfrak{A}$ *is a finite set, called the* **alphabet,** *having distinguished symbols* 0 *and* 1.
- S *is a finite set, called the* **collection of states.**
- s *is a distinguished state, called the* **initial state.**
- $T \colon L \times \mathfrak{P}(C) \times \mathfrak{A} \times S \to \mathfrak{P}(C) \times \mathfrak{A} \times S$ *is a function, called the* **lookup table,** *such that* $T(\ell, \emptyset, 0, s) = (\emptyset, 0, s)$ *for all* $\ell \in L$, *i.e., if any vertex is in the initial state, currently displays* 0, *and has received no pulses, then that vertex doesn't do anything in the next step. This lookup table can be thought of as specifying a* transition function.

A $\mathcal{G}$**-Turing machine** *(or simply a* $\mathcal{G}$*-machine) is a graph machine with underlying graph $\mathcal{G}$. A* **computable graph machine** *is a graph machine whose underlying graph is computable.*

If A is a subset of the underlying set of $\mathfrak{M}$, then $\mathfrak{M}|_A$ is the graph machine with underlying graph $\mathcal{G}|_A$ having the same alphabet, states, and lookup table as $\mathfrak{M}$.

The intuition is that a graph machine should consist of a graph where at each timestep, every vertex is assigned a state and an element of the alphabet, which it displays. To figure out how these assignments are updated over time, we apply the transition function determined by the lookup table which tells us, given the label of a vertex, its current state, and its currently displayed symbol, along with the colored *pulses* the vertex has most recently received, what state to set the vertex to, what symbol to display next, and what colored pulses to send to its neighbors.

For the rest of the paper, $\mathfrak{M} = (\mathcal{G}, (\mathfrak{A}, \{0, 1\}), (S, s), T)$ will denote a *computable* graph machine whose underlying (computable) graph is $\mathcal{G} = (G, (L, V), (C, E))$.

Definition 2.3. *A* **configuration** *of $\mathfrak{M}$ is a function $f \colon G \to \mathfrak{P}(C) \times \mathfrak{A} \times S$. A configuration f is a* **bounded configuration** *when* $|\{v \in G : f_{[2]}(v) \neq 0\}|$ *is finite. A bounded configuration f is a* **starting configuration** *when further* $f_{[1]}(v) = \emptyset$ *and* $f_{[3]}(v) = s$ *for all* $v \in G$.

In other words, a starting configuration is an assignment in which all vertices are in the initial state s, no pulses have been sent, and only finitely many vertices display a non-zero symbol.

Note that if A is a subset of the underlying set of $\mathfrak{M}$ and f is a starting configuration for $\mathfrak{M}$, then $f|_A$ is a starting configuration for $\mathfrak{M}|_A$.

Definition 2.4. *Given a configuration f for $\mathfrak{M}$, the **run** of $\mathfrak{M}$ on f is the function $\langle \mathfrak{M}, f \rangle \colon G \times \mathbb{N} \to \mathfrak{P}(C) \times \mathfrak{A} \times S$ satisfying, for all $v \in G$,*

- $\langle \mathfrak{M}, f \rangle(v, 0) = f(v)$ *and*
- $\langle \mathfrak{M}, f \rangle(v, n+1) = T(V(v), X, z, t)$ *for all $n \in \mathbb{N}$, where*
  - *$X = \bigcup_{w \in G}\left(E(w, v) \cap \langle \mathfrak{M}, f \rangle_{[1]}(w, n)\right)$,*
  - *$z = \langle \mathfrak{M}, f \rangle_{[2]}(v, n)$, and*
  - *$t = \langle \mathfrak{M}, f \rangle_{[3]}(v, n)$.*

*We say that a run **halts at stage** n if $\langle \mathfrak{M}, f \rangle(v, n) = \langle \mathfrak{M}, f \rangle(v, n+1)$ for all $v \in G$.*

A run of a graph machine is the function which takes a configuration for the graph machine and a natural number n, and returns the result of letting the graph machine process the configuration for n-many timesteps.

The following lemma is immediate from Definition 2.4.

Lemma 2.5. *Suppose f is a configuration for $\mathfrak{M}$, and for each $n \in \mathbb{N}$, define $f_n := \langle \mathfrak{M}, f \rangle(\,\cdot\,, n)$. Then for all $n, m \in \mathbb{N}$ and $v \in G$, the function f_n is a configuration for $\mathfrak{M}$ such that $f_{n+m}(v) = \langle \mathfrak{M}, f_n \rangle(v, m) = \langle \mathfrak{M}, f \rangle(v, n+m)$.* $\square$

We now describe how a graph machine defines a function.

Definition 2.6. *For $x \in \mathfrak{A}^{<G}$, let $\widehat{x}$ be the configuration such that $\widehat{x}(v) = (\emptyset, x(v), s)$ for all $v \in G$. Let $\{\mathfrak{M}\} \colon \mathfrak{A}^{<G} \to \mathfrak{A}^G$ be the partial function such that*

- *$\{\mathfrak{M}\}(x)\!\uparrow$, i.e., is undefined, if the run $\langle \mathfrak{M}, \widehat{x} \rangle$ does not halt, and*
- *$\{\mathfrak{M}\}(x) = y$ if $\langle T, \widehat{x} \rangle$ halts at stage n and $y(v) = \langle \mathfrak{M}, \widehat{x} \rangle_{[2]}(v, n)$ for all $v \in G$.*

Note that $\{\mathfrak{M}\}(x)$ is well defined as $\widehat{x}$ is always a starting configuration for $\mathfrak{M}$.

*The graph machine $\mathfrak{M}$ is **total bounded** if $\{\mathfrak{M}\}(\widehat{x}) \in \mathfrak{A}^{<G}$ for all $x \in \mathfrak{A}^{<G}$; in particular, the partial function $\{\mathfrak{M}\}$ is total. When $\mathfrak{M}$ is total bounded, we will sometimes write $\{\mathfrak{M}\} \colon \mathfrak{A}^{<G} \to \mathfrak{A}^{<G}$.*

While in general, the output of $\{\mathfrak{M}\}(x)$ might have infinitely many non-zero elements, for purposes of considering which Turing degrees are graph computable, we will mainly be interested in the case of total bounded machines.

When defining a function using a graph machine, it will often be convenient to have extra vertices whose labels don't affect the function being defined, but whose presence allows for a simpler definition. These extra vertices can be thought as "scratch paper" and play the role of extra tapes (beyond the main input/output tape) in a multi-tape Turing machine. We now make this precise.

Definition 2.7. *Let X be an infinite computable subset of G. A map $\zeta \colon \mathfrak{A}^{<X} \to \mathfrak{A}^X$ is $\langle \mathcal{G}, X \rangle$-**computable** via $\mathfrak{M}$ if*

(a) $\{\mathfrak{M}\}$ is total,
(b) for $x, y \in \mathfrak{A}^{<G}$, if $x|_X = y|_X$ then $\{\mathfrak{M}\}(x) = \{\mathfrak{M}\}(y)$, and
(c) for all $x \in \mathfrak{A}^{<G}$, for all $v \in G \setminus X$, we have $\{\mathfrak{M}\}(x)(v) = 0$, i.e., when $\{\mathfrak{M}\}(x)$ halts, v displays 0, and
(d) for all $x \in \mathfrak{A}^{<G}$, we have $\{\mathfrak{M}\}(x)|_X = \zeta(x|_X)$.

A function is $\mathcal{G}$-**computable via** $\mathfrak{M}$ if it is $\langle \mathcal{G}, X \rangle$-computable via $\mathfrak{M}$ for some infinite computable $X \subseteq G$. A function is $\mathcal{G}$-**computable** if it is $\mathcal{G}$-computable via $\mathfrak{M}^\circ$ for some computable $\mathcal{G}$-machine $\mathfrak{M}^\circ$. A function is **graph Turing computable**, or simply **graph computable**, when it is $\mathcal{G}^\circ$-computable for some computable graph $\mathcal{G}^\circ$.

The following easy lemma captures the sense in which functions that are $\langle \mathcal{G}, X \rangle$-computable via $\mathfrak{M}$ are determined by their restrictions to X.

Lemma 2.8. *Let X be an infinite computable subset of $\mathcal{G}$. There is at most one function $\zeta \colon \mathfrak{A}^{<X} \to \mathfrak{A}^X$ that is $\langle \mathcal{G}, X \rangle$-computable via $\mathfrak{M}$, and it must be Turing equivalent to $\{\mathfrak{M}\}$.* □

3   Arbitrary Graphs

In this section, we consider the possible Turing degrees of total graph computable functions. We begin with a bound for finite graphs.

Lemma 3.1. *Suppose G is finite. Let $\mathbf{h}$ be the map which takes a configuration f for $\mathfrak{M}$ and returns $n \in \mathbb{N}$ if $\langle \mathfrak{M}, f \rangle$ halts at stage n (and not earlier), and returns ∞ if $\langle \mathfrak{M}, f \rangle$ doesn't halt. Then $\langle \mathfrak{M}, f \rangle$ is computable and $\mathbf{h}$ is computable.*

Proof. Because G is finite, $\langle \mathfrak{M}, f \rangle$ is computable. Further, there are only finitely many configurations of $\mathfrak{M}$. Hence there must be some $n, k \in \mathbb{N}$ such that for all vertices v in the underlying set of $\mathfrak{M}$, we have $\langle \mathfrak{M}, f \rangle(v, n) = \langle \mathfrak{M}, f \rangle(v, n + k)$, and the set of such pairs (n, k) is computable. Note that $\langle \mathfrak{M}, f \rangle$ halts if and only if there is some n, less than or equal to the number of configurations for $\mathfrak{M}$, for which this holds for $(n, 1)$. Hence $\mathbf{h}$, which searches for the least such n, is computable. □

We next investigate which Turing degrees are achieved by arbitrary computable graph machines.

3.1   Upper Bound

We now show that every graph computable function is computable from $\mathbf{0}^{(\omega)}$.

Definition 3.2. *Let f be a configuration for $\mathfrak{M}$, and let A be a finite subset of G. We say that $(B_i)_{i \leq n}$ is an n-**approximation** of $\mathfrak{M}$ and f on A if*

- $A = B_0$,
- $B_i \subseteq B_{i+1} \subseteq G$ for all $i < n$, and
- if $B_{i+1} \subseteq B \subseteq G$ then $\langle \mathfrak{M}|_{B_{i+1}}, f_i|_{B_{i+1}} \rangle(v, 1) = \langle \mathfrak{M}|_B, f_i|_B \rangle(v, 1)$ for all $v \in B_{i+1}$, where again $f_i := \langle \mathfrak{M}, f \rangle(\,\cdot\,, i)$.

The following proposition (in the case where $\ell = n - n'$) states that if $(B_i)_{i \le n}$ is an n-approximation of $\mathfrak{M}$ and f on A, then as long as we are only running $\mathfrak{M}$ with starting configuration f for ℓ-many steps, and are only considering the states of elements within $B_{n'}$, then it suffices to restrict $\mathfrak{M}$ to B_n. It follows by a straightforward though technical induction.

Proposition 3.3. *The following claim holds for every $n \in \mathbb{N}$: For every configuration f for $\mathfrak{M}$, and finite $A \subseteq G$,*

- *there is an n-approximation of $\mathfrak{M}$ and f on A, and*
- *if $(B_i)_{i \le n}$ is such an approximation, then*

$$(\forall n' < n)(\forall \ell \le n - n')(\forall v \in B_{n'})\langle \mathfrak{M}|_{B_{n'+\ell}}, f|_{B_{n'+\ell}} \rangle(v, \ell) = \langle \mathfrak{M}, f \rangle(v, \ell). \quad \square$$

We now analyze the computability of approximations and of runs.

Proposition 3.4. *Let $n \in \mathbb{N}$. For all computable graph machines $\mathfrak{M}$ and configurations f for $\mathfrak{M}$, the following are $\mathbf{f}^{(n)}$-computable, uniformly in n, where $\mathbf{f}$ is the Turing degree of f.*

- *The collection $P_n(f) := \{(A, (B_i)_{i \le n}) \,:\, A \subseteq G$ is finite and $(B_i)_{i \le n}$ is an n-approximation of $\mathfrak{M}$ and f on $A\}$.*
- *The function $f_n := \langle \mathfrak{M}, f \rangle(\,\cdot\,, n)$.* $\quad \square$

Corollary 3.5. *If f is a configuration for $\mathfrak{M}$, then f_n is $\mathbf{f}^{(n)}$-computable and so $\langle \mathfrak{M}, f \rangle$ is $\mathbf{f}^{(\omega)}$-computable, where $\mathbf{f}$ is the Turing degree of f.*

Proof. By Proposition 3.3, for each $v \in G$ (the underlying set of $\mathfrak{M}$) and each $n \in \mathbb{N}$, there is an approximation of $\mathfrak{M}$ for f and $\{v\}$ up to n. Further, by Proposition 3.4 we can $\mathbf{f}^{(n)}$-compute such an approximation, uniformly in v and n. But if $(B_i^v)_{i \le n}$ is an approximation of $\mathfrak{M}$ for f and $\{v\}$ up to n then $\langle \mathfrak{M}|_{B_n^v}, f|_{B_n^v} \rangle(v, n) = \langle \mathfrak{M}, f \rangle(v, n)$. So $f_n = \langle \mathfrak{M}, f \rangle(\,\cdot\,, n)$ is $\mathbf{f}^{(n)}$-computable, uniformly in n. Hence $\langle \mathfrak{M}, f \rangle$ is $\mathbf{f}^{(\omega)}$-computable. $\quad \square$

Theorem 3.6. *Suppose that $\{\mathfrak{M}\}$ is a total function. Then $\{\mathfrak{M}\}$ is computable from $\mathbf{0}^{(\omega)}$.*

Proof. Let f be any starting configuration of $\mathfrak{M}$. Then f is computable. Hence by Corollary 3.5, $\langle \mathfrak{M}, f \rangle(v, n + 1)$ is $\mathbf{0}^{(n+1)}$-computable. This then implies that the function determining whether or not $\{\mathfrak{M}\}(x)$ halts after n steps is $\mathbf{0}^{(n+2)}$-computable.

But by assumption, $\{\mathfrak{M}\}(x)$ halts for every $x \in \mathfrak{A}^{<G}$, and so $\{\mathfrak{M}\}$ is $\mathbf{0}^{(\omega)}$-computable. $\quad \square$

3.2 Lower Bound

We have seen that every graph computable function is computable from $\mathbf{0}^{(\omega)}$. In this subsection, we will see that this bound can be obtained. We begin by showing that every arithmetical Turing degree has an element that is graph computable. From this we then deduce that there is a graph computable function Turing equivalent to $\mathbf{0}^{(\omega)}$.

We first recall the following standard result from computability theory (see [16, III.3.3]).

Lemma 3.7. *Suppose $n \in \mathbb{N}$ and $X \subseteq \mathbb{N}$. Then the following are equivalent.*

- $X \leq_{\mathrm{T}} \mathbf{0}^{(n)}$.
- *There is a computable function $g \colon \mathbb{N}^{n+1} \to \mathbb{N}$ such that*
  - $h(\,\cdot\,) := \lim_{x_0 \to \infty} \cdots \lim_{x_{n-1} \to \infty} g(x_0, \ldots, x_{n-1}, \,\cdot\,)$ *is total.*
  - $h \equiv_{\mathrm{T}} X$. $\qquad\qquad\square$

We now sketch the following construction.

Proposition 3.8. *Let $n \in \mathbb{N}$ and suppose $g \colon \mathbb{N}^{n+1} \to \mathbb{N}$ is computable such that*

$$h(\,\cdot\,) := \lim_{x_0 \to \infty} \cdots \lim_{x_{n-1} \to \infty} g(x_0, \ldots, x_{n-1}, \,\cdot\,)$$

is total. Then h is graph computable via a graph machine whose lookup table does not depend on n or g.

Proof sketch. We define a "subroutine" graph machine that, on its own, computes the limit of a computable binary sequence. We then embed n repetitions of this subroutine into a single graph machine that computes the n-fold limit of the $(n + 1)$-dimensional array given by g.

The subroutine graph machine has a countably infinite sequence (with one special vertex) as its underlying graph, in that every vertex is connected to all previous vertices (and all are connected to the special vertex). Each vertex first activates itself, setting its displayed symbol to the appropriate term in the sequence whose limit is being computed. Each vertex sends a pulse to every previous vertex, signaling its displayed state. Any vertex which receives both 0 and 1 from vertices later in the sequence knows that the sequence alternates at some later index. Finally, any vertex which only receives a 0 or 1 pulse, but not both, sends a pulse corresponding to the one it receives to the special vertex. This special vertex then knows the limiting value of the sequence. $\qquad\square$

This technical construction allows us to conclude the following.

Corollary 3.9. *Suppose $X \subseteq \mathbb{N}$ is such that $X \leq_{\mathrm{T}} \mathbf{0}^{(n)}$ for some $n \in \mathbb{N}$. Then X is Turing-equivalent to some graph computable function.*

Proof. By Lemma 3.7, X is Turing equivalent to the n-fold limit of some computable function. By Proposition 3.8, this n-fold limit is graph computable. $\qquad\square$

Not only are all graph computable functions Turing reducible to $\mathbf{0}^{(\omega)}$, but this bound can be achieved.

Theorem 3.10. *There is a graph computable function that is Turing equivalent to $\mathbf{0}^{(\omega)}$.* $\qquad\qquad\square$

4 Finite Degree Graphs

We have seen that every arithmetical function is graph computable. However, as we will see in this section, if we instead limit ourselves to graphs where each vertex has finite degree, then not only is every graph computable function computable from $\mathbf{0}'$, but also we can obtain more fine-grained control over the Turing degree of the function by studying the degree structure of the graph.

4.1 Upper Bound

Before we move to the specific case of graphs of finite degree (defined below), there is an important general result concerning bounds on graph computability and approximations to computations.

Definition 4.1. Let $\Theta \colon \mathfrak{P}_{<\omega}(G) \to \mathfrak{P}_{<\omega}(G)$. We say that Θ is a **uniform approximation** of $\mathfrak{M}$ if for all finite subsets $A \subseteq G$, we have $A \subseteq \Theta(A)$, and for any configuration f for $\mathfrak{M}$, the pair $(A, \Theta(A))$ is a 1-approximation of $\mathfrak{M}$ and f on A.

The following is an easy induction on n.

Lemma 4.2. Let $\Theta(A)$ be a uniform approximation of $\mathfrak{M}$. Then for any finite subset A of G, any configuration f for $\mathfrak{M}$, and any $n \in \mathbb{N}$, the tuple $(A, \Theta(A), \Theta^2(A), \ldots, \Theta^n(A))$ is an n-approximation of $\mathfrak{M}$ and f on A. □

Note that while we will be able to get even better bounds in the case of finite degree graphs, we do have the following bound on computability, as a straightforward consequence of Lemma 4.2.

Lemma 4.3. Let Θ be a uniform approximation of $\mathfrak{M}$. Then for any configuration f, the function $\langle \mathfrak{M}, f \rangle$ is computable from Θ and f (uniformly in f), and if $\{\mathfrak{M}\}$ is total, then $\{\mathfrak{M}\} \leq_{\mathrm{T}} \Theta'$. □

We now introduce the *degree function* of a graph.

Definition 4.4. For $v \in G$, define the **degree** of v to be the number of vertices incident with it, i.e., $\deg_{\mathcal{G}}(v) := |\{w : E(v,w) \cup E(w,v) \neq \emptyset\}|$, and call $\deg_{\mathcal{G}}(\cdot) \colon G \to \mathbb{N} \cup \{\infty\}$ the **degree function** of $\mathcal{G}$.
   We say that $\mathcal{G}$ has **finite degree** when $\mathrm{rng}(\deg_{\mathcal{G}}) \subseteq \mathbb{N}$, and say that $\mathcal{G}$ has **constant degree** when $\deg_{\mathcal{G}}$ is constant.

We will see that for a graph $\mathcal{G}$ of finite degree, its degree function bounds the computability of $\mathcal{G}$-computable functions.

The following easy lemma (using the fact that each vertex of a computable graph has a computably enumerable set of neighbors) allows us to provide a computation bound on graph Turing machines all vertices of whose underlying graph have finite degree.

Lemma 4.5. Suppose that $\mathcal{G}$ has finite degree. Then $\deg_{\mathcal{G}} \leq_{\mathrm{T}} \mathbf{0}'$. □

We next need the following definition.

Definition 4.6. *For each $A \subseteq G$ and $n \in \mathbb{N}$, the n-neighborhood of A, written $\mathbf{N}_n(A)$, is defined by induction as follows.*
Case 1 : The 1-neighborhood of A is $\mathbf{N}_1(A) := A \cup \{v \in G : (\exists a \in A) \ E(v, a) \cup E(a, v) \neq \emptyset\}$.
Case $k + 1$: The $k + 1$-neighborhood of A is $\mathbf{N}_{k+1}(A) = \mathbf{N}_1(\mathbf{N}_k(A))$.

Lemma 4.7. *Suppose that $\mathcal{G}$ has finite degree. Then*

(a) the 1-neighborhood map $\mathbf{N}_1$ is computable from $\deg_{\mathcal{G}}$, and
(b) for any $\mathcal{G}$-machine $\mathfrak{M}$, the map $\mathbf{N}_1$ is a uniform approximation to $\mathfrak{M}$.

Proof. Clause (a) follows from the fact that given the degree of a vertex one can search for all of its neighbors, as this set is computably enumerable (uniformly in the vertex) and of a known finite size.

Clause (b) follows from the fact that if a vertex receives a pulse, it must have come from some element of its 1-neighborhood. □

The following more precise upper bound on the computability of a graph computable function holds when the underlying graph has finite degree.

Theorem 4.8. *Suppose that $\mathcal{G}$ has finite degree and $\{\mathfrak{M}\}$ is total. Then $\mathfrak{M}$ is bounded and $\{\mathfrak{M}\}$ is computable from $\deg_{\mathcal{G}}$.* □

The following important corollaries are immediate from Lemma 4.5, Theorem 4.8, and the fact that if $\deg_{\mathcal{G}}$ is constant, it is computable.

Corollary 4.9. *Suppose that $\mathcal{G}$ has finite degree. Then any $\mathcal{G}$-computable function is $\mathbf{0}'$-computable.* □

Corollary 4.10. *Suppose that $\mathcal{G}$ has constant degree. Then any $\mathcal{G}$-computable function is computable (in the ordinary sense).* □

4.2   Lower Bound

Finally we consider the possible Turing degrees of graph computable functions where the underlying graph has finite degree. In particular, we show that every Turing degree below $\mathbf{0}'$ is the degree of some total graph computable function where the underlying graph has finite degree.

Recall from Lemma 3.7 (for $n = 1$) that a set $X \subseteq \mathbb{N}$ satisfies $X \leq_T \mathbf{0}'$ when the characteristic function of X is the limit of a 2-parameter computable function.

Theorem 4.11. *For every $X \colon \mathbb{N} \to \{0, 1\}$ such that $X \leq_T \mathbf{0}'$ there is a graph machine $\mathcal{N}_X$ such that every vertex of its underlying graph has degree at most 3, and $\{\mathcal{N}_X\}$ is total and Turing equivalent to X.* □

5   Representations of Other Computational Models via Graph Machines

Many other models of computation can be viewed as special cases of graph machines, providing further evidence for graph machines being a universal model of computation on graphs.

5.1   Ordinary Turing Machines

An ordinary Turing machine can be simulated by a graph Turing machine, where the tape of the Turing machine is encoded by an underlying graph that is a $\mathbb{Z}$-chain.

The doubly-infinite one-dimensional read/write tape of an ordinary Turing machine has cells indexed by $\mathbb{Z}$, the free group on one generator, and in each timestep the head moves according to the generator or its inverse. This interpretation of a Turing machine as a $\mathbb{Z}$-machine has been generalized to H-machines for arbitrary finitely generated groups H by [4], and the simulation mentioned above extends straightforwardly to this setting as well.

One might next consider how cleanly one might embed various extensions of Turing machines where the tape is replaced by a graph, such as Kolmogorov–Uspensky machines [13], Knuth's pointer machines [12, pp. 462–463], and Schönhage's storage modification machines [15]. For a further discussion of these and their relation to sequential abstract state machines, see [10,11].

5.2   Cellular Automata

Cellular automata (which we take to be finite-dimensional, finite-radius and with finitely-many states) can be naturally simulated by graph machines, moreover of constant degree. In particular, Corollary 4.10 applies to this embedding. For background on cellular automata, see, e.g., the book [18].

Cells of the automata are taken to be the vertices of the graph, and cells are connected to its "neighbors" (other cells within the given radius) by a collection of edges (of the graph) whose labels encode their relative position (e.g., "1 to the left of") and all possible cellular automaton states. (In particular, every vertex has the same finite degree.) The displayed symbol of each vertex encodes the cellular automaton state of that cell. The rule of the cellular automaton is encoded in the lookup table so as to provide a bisimulation between the original cellular automaton and the graph machine built based on it.

Not only is the evolution of every such cellular automata computable (moreover via this embedding as a graph machine), but there are particular automata whose evolution encodes the behavior of a universal Turing machine [9,19]. Several researchers have also considered the possibility of expressing intermediate Turing degrees via this evolution [5,8,17]. Analogously, one might ask which Turing degrees can be expressed in the evolution of graph machines.

5.3 Parallel Graph Dynamical Systems

Parallel graph dynamical systems [2] can be viewed as essentially equivalent to the finite case of graph Turing machines, as we now describe. Finite cellular automata can also be viewed as a special case of parallel graph dynamical systems, as can finite boolean networks, as noted in [2, Sect. 2.2]. For more on parallel graph dynamical systems, see [1,2,6].

In contrast, it is not immediately clear how best to encode an arbitrary (parallel) abstract state machine [7] as a graph Turing machine (due to the higher arity relations of the ASM).

Acknowledgements. The authors would like to thank Tomislav Petrović, Linda Brown Westrick, and the anonymous referees of earlier versions for helpful comments.

References

1. Aledo, J.A., Martinez, S., Valverde, J.C.: Graph dynamical systems with general Boolean states. Appl. Math. Inf. Sci. **9**(4), 1803–1808 (2015)
2. Aledo, J.A., Martinez, S., Valverde, J.C.: Parallel dynamical systems over graphs and related topics: a survey. J. Appl. Math. (2015). Article no. 594294
3. Angluin, D., Aspnes, J., Bazzi, R.A., Chen, J., Eisenstat, D., Konjevod, G.: Effective storage capacity of labeled graphs. Inform. Comput. **234**, 44–56 (2014)
4. Aubrun, N., Barbieri, S., Sablik, M.: A notion of effectiveness for subshifts on finitely generated groups. Theor. Comput. Sci. **661**, 35–55 (2017)
5. Baldwin, J.: Review of A New Kind of Science by Stephen Wolfram. Bull. Symb. Logic **10**(1), 112–114 (2004)
6. Barrett, C.L., Chen, W.Y.C., Zheng, M.J.: Discrete dynamical systems on graphs and Boolean functions. Math. Comput. Simul. **66**(6), 487–497 (2004)
7. Blass, A., Gurevich, Y.: Abstract state machines capture parallel algorithms. ACM Trans. Comput. Log. **4**(4), 578–651 (2003)
8. Cohn, H.: Review of A New Kind of Science by Stephen Wolfram. MAA Reviews, Washington D.C. (2002)
9. Cook, M.: Universality in elementary cellular automata. Complex Syst. **15**(1), 1–40 (2004)
10. Gurevich, Y.: Kolmogorov machines and related issues. In: Current Trends in Theoretical Computer Science. World Scientific Series in Computer Science, vol. 40, pp. 225–234. World Scientific (1993)
11. Gurevich, Y.: Sequential abstract-state machines capture sequential algorithms. ACM Trans. Comput. Log. **1**(1), 77–111 (2000)
12. Knuth, D.E.: The Art of Computer Programming: Volume 1: Fundamental Algorithms. Addison-Wesley, Boston (1968)
13. Kolmogorov, A.N., Uspensky, V.A.: On the definition of an algorithm. Uspekhi Mat. Nauk **13**(4), 3–28 (1958)
14. Lovász, L.: Very large graphs. In: Current Developments in Mathematics, vol. 2008, pp. 67–128. International Press, Somerville (2009)
15. Schönhage, A.: Storage modification machines. SIAM J. Comput. **9**(3), 490–508 (1980)
16. Soare, R.I.: Recursively Enumerable Sets and Degrees. Perspectives in Mathematical Logic. Springer, Berlin (1987)

17. Sutner, K.: Cellular automata and intermediate degrees. Theoret. Comput. Sci. **296**(2), 365–375 (2003)
18. Toffoli, T., Margolus, N.: Cellular Automata Machines: A New Environment for Modeling. MIT Press, Cambridge (1987)
19. Woods, D., Neary, T.: The complexity of small universal Turing machines: a survey. Theoret. Comput. Sci. **410**(4–5), 443–450 (2009)

Independence-Friendly Logic Without Henkin Quantification

Fausto Barbero[1]([⊠]), Lauri Hella[2], and Raine Rönnholm[2]

[1] Philosophy, Faculty of Arts, University of Helsinki, Helsinki, Finland
fausto.barbero@helsinki.fi
[2] Mathematics, Faculty of Natural Sciences, University of Tampere,
Tampere, Finland
{lauri.hella,raine.ronnholm}@uta.fi

Abstract. We analyze from a global point of view the expressive resources of IF logic that do not stem from Henkin (partially-ordered) quantification. When one restricts attention to regular IF sentences, this amounts to the study of the fragment of IF logic which is individuated by the game-theoretical property of Action Recall. We prove that the fragment of Action Recall can express all existential second-order (ESO) properties. This can be accomplished already by the prenex fragment of Action Recall, whose only second-order source of expressiveness are the so-called signalling patterns. The proof shows that a complete set of Henkin prefixes is explicitly definable in the fragment of Action Recall. In the more general case, in which also irregular IF sentences are allowed, we show that full ESO expressive power can be achieved using neither Henkin nor signalling patterns.

Keywords: Independence-Friendly logic · Game-theoretical semantics · Henkin quantification · Action recall · Signalling · Existential second-order logic · Expressive power

1 Introduction

Independence-Friendly logic [10,15] is one of a number of formalisms that have been developed in order to make various notions of *dependence* and *independence* accessible to the instruments of logical investigation. Independence-Friendly (IF) logic and similar formalisms (Dependence-Friendly logic, Dependence logic [18]), in particular, were developed as a more flexible approach to the logic of *Henkin quantifiers* [9]. The Henkin quantifier H_k^n is a matrix

$$\begin{pmatrix} \forall x_1^1 \ \forall x_1^2 \ \ldots \ \forall x_1^n \ \exists y_1 \\ \vdots \quad \vdots \quad \ddots \quad \vdots \quad \vdots \\ \forall x_k^1 \ \forall x_k^2 \ \ldots \ \forall x_k^n \ \exists y_k \end{pmatrix}$$

which, differently from a linear sequence of the same quantifiers, is meant to state that each y_i is supposed to be chosen as a function of $x_i^1, \ldots, x_i^n$ only. In IF

© Springer-Verlag GmbH Germany 2017
J. Kennedy and R.J.G.B. de Queiroz (Eds.): WoLLIC 2017, LNCS 10388, pp. 14–30, 2017.
DOI: 10.1007/978-3-662-55386-2_2

logic, the same is achieved by means of a linear prefix, together with a slashing device. For example, the Henkin quantifier H_2^1 is expressed in IF logic by the sequence of quantifiers

$$\forall x_1^1 \exists y_1 \forall x_2^1 (\exists y_2 / \{x_1^1, y_1\}).$$

The slashed quantifier $(\exists y_2 / \{x_1^1, y_1\})$ expresses the fact that y_2 is *independent* from x_1^1 and y_1.

It has been gradually realized that, in spite of the fact that it stems from the study of Henkin quantifiers, IF logic derives its expressiveness also from other sources. Henkin quantifiers are partial orderings of first-order quantifiers; but in IF logic also intransitive (thus not ordered) dependence sequences are allowed, for example

$$\forall x \exists y (\exists z / \{x\}).$$

Here y depends on x, z depends on y, but z does not depend on x. It is known that such quantifier sequences, also known as *signalling* sequences (or patterns) can be used to express higher-order concepts [4,7,17]; for example, the IF sentence $\exists v \forall x \exists y (\exists z / \{x\})(x = z \wedge y \neq v)$ is known to characterize the class of all infinite structures (this idea is attributed, in [7], to Fred Galvin).

Henkin and signalling patterns are known to exhaust the higher-order expressive power of *prenex regular*[1] IF logic: if a regular sentence is in prenex normal form and does not contain Henkin or signalling patterns, then it is equivalent to some first-order sentence [17]. Non-prenex, regular IF logic is known to contain further expressive synctactical patterns (involving the interaction of quantifiers and disjunctions) that are neither of the Henkin nor the signalling type, yet allow describing NP-complete problems such as SAT and SET PARTITIONING [2]; the problem of a complete classification of such patterns is still open. Less is known of irregular IF logic, which will be addressed here in Sect. 5. The aim of the present paper is a better understanding of the resources of IF logic that do *not* stem from Henkin quantification.

A peculiarity of Independence-Friendly logic is the close link between its syntax and the theory of extensive games of imperfect information. The link is given by the so-called Game-Theoretical Semantics, that we will review in Sect. 2. Through this connection, game-theoretical concepts throw light on peculiarities of the logic; and vice versa, the study of logical phenomena can cast new light on the foundations of game theory.

It is well known, through the works of Henkin, Hintikka and others, that it is possible to define a notion of truth for first-order languages in terms of certain games of perfect information, which involve two players called Verifier ("Eloise") and Falsifier ("Abelard"), who take it in turns to point out evidence for or against the truth of a given sentence φ in a given structure M. The resulting Game-Theoretical Semantics (GTS) is equivalent to the usual Tarskian one.

---

[1] The notion of regularity will be defined in Sect. 2.

When moving from first-order to IF languages, extending the Tarskian semantics is not straightforward;[2] instead, it is quite natural to generalize the semantic games by allowing *imperfect information*, in a way that the independence constraints expressed by syntax correspond (roughly speaking) to the fact that a player is forced to make his/her choices in ignorance of the outcomes of some earlier moves [10]. This generalization allows new complex possibilities. Many IF games are actually games of *imperfect recall*: the players may forget what they knew at earlier stages of the game.

In this paper, we will be particularly interested in a game-theoretical property called *action recall*. Eloise has action recall if she cannot forget her own moves; assuming regularity, an IF sentence has action recall (i.e., all its corresponding games have action recall) for Eloise if its sets of slashed variables associated to existential quantifiers contain no existentially quantified variables. Thus for example $\forall x(\exists y/\{x\})R(x,y)$ has action recall, while $\exists x(\exists y/\{x\})R(x,y)$ does not.

The fragment of sentences with action recall for Eloise is particularly important, because, in it, it is impossible to write the usual IF translations of Henkin prefixes,[3] and yet, it is a highly expressive fragment. Therefore, it is natural to wonder to what degree the IF-definable concepts are expressible under the restriction of action recall. IF logic is known to capture exactly the existential second-order (ESO) definable classes. In Sect. 4 we will show that the Henkin prefixes H_2^n are explicitly definable in the *prenex, regular* fragment of action recall (therefore, by means of signalling). The H_2^n prefixes, taken together, are known to capture all ESO definable concepts [14]; therefore, the prenex, regular fragment of action recall suffices for full IF expressive power.

When *irregular* IF sentences are allowed, instead, it becomes possible to violate action recall in new ways; in Sect. 5 we use this fact to show, by means of a translation procedure, that it is possible to express all IF properties using neither Henkin *nor signalling* quantifier patterns. Section 2 reviews preliminary notions about game-theoretical semantics and action recall, while Sect. 3 presents some significant examples.

2   Preliminaries

Notation. Structures are denoted by capital italic letters. To keep the notation simple, we do not introduce a separate symbol for the domain of a structure; thus if M is a structure, $a \in M$ and $R \subseteq M^2$ mean that a is an element of the domain of M and R is a binary relation on the domain of M, respectively.

An *assignment* of variables on a structure M is a function $s : V \to M$, where the domain V of s is a finite set of variables. We denote the set of all assignments on M with domain V by $\mathrm{As}(V, M)$. Given an assignment $s \in \mathrm{As}(V, M)$ and an element $a \in M$, we write $s(a/v)$ for the assignment with domain $V \cup \{v\}$ such

---

[2] It can be done, at the cost of defining a notion of satisfaction by sets of assignment, instead of the usual single assignments. See e.g. [5,11,12,15,18].

[3] This point is exemplified by the IF rendition of the H_2^1 prefix, shown above: its "slash set" $\{x_1^1, y_1\}$ contains an existentially quantified variable, y_1.

that $s(a/v)(v) = a$ and $s(a/v)(u) = s(u)$ for $u \in V \setminus \{v\}$. If $\overline{x} = (x_1, \ldots, x_n)$ is a tuple of variables, we use the shorthand notation $s(\overline{x})$ for $(s(x_1), \ldots, s(x_n))$.

Game-Theoretical Semantics. The syntax of IF logic is a restriction of the usual first-order syntax, to which we add quantifiers of the forms $(\exists v/V)$ and $(\forall v/V)$, where V is a finite set of variables, called the *slash set* of the quantifier. When $V = \emptyset$, we use the abbreviation $Qx := (Qx/V)$. The syntax is restricted, with regards to usual first-order languages, in that

- we only allow the connectives $\wedge$, $\vee$ and $\neg$, and
- for simplicity, we only allow $\neg$ to occur in front of atomic formulae.

The set Free(φ) of free variables of a formula φ is defined as usual, with the proviso that also variables from slash sets can be either free or bound. For example, in $\forall x (\exists y / \{x, y, z\}) \psi$, the occurrence of x in the slash set is bound, while the occurrences of y, z are free.

A further restriction on the syntax of IF logic that is often assumed in the literature (see, e.g., [3,15]) is that variables are not requantified:

- A sentence is *regular* if no quantifier (Qv/V) occurs in the scope of another quantifier $(Q'v/W)$ over the same variable v.

We denote IF logic with this regularity restriction by IF^r. We will mostly restrict our studies to IF^r, but in Sect. 5 we will also consider irregular sentences.

Game-Theoretical Semantics (GTS) associates to each triple (φ, M, s), where φ is an IF formula, M is a structure, and $s \in \text{As}(V, M)$ for a set V of variables such that Free(φ) $\subseteq V$, a 2-player win-lose extensive game of imperfect information $G(\varphi, M, s)$. In case φ is a sentence and $s = \emptyset$, we simply write $G(\varphi, M)$. The two players, usually called Eloise and Abelard, can be thought of as trying to verify, respectively falsify, the sentence φ on the structure M. Their moves are triggered by the most external logical operator of φ:

- in $G(\psi_1 \vee \psi_2, M, s)$, Eloise chooses a disjunct ψ_i and then $G(\psi_i, M, s)$ is played;
- in $G(\psi_1 \wedge \psi_2, M, s)$, the same kind of move is performed by Abelard;
- in $G((\exists v/V)\psi, M, s)$, Eloise picks an element $a \in M$ and then the game $G(\psi, M, s(a/v))$ is played;
- in $G((\forall v/V)\psi, M, s)$ the same kind of move is performed by Abelard;
- in $G(\alpha, M, s)$, with α a literal (i.e., an atomic formula or the negation of an atomic formula), the winner is decided: it is Eloise in case $M, s \models \alpha$ (in the usual first-order sense), and Abelard otherwise.

Imperfect information manifests itself in that some histories of the game are considered indistinguishable for the player who has the turn to move at the end of them. If two histories h and h' both end with the choice of a subgame associated with the same occurrence of a subformula $(Qv/V)\psi$ with assignments

$s_h, s_{h'} \in \mathrm{As}(W, M)$ such that $s_h(w) = s_{h'}(w)$ for every $w \in W \setminus V$, then h and h' are indistinguishable for the player associated to (Qv/V), and we write $h \sim_V h'$.

A *strategy* for Eloise in game $G(\varphi, M, s)$ is a function associating, to each history ending in a subgame $G((\exists v/V)\psi, M, s')$, an element $a \in M$; and, to every history ending in a subgame $G(\psi_1 \vee \psi_2, M, s')$, either ψ_1 or ψ_2. Strategies for Abelard can be similarly defined.

A strategy of Eloise is *winning* if, playing according to it, Eloise wins, whatever moves Abelard makes. Winning strategies for Abelard are defined dually.

A strategy σ is *uniform* if, whenever two histories h, h' are in its domain and $h \sim_V h'$ (for the only appropriate V), then $\sigma(h) = \sigma(h')$.

With this game-theoretical apparatus, it is possible to define the notions of *truth* and *falsity* for IF sentences as the existence of appropriate strategies:

$M \models \varphi$ if Eloise has a uniform winning strategy in $G(\varphi, M)$

$M \models^- \varphi$ if Abelard has a uniform winning strategy in $G(\varphi, M)$.

There is also a third possibility: it may happen that neither player has a uniform winning strategy (consider, e.g., the sentence $\forall x (\exists y/\{x\}) x = y$). In that case, the game and the truth value of the sentence on M are said to be *undetermined*. In this paper, we only focus on the truth/nontruth distinction. Accordingly, we say that a class K of structures is definable in IF, if there is an IF sentence φ such that for all structures M, we have $M \in K \Leftrightarrow M \models \varphi$. As was already shown in [10], IF logic has the same expressive power as existential second-order logic ESO: a class of structures is definable in IF if and only if it is definable in ESO.

Signalling and Henkin Patterns. We have been talking informally of Henkin and signalling patterns of quantifiers. Exact definitions were given in [17]; We extend these definitions for irregular sentences.

- Let (Qv/V) and $(Q'u/U)$ be quantifiers occurring in a prefix or in a sentence. We use the following terminology:
  - (Qv/V) *is in the effective scope* of $(Q'u/U)$, if (Qv/V) is in the scope of $(Q'u/U)$ and there is no quantifier $(Q''u/W)$ in the scope of $(Q'u/U)$ such that (Qv/V) in the scope of $(Q''u/W)$. We write $(Qv/V) \in \mathrm{Es}(Q'u/U)$ if this is the case.
  - (Qv/V) *depends on* $(Q'u/U)$ if $(Qv/V) \in \mathrm{Es}(Q'u/U)$ and $u \notin V$.
- A *signalling pattern* in a sentence consists of three quantifiers $(\forall x/X)$, $(\exists y/Y)$, $(\exists z/Z)$ such that x, y and z are distinct, $(\exists z/Z) \in \mathrm{Es}(\exists y/Y) \cap \mathrm{Es}(\forall x/X)$, $(\exists y/Y) \in \mathrm{Es}(\forall x/X)$ and
  - $(\exists y/Y)$ depends on $(\forall x/X)$;
  - $(\exists z/Z)$ depends on $(\exists y/Y)$, but not on $(\forall x/X)$.
- A *Henkin pattern* in a sentence consists of four quantifiers $(\forall x/X)$, $(\exists y/Y)$, $(\forall z/Z)$, $(\exists w/W)$ s.t. x, y, z, w are distinct, $(\exists w/W) \in \mathrm{Es}(\exists y/Y) \cap \mathrm{Es}(\forall z/Z)$, $(\exists w/W), (\forall z/Z), (\exists y/Y) \in \mathrm{Es}(\forall x/X)$, and
  - $(\exists y/Y)$ depends on $(\forall x/X)$, but not on $(\forall z/Z)$;
  - $(\exists w/W)$ depends on $(\forall z/Z)$, but not on $(\forall x/X)$ or $(\exists y/Y)$.

Note that the last condition holds only if the existentially quantified variable y is in the slash set W. Also note that in the case of regular sentences, we may simply talk about scopes instead of effective scopes in the definitions above.

Action Recall Fragment of IF Logic. Even though many different games are associated to each single sentence (one game for each sentence-structure pair), some interesting properties of the games are characterized by synctactical properties of the associated sentences; they are invariants of the sentence alone. As a consequence, such game-theoretical properties define associated fragments of IF logic. In particular, in the literature ([13, 15, 16] sect. 6.4, [1]) there has been some interest in properties that limit the ability of players to forget. Considering for example the role of Eloise:

- Eloise has *action recall* if she cannot forget her own moves.
- Eloise has *knowledge memory* if she cannot forget what she knew at earlier stages of the game.
- Eloise has *perfect recall* if she has both action recall and knowledge memory.

Under the assumption of regularity, each of these properties has been given a syntactical characterization in the literature (see e.g., [15], [1]). For action recall the characterization is as follows: Assume that φ is a regular IF sentence. Then Eloise has action recall in the game $G(\varphi, M, s)$ if and only if φ satisfies the following *restriction on slash sets*:

(RS) If an existential quantifier $(\exists v/V)$ occurs in the scope of another existential quantifier $(\exists u/U)$, then $u \notin V$.

It should be noted that the condition (RS) does not guarantee action recall for Eloise on irregular sentences. For example, the formula $\exists x \exists x \exists y (x = y)$ violates action recall: in her third move, Eloise forgets the value chosen in the first move, because it has been overwritten by the second move.

We denote the fragment of IF consisting of all regular formulae that satisfy (RS) by $IF^r_{AR(\exists)}$ (here AR($\exists$) stands for Action Recall for $\exists$loise). Note that it is impossible to write a Henkin quantifier in $IF^r_{AR(\exists)}$; for this reason, the fragment $IF^r_{AR(\exists)}$ will be one of the main objects of study in this paper. We will also be interested in the set $IF^{p,r}_{AR(\exists)}$ of *prenex* sentences of $IF^r_{AR(\exists)}$.

The fragments of perfect recall and knowledge memory are relatively well-understood; truth, in both of them, can only capture first-order concepts (for the former fragment, the result was anticipated in [13, 16] and adequately proved in [15]; the latter fragment was addressed in [1]). The regular action recall fragment $IF^r_{AR(\exists)}$ is by far less understood; some examples in the literature show that it is capable of expressing higher-order concepts, such as infinity over the empty signature, and some NP-complete problems (see Sect. 3). But a general understanding of its expressive power is lacking, and will be addressed in the present paper.

3   Examples

The main result that will be proved in Sect. 4 implies that any ESO concept can be expressed by some regular, prenex action recall formula (therefore, by means of signalling). However, the defining sentences provided by the theorem are often unnecessarily complicated. We give here some examples of NP-complete problems that can be expressed by relatively simple sentences of $\text{IF}^{\text{p,r}}_{\text{AR}(\exists)}$.

Example 1. In [17], it was shown that the EXACT COVER BY 3-SETS problem can be defined by an $\text{IF}^{\text{p,r}}_{\text{AR}(\exists)}$ sentence. This problem consists in deciding, given a set U of $3k$ elements and a family C of 3-element subsets of U, whether there is a subfamily of C which is a partition of U. It is defined by the sentence

$$\forall x \exists y (\exists z/\{x\})(U(x) \to (K(y) \land E(x,z)))$$

on finite structures M of domain $U \cup C$ (where $U \cap C = \emptyset$), such that $U^M = U$, $\text{Card}(K^M) = k$ and $E^M = \{\langle a, B\rangle \mid a \in U, B \in C, a \in B\}$. We wish to point out that, if we restrict, w.l.o.g., the class of structures by the additional constraint that $K^M \subseteq M \setminus U^M$, then the condition above can be shown (see Appendix) to be equivalent to an ESO sentence of prefix $\exists f \forall x$:

$$\varphi^* = \exists f \forall x (U(x) \to (K(f(x)) \land E(x, f(f(x))))).$$

$\exists f \forall x$ is the simplest non-trivial prefix of functional ESO. The fact that it can capture NP-complete problems was shown by Grandjean [8]; he applied this prefix to a conjunction of *twenty-one* clauses to define the HAMILTON PATH problem.

Example 2. We consider another NP-complete problem, DOMINATING SET: the problem of deciding, given an integer k and a graph $G = (V, E^G)$ as input, whether there is a set $D \subseteq V$ of vertices of size at most k, such that, for every vertex $x \in V$, either $x \in D$ or $(y, x) \in E^G$ for some $y \in D$. Assuming that the intended structures encode k by an interpreted unary predicate P^G of cardinality k, the problem is described (see Appendix) by the $\text{IF}^{\text{p,r}}_{\text{AR}(\exists)}$ sentence

$$\forall x \exists z (\exists y/\{x\})((E(y,x) \lor y = x) \land P(z)).$$

This description is based on an analogous result for Dependence logic [19].

Example 3. Also the problem SAT is expressible by means of signalling. SAT is stated as follows: given a propositional formula π in conjunctive normal form, is π satisfiable? The problem can be modeled over structures M of signature $P, N, C, 0, 1$, with $0^M, 1^M$ distinct constants; $C^M \subseteq M$ representing the set of clauses; $P^M, N^M \subseteq (M \setminus C^M) \times C^M$, representing the fact that the first argument (thought of as a propositional variable) occurs positively, respectively negatively, in the second argument (thought of as a clause). In this class of structures, SAT is described (see Appendix) by the following $\text{IF}^{\text{p,r}}_{\text{AR}(\exists)}$ sentence:

$$\forall x \exists y (\exists z/\{x\})(C(x) \to ((P(y,x) \land z = 1) \lor (N(y,x) \land z = 0))).$$

Since this specific form of SAT is known to be NP-complete under quantifier-free reductions ([6]), we could give an argument based on standard tools to show that $\text{IF}^{\text{p,r}}_{\text{AR}(\exists)}$ captures NP. In principle, we could extend this argument to show that an "infinite" version of SAT is complete for ESO under quantifier-free reductions, and thus $\text{IF}^{\text{p,r}}_{\text{AR}(\exists)}$ captures ESO. However, we will prove this result in the next section with a more direct argument.

4   Explicit Definition of Henkin Quantifiers by Signalling

In this section we show that the prenex action recall fragment $\text{IF}^{\text{p,r}}_{\text{AR}(\exists)}$ has the same expressive power as the full IF logic. In the proof of this result we exploit the fact that ESO is captured by Henkin quantifiers with two rows:

Theorem 1 [14]. *For any* ESO *sentence there is an equivalent sentence of the form*

$$\begin{pmatrix} \forall x_1 \dots \forall x_n \ \exists u \\ \forall y_1 \dots \forall y_n \ \exists v \end{pmatrix} \psi$$

for some $n \in \mathbb{N}$, and some quantifier-free formula ψ.

By this result, it suffices to prove that, for any n, any sentence that is obtained by applying the Henkin quantifier H^n_2 to a quantifier-free formula, is expressible in $\text{IF}^{\text{p,r}}_{\text{AR}(\exists)}$. Since $\text{IF}^{\text{p,r}}_{\text{AR}(\exists)}$ is a fragment of IF, and IF is expressively equivalent to ESO, it follows then that the expressive powers of all the three logics $\text{IF}^{\text{p,r}}_{\text{AR}(\exists)}$, IF and ESO coincide.

Thus, we consider a sentence starting with the Henkin quantifier H^n_2; let

$$\varphi := \begin{pmatrix} \forall x_1 \dots \forall x_n \ \exists u \\ \forall y_1 \dots \forall y_n \ \exists v \end{pmatrix} \psi(x_1, \dots, x_n, u, y_1, \dots, y_n, v),$$

where ψ is a quantifier-free formula. In order to make the argument below more transparent, we formulate the truth condition of φ in a slightly non-standard way: $M \models \varphi$ if and only if there are relations $F_a, F_b \subseteq M^{n+1}$ such that

(a) $(M, F_a) \models \forall \overline{z} \exists w \, F_a(\overline{z}, w)$,
(b) $(M, F_b) \models \forall \overline{z} \exists w \, F_b(\overline{z}, w)$,
(c) $(M, F_a, F_b) \models \forall \overline{x} \forall u \forall \overline{y} \forall v (\neg F_a(\overline{x}, u) \vee \neg F_b(\overline{y}, v) \vee \psi(\overline{x}, u, \overline{y}, v))$.

Here, and in the sequel, $\overline{z}$ denotes a tuple $(z_1, \dots, z_n)$ of distinct variables; similarly, $\overline{x} = (x_1, \dots, x_n)$ and $\overline{y} = (y_1, \dots, y_n)$.

We will now build a sentence θ of $\text{IF}^{\text{p,r}}_{\text{AR}(\exists)}$ that expresses the three conditions above. The idea is to use the variables $\overline{z}$ and w for expressing conditions (a) and (b), and the variables $\overline{x}$, u, $\overline{y}$ and v for expressing (c). In addition we use an "index variable" i that Abelard will use in the game $G(\theta, M)$ to separate the conditions (a), (b) and (c) from each other, and another "index variable" j that Eloise uses either to signal the value of i, or to choose a disjunct of the quantifier-free part in (c).

To simplify the presentation, we assume first that the signature contains three constants, **a**, **b** and **c**, and consider only structures in which they are interpreted by distinct elements. In this case, the sentence θ is defined as follows:

$$\theta := \forall \overline{x}\, \forall u \forall \overline{y}\, \forall v \forall \overline{z}\, \forall i\, (\exists j/Z)(\exists w/X)\, \eta,$$

where Z is the set $\{z_1, \ldots, z_n\}$, X is the set $\{x_1, \ldots, x_n, u, y_1, \ldots, y_n, v, i\}$ and η is the following quantifier-free formula

$$(i = \mathbf{a} \rightarrow j = \mathbf{a}) \tag{1}$$
$$\wedge\, (i = \mathbf{b} \rightarrow j = \mathbf{b}) \tag{2}$$
$$\wedge\, (i = \mathbf{c} \wedge \overline{z} = \overline{x} \wedge j = \mathbf{a} \rightarrow w \neq u) \tag{3}$$
$$\wedge\, (i = \mathbf{c} \wedge \overline{z} = \overline{y} \wedge j = \mathbf{b} \rightarrow w \neq v) \tag{4}$$
$$\wedge\, (i = \mathbf{c} \wedge j = \mathbf{c} \rightarrow \psi(\overline{x}, u, \overline{y}, v)) \tag{5}$$
$$\wedge\, (i = \mathbf{c} \rightarrow (j = \mathbf{a} \vee j = \mathbf{b} \vee j = \mathbf{c})) \tag{6}$$

Lemma 1. *Let M be a structure such that $a \neq b \neq c \neq a$, where $a = \mathbf{a}^M$, $b = \mathbf{b}^M$ and $c = \mathbf{c}^M$. Then $M \models \varphi$ if and only if $M \models \theta$.*

Proof. Assume first that $M \models \varphi$. Let F_a and F_b be relations satisfying the conditions (a), (b) and (c). Without loss of generality, we can assume that F_a and F_b are actually functions. We describe a winning strategy σ for Eloise in the semantic game $G(\theta, M)$. In the first six moves of the game, Abelard chooses interpretations for the variables $\overline{x}$, u, $\overline{y}$, v, $\overline{z}$ and i; let s be the assignment formed during these moves. Then Eloise answers by choosing a value d_s for the variable j as follows:

- If $s(i) = a$, then Eloise sets $d_s = a$,
- If $s(i) = b$, then Eloise sets $d_s = b$,
- Assume then that $s(i) = c$. By condition (c), one of the following holds:
  (i) $s(u) \neq F_a(s(\overline{x}))$, (ii) $s(v) \neq F_b(s(\overline{y}))$, or (iii) $M, s \models \psi$.
  In case (i), Eloise sets $d_s = a$; in case (ii), Eloise sets $d_s = b$; in case (iii), Eloise sets $d_s = c$.

In the next move, Eloise chooses a value e_s for the variable w. If $d_s \in \{a, b\}$, she sets $e_s = F_{d_s}(s(\overline{z}))$; in case $d_s = c$, she chooses an arbitrary $e_s \in M$.

Note that the choice of d_s above does not depend on $s(z_l)$ for any l. Similarly, the choice of e_s is independent of $s(x_1), \ldots, s(x_n), s(u), s(y_1), \ldots, s(y_n), s(v)$ and $s(i)$. Thus, the strategy of Eloise described above is uniform. Furthermore, it is straightforward to verify that Eloise has a winning strategy in $G(\eta, M, s')$, where $s' = s(d_s/j, e_s/w)$. Thus, we see that $M \models \theta$.

Assume then for the other direction that $M \models \theta$. Then, given any assignment $s \in \mathrm{As}(X \cup Z, M)$, Eloise can choose values d_s and e_s for the variables j and w such that d_s does not depend on $s(\overline{z})$, e_s does not depend on $s(\overline{x}u\overline{y}vi)$ (but may depend on d_s), and Eloise has a winning strategy in the game $G(\eta, M, s(d_s/j, e_s/w))$.

We define now relations $F_a, F_b \subseteq M^{n+1}$ as follows:

- $F_a := \{(s(\overline{z}), e_s) \mid s \in \mathrm{As}(X \cup Z, M), d_s = a\}$,
- $F_b := \{(s(\overline{z}), e_s) \mid s \in \mathrm{As}(X \cup Z, M), d_s = b\}$.

It suffices to show that conditions (a), (b) and (c) hold for these relations. In order to prove (a), let $\overline{m} = (m_1, \ldots, m_n) \in M^n$, and consider an assignment $s \in \mathrm{As}(X \cup Z, M)$ such that $s(\overline{z}) = \overline{m}$ and $s(i) = a$. Then, d_s is necessarily a, since otherwise Eloise would lose the game $G(\eta, M, s(d_s/j, e_s/w))$ if Abelard chooses the first conjunct $i = \mathbf{a} \rightarrow j = \mathbf{a}$ of η. Thus, by the definition above, $(\overline{m}, e_s) \in F_a$. Condition (b) is proved symmetrically by using conjunct (2) of η.

Note that since the choice of e_s depends only on $s(\overline{z})$ and d_s, we can assume that the relations F_a and F_b are actually functions $M^n \rightarrow M$.

To prove (c), let s_0 be an assignment with domain $X \setminus \{i\}$. We need to show that $F_a(s_0(\overline{x})) \neq s_0(u)$, $F_b(s_0(\overline{y})) \neq s_0(v)$ or $M, s_0 \models \psi$. Let s be an extension of s_0 to domain $X \cup Z$ such that $s(i) = c$. Then necessarily $d_s \in \{a, b, c\}$, since otherwise Eloise would lose the game $G(\eta, M, s(d_s/j, e_s/w))$ if Abelard chooses the last conjunct (6) of η.

Assume first that $d_s = a$. Since the choice of d_s does not depend on $s(\overline{z})$, we have $d_{s'} = d_s = a$, where $s' = s(s(\overline{x})/\overline{z})$. Then by the definition of F_a, we have $e_{s'} = F_a(s'(\overline{z})) = F_a(s_0(\overline{x}))$. On the other hand, it must be the case that $e_{s'} \neq s'(u) = s_0(u)$, since otherwise Eloise would lose the game $G(\eta, M, s'(d_{s'}/j, e_{s'}/w))$ if Abelard chooses conjunct (3) of η. Thus, we see that $F_a(s_0(\overline{x})) \neq s_0(u)$. In the case $d_s = b$, we can prove in the same way that $F_b(s_0(\overline{y})) \neq s_0(v)$, by using conjunct (4) of η.

Assume finally, that $d_s = c$. Then it follows immediately that $M, s_0 \models \psi$. This is because otherwise Eloise would lose the game $G(\eta, M, s(d_s/j, e_s/w))$ if Abelard chooses conjunct (5) of η. $\qquad\square$

We will next eliminate the assumption of three constants with distinct interpretations. On structures with at least two different elements, this is done by replacing the quantifiers $\forall i$ and $(\exists j/Z)$ in θ by the sequences $\forall i \forall i' \forall i''$ and $(\exists j/Z)(\exists j'/Z)(\exists j''/Z)$, respectively. Furthermore, the subformulae $i = \mathbf{a}$, $i = \mathbf{b}$ and $i = \mathbf{c}$ of η are replaced by $i = i' \wedge i \neq i''$, $i = i'' \wedge i \neq i'$ and $i' = i'' \wedge i \neq i'$, and similarly for the subformulae $j = \mathbf{a}$, $j = \mathbf{b}$ and $j = \mathbf{c}$. Let θ' be the formula obtained from θ by performing these changes. By a straightforward modification of the proof of Lemma 1, we see that $M \models \varphi \Leftrightarrow M \models \theta'$ holds for all structures M with at least two elements.

If M has only one element, then clearly $M \models \varphi \Leftrightarrow M \models \forall \overline{x} \forall u \forall \overline{y} \forall v \, \psi$. Furthermore, the implication $M \models \forall \overline{x} \forall u \forall \overline{y} \forall v \, \psi \Rightarrow M \models \varphi$ holds for all structures. Thus, we see that φ is equivalent to θ^* on all structures, where θ^* is obtained from θ' by adding (in the end of the prefix) the sequence $\forall \overline{x}' \forall u' \forall \overline{y}' \forall v'$ of universal quantifiers and the disjunct $\psi(\overline{x}', u', \overline{y}', v')$ to the quantifier-free part, for some fresh variables $\overline{x}' = (x_1', \ldots, x_n'), \overline{y}' = (y_1', \ldots, y_n'), u'$ and v'. This completes the proof of the main result in this section:

Theorem 2. $\mathrm{IF}^{\mathrm{p,r}}_{\mathrm{AR}(\exists)}$ *has the same expressive power as* ESO. *In particular, any class definable in* IF *is already definable in* $\mathrm{IF}^{\mathrm{p,r}}_{\mathrm{AR}(\exists)}$.

Note that the length of the $\mathrm{IF^r_{AR(\exists)}}$ translation θ^* given in the proof of Theorem 2 is only linear with respect to the length of the original H_2^n formula φ. Another interesting observation that follows from the proof is that there is no hierarchy of expressive power based on the number or length of signalling patterns: the signalling pattern in θ^* is independent of n.

5   No-Henkin, No-Signalling Irregular IF Logic

In Sect. 4, we showed that all ESO properties can be defined by regular, prenex IF sentences of action recall. Such sentences are of the signalling, but not of the Henkin kind. We show now that, if the restrictions of regularity and prenex form are abandoned, then all ESO properties can be expressed by IF sentences which contain neither Henkin nor signalling patterns.

Eliminating Henkin and Signalling Sequences by Requantification

In order to create a Henkin or a signalling pattern, we need to have two existential quantifiers that have certain dependencies to each other. In order to eliminate these patterns, we now attempt to replace existential quantifiers with universal quantifiers that have essentially the same role in the semantic game. This can be done by a simple trick if we allow requantification of variables.

Let φ be a regular IF formula and let y be a fresh variable. Suppose that φ has a subformula of the form $(\exists x/V)\psi$. Now $(\exists x/V)\psi$ is equivalent to the irregular IF formula $(\exists y/V)\forall x(x \neq y \vee \forall y\,\psi)$. The truth of this equivalence can be seen by the following game-theoretical reasoning: After the quantification for $(\exists y/V)$, Abelard has to choose the same value for x as Eloise chose for y, since else he loses the game (when Eloise chooses the left disjunct). Hence we see that Eloise can indirectly "force" Abelard to choose a value for x in any (V-uniform) way she wishes. And since Abelard may then choose a new value for y, Eloise cannot use its value for signalling later in the game.

By replacing $(\exists x/V)\psi$ with $(\exists y/V)\forall x(x \neq y \vee \forall y\,\psi)$ in the sentence φ, we obtain an equivalent sentence φ^*. If the existential quantifier $(\exists x/V)$ created Henkin or signalling patterns in φ with some other quantifiers, these patterns are broken, since x is not existentially quantified anymore in φ^*. The new existential quantifier $(\exists y/V)$, that was introduced, cannot create any new Henkin patterns since the fresh variable y is not in any slash set of φ^* and $(\exists y/V)$ has no existential quantifiers in its effective scope.

By iterating this translation process to every existential quantifier in φ, we obtain an equivalent formula φ' which has no Henkin patterns. But φ' has no signalling patterns either, since no existential quantifier in φ' is in the effective scope of any other existential quantifier.

Theorem 3. *Every IF sentence φ is equivalent to an irregular IF sentence φ' that has neither Henkin nor signalling patterns.*

Proof. Let φ be any IF-sentence. Let $\{x_1, \ldots, x_k\}$ be the set of (distinct) variables that are existentially quantified in φ. Let $\{y_1, \ldots, y_k\}$ be a set of distinct variables that do not occur in φ. We define φ' recursively as follows:

$$\varphi' = \varphi, \quad \text{if } \varphi \text{ is a literal}$$
$$(\psi \vee \theta)' = \psi' \vee \theta', \ (\psi \wedge \theta)' = \psi' \wedge \theta'$$
$$((\forall x/V)\psi)' = (\forall x/V)\psi'$$
$$((\exists x_i/V)\psi)' = (\exists y_i/V)\forall x_i(x_i \neq y_i \vee \forall y_i \psi').$$

By the observations that we made above, φ' has neither Henkin nor signalling patterns. Thus it suffices to show that $M \models \varphi$ if and only if $M \models \varphi'$. This is quite easy to see by the game-theoretical intuition that we gave above. We give a formal proof for this equivalence in the appendix. □

Note that our translation above can be applied for any IF formula – including non-prenex and irregular formulae. Furthermore, the translation increases the length of a given formula only by a small constant for each existential quantifier in it. If a sentence φ in a prenex form is translated to φ' as above, the prenex form is lost. However, φ' is still "almost prenex" since only disjunctions with a literal as the left disjunct are created within the quantifier prefix. See the following example for explicitly expressing Henkin prefix H_2^1 without using Henkin or signalling patterns.

Example 4. Consider the IF sentence $\forall x \exists y \forall z (\exists w/\{x, y\})\psi$, where ψ is quantifier free, and suppose that y' does not occur in ψ. By applying our translation procedure to the most external occurrence of an existential quantifier, $\exists y$, we obtain the formula $\xi := \forall x \exists y' \forall y (y \neq y' \vee \forall y' \forall z (\exists w/\{x, y\})\psi)$. Note here that there is no need to apply the translation procedure to the innermost existential quantifier $(\exists w/\{x, y\})$. What happens to the flow of information in ξ? In the right disjunct, the variables y and y' carry the same value; as a signal, y is blocked by the slash set of $\exists w$ and, as a signal, y' is blocked by $\forall y'$; but the value of y (equal to the old value of y') can still be used within ψ.

Action Recall for Irregular Formulae

As we discussed after defining the restriction (RS), there are irregular formulae which satisfy it, but for which Eloise does not have action recall in the corresponding semantic game. We analyse here what would be the proper characterization of the syntactical fragment of action recall for irregular formulae.

Let φ be an irregular IF sentence in which a variable x is quantified twice. If x is first universally quantified, this requantification does not violate action recall for Eloise, since the first value for x is chosen by Abelard. But if x is first existentially quantified, then there is a play of the semantic game for φ in which Eloise chooses the value for x and then forgets that value when x is requantified, and thus she does not have action recall, supposing that Eloise has at least one action to perform in the game after x has been requantified.

By the observation above, the action recall fragment for Eloise for all (including irregular) formulae, denoted by $\mathrm{IF}_{\mathrm{AR}(\exists)}$, is obtained simply by requiring the following condition in addition to (RS):

– No quantifier (Qx/V) is in the scope of a quantifier $(\exists x/W)$ such that a formula of the form $\psi \vee \theta$ or $(\exists y/U)\psi$ is in the scope of (Qx/V).

6   Conclusions

We have shown that full IF (i.e., ESO) expressive power can be achieved, without the use of Henkin prefixes, already within each of the two following fragments of IF logic: (1) prenex, regular IF logic with action recall ($\mathrm{IF}^{\mathrm{r}}_{\mathrm{AR}(\exists)}$), and (2) non-prenex, irregular IF logic without Henkin and signalling patterns. The proof of the first result shows that the H^n_2 Henkin prefixes are explicitly definable by means of signalling prefixes with a constant number of signalling variables. Consequently, there are no hierarchies based on the number or length of signalling patterns in $\mathrm{IF}^{\mathrm{r}}_{\mathrm{AR}(\exists)}$.

These results extend the analysis of the expressive resources of IF logic which was initiated in [2,17], and they raise a number of questions to be further investigated:

– Is it possible to capture ESO within $\mathrm{IF}^{\mathrm{r}}_{\mathrm{AR}(\exists)}$ without the use of signalling? Note that this is not possible for $\mathrm{IF}^{\mathrm{p,r}}_{\mathrm{AR}(\exists)}$, since prenex, regular IF logic without Henkin and signalling patterns collapses to first-order logic [17].
– When considering irregular prenex sentences, are there other sources of second-order expressive power, besides Henkin and signalling patterns?
– Are there interesting hierarchies of signalling prefixes, e.g. based on the number of universal or existential quantifiers?

7   Appendix

Example 1, and the Prefix $\exists f \forall x$

By applying so-called *Skolemization*, we can translate any IF sentence φ to an equivalent ESO sentence of the form $\exists f_1 \ldots \exists f_n \psi$, where ψ is a first order formula that contains no existential quantifiers. (The functions f_i here correspond to Eloise's "choice functions" for choosing values for the existentially quantified variables in φ.) For more details on Skolemization, see [15].

By applying Skolemization to $\forall x \exists y (\exists z/\{x\})(U(x) \rightarrow (K(y) \wedge E(x,z)))$, we obtain an equivalent ESO sentence $\varphi := \exists h \exists g \psi$, where

$$\psi := \forall x(U(x) \rightarrow (K(h(x)) \wedge E(x, g(h(x))))).$$

A proof that this ESO sentence defines the problem EXACT COVER BY 3-SETS on appropriate structures can be found in [17]. Instead, we prove here that, if we restrict the class of adequate structures for the problem EXACT COVER

BY 3-SETS by the additional constraint $K^M \subseteq M \setminus U^M$, then φ captures the same class of structures as

$$\varphi^* := \exists f \forall x (U(x) \to (K(f(x)) \wedge E(x, f(f(x))))).$$

This, together with the arguments in [17], gives a new proof that the NP-complete problem EXACT COVER BY 3-SETS is expressible by φ^*, that is, by the quantifier prefix $\exists f \forall x$. (Note that the additional constraint $K^M \subseteq M \setminus U^M$ does not decrease the generality of the problem.)

In one direction, it is apparent that φ^* logically implies φ. Suppose instead that φ holds in an appropriate structure M. Let $g, h : M \to M$ be two functions that satisfy ψ. Define

$$f(x) = \begin{cases} h(x) \text{ if } x \in U^M \\ g(x) \text{ if } x \in M \setminus U^M \end{cases}$$

Then, for all $a \in U^M$, we have $f(a) = h(a)$ and so from $h(a) \in K^M$ we obtain $f(a) \in K^M$; from our assumption that $K^M \subseteq M \setminus U^M$ we get $f(a) \in M \setminus U^M$; so, $g(h(a)) = g(f(a)) = f(f(a))$; then, from the fact that $(a, g(h(a))) \in E^M$ we deduce $(a, f(f(a))) \in E^M$. Therefore $M \models \varphi^*$.

Example 2, DOMINATING SET

We need to prove that the DOMINATING SET problem is captured by the sentence $\forall x \exists z (\exists y / \{x\})((E(y, x) \vee y = x) \wedge P(z))$. By using Skolemization, it suffices to prove this claim for the ESO sentence

$$\zeta := \exists f \exists g \forall x ((E(g(f(x)), x) \vee g(f(x)) = x) \wedge P(f(x))).$$

Fix an integer k. Let $G = (V, E^G, P^G)$ be any structure such that (V, E^G) is a graph, and such that $P^G = \{d_1, \ldots, d_k\}$ is a subset of V of cardinality k. Suppose first that G has a dominating set D of cardinality k. Enumerate D as $\{c_1, \ldots, c_k\}$. Since D is a dominating set, to each $a \in V$ we can associate a $b_a \in D$ such that either $(b_a, a) \in E^G$ or $b_a = a$. Now, define $f : V \to P^G$ as follows: if $b_a = c_i$, then set $f(a) := d_i$. Define $g : V \to V$ as follows: $g(d_i) = c_i$; for $a \in V \setminus P^G$, $g(a)$ takes an arbitrary value. Note then that, by the definitions, for every $a \in V$, $g(f(a)) = b_a$. Therefore, $(G, f, g) \models E(g(f(x)), x) \vee g(f(x)) = x$. And the definition of f implies that $(G, f, g) \models P(f(x))$.

Suppose instead that $G \models \zeta$. Then, there are functions $f : V \to P^G$ and $g : V \to V$ such that, for every $a \in V$, either $(g(f(a)), a) \in E^G$ or $g(f(a)) = a$. Define $D := g[P^G] = \{g(a) \mid a \in P^G\}$. Clearly $\mathrm{Card}(D) \leq \mathrm{Card}(P^G) = k$, and since $g(f(a)) \in D$ for every $a \in V$, D is a dominating set.

Example 3, SAT by Signalling

For proving that $\forall x \exists y (\exists z / \{x\})(C(x) \to ((P(y, x) \wedge z = 1) \vee (N(y, x) \wedge z = 0)))$ captures SAT, we apply Skolemization again and prove the claim for the ESO sentence $\xi := \exists f \exists g \psi$, where

$$\psi := \forall x (C(x) \to ((P(f(x), x) \wedge g(f(x)) = 1) \vee (N(f(x), x) \wedge g(f(x)) = 0))).$$

Let M be an appropriate structure, and π the propositional formula encoded by it. Suppose first that M is a "yes" instance of SAT; then there is a truth assignment T such that each clause c of π contains a literal α_c for which we have $T(\alpha_c) = 1$. A literal α_c can either be of the form p_c or $\neg p_c$, with p_c a proposition symbol. In the first case, we then have $T(p_c) = 1$, while in the second $T(p_c) = 0$. Let $f : M \to M$ be the function that maps c to p_c (define it arbitrarily on elements that are not clauses); let $g : M \to M$ be defined by $g(p) := T(p)$ if p is a proposition symbol, and an arbitrary constant otherwise. With these f and g, $(M, f, g) \models \psi$.

Vice versa, suppose $M \models \xi$. Let f, g be two functions that satisfy ψ. Let T be a truth assignment such that $T(p) = g(p)$ for all the proposition symbols in π. Now for any $c \in C^M$, either $(f(c), c) \in P^M$ and $g(f(c)) = 1$, or $(f(c), c) \in N^M$ and $g(f(c)) = 0$. In the former case $f(c)$ is a proposition symbol occurring positively in c, to which T assigns truth value 1. Similarly, in the second case $f(c)$ is a proposition symbol which occurs negatively in c, to which T assigns truth value 0. These remarks show that T satisfies π.

Theorem 3, a Proof of the Equivalence

We argue by using the so-called *team semantics* for IF logic. For the definition of this semantics – and the notation and terminology that we use here – refer to [15]. In the proof we also use the *downwards closure property* of IF logic. That is, if $M, X \models \varphi$ and $Y \subseteq X$, then $M, Y \models \varphi$.

Let μ be a subformula of φ and let X be a team for which $\mathrm{dom}(X) = \mathrm{Free}(\mu)$. We show by the structural induction on φ that the following equivalence holds for any team Y for which $\mathrm{dom}(Y) = \{y_1, \ldots, y_k\}$:

$$M, X \models \mu \quad \text{iff} \quad M, X \times Y \models \mu', \tag{$\star$}$$

where $X \times Y := \{s \cup s' \mid s \in X \text{ and } s' \in Y\}$. (Note that $\mathrm{dom}(X) \cap \mathrm{dom}(Y) = \emptyset$.)

- The case when μ is a literal holds trivially since then $\mu = \mu'$ and the values of the variables y_i in the team do not affect the truth of μ.
- The cases $\mu = \psi \wedge \theta$ and $\mu = (\forall x_j / V)\psi$ are proven easily by just applying the inductive hypothesis.
- Let $\mu = \psi \vee \theta$. Suppose first that $M, X \models \psi \vee \theta$ and let Y be a team. Now there are $X_1, X_2 \subseteq X$ s.t. $X_1 \cup X_2 = X$, $M, X_1 \models \psi$ and $M, X_2 \models \theta$. By the inductive hypothesis we have $M, X_1 \times Y \models \psi'$ and $M, X_2 \times Y \models \theta'$. Since $X_1 \times Y \cup X_2 \times Y = X \times Y$, we have $M, X \times Y \models \psi' \vee \theta'$, i.e. $M, X \times Y \models \mu'$. Suppose then that $M, X \times Y \models \mu'$ for any Y. In particular $M, X \times \{r\} \models \mu'$ for some singleton $\{r\}$. Now there are $X_1, X_2 \subseteq X$ s.t. $X_1 \cup X_2 = X$, such that $M, X_1 \times \{r\} \models \psi'$ and $M, X_1 \times \{r\} \models \theta'$. By the inductive hypothesis we have $M, X_1 \models \psi$ and $M, X_2 \models \theta$. Therefore $M, X \models \psi \vee \theta$, i.e. $M, X \models \mu$.
- Let $\mu = (\exists x_j / V)\psi$ for some $j \leq k$. Suppose first that $M, X \models \varphi$ and let Y be a team. Hence there is a V-uniform $f : X \to M$ s.t. $M, X[f/x_j] \models \psi$. Let $g : X \times Y \to M$ s.t. $g(s) = f(s \upharpoonright \mathrm{dom}(X))$ for every $s \in X \times Y$. We define the teams $X' := (X \times Y)[g/y_j, M/x_j]$, $X_1 := \{s \in X' \mid x_j \neq y_j\}$ and

$X_2 := \{s \in X' \mid x_j = y_j\}$. Now clearly $X_1 \cup X_2 = X'$ and $M, X_1 \models x_j \neq y_j$. Since $M, X_2 \models x_j = y_j$, by the definition of g it is quite easy to see that

$$X_2[M/y_j] = (X[f/x_j]) \times (Y[M/y_j])$$

By the inductive hypothesis we have $M, (X[f/x_j]) \times (Y[M/y_j]) \models \psi'$ and therefore $M, X_2[M/y_j] \models \psi'$. Furthermore we have $M, X_2 \models \forall y_j \psi'$ and thus $M, X' \models x_j \neq y_j \vee \forall y_j \psi'$. Since f was V-uniform also g is V-uniform and thus $M, X \times Y \models (\exists y_j/V) \forall x_j (x_j \neq y_j \vee \forall y_j \psi')$, i.e. $M, X \times Y \models \mu'$.

Suppose then that $M, X \times Y \models \mu'$ for any Y. In particular $M, X \times \{r\} \models \mu'$ for some singleton $\{r\}$. Now there is a V-uniform function $g : X \times \{r\} \to M$ s.t. $M, X' \models x_j \neq y_j \vee \forall y_j \psi'$, where $X' = (X \times \{r\})[g/y_j, M/x_j]$. Hence there are $X_1, X_2 \subseteq X'$ such that $M, X_1 \models x_j \neq y_j$ and $M, X_2 \models \forall y_j \psi'$. Let $Z := \{s \in X' \mid s(x_j) = s(y_j)\}$. Since $M, X_1 \models x_j \neq y_j$ we must have $Z \subseteq X_2$. Thus by downwards closure $M, Z \models \forall y_j \psi'$, i.e. $M, Z[M/y_j] \models \psi'$.

Let $f : X \to M$ be such that $f(s) = g(s \cup r)$. Since $M, Z \models x_j = y_j$, by the definition of f it is quite easy to see that

$$(X[f/x_j]) \times \{r\} \subseteq Z[M/y_j].$$

Therefore, again by downwards closure, $M, (X[f/x_j]) \times \{r\} \models \psi'$. By the inductive hypothesis $M, X[f/x_j] \models \psi$. Since g was V-uniform also f is V-uniform and thus we have $M, X \models (\exists x_j/V)\psi$, i.e. $M, X \models \mu$.

By Theorem 5.2 of [3] the values of variables in a team cannot affect the truth of an IF *sentence*. Therefore, when $\mu = \varphi$ and $X = \{\emptyset\}$, the equivalence in the proof of Theorem 3 follows from $(\star)$. This concludes the proof.   □

References

1. Barbero, F.: On existential declarations of independence in IF logic. Rev. Symb. Log. **6**, 254–280 (2013)
2. Barbero, F.: Complexity of syntactical tree fragments of Independence-Friendly logic. pre-print arXiv:1610.03406
3. Caicedo, X., Dechesne, F., Janssen, T.M.V.: Equivalence and quantifier rules for logic with imperfect information. Log. J. IGPL **17**, 91–129 (2009)
4. Caicedo, X., Krynicki, M.: Quantifiers for reasoning with imperfect information and Σ_1^1-logic. In: Carnielli, W.A., D'Ottaviano, I.M.L. (eds.) Contemporary Mathematics, vol. 235, pp. 17–31. American Mathematical Society (1999)
5. Cameron, P., Hodges, W.: Some combinatorics of imperfect information. J. Symb. Log. **66**, 673–684 (2001)
6. Dahlhaus, E.: Reduction to NP-complete problems by interpretations. In: Proceedings of the Symposium "Rekursive Kombinatorik" on Logic and Machines: Decision Problems and Complexity, pp. 357–365 (1983)
7. Enderton, H.B.: Finite partially ordered quantifiers. Math. Log. Q. **16**(8), 393–397 (1970)
8. Grandjean, E.: First-order spectra with one variable. J. Comput. Syst. Sci. **40**, 136–153 (1990)

9. Henkin, L.: Some remarks on infinitely long formulas. In: Infinitistic Methods. Pergamon Press, Oxford, New York (1961)
10. Hintikka, J., Sandu, G.: Informational independence as a semantical phenomenon. In: Fenstad, J.E., et al. (eds.) Logic, Methodology and Philosophy of Science VIII, pp. 571–589. Elsevier Science Publishers B.V, Amsterdam (1989)
11. Hodges, W.: Compositional semantics for a language of imperfect information. Log. J. IGPL **5**, 539–563 (1997)
12. Hodges, W.: Some strange quantifiers. In: Mycielski, J., Rozenberg, G., Salomaa, A. (eds.) Structures in Logic and Computer Science. LNCS, vol. 1261, pp. 51–65. Springer, Heidelberg (1997). doi:10.1007/3-540-63246-8_4
13. Hyttinen, T., Tulenheimo, T.: Decidability of IF modal logic of perfect recall. In: Advances in Modal Logic, vol. 5 (2005)
14. Krynicki, M.: Hierarchies of partially ordered connectives and quantifiers. Math. Log. Q. **39**, 287–294 (1993)
15. Mann, A.L., Sandu, G., Sevenster, M.: Independence-Friendly Logic - a Game-Theoretic Approach. London Mathematical Society Lecture Note Series, vol. 386. Cambridge University Press, Cambridge (2011)
16. Sevenster, M.: Branches of imperfect information: logic, games, and computation. Ph.D. thesis, ILLC, Universiteit van Amsterdam (2006)
17. Sevenster, M.: Dichotomy result for independence-friendly prefixes of generalized quantifiers. J. Symbol. Log. **79**(04), 1224–1246 (2014)
18. Väänänen, J., Logic, D.: A New Approach to Independence Friendly Logic. London Mathematical Society Student Texts, vol. 70. Cambridge University Press, Cambridge (2007)
19. Virtema, J.: Approaches to finite variable dependence. Ph.D. thesis, University of Tampere (2014)

Total Search Problems in Bounded Arithmetic and Improved Witnessing

Arnold Beckmann and Jean-José Razafindrakoto[✉]

Department of Computer Science, College of Science,
Swansea University, Swansea, UK
a.beckmann@swansea.ac.uk, jjrazaf@icloud.com

Abstract. We define a new class of total search problems as a subclass of Megiddo and Papadimitriou's class of total NP search problems, in which solutions are verifiable in AC^0. We denote this class $\forall\exists AC^0$. We show that all total NP search problems are equivalent, w.r.t. AC^0-many-one reductions, to search problems in $\forall\exists AC^0$. Furthermore, we show that $\forall\exists AC^0$ contains well-known problems such as the Stable Marriage and the Maximal Independent Set problems. We introduce the class of Inflationary Iteration problems in $\forall\exists AC^0$, and show that it characterizes the provably total NP search problems of the bounded arithmetic theory corresponding to polynomial-time. Cook and Nguyen introduced a generic way of defining a bounded arithmetic theory VC for complexity classes C which can be obtained using a complete problem. For such C we will define a new class KPT[C] of $\forall\exists AC^0$ search problems based on Student-Teacher games in which the student has computing power limited to AC^0. We prove that KPT[C] characterizes the provably total NP search problems of the bounded arithmetic theory corresponding to C. All our characterizations are obtained via "new-style" witnessing theorems, where reductions are provable in a theory corresponding to AC^0.

1 Introduction

The two-sorted bounded arithmetic theories VC [8] are well-known for their proof theoretic strength corresponding to complexity classes C, for many C between AC^0 and PH. It is a fundamental open question in computer science whether any two complexity classes within the following sequence

$$AC^0(6) \subseteq TC^0 \subseteq NC^1 \subseteq L \subseteq NL \subseteq NC \subseteq P \subseteq NP \subseteq PH,$$

are equal or not, a question which is a weaker version of the P versus NP question. Likewise, it is a fundamental open problem whether any of the corresponding bounded arithmetic theories are distinct. The difference in working with bounded arithmetic theories instead directly with computational classes is that the theories may possibly be shown to be distinct by combining logical considerations of provability along with computational complexity considerations. Another motivation for studying bounded arithmetic theories lies in their relation to propositional proof complexity, in that proving in bounded arithmetic

© Springer-Verlag GmbH Germany 2017
J. Kennedy and R.J.G.B. de Queiroz (Eds.): WoLLIC 2017, LNCS 10388, pp. 31–47, 2017.
DOI: 10.1007/978-3-662-55386-2_3

theories corresponds to uniform provability in corresponding propositional proof systems [8]. In this paper, we give characterizations of the total search problems with AC^0 graphs which are definable in bounded arithmetic theories VC for many C between AC^0 and P, where necessary reductions are proven in the weakest theory of bounded arithmetic V^0 related to AC^0 reasoning. In particular, we give improved "new-style" witnessing theorems for such theories.

A classical way to associate a theory T with a complexity class C is to show that the provably total functions in T are precisely the functions in the function class FC associated with C. This assertion splits into two parts: the first, usually easier part shows that all function in FC can be suitably defined and proven total in T (and thus are called provably total); the second, usually more involved part often employs a witnessing theorem. Witnessing theorems in their original form were introduced by Buss [5] to show that existential statements with parameters provable in a bounded arithmetic theory T can be witnessed by functions from a corresponding function class, and that this witnessing property is provable in T. For example, one result of Buss [5], adapted to the two-sorted bounded arithmetic theory V^1, shows that given a $\forall \Sigma_1^B$-consequence of V^1 we can find a polynomial time computable function witnessing the existential quantifier, where the correctness of the witnessing function is provable in V^1. Σ_1^B formulas have a certain syntactic form starting with a bounded existential quantifier—such formulas express exactly NP properties over the domain of natural numbers. We will denote the set of $\forall \Sigma_1^B$-consequences of a theory T by $\forall \Sigma_1^B(T)$.

Cook and Nguyen [8] have a generic way of defining a bounded arithmetic theory VC for those complexity classes C which can be obtained using a complete problem. They show that the set of provably total functions in VC corresponds to FC. Their approach is to construct a universal conservative extension $\overline{VC}$ of VC, where the terms of $\overline{VC}$ represent precisely functions in FC. They then apply Herbrand's theorem to obtain their desired correspondence. The correctness of witnessing functions is proved in VC.

Recently the focus has turned to "new-style" witnessing theorems, in which the correctness of the witnessing function is proved in a weaker theory than the one proving the $\forall \Sigma_1^B$-statement [2–4,15,24]. Furthermore, the focus has shifted to search problems, i.e. multifunctions, instead of functions. The class TFNP [20] of total NP search problems, whose solutions are verifiable in polynomial-time, has been extensively studied from the point of view of complexity theory and contains a host of important problems like the Polynomial Local Search problems PLS [13]. For the theories V^i corresponding to the i-th level of the polynomial time hierarchy PH, a host of characterizations of $\forall \Sigma_1^B(V^i)$ have been given in terms of subclasses of TFNP, using V^1-provability for correctness of witnessing functions. For instance, Buss and Krajíček [6] characterized $\forall \Sigma_1^B(V^2)$ in terms of PLS; Krajíček et al. [17] characterized $\forall \Sigma_1^B(V^3)$ in terms of colored PLS (denoted CPLS), and, for $0 < i$, Beckmann and Buss [2,3] characterized $\forall \Sigma_1^B(V^{i+1})$ in terms of some relativized notion of PLS called Π_i^p-PLS with Π_i^p-goals, which we denote Π_i^p-PLS for the purpose of this introduction.

The aim of this paper is to provide characterizations of $\forall \Sigma_1^B(\mathsf{VC})$, for C below P, and $\forall \Sigma_1^B(\mathsf{V}^1)$ in terms of subclasses of TFNP, using new-style witnessing theorems in which the correctness of witnessing functions is provable in V^0—these new-style witnessing theorems are similar to the ones in [2–4,15,24], except for the correctness of witnessing functions that is now proved over a weaker theory. To achieve our aim, we define the class of *total* $\mathsf{N\text{-}AC}^0$ *search problem* as those total NP search problems for which solutions are verifiable in AC^0 rather than in P. We denote this class as $\forall \exists \mathsf{AC}^0$. From the point of view of bounded arithmetic, $\forall \exists \mathsf{AC}^0$ can be identified with the set of all true $\forall \Sigma_1^B$-sentences. We will show that $\forall \exists \mathsf{AC}^0$ is equivalent to TFNP under AC^0-many-one reductions, and that it contains many well-known problems like the problem of finding an inverse of a square matrix, the Stable Marriage problem, or the Maximal Independent Set problem.

Each known characterization of $\forall \Sigma_1^B(\mathsf{V}^i)$ as a subclass $\mathcal{S}$ of TFNP, which can be found in the literature, is given in the form of a generic search principle $\mathcal{S}'(F_1, \ldots, F_n)$ such that $\mathcal{S}$ is obtained by instantiating $F_1, \ldots, F_n$ in $\mathcal{S}'$ with all possible choices of functions from FP. It is then natural to consider $\mathsf{AC}^0\text{-}\mathcal{S}$ obtained by instantiating $\mathcal{S}'$ with functions from FAC^0, and study the question whether $\mathsf{AC}^0\text{-}\mathcal{S}$ still characterizes $\forall \Sigma_1^B(\mathsf{V}^i)$ under AC^0-many-one reducibility, provable in V^0. For many characterisations, it is the case: Cook and Nguyen [8] showed that $\mathsf{AC}^0\text{-}\mathsf{PLS}$ characterizes $\forall \Sigma_1^B(\mathsf{V}^2)$ under AC^0-many-one reducibility, provable in V^0. Furthermore, it is shown in [23] that $\mathsf{AC}^0\text{-}(\Pi_i^p\text{-}\mathsf{PLS})$ characterizes $\forall \Sigma_1^B(\mathsf{V}^{i+1})$ under AC^0-many-one reducibility, provable in V^0. From that latter result and the fact that CPLS characterizes $\forall \Sigma_1^B(\mathsf{V}^3)$, it follows directly that $\mathsf{AC}^0\text{-}\mathsf{CPLS}$ is AC^0-many-one reducible to $\mathsf{AC}^0\text{-}(\Pi_i^p\text{-}\mathsf{PLS})$. However, it is an open problem whether the other direction holds. We conjecture that $\mathsf{AC}^0\text{-}\mathsf{CPLS}$, based on CPLS in its literal form as defined in [17], is not AC^0-many-one reducible to $\mathsf{AC}^0\text{-}(\Pi_1^p\text{-}\mathsf{PLS})$—we note here that proving this conjecture implies $\mathsf{P} \neq \mathsf{NP}$.

The outline of the paper is as follows: The next section is a preliminary section providing the necessary background. In Sect. 3, we introduce the class $\forall \exists \mathsf{AC}^0$ as a subclass of TFNP, and show that it is equivalent to TFNP w.r.t. AC^0-many-one reducibility, and that it contains a variety of well-known problems.

In Sect. 4, we define a class of total $\mathsf{N\text{-}AC}^0$ search problems which we call $\mathsf{KPT}[\mathsf{C}]$. The class $\mathsf{KPT}[\mathsf{C}]$ is a class of total search problems motivated by the KPT witnessing theorem [16], where the process of finding a solution to an instance of a problem in $\mathsf{KPT}[\mathsf{C}]$ is carried out cooperatively between a student S and a teacher T: the student computes a potential solution, that either T accepts or rejects, and in the case that T rejects, then T must come up with a counterexample that S can then use in order to compute the next candidate solution. We use $\mathsf{KPT}[\mathsf{C}]$ in order to characterize $\forall \Sigma_1^B(\mathsf{VC})$, where the reduction is provable in V^0, using a new-style witnessing theorem for VC.

For $\forall \Sigma_1^B(\mathsf{V}^1)$, we introduce, in Sect. 5, a class of total $\mathsf{N\text{-}AC}^0$ search problems that we call *Inflationary Polynomial Local Search* (IPLS). The class IPLS is $\mathsf{AC}^0\text{-}\mathsf{PLS}$, but with some restriction on its neighborhood function in that this function must be inflationary. We show that IPLS has a complete problem class

that we call Inflationary Iteration (IITER), which is based on the iteration principle [7] (which can be viewed as the problem of finding a sink in an exponentially
large directed acyclic graph). We show that IITER characterizes $\forall \Sigma_1^B(V^1)$, where
the reduction is provable in V^0, using a new-style witnessing theorem for V^1.

2   Preliminaries

We assume familiarity with bounded arithmetic in either its one-sorted [5] or
two-sorted [8] setting, but we will quickly review all necessary notation and
results used in this paper. We assume a basic understanding of complexity classes
between AC^0 and P. For circuit complexity classes covered here, the uniformity
we implicitly use is first-order uniformity [12,21]. Overall, our exposition follows [8].

The Language of Two-Sorted Bounded Arithmetic. In the two-sorted setting,
there are two kinds of variables: *number variables* $x, y, z, \ldots$ of the first sort,
intended to range over $\mathbb{N}$, and *string variables* $X, Y, Z, \ldots$ of the second sort,
intended to range over finite subsets of $\mathbb{N}$. We interpret finite subsets of $\mathbb{N}$ as
bit strings. The base language $\mathcal{L}_A^2$ consists of the usual symbols $0, 1, +, \cdot, \leq$ of
arithmetic on $\mathbb{N}$, the function $|X|$ (whose intended meaning is 0 if X is empty,
and 1 plus the maximal element in X, otherwise), the set membership relation $\in$,
and the relations $=_1$ and $=_2$, which are intended to be the equality on numbers
and strings respectively. Since there will be no confusion, the subscripts in $=_i$
will often be omitted. We will usually write $X(i)$ for $i \in X$ and this is understood
to denote the i-th bit in X.

   Terms over $\mathcal{L}_A^2$ are built in the usual way. Note that the only string terms are
string variables. If $\mathcal{L}_A^2$ is extended with additional string function symbols, then
other string terms are built as usual. *Formulae* over $\mathcal{L}_A^2$ are built using $\wedge, \vee, \neg$,
number quantifiers (i.e., $\exists x$ and $\forall x$) and string quantifiers (i.e., $\exists X$ and $\forall X$).
Bounded number quantifiers are defined as usual, whereas the *bounded string
quantifier* $(\exists X \leq t)\varphi$ stands for $\exists X(|X| \leq t \wedge \varphi)$ and $(\forall X \leq t)\varphi$ stands for
$\forall X(|X| \leq t \supset \varphi)$, where $\varphi \supset \psi$ stands for $\neg \varphi \vee \psi$, and where X does not appear
in t.

   The class Σ_0^B (or Π_0^B) consists of those $\mathcal{L}_A^2$-formulae with no string quantifiers and only bounded number quantifiers. Inductively, Σ_{i+1}^B consists of those
formulae of the form $(\exists X_1 \leq t_1) \ldots (\exists X_k \leq t_k)\varphi$, where $\varphi \in \Pi_i^B$, and Π_{i+1}^B consists of those formulae of the form $(\forall X_1 \leq t_1) \ldots (\forall X_k \leq t_k)\varphi$, where $\varphi \in \Sigma_i^B$.
In general, we write $\Sigma_i^B(\mathcal{L})$ to denote the class Σ_i^B that allows function and
predicate symbols from $\mathcal{L} \cup \mathcal{L}_A^2$. Finally, a formula is in Σ_1^1 if it is of the form
$(\exists X_1) \ldots (\exists X_k)\varphi$, where $\varphi \in \Sigma_0^B$.

Two-Sorted Complexity Classes. Two-sorted complexity classes consist of rela-
tions $R(\boldsymbol{x}, \boldsymbol{X})$ that are taking arguments of both sorts, where the string arguments $\boldsymbol{X}$ are the main inputs and $\boldsymbol{x}$ only play an auxiliary role. However, for our
purpose, it is convenient to assume that R only takes a single string argument,

as we can always pair x, X into one single string X. The following fact will be frequently used:

Theorem 1 (Σ_0^B Representation Theorem [25]**).** *A relation is in* AC^0 *if, and only if, it is represented by some Σ_0^B-formula.*

For each two-sorted complexity C of interest, there is a corresponding function class FC. For a string function $F(X)$ to be in FC, $F(X)$ needs to be *p-bounded* (i.e., $|F(X)|$ is bounded by some polynomial in $|X|$) and its *bit graph* (i.e., the relation $B_F(i, X)$ that holds if, and only if, the i-th bit of $F(X)$ is 1) is in C.

Two-Sorted Bounded Arithmetic Theories. The theory BASIC consists of some finite set of axioms defining the non-logical symbols in $\mathcal{L}_A^2$. Then, for $i = 0, 1$, the theory V^i is BASIC plus the Σ_i^B-*comprehension axiom scheme*, denoted Σ_i^B-COMP, which is $(\exists X \leq y)(\forall z < y)[X(z) \leftrightarrow \varphi(z)]$, where $\varphi \in \Sigma_i^B$ and X does not occur free in φ. For $\Phi = \Sigma_i^B$, the following axiom schemes are provable in V^i:

Φ-IND: $[\varphi(0) \wedge \forall x(\varphi(x) \supset \varphi(x + 1))] \supset \forall x \varphi(x)$,
Φ-MAX: $\varphi(0) \supset (\exists x \leq y)(\varphi(x) \wedge (\forall z \leq y)(x < z \supset \neg \varphi(z)))$.

We will usually be working with a universal conservative extension $\overline{\mathsf{V}}^0$ of V^0, whose language $\mathcal{L}_{\overline{\mathsf{V}}^0}$ has a symbol for each function in FAC^0.

A string function $F(X)$ is *provably total in a theory* $\mathcal{T}$ if its graph $Y = F(X)$ is represented by a Σ_1^B-formula $\varphi(X, Y)$ and $\mathcal{T}$ proves $\forall X \exists! Y \varphi(X, Y)$.

For certain complexity classes C within P, Cook and Nguyen [8] showed how to construct a theory VC corresponding to FC (i.e., the provably total functions in VC are precisely those in FC). Before we give the definition of VC, let us first review the notion of AC^0-*reduction*.

A relation R is AC^0-*reducible* to a collection $\mathcal{L}$ of functions if there is a sequence of string functions $G_1, \ldots, G_n$ such that each G_i is p-bounded and its bit graph is represented by a $\Sigma_0^B(\mathcal{L} \cup \{G_1, \ldots, G_{i-1}\})$-formula and R is represented by a $\Sigma_0^B(\mathcal{L} \cup \{G_1, \ldots, G_n\})$-formula.

For a two-sorted complexity class C of interest, fix a function F so that C is the class of all relations that are AC^0-reducible to $\{F\}$ (we keep F fixed in what follows) and so that there is a Σ_0^B-formula $\delta_F(X, Y)$ and some $\mathcal{L}_A^2$-term $t(X)$ such that the graph $Y = F(X)$ of F is represented by $|Y| \leq t(X) \wedge \delta_F(X, Y)$. Furthermore, assume that V^0 proves the uniqueness of the value of F. Let the *aggregate function* $F^*(b, X)$ of $F(X)$ be the function that gathers the values of F for a polynomially long sequence of arguments. Thus, F^* is defined so that

$$\forall i < b, F^*(b, X)^{[i]} = F(X^{[i]}),$$

where $X^{[i]}(j)$ holds if and only if $j < |X| \wedge X(i, j)$ holds—we obtain arrays of more than one dimension by using a suitable pairing function $\langle x, y \rangle$ on numbers x, y, e.g. $X(i, j)$ stands for $X(\langle i, j \rangle)$. Let $G_F(b, X, Y)$ be a Σ_0^B-formula that represents the graph of $F^*(b, X)$. The theory VC is then V^0 plus the Σ_1^B-statement $(\exists Y \leq \langle b, t \rangle) G_F(b, X, Y)$.

Two-Sorted Search Problems. A *total search problem* (or simply a *search problem*) is a binary relation $R(X, Y)$ such that $\forall X \exists Y R(X, Y)$ holds (we also call R the *graph* of the search problem). The search task associated with R is the following: given an *instance* X of R, find a *solution* Y such that $R(X, Y)$ holds.

The class TFNP [20] consists of those search problems $R(X, Y)$ such that R is polynomial-time computable and $|Y|$ is bounded by a polynomial in $|X|$.

Let R be a search problem. R is *provably total in a theory* T if the graph of R is represented by a Σ_1^B-formula $\varphi(X, Y)$ and T proves $\forall X \exists Y \varphi(X, Y)$.

Let C be a complexity class. Then a search problem R is C-*many-one reducible* to a search problem Q, denoted $R \leq_m^C Q$, if there are functions $F, G \in$ FC such that $Q(F(X), Y)$ implies $R(X, G(X, Y))$, for all X, Y. For two classes Γ and Δ of search problems, we say that Γ is C-many-one reducible to Δ, denoted $\Gamma \leq_m^C \Delta$, if for all $R \in \Gamma$, there is some $Q \in \Delta$ such that $R \leq_m^C Q$. We say that Γ and Δ are C-*equivalent* if $\Gamma \leq_m^C \Delta$ and $\Delta \leq_m^C \Gamma$. Finally, we say that Γ is C-many-one complete for Δ if $\Gamma \subseteq \Delta$ and $\Delta \leq_m^C \Gamma$.

3   The Class $\forall \exists \mathbf{AC}^0$

Definition 2. *A search problem R is said to be in $\forall \exists AC^0$ if R can be expressed as a TFNP problem with AC^0 graph.*

We observe that $\forall \exists AC^0$ is AC^0-many-one equivalent to TFNP. To see this, note that the statement "string W is a valid encoding of the full computation of a fixed polynomial-time Turing machine on a given input" can be expressed by a Σ_0^B-formula. From that, and the Σ_0^B representation theorem, we can turn R into a $\forall \exists AC^0$ problem Q, whose solution can then be mapped into a solution for R.

Another motivation for studying $\forall \exists AC^0$ is the fact that it contains a host of well-known problems. As already noted in the introduction, Cook and Nguyen [8] show that PLS is equivalent to AC^0-PLS. Another example [8] stems from linear algebra: $(\star)$ given an $n \times n$ matrix A over some field, find an $n \times n$ matrix $B \neq 0$ such that $AB = I \vee AB = 0$. Observe that the provability of $(\star)$ in VNC^1 is still an open problem.

In what follows, we demonstrate that the Stable Marriage problem and the Maximal Independent Set problem are $\forall \exists AC^0$ problems.

The Stable Marriage Problem. The Stable Marriage problem (SM) was first introduced by Gale and Shapley [11]. Besides having practical applications, SM is of importance for the NC vs P question: It has been shown that SM is complete for Subramanian's complexity class CC [19], a subclass of P based on comparator circuits. Furthermore, Cook et al. [9] gave strong evidence that CC and NC, which is also a subclass of P, are incomparable.

An instance of size n of SM involves two sets of n men and n women. Associated with each person p is a strictly ordered preference list $l = q_1, \ldots, q_n$ containing all the members of the opposite sex: person p prefers person q to r if, and only if, there is a q_i and a q_j in l such that $q_i = q$ and $q_j = r$ and $i < j$.

Given an instance of SM, a *matching* M is a bijection between the sets of men and women. A man m and a woman w are called *partners in* M if, and only if, they are matched in M; we write $p_M(m)$ to denote the partner of m in M (similarly for $p_M(w)$). A matching M is called *unstable* if there is a man m and a woman w such that m and w are not partners in M, but m prefers w to $p_M(m)$ and w prefers m to $p_M(w)$; otherwise, M is called *stable*.

The search task associated with SM is as follows: given an instance of SM, find a matching that is stable. Gale and Shapley showed that such a stable matching always exists. Hence, SM is a total search problem.

We argue that the SM search problem is in $\forall\exists\mathsf{AC}^0$. Let $\{0,1,\ldots,n-1\}$ corresponds to the set of men and $\{n, n+1,\ldots,2n-1\}$ to the set of women. Then a preference list for a person p can be encoded in bounded arithmetic as a three-dimensional array $L(p, j, q_j)$, which holds if and only if q_j sits at j-th position in person p's preference list. A matching can be encoded as a two-dimensional array $M(p, q)$ with size bounded by $\langle n, n\rangle$. It is easy to see that the statement "M is a stable matching for (n, L)" can be expressed as a Σ_0^B-formula. Thus, by the representation theorem for Σ_0^B, SM is a $\forall\exists\mathsf{AC}^0$ search problem.

The Maximal Independent Set Problem. Another example for a problem in $\forall\exists\mathsf{AC}^0$ is the Maximal Independent Set problem (MIS), which is a fundamental problem in Graph Theory since several important problems can be reduced to it. For instance, Karp and Widgerson [14] show that the maximal set packing and the maximal matching problems are NC^1-reducible to MIS, and that the 2-satisfiability problem is NC^2-reducible to MIS. In terms of its complexity, Luby [18] and, independently, Alon et al. [1] proved the existence of NC^2-algorithms that solve MIS. However, it is still open whether MIS can be solved by an NC^1-algorithm.

Let G be a graph. An *independent set* in G is a set of vertices such that no two of them are adjacent. A *maximal independent set* I in G is an independent set such that for every vertex v in G, either v belongs to I or v has at least one neighbor vertex that belongs to I.

The MIS problem is the following computational problem: given a graph G, find a maximal independent set in G. MIS is a total search problem, since for a given graph G, a maximal independent set I is always guaranteed to exist.

The MIS problem is in $\forall\exists\mathsf{AC}^0$. For that, we specify a graph G by a pair (n, E), where $0, 1, \ldots, n-1$ are the vertices in G and $E(u, v)$ holds if, and only if, there is an edge between vertex u and v in G. Then the statement "U is a maximal independent set in G" can be written as a Σ_0^B-formula. Note that the size of U is bounded by n.

4   The Class KPT[C] and VC

For this section, we fix a function $F(X)$ in FC so that C is the AC^0-closure of F. Let $G_F(b, X, Y)$ be a Σ_0^B-formula that states that Y is the value for the aggregate function $F^*(b, X)$ of $F(X)$. In the following we will identify F with F^*—it will be clear from the context which of the two is meant.

The following lemma is an application of the KPT witnessing theorem [16]. It says that if the theory VC proves $\forall X \exists Y \varphi(X,Y)$, where φ is a Σ_0^B-formula, then for a given X, we can construct a witness for $\exists Y \varphi(X,Y)$ in a collaborative fashion by using F and some AC^0-functions $F_1(X), \ldots, F_k(X, Z_1, \ldots, F_{k-1})$.

Lemma 3. *Let $\varphi(X,Y)$ be a Σ_0^B-formula and $\theta(X,Y,Z)$ denote*

$$G_F(|Y^{[1]}|, Y^{[2]}, Z) \supset \varphi(X, Y^{[0]}).$$

Suppose that the theory VC proves $\forall X \exists Y \varphi(X,Y)$. Then there exist some AC^0-functions $F_1(X), \ldots, F_k(X, Z_1, \ldots, Z_{k-1})$ such that $\overline{V}^0$ proves

$$\bigvee_{i=1}^{k} \theta(X, F_i(X, Z_1, \ldots, Z_{i-1}), Z_i). \tag{1}$$

Proof. The theory VC is defined as V^0 plus a $\forall \Sigma_1^B$ sentence expressing the existence of a solution of a complete problem in C. Applying the deduction theorem of first-order logic to a VC proof of $\forall X \exists Y \varphi(X,Y)$ and working in a conservative extension $\overline{V}^0$ of V^0, we obtain a $\overline{V}^0$ proof of a statement to which the KPT witnessing theorem is applicable. □

We can think of Lemma 3 as a game about the formula

$$\exists Y \forall Z \theta(X, Y, Z) \tag{2}$$

between a student E and a teacher U, where E's role is to find a witness Y to the existential quantifier in (2), but has computing power limited to FAC^0, whereas U's role is to find a counterexample Z to the universal quantifier in (2), if it exists. More precisely, the game starts with E producing a potential witness $Y_1 = F_1(X)$, which U either approves or rejects – U approves Y_1 if $\forall Z \theta(X, Y_1, Z)$ is true, otherwise U rejects Y_1 and has to provide a counterexample Z_1 such that $\neg\theta(X, Y_1, Z_1)$ holds, that is to say, $Z_1 = F(|Y_1^{[1]}|, Y_1^{[2]})$ and $\neg\varphi(X, Y_1^{[0]})$ is true. If U rejects Y_1 by producing a counterexample Z_1, then E can use Z_1 in order to compute the next potential witness $Y_2 = F_2(X, Z_1)$. Again, either U approves or rejects Y_2. As before, if U rejects Y_2, then he has to provide E with a counterexample Z_2. This process will continue for at most k steps, after which E finds a witness to the existential quantifier in (2). Note that E cannot compute F, since F is beyond E's computing power.

In the student-teacher game interpretation of Lemma 3, the student is always guaranteed to find a value Y such that $\forall Z \theta(X, Y, Z)$ holds after at most k steps. However, if $\varphi(X,Y)$ and $F_1(X), \ldots, F_k(X, Z_1, \ldots, Z_k)$ were to be picked arbitrarily, then there is no guarantee that the student would still win, that is to say that he would find a value Y that satisfies $\forall Z \theta(X, Y, Z)$. This is because, for an arbitrary X, it is not always the case that there is a Y such that $\varphi(X,Y)$ is true. Also, even if $\forall X \exists Y \varphi(X,Y)$ happened to be true, nothing tells us that $\forall Z \theta(X, F_j(X, Z_1, \ldots, Z_{j-1}), Z)$ will hold, for some F_j in $F_1, \ldots, F_k$.

The class KPT[C] will be defined with the student-teacher game interpretation of Lemma 3 in mind, but where φ and $F_1, \ldots, F_k$ are given arbitrarily. Therefore, some care needs to be taken when defining KPT[C] in order to ensure its totality. More precisely, if in case there is no F_j in $F_1, \ldots, F_k$ such that $\forall Z \theta(X, F_j(X, Z_1, \ldots, Z_{j-1}), Z)$ holds, then we will just force part of the formula that defines the graph of a KPT[C] search problem to be trivially true.

In the following, we write $\hat{F}_i(X, W)$ for $F_i(X, W^{[1]}, \ldots, W^{[i-1]})$.

Definition 4. *A* KPT[C] *search problem* $Q(X, W)$ *is specified by a* $k \in \mathbb{N}$, *a* Σ_0^B-*formula* $\varphi(X, Y)$ *and* AC^0-*functions* $F_1(X), \ldots, F_k(X, Z_1, \ldots, Z_{k-1})$. *A string* W *is a solution to an instance* X *of* Q *if, and only if, the following hold:*

1. For all i *from* 1 *to* k,

$$G_F(|\hat{F}_i(X, W)^{[1]}|, \hat{F}_i(X, W)^{[2]}, W^{[i]}). \tag{3}$$

2. There exists an i *between* 1 *and* k *such that the following holds:*

$$[W^{[0]} = \hat{F}_i(X, W)^{[0]} \wedge [i < k \supset \varphi(X, W^{[0]})] \wedge \bigwedge_{j < i} \neg\varphi(X, \hat{F}_j(X, W)^{[0]}). \tag{4}$$

We will call φ *and* $F_1, \ldots, F_k$ *the* components *of* Q.

We will explain (3) and (4) here. The formula in (3) says that $W^{[i]}$ stores the counterexample $F(|\hat{F}_i(X, W)^{[1]}|, \hat{F}_i(X, W)^{[2]})$ given by the teacher to the student – in fact, note that even if $\varphi(X, \hat{F}_i(X, W)^{[0]})$ is true, then $W^{[i]}$ always stores $F(|\hat{F}_i(X, W)^{[1]}|, \hat{F}_i(X, W)^{[2]})$. Next, the formula in (4) guarantees the totality of Q. If there is no F_j in $F_1, \ldots, F_k$ such that $\varphi(X, \hat{F}_j(X, W)^{[0]})$ is true, then the above formula trivially holds by taking $i = k$ and $W^{[0]}$ to be equal to $\hat{F}_i(X, W)^{[0]}$, and in case there is an F_i in $F_1, \ldots, F_k$ such that $\varphi(X, \hat{F}_i(X, W)^{[0]})$ is true, then the formula in (4) tells us that i is the least value in $\{1, \ldots, k\}$ such that $\varphi(X, \hat{F}_i(X, W)^{[0]})$ is true. Finally, using the Σ_0^B representation theorem, it is easy to see that the graph of a KPT[C] search problem is in AC^0.

Lemma 5. *Let* Q *be a* KPT[C] *search problem. Then* VC *proves that* Q *is total.*

Proof. The proof is a straightforward case analysis. $\qquad\square$

The next theorem is a converse of Lemma 5.

Theorem 6 (New-style Witnessing Theorem for VC**).** *Let* $\varphi(X, Y)$ *be a* Σ_1^1-*formula such that* VC *proves* $\forall X \exists Y \varphi(X, Y)$. *Then there is a* KPT[C] *search problem* Q *and an* AC^0-*function* H *such that* $\overline{V}^0$ *proves*

$$Q(X, W) \supset \varphi(X, H(X, W)). \tag{5}$$

Proof. W.l.o.g. we can assume that $\varphi \in \Sigma_0^B$. By Lemma 3, we obtain some AC^0-functions $F_1(X), \ldots, F_k(X, Z_1, \ldots, Z_k)$ such that $\overline{V}^0$ proves

$$\forall X \forall Z_1 \ldots \forall Z_k \bigvee_{i=1}^{k} \theta(X, F_i(X, Z_1, \ldots, Z_{i-1}), Z_i), \tag{6}$$

where $\theta(X, Y, Z)$ is the formula $G_F(|Y^{[1]}|, Y^{[2]}, Z) \supset \varphi(X, Y^{[0]})$. Define a KPT[C] search problem Q using φ and $F_1, \ldots, F_k$.

Arguing in $\overline{V}^0$, we want to show (5). Suppose that $Q(X, W)$ holds. Then (4) is true for some $i \leq k$. If $i < k$, then $\varphi(X, W^{[0]})$ follows directly. Otherwise, $i = k$, and we have that

$$\bigwedge_{j < k} \neg\varphi(X, F_j(X, W^{[1, \ldots, j-1]})^{[0]})$$

holds. Combining this with (6), it is easy to see that $\varphi(X, F_k(X, W^{[1, \ldots, k-1]})^{[0]})$ and $W^{[0]} = F_k(X, W^{[1, \ldots, k-1]})^{[0]}$. By letting $H(X, W) = W^{[0]}$ the assertion follows. □

Combining Lemma 5, Theorem 6 and the fact that $\overline{V}^0$ is a universal conservative extension of V^0, we obtain the following theorem:

Theorem 7. KPT[C] *is* AC^0-*many-one complete for the provably total* NP *search problems in* VC. *Furthermore, the reduction is provable in the theory* V^0.

5 The Class of Inflationary Iteration Problems and V^1

Finite subsets of $\mathbb{N}$ can be viewed as finite binary strings with no leading zeros by letting an element in the set indicate whether the corresponding bit in the string is set to one. Using this identification of strings with finite sets, we define the notion of an "inflationary" string function:

Definition 8. *A string function* $F(X, Z)$ *is said to be* inflationary *if, and only if, for all* X, Z, *we have that* $Z \subseteq F(X, Z)$.

The complexity class PLS [13] is based on the principle that every finite directed acyclic graph has a sink. Additionally, if the local search function is given by an inflationary FAC^0-function, then we obtain the class IPLS:

Definition 9. *An* IPLS *problem* $Q(X, Y)$ *is specified by the following:*

1. *An* AC^0-*relation* $F_Q(X, Y)$ *and an* $\mathcal{L}_A^2$-*term* $t(X)$ *such that the following conditions hold:*

$$F_Q(X, \emptyset),$$
$$F_Q(X, Z) \supset |Z| \leq t(X).$$

The set of all Y *with* $F_Q(X, Y)$ *is the set of all* candidate solutions *for* Q *on instance* X.

2. *An* FAC^0-*function* $P_Q(X, Y)$, *which computes the* profit *of* Y, *and an inflationary* FAC^0-*function* $N_Q(X, Y)$, *which computes the* neighbor *of* Y, *such that for any* Y *that satisfies* $F_Q(X, Y)$, *the following holds:*

$$[N_Q(X, Y) = Y] \vee [F_Q(X, N_Q(X, Y)) \wedge P_Q(X, Y) < P_Q(X, N_Q(X, Y))].$$

where $X < Y$ is the less than relation on strings. A solution to an instance X of Q is any string Y such that

$$F_Q(X,Y) \wedge N_Q(X,Y) = Y$$

holds. We will usually refer to F_Q, P_Q, N_Q and t as the components of Q.

Any IPLS problem is a total search problem. Moreover, checking if a string is a solution to an instance of an IPLS problem is an AC^0-property. Thus every IPLS problem is a $\forall\exists AC^0$ search problem.

We will next introduce the class IITER, which is based on the iteration principle [7]. The iteration principle is also based on the fact that every finite directed acyclic graph $\mathcal{G} = (\mathcal{V}, \mathcal{E})$ has a sink. In an exponential sized graph $\mathcal{G}$, it may take exponentially many steps to find a sink following a path through the graph. However, if the edge relation is given by an inflationary function, paths are bound to be of polynomial length.

Definition 10. An IITER $Q_F(X,Y)$ is specified by an inflationary FAC^0-function $F(X,Y)$ and an $\mathcal{L}_A^2$-term $t(X)$. A solution to an instance X of Q_F is a string Y satisfying the formula $\psi_F(X,Y)$, which is (omitting the parameter X) given as follows:

$$[Y = \emptyset \wedge F(Y) = Y] \vee$$
$$[|Y| \le t \wedge Y < F(Y) \wedge [t < |F(Y)| \vee F(F(Y)) \le F(Y)]]. \quad (7)$$

We will usually refer to F and t as the components of Q_F. We say that a string Y is a candidate solution to Q_F on instance X if Y satisfies the following condition:

$$|Y| \le t \wedge (Y = \emptyset \vee Y < F(X,Y)). \quad (8)$$

It is known that the iteration principle is AC^0-many-one complete for PLS [8,22]. In what follows, we show that IITER is AC^0-many-one complete for IPLS.

Lemma 11. Every IITER problem is an IPLS problem.

Proof. The proof is a direct adaptation of the one for [8, Lemma VIII.5.7]. □

Lemma 12. Every IPLS problem is AC^0-many-one reducible to an IITER problem.

Proof. The proof is easier than the one for [8, Theorem VIII.5.8]. Observe that $X \subseteq Y$ implies $X \le Y$. Given a IPLS problem Q with components F_Q, N_Q, P_Q and t, we can define an IITER problem Q_F using N_Q on F_Q and t. Given an instance X, it is easy to see that a solution Y to Q_F is one step beyond a solution to Q, the latter being given by $N_Q(X,Y)$. □

From Lemmas 11 and 12, we immediately obtain the following corollary:

Corollary 13. IITER is AC^0-many-one complete for IPLS. □

Theorem 14. *Let Q be an IITER problem. Then Q is provably total in V^1.*

Proof. Let Q be an IITER problem with components F and t. Let numones(y, Y) be the function that computes the total number of elements in Y that are strictly less than y. The function numones is a polytime function definable in V^1. Consider $\eta(X, Z)$ to be the formula $Z = \emptyset \vee Z < F(X, Z)$ and $\bar{\eta}(X, z)$ to be

$$\exists Z \leq t(X)[z = \text{numones}(Z) \wedge \eta(X, Z)].$$

Then η is in Σ_0^B, and $\bar{\eta}$ equivalent to a formula in Σ_1^B. Using maximisation on z, which is available in V^1, we obtain a Z with maximal number of elements amongst those satisfying η. It is easy to see that this Z is a solution to Q.   □

The converse of Theorem 14 is the new-style witnessing theorem for V^1.

Theorem 15 (New-style Witnessing Theorem for V^1). *Suppose that $\varphi(X, Y)$ is a Σ_1^1-formula such that*

$$\mathsf{V}^1 \vdash \forall X \exists Y \varphi(X, Y).$$

Then there is an IITER problem Q_F with graph $\psi_F(X, Y)$ (as in (7)), and an FAC^0-function $G(X, Y)$, such that

$$\overline{\mathsf{V}}^0 \vdash \psi_F(X, Y) \supset \varphi(X, G(X, Y)). \tag{9}$$

Proof (Proof Idea). The idea of this proof is to construct the required search problem by induction on an appropriate sequent calculus derivation of the original statement. For this we will have to redefine V^1 in terms of an appropriate induction scheme, and use a corresponding inference rule in the definition of the sequent calculus. The main step in the construction is to deal with applications of this induction rule. From an IITER problem given for the premise of the induction rule, we obtain one for the conclusion by iterating the former polynomially many times, creating in each step an additional entry in a polynomially long board in order to guarantee the result to be inflationary.

Further details can be found in Appendix A.   □

Combining Theorems 14 and 15 and the fact that $\overline{\mathsf{V}}^0$ is a universal conservative extension of V^0, we obtain the following corollary:

Corollary 16. *IITER is AC^0-many-one complete for the provably total NP search problems in V^1. Furthermore, the reduction is provable in the theory V^0.*   □

Acknowledgement. We would like to thank Noahi Eguchi. Our characterisation of $\forall \Sigma_1^B(\mathsf{V}^1)$ using inflationary iteration grew out of discussions with him on his attempt to capture P via a two-sorted theory using axioms on inductive definitions [10].

Appendix A: Proof of Theorem 15

In what follows, when we say that a theory $\mathcal{T}$ proves a sequent

$$\varphi_1, \ldots, \varphi_k \longrightarrow \psi_1, \ldots \psi_l,$$

we mean that $\mathcal{T}$ proves

$$\bigwedge_{i=1}^{k} \varphi_i \supset \bigvee_{j=1}^{l} \psi_j.$$

Buss [5] originally proved his witnessing theorem for V^1 via a witnessing lemma. Here, we do the same; that is to say, we use a new-style witnessing lemma in order to prove Theorem 15.

Lemma 17 (New-style Witnessing Lemma for V^1). *Suppose that the theory V^1 proves a sequent $\Gamma(A) \longrightarrow \Delta(A)$ of the form*

$$\ldots, \exists X_i \phi_i'(X_i), \ldots, \Lambda \longrightarrow \Pi, \ldots, \exists Y_j \psi_j'(Y_j), \ldots \tag{10}$$

where $\phi_i', \psi_j', \Lambda$ and Π are Σ_0^B-formulae. Then there is an IITER problem Q_F with graph ψ_F and FAC^0-functions G such that $\overline{V}^0$ proves the sequent $\Gamma' \longrightarrow \Delta'$, which is

$$\ldots, \phi_i'(\beta_i), \ldots, \Lambda, \psi_F(A, \beta, \gamma) \longrightarrow \Pi, \ldots, \psi_j'(G_j(A, \beta, \gamma)), \ldots \tag{11}$$

We will use a version of the sequent calculus to prove this lemma. Given a sequent calculus proof π of (10) we try to show the conclusion of Lemma 17 by structural induction on the depth of a sequent S in π. If we use directly a sequent calculus for V^1, we have the issue that the Σ_1^B-COMP axiom is in general not equivalent to a Σ_1^B-formula. As a result, the proof π may contain formulae that are not Σ_1^1. To circumvent this obstacle, we need to work with a slightly different theory $\widetilde{V}^1$ equivalent to V^1. For that, first consider the following definition:

Definition 18 (Cook and Nguyen [8]). *Let $\psi(X)$ be an $\mathcal{L}_A^2$-formula. Then ψ is a single-Σ_1^1-formula if ψ is of the form $\exists Y \varphi(X, Y)$, where φ is a Σ_0^B-formula. If ψ is of the form $(\exists Y \leq t)\varphi(X, Y)$, where φ is a Σ_0^B-formula and t is an $\mathcal{L}_A^2$-term not involving Y, then ψ is a single-Σ_1^B-formula.*

Definition 19 (Cook and Nguyen [8]). *The theory $\widetilde{V}^1$ is axiomatized by the axioms of V^0 plus the single-Σ_1^B-IND axiom scheme.*

Below, we merely state that $\widetilde{V}^1 = V^1$ without proof. A full proof of it can be found in [8, Theorem VI.4.8].

Theorem 20 (Cook and Nguyen [8]). *The theories $\widetilde{V}^1$ and V^1 are the same.*

The sequent calculus $\mathsf{LK}\text{-}\tilde{V}^1$ for $\tilde{V}^1$ is essentially the sequent calculus $\mathsf{LK}\text{-}V^0$ for V^0 (c.f. [8]) augmented with the *single-Σ_1^B-IND rule*, which is

$$\frac{\chi(b), \Gamma \longrightarrow \Delta, \chi(b+1)}{\chi(0), \Gamma \longrightarrow \Delta, \chi(t)},$$

where $\chi \in \Sigma_1^B$, and b is an eigenvariable and cannot appear in the lower sequent.

The sequent calculus $\mathsf{LK}\text{-}\tilde{V}^1$ satisfies the following property, whose proof can be found in [8]:

Theorem 21 (Cook and Nguyen [8]). *Suppose that $\tilde{V}^1$ proves a sequent $\Gamma \longrightarrow \Delta$ consisting only of single-$\tilde{\Sigma}_1^1$-formulae. Then there is an $\mathsf{LK}\text{-}\tilde{V}^1$ proof π of $\Gamma \longrightarrow \Delta$ such that every formula in π is a single-$\tilde{\Sigma}_1^1$-formula.*

We are now ready to prove Lemma 17. The proof technique we use to prove Lemma 17 is similar to the one used for Theorem VI.4.1 in [8, p. 154] (which is a witnessing theorem for V^1), which adopts the same proof technique as Buss (cf. [5, Theorem 5]).

Proof (of the New-style Witnessing Lemma for V^1, *Lemma 17).* Since $\tilde{V}^1$ and V^1 are the same, it follows that $\tilde{V}^1$ proves (10). By Theorem 21, let π be an $\mathsf{LK}\text{-}\tilde{V}^1$ proof of (10) such that every formula in π is a single-$\tilde{\Sigma}_1^1$-formula. We show that $\overline{V}^0$ proves the conclusion of Lemma 17 by induction on the depth of a sequent S in π. The inductive proof splits into cases, depending on whether S is an initial sequent or generated by the use of an inference rule. The most crucial case is the case of the single-Σ_1^B-IND rule.

Suppose that S is obtained by the application of the single-Σ_1^B-IND rule. Then S is the bottom sequent of

$$\frac{\psi(b), \Lambda \longrightarrow \Pi, \psi(b+1)}{\psi(0), \Lambda \longrightarrow \Pi, \psi(t)}$$

where (omitting the parameters A) $\psi(b)$ is of the form $(\exists X \leq r(b))\psi_0(b, X)$ and

$$\Pi = \Pi', \exists Y_1 \psi_1'(Y_1), \ldots, \exists Y_l \psi_l'(Y_l).$$

Here $\Pi', \psi_1', \ldots, \psi_l'$ is a sequence of Σ_0^B-formulae. Let $\eta(b, \beta)$ denote the formula $|\beta| \leq r(b) \wedge \psi_0(b, \beta)$. By the induction hypothesis, let Q_{F_1} be an IITER problem specified by F_1 and t_1, with graph ψ_{F_1}, and $G_1^1, \ldots, G_l^1$ and G_{l+1}^1 be the witnessing functions for the formulae in $\Pi, \psi(b+1)$ such that $\overline{V}^0$ proves the following (omitting the parameters A, $\boldsymbol{\lambda}$, where $\boldsymbol{\lambda}$ are witnesses for the formulae in Λ):

$$\eta(b, \beta), \Lambda', \psi_{F_1}(b, \beta, \gamma) \longrightarrow \Pi''(G_j^1(b, \beta, \gamma)), \eta(b+1, G_{l+1}^1(b, \beta, \gamma)) \qquad (12)$$

where Λ' is the result of witnessing $\tilde{\Sigma}_1^1$-formulae in Λ and leaving the rest unchanged and $\Pi''(G_j^1(b, \beta, \gamma)) = \Pi', \psi_1'(G_1^1(b, \beta, \gamma)), \ldots, \psi_l'(G_l^1(b, \beta, \gamma))$. Our goal is to construct an IITER problem Q_F (with graph ψ_F) and FAC^0-functions $G_1, \ldots, G_l$ and G_{l+1} such that $\overline{V}^0$ proves the following:

$$\eta(0, \beta_0), \Lambda', \psi_F(\beta_0, \gamma) \longrightarrow \Pi''(G_j(\beta_0, \gamma)), \eta(t, G_{l+1}(\beta_0, \gamma)). \tag{13}$$

The intuitive idea behind the definition of Q_F is that, assuming that $\eta(0, \beta_0)$ is true, we will repeatedly use Q_{F_1} and G_{l+1}^1 in order to generate witnesses $\beta_1, \ldots, \beta_n$ for $\psi(1), \ldots, \psi(n)$, respectively, for $n \leq t$. If $n < t$, then Q_{F_1} failed to generate a witness to $\psi(n+1)$. Therefore, assuming that the hypothesis for (13) is true and using (12), we obtain our desired goal.

In what follows, the *string concatenation* function $X *_z Y$ is an FAC^0 string function that concatenates the first z bits of X with Y and can be recursively extended in the natural way. Omitting the subscripts to $*$, we write $Y_0 * \ldots * y * \ldots * Y_k$ for $Y_0 * \ldots * Y * \ldots * Y_k$, where Y is the string representing the unary notation of the number value y.

We assume that the search variable for Q_F is of the form

$$\gamma = \langle A, \beta_0, \boldsymbol{\lambda} \rangle *_s S_0 *_{2s} S_1 *_{3s} \cdots *_{(m+1)s} S_m,$$

where s (s is obtained from t and the bounding term r, in the induction-formula ψ, and the bounding term t_1 for Q_{F_1}) is a suitable $\mathcal{L}_A^2$-term that bounds $|\langle A, \beta_0, \boldsymbol{\lambda} \rangle|, |S_0|, \ldots, |S_m|$; the symbol S_i denotes $i * \beta_i * \gamma_i * 1$ and $m \leq t$. Note here that, even though we omitted the subscripts to $*$ in S_i, they are somehow implicit. Let us now define the transition function F for Q_F. In the following, we again omit the parameters $A, \boldsymbol{\lambda}$ for F. As usual, we will drop the subscripts to $*$ in $F(\beta_0, \gamma)$. If $\gamma = \emptyset$, then

$$F(\beta_0, \gamma) = \langle A, \beta_0, \boldsymbol{\lambda} \rangle * 0 * \beta_0 * \emptyset * 1. \tag{14}$$

Assume now that $\gamma \neq \emptyset$ and suppose that $m < t$ and $\eta(m, \beta_m)$ is true. Then there are two cases to consider. First, if $|\gamma_m| \leq t_1 \wedge \gamma_m < F_1(m, \beta_m, \gamma_m) \wedge \neg\psi_{F_1}(m, \beta_m, \gamma_m)$ is true, then

$$F(\beta_0, \gamma) = \langle A, \beta_0, \boldsymbol{\lambda} \rangle * S_0 * \ldots * S_{m-1} * m * \beta_m * F_1(m, \beta_m, \gamma_m) * 1. \tag{15}$$

Second, if $|\gamma_m| \leq t_1 \wedge \gamma_m < F_1(m, \beta_m, \gamma_m) \wedge \psi_{F_1}(m, \beta_m, \gamma_m)$, then

$$F(\beta_0, \gamma) = \langle A, \beta_0, \boldsymbol{\lambda} \rangle * S_0 * \ldots * S_m * (m+1) * G_{l+1}^1(m, \beta_m, \gamma_m) * \emptyset * 1. \tag{16}$$

In all other cases, $F(\beta_0, \gamma) = \gamma$. Let t_{Q_F} be $(t+2) \cdot s$ and Q_F be specified by F and t_{Q_F}. Finally, we define the FAC^0-functions G_i, for $i = 1, \ldots, l+1$, as follows:

$$G_j(\beta_0, \gamma) = \begin{cases} \beta_0 & \text{if } t = 0 \\ G_j^1(m, \beta_m, \gamma_m) & \text{otherwise,} \end{cases}$$

The fact that $\overline{V}^0$ proves (13) follows from (13)'s assumptions, from the following claim, the induction hypothesis and the definition of G_j above. As a side remark, note that if $t = 0$, then $\overline{V}^0$ proves (13) trivially.

Claim. We reason in $\overline{V}^0$. Suppose that $t \neq 0$, $\eta(0, \beta_0)$ is true and $\gamma = \langle A, \beta_0, \lambda \rangle *$ $S_0 * \ldots * S_m$ is a solution to $Q_F(\beta_0)$, where S_i is again of the form $i * \beta_i * \gamma_i * 1$. Then $\eta(m, \beta_m)$ is true; γ_m is a solution to $Q_{F_1}(m, \beta_m)$; and either $\neg\eta(m + 1, G^1_{l+1}(m, \beta_m, \gamma_m))$ or $\eta(t, G_{l+1}(\beta_0, \gamma))$ is true.

Proof of Claim. Since γ is a solution to $Q_F(\beta_0)$, then we have two possibilities: either $\gamma = \emptyset$ and $F(\beta_0, \gamma) = \gamma$, or

$$|\gamma| \leq t_{Q_F} \wedge \gamma < F(\beta_0, \gamma) \wedge [|F(\beta_0, \gamma)| > t_{Q_F} \vee F(\beta_0, F(\beta_0, \gamma)) = F(\beta_0, \gamma)].$$

Note that, by the definition of F, $\emptyset$ cannot be a solution to $Q_F(\beta_0)$ and $|F(\beta_0, \gamma)| \leq t_{Q_F}$. Therefore, we have that

$$\gamma \neq \emptyset \wedge \gamma < F(\beta_0, \gamma) = F(\beta_0, F(\beta_0, \gamma)). \tag{17}$$

The only way for (17) to hold is if (16) is true. This implies that $\eta(m, \beta_m)$ holds and $\psi_{F_1}(m, \beta_m, \gamma_m)$ is true; that is to say, γ_m is a solution $Q_{F_1}(m, \beta_m)$. Hence, we are left with proving the following:

$$\neg\eta(m + 1, G^1_{l+1}(m, \beta_m, \gamma_m)) \vee \eta(t, G_{l+1}(\beta_0, \gamma)).$$

If $m + 1 = t$, then we are done. So, assume that $m + 1 < t$. For the sake of contradiction, assume that $\eta(m + 1, G^1_{l+1}(m, \beta_m, \gamma_m))$ holds. This means that $F(\beta_0, \gamma) < F(\beta_0, F(\beta_0, \gamma))$, which is a contradiction. Thus, we are done with the proof of the claim. $\square$

This finishes the proof of Lemma 17.

References

1. Alon, N., Babai, L., Itai, A.: A fast and simple randomized parallel algorithm for the maximal independent set problem. J. Algorithms **7**(4), 567–583 (1986)
2. Beckmann, A., Buss, S.R.: Polynomial local search in the polynomial hierarchy and witnessing in fragments of bounded arithmetic. J. Math. Log. **9**(1), 103–138 (2009). doi:10.1142/S0219061309000847
3. Beckmann, A., Buss, S.R.: Characterising definable search problems in bounded arithmetic via proof notations. In: Ways of Proof Theory. Ontos Series on Mathematical Logic, vol. 2, pp. 65–133. Ontos Verlag, Heusenstamm (2010)
4. Beckmann, A., Buss, S.R.: Improved witnessing and local improvement principles for second-order bounded arithmetic. ACM Trans. Comput. Log. **15**(1), 35 (2014). doi:10.1145/2559950. (Art. 2)
5. Buss, S.R.: Bounded Arithmetic. Studies in Proof Theory. Lecture Notes, vol. 3. Bibliopolis, Naples (1986)
6. Buss, S.R., Krajíček, J.: An application of Boolean complexity to separation problems in bounded arithmetic. Proc. Lond. Math. Soc. (3) **69**(1), 1–21 (1994). doi:10.1112/plms/s3-69.1.1
7. Chiari, M., Krajíček, J.: Witnessing functions in bounded arithmetic and search problems. J. Symb. Log. **63**(3), 1095–1115 (1998). doi:10.2307/2586729

8. Cook, S., Nguyen, P.: Logical Foundations of Proof Complexity, 1st edn. Cambridge University Press, New York (2010)

9. Cook, S.A., Filmus, Y., Lê, D.T.M.: The complexity of the comparator circuit value problem. ACM Trans. Comput. Theory **6**(4), 44 (2014). doi:10.1145/2635822

10. Eguchi, N.: Characterising complexity classes by inductive definitions in bounded arithmetic, June 2013. ArXiv e-prints

11. Gale, D., Shapley, L.S.: College admissions and the stability of marriage. Am. Math. Mon. **120**(5), 386–391 (2013). doi:10.4169/amer.math.monthly.120.05.386. (Reprint of MR1531503)

12. Immerman, N.: Descriptive Complexity. Graduate Texts in Computer Science. Springer, New York (1999). doi:10.1007/978-1-4612-0539-5

13. Johnson, D.S., Papadimitriou, C.H., Yannakakis, M.: How easy is local search? J. Comput. Syst. Sci. **37**(1), 79–100 (1988). doi:10.1016/0022-0000(88)90046-3. (In: 26th IEEE Conference on Foundations of Computer Science, Portland, OR (1985))

14. Karp, R.M., Wigderson, A.: A fast parallel algorithm for the maximal independent set problem. J. Assoc. Comput. Mach. **32**(4), 762–773 (1985). doi:10.1145/4221. 4226

15. Kołodziejczyk, L.A., Nguyen, P., Thapen, N.: The provably total NP search problems of weak second order bounded arithmetic. Ann. Pure Appl. Log. **162**(6), 419–446 (2011). doi:10.1016/j.apal.2010.12.002

16. Krajíček, J., Pudlák, P., Takeuti, G.: Bounded arithmetic and the polynomial hierarchy. Ann. Pure Appl. Logic **52**(1–2), 143–153 (1991). doi:10.1016/ 0168-0072(91)90043-L. (In: International Symposium on Mathematical Logic and Its Applications, Nagoya (1988))

17. Krajíček, J., Skelley, A., Thapen, N.: NP search problems in low fragments of bounded arithmetic. J. Symb. Log. **72**(2), 649–672 (2007). doi:10.2178/jsl/ 1185803628

18. Luby, M.: A simple parallel algorithm for the maximal independent set problem. In: Proceedings of the Seventeenth Annual ACM Symposium on Theory of Computing, STOC 1985, pp. 1–10. ACM, New York (1985). doi:10.1145/22145.22146

19. Mayr, E.W., Subramanian, A.: The complexity of circuit value and network stability. J. Comput. Syst. Sci. **44**(2), 302–323 (1992). doi:10.1016/ 0022-0000(92)90024-D

20. Megiddo, N., Papadimitriou, C.H.: On total functions, existence theorems and computational complexity. Theoret. Comput. Sci. **81**(2), 317–324 (1991). doi:10. 1016/0304-3975(91)90200-L. (Algorithms, Automata, Complexity and Games)

21. Mix-Barrington, D.A., Immerman, N., Straubing, H.: On uniformity within NC1. J. Comput. Syst. Sci. **41**(3), 274–306 (1990). doi:10.1016/0022-0000(90)90022-D

22. Morioka, T.: Classification of search problems and their definability in bounded arithmetic (2001)

23. Razafindrakoto, J.J.: Witnessing theorems in bounded arithmetic and applications. Ph.D. thesis, Swansea University (2016). http://cs.swan.ac.uk/~csjjr/ Papers/thesis.pdf

24. Thapen, N.: Higher complexity search problems for bounded arithmetic and a formalized no-gap theorem. Arch. Math. Log. **50**(7–8), 665–680 (2011). doi:10. 1007/s00153-011-0240-0

25. Zambella, D.: Notes on polynomially bounded arithmetic. J. Symb. Log. **61**(3), 942–966 (1996). doi:10.2307/2275794

On the Reflection Calculus with Partial Conservativity Operators

Lev D. Beklemishev[✉]

Steklov Mathematical Institute of the Russian Academy of Sciences,
Moscow, Russia
bekl@mi.ras.ru

Abstract. Strictly positive logics recently attracted attention both in
the description logic and in the provability logic communities for their
combination of efficiency and sufficient expressivity. The language of
Reflection Calculus RC consists of implications between formulas built
up from propositional variables and the constant 'true' using only conjunction and the diamond modalities which are interpreted in Peano
arithmetic as restricted uniform reflection principles.

We extend the language of RC by another series of modalities representing the operators associating with a given arithmetical theory T
its fragment axiomatized by all theorems of T of arithmetical complexity
Π_n^0, for all $n > 0$. We note that such operators, in a precise sense, cannot
be represented in the full language of modal logic.

We formulate a formal system extending RC that is sound and, as
we conjecture, complete under this interpretation. We show that in this
system one is able to express iterations of reflection principles up to
any ordinal $< \varepsilon_0$. On the other hand, we provide normal forms for its
variable-free fragment. Thereby, the variable-free fragment is shown to
be algorithmically decidable and complete w.r.t. its natural arithmetical
semantics.

Keywords: Strictly positive logics · Reflection principle · Provability ·
GLP

1 Introduction

A system, called *Reflection Calculus* and denoted RC, was introduced in [8]
and, in a slightly different format, in [13]. From the point of view of modal logic,
RC can be seen as a fragment of Japaridze's polymodal provability logic GLP
[12,18,24] consisting of the implications of the form $A \rightarrow B$, where A and B
are formulas built-up from $\top$ and propositional variables using just $\wedge$ and the
diamond modalities. We call such formulas A and B *strictly positive.*

Strictly positive modal logics independently and earlier appeared in the work
on description logic, see [21] for some results and further references. In particular,

This work is supported by the Russian Science Foundation under grant 16–11–10252.

© Springer-Verlag GmbH Germany 2017
J. Kennedy and R.J.G.B. de Queiroz (Eds.): WoLLIC 2017, LNCS 10388, pp. 48–67, 2017.
DOI: 10.1007/978-3-662-55386-2_4

the strictly positive language corresponds to the OWL2EL profile of the OWL web ontology language.

Reflection calculus RC is much simpler than its modal companion GLP yet expressive enough to retain its main proof-theoretic applications. It has been outlined in [8] that RC allows one to define a natural system of ordinal notations up to ε_0 and serves as a convenient basis for a proof-theoretic analysis of Peano Arithmetic in the style of [5,6]. This includes a consistency proof for Peano arithmetic based on transfinite induction up to ε_0, a characterization of its Π_n^0-consequences in terms of iterated reflection principles, a slowly terminating term rewriting system [2] and a combinatorial independence result [7].

An axiomatization of RC (as an equational calculus) has been found by Evgeny Dashkov in his paper [13] which initiated the study of strictly positive fragments of provability logics. Dashkov proved two important further facts about RC which sharply contrast with the corresponding properties of GLP. Firstly, RC is complete with respect to a natural class of finite Kripke frames. Secondly, RC is decidable in polynomial time, whereas most of the standard modal logics are PSPACE-complete and the same holds for the variable-free fragment of GLP [22].

Another advantage of going to a strictly positive language is exploited in the present paper. Strictly positive modal formulas allow for more general arithmetical interpretations than those of the standard modal logic language. In particular, propositional formulas can now be interpreted as arithmetical *theories* rather than individual *sentences*. (Notice that the 'negation' of a theory would not be well-defined.)

Any monotone operator acting on the semilattice of arithmetical theories can be considered as a modality in strictly positive logic. One such operation is particularly attractive from the point of view of proof-theoretic applications, namely the map associating with a theory T its fragment $\Pi_{n+1}(T)$ axiomatized by all theorems of T of arithmetical complexity Π_{n+1}^0. Since the Π_{n+1}^0-conservativity relation of T over S can be expressed by $S \vdash \Pi_{n+1}(T)$, we call such operators Π_{n+1}^0-*conservativity operators*.

This relates our study to the fruitful tradition of research on conservativity and interpretability logics, see e.g. [14,16,27]. Our framework happens to be both weaker and stronger than the traditional one: in our system we are able to express the conservativity relations for each class Π_{n+1}^0 and are able to relate not only sentences but theories. However, in this framework negation is lacking and the conservativity is not a binary modality and cannot be iterated. Yet, we believe that the strictly positive language is both simpler and better tuned to the needs of proof-theoretic analysis of formal systems of arithmetic.

We introduce the system RC^∇ with modalities $\Diamond_n$ representing uniform reflection principles of arithmetical complexity Σ_n^0, and ∇_n representing Π_{n+1}^0-conservativity operators. We provide an adequate semantics of RC^∇ in terms of the semilattice $\mathfrak{G}_{\mathrm{EA}}$ of (numerated) arithmetical r.e. theories extending elementary arithmetic EA. Further, we introduce transfinite iterations of monotone semi-idempotent operators along elementary well-orderings, somewhat generalizing the notion of a Turing–Feferman recursive progression of axiomatic systems but mainly following the same development as in [4]. Our first result shows that RC^∇ can express α-iterations of modalities $\Diamond_n$, for each $n < \omega$ and ordinals $\alpha < \varepsilon_0$. This result requires some arithmetical prerequisites and is postponed until the Appendix. A variable-free strictly positive logic where such iterations are explicitly present in the language has been introduced by Joosten and Reyes [19] which is, thereby, contained in RC^∇.

Then we turn to a purely syntactic study of the variable-free fragment of the system RC^∇ and provide unique normal forms for its formulas. A corollary is that the variable-free fragment of RC^∇ is decidable and arithmetically complete.

Whereas the normal forms for the variable-free formulas of RC correspond in a unique way to ordinals below ε_0, the normal forms of RC^∇ are more general. It turns out that they are related in a canonical way to the collections of proof-theoretic ordinals of (bounded) arithmetical theories for each complexity level Π^0_{n+1}, as defined in [4].

Studying the collections of proof-theoretic ordinals corresponding to several levels of logical complexity as single objects seems to be a rather recent and interesting develdpoment. Such collections appeared for the first time in the work of Joosten [20]. He established a one-to-one correspondence between such collections (for a certain class of theories) and the points of the universal model for the variable-free fragment of GLP due to Ignatiev [17]. We call such collections *conservativity spectra* of arithmetical theories. Our results show that RC^∇ provides a way to syntactically represent and conveniently handle such conservativity spectra. The study of the universal model for the variable-free fragment of RC^∇ [3] highlights the underlying algebraic and order structures.[1] The work of Pakhomov [23] shows that the elementary theory of the system of ordinal notation for ε_0 viewed as a lower semilattice with operations $\Diamond_n$, for each $n < \omega$, is undecidable.

Thanks are due to Albert Visser for suggesting many improvements including Lemma 1.

2   The Lattice of Arithmetical Theories

We define the intended arithmetical interpretation of the strictly positive modal language. Propositional variables (and strictly positive formulas) will now denote possibly infinite theories rather than individual sentences. We deal with r.e. theories formulated in the language of *elementary arithmetic* EA and containing the axioms of EA. To avoid well-known problems with the representation of theories in arithmetic, we assume that each theory S comes equipped with an *elementary recursive numeration*, that is, a bounded formula $\sigma(x)$ in the language of EA defining the set of axioms of S in the standard model of arithmetic $\mathbb{N}$.

Given such a σ, we have a standard arithmetical Σ^0_1-formula $\Box_\sigma(x)$ expressing the provability of x in S (see [15]). We often write $\Box_\sigma \varphi$ for $\Box_\sigma(\ulcorner \varphi \urcorner)$. The

---

[1] The paper [3] is a longer version of the present one containing additional material and more detailed proofs of some results.

expression $\bar{n}$ denotes the numeral $0'^{\cdots\prime}$ (n times). If $\varphi(v)$ contains a parameter v, then $\Box_\sigma \varphi(\bar{x})$ denotes a formula (with a parameter x) expressing the provability of the sentence $\varphi(\bar{x}/v)$ in S.

Given two numerations σ and τ, we write $\sigma \leqslant_{\mathrm{EA}} \tau$ if

$$\mathrm{EA} \vdash \forall x \, (\Box_\tau(x) \rightarrow \Box_\sigma(x)).$$

We will only consider the numerations σ such that $\sigma \leqslant_{\mathrm{EA}} \sigma_{\mathrm{EA}}$, where σ_{EA} is some standard numeration of EA. We call such numerated theories *Gödelian extensions* of EA.

The relation $\leqslant_{\mathrm{EA}}$ defines a natural preorder on the set $\mathfrak{G}_{\mathrm{EA}}$ of Gödelian extensions of EA. Let $\overline{\mathfrak{G}}_{\mathrm{EA}}$ denote the quotient by the associated equivalence relation $=_{\mathrm{EA}}$, where by definition $\sigma =_{\mathrm{EA}} \tau$ iff both $\sigma \leqslant_{\mathrm{EA}} \tau$ and $\tau \leqslant_{\mathrm{EA}} \sigma$. $\overline{\mathfrak{G}}_{\mathrm{EA}}$ is a lattice with $\wedge_{\mathrm{EA}}$ corresponding to the union of theories and $\vee_{\mathrm{EA}}$ to their intersection. These operations are defined on elementary numerations as follows:

$$\sigma \wedge_{\mathrm{EA}} \tau := \sigma(x) \vee \tau(x),$$

$$\sigma \vee_{\mathrm{EA}} \tau := \exists x_1, x_2 \leqslant x \, (\sigma(x_1) \wedge \tau(x_2) \wedge x = \mathrm{disj}(x_1, x_2)),$$

where $\mathrm{disj}(x_1, x_2)$ is an elementary term computing the Gödel number of the disjunction of formulas given by Gödel numbers x_1 and x_2.

We will only be concerned with the operation $\wedge_{\mathrm{EA}}$, that is, with the structure of lower semilattice $(\overline{\mathfrak{G}}_{\mathrm{EA}}, \wedge_{\mathrm{EA}})$. Notice that the top element 1_{EA} corresponds to (the equivalence class of) EA, whereas 0_{EA} is the class of all inconsistent extensions of EA.

An operator $R : \mathfrak{G}_{\mathrm{EA}} \rightarrow \mathfrak{G}_{\mathrm{EA}}$ is called *extensional* if $\sigma =_{\mathrm{EA}} \tau$ implies $R(\sigma) =_{\mathrm{EA}} R(\tau)$. Similarly, R is called *monotone* if $\sigma \leqslant_{\mathrm{EA}} \tau$ implies $R(\sigma) \leqslant_{\mathrm{EA}} R(\tau)$. Clearly, each monotone operator is extensional and each extensional operator correctly acts on the quotient lattice $\overline{\mathfrak{G}}_{\mathrm{EA}}$. An operator R is called *semi-idempotent* if $R(R(\sigma)) \leqslant_{\mathrm{EA}} R(\sigma)$. R is a *closure* operator if it is monotone, semi-idempotent and, in addition, $\sigma \leqslant_{\mathrm{EA}} R(\sigma)$. Operators considered in this paper will usually be at least monotone and semi-idempotent.

Meaningful monotone operators abound in arithmetic. Typical examples are the uniform Σ_n-reflection principles $R_n(\sigma)$ associating with σ the extension of EA by the schema $\{\forall x \, (\Box_\sigma \varphi(\bar{x}) \rightarrow \varphi(x)) : \varphi \in \Sigma_n\}$ taken with its natural elementary numeration that we denote $x \in R_n(\sigma)$. It is known that the theory $R_n(\sigma)$ is finitely axiomatizable. Moreover, $R_0(\sigma)$ is equivalent to Gödel's consistency assertion $\mathrm{Con}(\sigma)$ for σ.

In this paper we will study another series of monotone operators. Given a theory S numerated by σ, let $\Pi_n(S)$ denote the extension of EA by all theorems of S of complexity Π_n^0. The set $\Pi_n(S)$ is r.e. but in general not elementary recursive. In order to comply with our definitions we apply a form of Craig's trick that yields an elementary axiomatization of $\Pi_n(S)$.[2] Let $\Pi_n(\sigma)$ denote the

---

[2] Over $\mathrm{EA} + \mathrm{B}\Sigma_1$ one can work with a natural r.e. axiomatization of $\Pi_n(S)$.

elementary formula

$$\exists y, p \leqslant x \, (\mathrm{Prf}_\sigma(y,p) \wedge y \in \Pi_n^0 \wedge x = \mathrm{disj}(y, \ulcorner \bar{p} \neq \bar{p} \urcorner))$$

and the theory numerated by this formula over EA. Here, $\mathrm{Prf}_\sigma(y,p)$ is an elementary formula expressing that p is the Gödel number of a proof of y, so that $\exists p \, \mathrm{Prf}_\sigma(y,p)$ is $\Box_\sigma(y)$; and $x \in \Pi_n^0$ is an elementary formula expressing that x is the Gödel number of a Π_n^0-sentence. Then it is easy to see that the theory $\Pi_n(\sigma)$ is (externally) deductively equivalent to $\Pi_n(S)$.

We will implicitly rely on the following characterization.

Lemma 1. *It is provable in EA that*

$$\forall x \, (\Box_{\Pi_n(\sigma)}(x) \leftrightarrow \exists \pi \in \Pi_n^0 \, (\Box_\sigma(x) \wedge \Box_{\mathrm{EA}}(\pi \to x))).$$

Proof. The implication from right to left is easy, we sketch a proof of ($\to$). Reason within EA. Suppose p is a $\Pi_n(\sigma)$-proof of x. It is a EA-proof of x from some assumptions $\pi_1', \pi_2', \ldots, \pi_k'$ such that each π_i' has the form $\pi_i \vee \overline{p_i} \neq \overline{p_i}$ where $\pi_i \in \Pi_n^0$ and $\mathrm{Prf}_\sigma(\pi_i, p_i)$. Since p contains witnesses for all the proofs p_i, from p one can construct in an elementary way a sentence $\pi \in \Pi_n^0$ equivalent to $\pi_1 \wedge \cdots \wedge \pi_k$ together with its σ-proof and an EA-proof of $\pi \to x$, using a formalization of the deduction theorem in EA. A verification that it is, indeed, the required proof goes by an elementary induction on the length of p.

Using Lemma 1 one can naturally infer that all the operators R_n and Π_n are monotone and semi-idempotent, moreover Π_n is a closure. Moreover, EA can be replaced in all the previous considerations by any of its Gödelian extensions T.

The main source of interest for us in this paper will be the structure of semilattice with operators

$$(\overline{\mathfrak{G}}_T, \wedge_T, \{\mathrm{R}_n, \Pi_{n+1} : n < \omega\}).$$

We call it *the RC^∇-algebra of Gödelian extensions of T.* The term RC^∇-*algebra* will be explained below.

3   Strictly Positive Logics and Reflection Calculi

We refer the reader to a note [1] for a short introduction to strictly positive logic and to [21] for more information from the description logic perspective.

3.1   Normal Strictly Positive Logics

Consider a modal language $\mathcal{L}_\Sigma$ with propositional variables $p, q, \ldots$, a constant $\top$, conjunction $\wedge$, and a possibly infinite set of symbols $\Sigma = \{a_i : i \in J\}$ understood as diamond modalities. The family Σ is called the *signature* of the language $\mathcal{L}_\Sigma$. Strictly positive formulas (or simply *formulas*) are built up by the grammar:

$$A ::= p \mid \top \mid (A \wedge A) \mid aA, \quad \text{where } a \in \Sigma.$$

Sequents are expressions of the form $A \vdash B$ where A, B are strictly positive formulas.

Basic sequent-style system, denoted K^+, is given by the following axioms and rules:

1. $A \vdash A$; $A \vdash \top$;   if $A \vdash B$ and $B \vdash C$ then $A \vdash C$;
2. $A \wedge B \vdash A$; $A \wedge B \vdash B$;   if $A \vdash B$ and $A \vdash C$ then $A \vdash B \wedge C$;
3. if $A \vdash B$ then $aA \vdash aB$, for each $a \in \Sigma$.

It is well-known that K^+ axiomatizes the strictly positive fragment of a poly-modal version of basic modal logic K. All our systems will also contain the following principle corresponding to the transitivity axiom in modal logic:

4. $aaA \vdash aA$.

The extension of K^+ by this axiom will be denoted $K4^+$.

Let $C[A/p]$ denote the result of replacing in C all occurrences of a variable p by A. A set of sequents L is called a *normal strictly positive logic* if it contains the axioms and is closed under the rules of K^+ and under the following *substitution rule*: if $(A \vdash B) \in L$ then $(A[C/p] \vdash B[C/p]) \in L$. We will only consider normal strictly positive logics below. We write $A \vdash_L B$ for the statement that $A \vdash B$ is provable in L (or belongs to L). $A =_L B$ means $A \vdash_L B$ and $B \vdash_L A$.

Any normal strictly positive logic L satisfies the following simple *positive replacement lemma* that we leave without proof.

Lemma 2. *Suppose $A \vdash_L B$, then $C[A/p] \vdash_L C[B/p]$, for any formula C.*

3.2   The System RC

Reflection calculus RC is a normal strictly positive logic formulated in the signature $\{\Diamond_n : n \in \omega\}$. It is obtained by adjoining to the axioms and rules of $K4^+$ (stated for each $\Diamond_n$) the following principles:

5. $\Diamond_n A \vdash \Diamond_m A$, for all $n > m$;
6. $\Diamond_n A \wedge \Diamond_m B \vdash \Diamond_n(A \wedge \Diamond_m B)$, for all $n > m$.

We notice that RC proves the following *polytransitivity* principles:

$$\Diamond_n \Diamond_m A \vdash \Diamond_m A, \quad \Diamond_m \Diamond_n A \vdash \Diamond_m A, \quad \text{for each } m \leqslant n.$$

Also, the converse of Axiom 6 is provable in RC, so that in fact we have

$$\Diamond_n(A \wedge \Diamond_m B) =_{\mathrm{RC}} \Diamond_n A \wedge \Diamond_m B. \tag{1}$$

The system RC was introduced in an equational logic format by Dashkov [13], the present formulation is from [8]. Dashkov showed that RC axiomatizes the set of all sequents $A \vdash B$ such that the implication $A \rightarrow B$ is provable in the polymodal logic GLP. Moreover, unlike GLP itself, RC is polytime decidable (whereas GLP is PSPACE-complete [25]) and enjoys the finite model property (whereas GLP is Kripke incomplete).

We recall a correspondence between variable-free RC-formulas and ordinals [5]. Let $\mathbb{F}$ denote the set of all variable-free RC-formulas, and let $\mathbb{F}_n$ denote its restriction to the signature $\{\Diamond_i : i \geqslant n\}$, so that $\mathbb{F} = \mathbb{F}_0$. For each $n \in \omega$ we define binary relations $<_n$ on $\mathbb{F}$ by

$$A <_n B \stackrel{\text{def}}{\Longleftrightarrow} B \vdash_{\text{RC}} \Diamond_n A.$$

Obviously, $<_n$ is a transitive relation invariantly defined on the equivalence classes w.r.t. provable equivalence in RC (denoted $=_{\text{RC}}$). Since RC is polytime decidable, so are both $=_{\text{RC}}$ and all of $<_n$.

An RC-formula without variables and $\wedge$ is called a *word*. In fact, any such formula syntactically is a finite sequence of letters $\Diamond_i$ (followed by $\top$). If A, B are words then AB will denote $A[\top/B]$, that is, the word corresponding to the concatenation of these sequences. $A \stackrel{\circ}{=} B$ denotes the graphical identity of formulas (words).

The set of all words will be denoted $\mathbb{W}$, and $\mathbb{W}_n$ will denote its restriction to the signature $\{\Diamond_i : i \geqslant n\}$. The following facts are from [5,8]:

- Every $A \in \mathbb{F}_n$ is RC-equivalent to a word in $\mathbb{W}_n$;
- $(\mathbb{W}_n/=_{\text{RC}}, <_n)$ is isomorphic to $(\varepsilon_0, <)$.

Here, ε_0 is the first ordinal α such that $\omega^\alpha = \alpha$. Thus, the set $\mathbb{W}_n/=_{\text{RC}}$ is well-ordered by the relation $<_n$. The isomorphism can be established by an onto and order preserving function $o_n : \mathbb{W}_n \to \varepsilon_0$ such that, for all $A, B \in \mathbb{W}_n$,

$$A =_{\text{RC}} B \iff o_n(A) = o_n(B).$$

Then $o_n(A)$ is the order type of $\{B \in \mathbb{W}_n : B <_n A\}/=_{\text{RC}}$.

The function $o(A) := o_0(A)$ can be inductively calculated as follows: If $A \stackrel{\circ}{=} \Diamond_0^k \top$ then $o(A) = k$. If $A \stackrel{\circ}{=} A_1 \Diamond_0 A_2 \Diamond_0 \cdots \Diamond_0 A_n$, where all $A_i \in \mathbb{W}_1$ and not all of them are empty, then

$$o(A) = \omega^{o(A_n^-)} + \cdots + \omega^{o(A_1^-)}.$$

Here, B^- is obtained from $B \in \mathbb{W}_1$ replacing every $\Diamond_{m+1}$ by $\Diamond_m$. For $n > 0$ and $A \in \mathbb{W}_n$ we let $o_n(A) = o_{n-1}(A^-)$.

3.3 The System RC^∇

Definition 1. The signature of RC^∇ consists of modalities $\Diamond_n$ and ∇_n, for each $n < \omega$. The system RC^∇ is a normal strictly positive logic given by the following axioms and rules, for all $m, n < \omega$:

1. RC for $\Diamond_n$; RC for ∇_n;
2. $A \vdash \nabla_n A$;
3. $\Diamond_n A \vdash \nabla_n A$;

4. $\Diamond_m \nabla_n A \vdash \Diamond_m A$; $\nabla_n \Diamond_m A \vdash \Diamond_m A$ if $m \leqslant n$.

As a basic syntactic fact about RC^∇ we mention the following useful lemma. We often write $=$ for $=_{\mathrm{RC}^\nabla}$ and $\vdash$ for $\vdash_{\mathrm{RC}}$.

Lemma 3. *The following are theorems of* RC^∇, *for all* $m < n$:

(i) $\Diamond_n(A \wedge \nabla_m B) = \Diamond_n A \wedge \Diamond_m B$;
(ii) $\nabla_n(A \wedge \Diamond_m B) = \nabla_n A \wedge \Diamond_m B$.

Proof.

(i) Part ($\vdash$) follows from $\Diamond_n \nabla_m B \vdash \Diamond_m B$. Part ($\dashv$) follows from $\Diamond_n A \wedge \Diamond_m B \vdash \Diamond_n(A \wedge \Diamond_m B) \vdash \Diamond_n(A \wedge \nabla_m B)$ using positive replacement.
(ii) Part ($\vdash$) follows from $\nabla_n \Diamond_m B \vdash \Diamond_m B$. Part ($\dashv$) follows from $\nabla_n A \wedge \Diamond_m B \vdash \nabla_n A \wedge \nabla_m \Diamond_m B \vdash \nabla_n(A \wedge \nabla_m \Diamond_m B) \vdash \nabla_n(A \wedge \Diamond_m B)$ using Axiom 3.6 for ∇ modalities, the fact that $\Diamond_m B = \nabla_m \Diamond_m B$ and positive replacement.

A formula A is called *ordered* if no modality with a smaller index (be it $\Diamond_i$ or ∇_i) occurs in A within the scope of a modality with a larger index.

Lemma 4. *Every formula* A *of* RC^∇ *is equivalent to an ordered one.*

Proof. Apply Eq. (1) of RC for $\Diamond$ and for ∇ modalities, and the identities of Lemma 3 from left to right, until the rules are not applicable to any of the subformulas of A.

The intended arithmetical interpretation of RC^∇ maps strictly positive formulas to Gödelian theories in $\mathfrak{G}_T$ in such a way that $\top$ corresponds to T, $\wedge$ corresponds to the union of theories, $\Diamond_n$ corresponds to R_n and ∇_n corresponds to Π_{n+1}, for each $n \in \omega$.

Definition 2. An *arithmetical interpretation in* $\overline{\mathfrak{G}}_T$ is a map $*$ from strictly positive modal formulas to $\overline{\mathfrak{G}}_T$ satisfying the following conditions for all $n \in \omega$:

- $\top^* = 1_T$; $(A \wedge B)^* = (A^* \wedge_T B^*)$;
- $(\Diamond_n A)^* = \mathrm{R}_n(A^*)$; $(\nabla_n A)^* = \Pi_{n+1}(A^*)$.

The following result shows, as expected, that every theorem of RC^∇ represents an identity of the structure $(\overline{\mathfrak{G}}_T, \wedge_T, \{\mathrm{R}_n, \Pi_{n+1} : n < \omega\})$.

Theorem 1. *For any formulas* A, B *of* RC^∇, *if* $A \vdash_{\mathrm{RC}^\nabla} B$ *then* $A^* \leqslant_T B^*$, *for all arithmetical interpretations* $*$ *in* $\overline{\mathfrak{G}}_T$.

Proof. A proof of Theorem 1 is routine. For the axioms and rules of RC for the $\Diamond$-fragment it has been carefully verified in [9]. Of the remaining axioms and rules we only treat Axiom 3.6 for the ∇-fragment, that is, the principle

$$\nabla_n A \wedge \nabla_m B \vdash \nabla_n(A \wedge \nabla_m B). \tag{2}$$

Consider any arithmetical interpretation $*$, and let $S = A^*$ and $U = B^*$ be the corresponding Gödelian theories (with the associated numerations σ and τ, respectively). We rely on Lemma 1. The principle (2) is the formalization in EA of the following assertion: *For any sentence $\pi \in \Pi_{n+1}^0$, if $S \cup \Pi_{m+1}(U) \vdash \pi$ then $\Pi_{n+1}(S) \cup \Pi_{m+1}(U) \vdash \pi$.* Reasoning in EA, consider a sentence $\varphi \in \Pi_{m+1}(U)$ such that $S, \varphi \vdash \pi$. Then $S \vdash \varphi \to \pi$ and, since $\varphi \to \pi$ is logically equivalent to a Π_{n+1}^0-sentence, conclude $\Pi_{n+1}(S) \vdash \varphi \to \pi$. Thus, $\Pi_{n+1}(S) \cup \Pi_{m+1}(U) \vdash \pi$.

Theorem 1, together with Gödel's second incompleteness theorem, has as its corollary the following property of the logic RC^∇.

Corollary 1. *For all RC^∇ formulas A, $A \nvdash_{\mathrm{RC}^\nabla} \Diamond_n A$.*

A similar fact is known for GLP and can also be proved by purely modal logic means [11]. A simpler argument for RC is given in [3]. We will make use of the latter in the normal form theorems below.

Conjecture 1. RC^∇ is arithmetically complete, that is, the converse of Theorem 1 also holds, provided T is arihmetically sound.

Finally, we remark that RC^∇ is not Kripke complete. In fact, the sequent

$$\Diamond_1 A \wedge \nabla_0 B \vdash \Diamond_1 (A \wedge \nabla_0 B)$$

is valid in every Kripke frame satisfying RC^∇. However, it is unprovable in RC^∇ and even arithmetically invalid. This can be established by standard arguments, see [3] for some details. By Theorem 3 of [1] this has a noteworthy corollary that RC^∇ is not a strictly positive fragment of any normal polymodal logic. In this strong sense, the conservativity operators are not representable in a standard modal logic context.

4   The Variable-Free Fragment of RC^∇

Let $\mathbb{F}_n^\nabla$ denote the set of all variable-free strictly positive formulas in the language of RC^∇ with the modalities $\{\Diamond_i, \nabla_i : i \geqslant n\}$ only. We abbreviate $F \vdash_{\mathrm{RC}^\nabla} \nabla_n G$ by $F \vdash_n G$ and $\nabla_n F =_{\mathrm{RC}^\nabla} \nabla_n G$ by $F \equiv_n G$.

Lemma 5.

(i) *If $A \vdash_n B$ and $m < n$, then $A \wedge \Diamond_m C \vdash_n B \wedge \Diamond_m C$;*
(ii) *If $A \vdash_n B$ and $B \vdash \nabla_n C$, then $A \vdash \nabla_n C$;*
(iii) *If $A \vdash_n B$ and $B \vdash \Diamond_n C$, then $A \vdash \Diamond_n C$.*

Proof.

(i) $A \wedge \Diamond_m C \vdash \nabla_n B \wedge \Diamond_m C \vdash \nabla_n (B \wedge \Diamond_m C)$.
(ii) $A \vdash \nabla_n B \vdash \nabla_n \nabla_n C \vdash \nabla_n C$;
(iii) $A \vdash \nabla_n B \vdash \nabla_n \Diamond_n C \vdash \Diamond_n C$.

Lemma 6.

(i) $\Diamond_i A \wedge B = \nabla_i(\Diamond_i A \wedge B) \wedge B$;
(ii) $\nabla_i A \wedge B = \nabla_i(\nabla_i A \wedge B) \wedge B$.

Proof. In both (i) and (ii) the implication ($\vdash$) follows from the axiom $C \vdash \nabla_i C$. For ($\dashv$) we obtain $\nabla_i(\Diamond_i A \wedge B) \vdash \nabla_i \Diamond_i A = \Diamond_i A$ for (i) and simlarly $\nabla_i(\nabla_i A \wedge B) \vdash \nabla_i \nabla_i A = \nabla_i A$ for (ii).

Lemma 7. *The set of all formulas $\{\Diamond_n F, \nabla_n G : F, G \in \mathbb{W}_n\}$ is linearly ordered by $\vdash_{\mathrm{RC}^\nabla}$.*

Proof. For any $F, G \in \mathbb{W}_n$ we know that either $F \vdash_{\mathrm{RC}} \Diamond_n G$ or $G \vdash_{\mathrm{RC}} \Diamond_n F$ or $F =_{\mathrm{RC}} G$. In the first case we obtain provably in RC^∇: $\Diamond_n F \vdash \nabla_n F \vdash \Diamond_n G \vdash \nabla_n G$. The second case is symmetrical. In the third case we obtain $\Diamond_n F = \Diamond_n G \vdash \nabla_n F = \nabla_n G$.

Theorem 2. *For each $A \in \mathbb{F}_n^\nabla$ there is a word $W \in \mathbb{W}_n$ such that $A \equiv_n W$.*

Proof. By Lemma 4 it is sufficient to prove the theorem for ordered formulas A. The proof goes by induction on the length of ordered A. We can also assume that the minimal modality occurring in A is $\Diamond_n$ or ∇_n. (Otherwise, prove it for the minimum $m > n$ and infer $A \equiv_n W$ from $A \equiv_m W$.) The basis of induction is trivial, consider the induction step.

Assume that the induction hypothesis holds for all formulas shorter than A. Since A is ordered, A can be written in the form

$$A = \Diamond_n A_1 \wedge \cdots \wedge \Diamond_n A_k \wedge \nabla_n B_1 \wedge \ldots \nabla_n B_l \wedge D,$$

where $D \in \mathbb{F}_{n+1}^\nabla$ and $A_i, B_j \in \mathbb{F}_n^\nabla$. Since $\Diamond_n$ or ∇_n must occur in A, we know that D and each A_i, B_j are strictly shorter than A. By the induction hypothesis and Lemma 7 we can delete from the conjunction all but one members of the form $\Diamond_n A_i, \nabla_n B_j$. Thus, $A = D \wedge \Diamond_n A'$ or $A = D \wedge \nabla_n B'$, for some words $A', B' \in \mathbb{W}_n$.

Now we apply the induction hypothesis to D and obtain a word $V \in \mathbb{W}_{n+1}$ such that $V \equiv_{n+1} D$. It follows that $D \wedge \Diamond_n A' \equiv_{n+1} V \wedge \Diamond_n A'$ and $D \wedge \nabla_n B' \equiv_{n+1} V \wedge \nabla_n B'$, by Lemma 5. Hence, it is sufficient to prove that, for some $W \in \mathbb{W}_n$, $V \wedge \Diamond_n A' \equiv_n W$ and similarly, for some $W \in \mathbb{W}_n$, $V \wedge \nabla_n B' \equiv_n W$.

In the first case we actually have $V \wedge \Diamond_n A' =_{\mathrm{RC}} W$, for some word W, which immediately yields the claim.

In the second case we write $B' = B_1 \Diamond_n B_2$ where $B_1 \in \mathbb{W}_{n+1}$. There are three cases to consider: (a) $B_1 \vdash \Diamond_{n+1} V$, (b) $V \vdash \Diamond_{n+1} B_1$, (c) $V = B_1$.

In case (c) by Lemma 6 we obtain:

$$V \wedge \nabla_n B_1 \Diamond_n B_2 = V \wedge \nabla_n(V \wedge \Diamond_n B_2) = V \wedge \Diamond_n B_2 = V \Diamond_n B_2.$$

In case (a) we show $\nabla_n(V \wedge \nabla_n B') = \nabla_n B'$. Firstly,

$$B' \vdash \Diamond_{n+1} V \wedge \nabla_n B' \vdash \nabla_{n+1} V \wedge \nabla_n B' = \nabla_{n+1}(V \wedge \nabla_n B').$$

Hence, $\nabla_n B' \vdash \nabla_n \nabla_{n+1}(V \wedge \nabla_n B') = \nabla_n(V \wedge \nabla_n B')$. On the other hand,

$$\nabla_n(V \wedge \nabla_n B') \vdash \nabla_n \nabla_n B' \vdash \nabla_n B'.$$

In case (b) we show $\nabla_n(V \wedge \nabla_n B') = \nabla_n(V \wedge \Diamond_n B_2)$ so that one can infer $\nabla_n(V \wedge \nabla_n B') = \nabla_n V \Diamond_n B_2$. On the one hand, we have

$$\nabla_n B' = \nabla_n(B_1 \wedge \Diamond_n B_2) \vdash \nabla_n \Diamond_n B_2 = \Diamond_n B_2,$$

which implies $\nabla_n(V \wedge \nabla_n B') \vdash \nabla_n(V \wedge \Diamond_n B_2)$. On the other hand,

$$V \wedge \Diamond_n B_2 = V \wedge \Diamond_{n+1} B_1 \wedge \Diamond_n B_2 = V \wedge \Diamond_{n+1}(B_1 \wedge \Diamond_n B_2) = V \wedge \Diamond_{n+1} B' \vdash V \wedge \nabla_n B'.$$

Hence, $\nabla_n(V \wedge \Diamond_n B_2) \vdash \nabla_n(V \wedge \nabla_n B')$.

From Theorem 2 we obtain the following strengthening of Lemma 7.

Corollary 2. *The set of all formulas* $\{\Diamond_n F, \nabla_n G : F, G \in \mathbb{F}_n^\nabla\}$ *is linearly ordered by* $\vdash_{RC^\nabla}$.

Corollary 3. *For all formulas* $A, B \in \mathbb{F}_n^\nabla$, *either* $A \vdash \Diamond_n B$, *or* $B \vdash \Diamond_n A$, *or* $A \equiv_n B$.

Proof. Consider the words $A_1 \equiv_n A$ and $B_1 \equiv_n B$. By the linearity property for words either $A_1 \vdash \Diamond_n B_1$ or $B_1 \vdash \Diamond_n A_1$ or $A_1 = B_1$. In the first case we obtain $A \vdash \nabla_n A_1 \vdash \nabla_n \Diamond_n B_1 \vdash \Diamond_n B_1 \vdash \Diamond_n \nabla_n B \vdash \Diamond_n B$. The second case is symmetrical, the third one implies $A \equiv_n B$ immediately.

Corollary 4. *For all* $A, B \in \mathbb{F}_n^\nabla$, $\Diamond_n A \vdash \Diamond_n B$ *iff* $A \vdash \nabla_n B$.

Proof. Assume $\Diamond_n A \vdash \Diamond_n B$. By Corollary 3, either $A \vdash \Diamond_n B$, or $B \vdash \Diamond_n A$, or $A \equiv_n B$. In the first and the third cases we immediately have $A \vdash \nabla_n B$. In the second case we obtain $\Diamond_n A \vdash \Diamond_n B \vdash \Diamond_n \Diamond_n A$ contradicting Corollary 1.

In the opposite direction, if $A \vdash \nabla_n B$ then $\Diamond_n A \vdash \Diamond_n \nabla_n B \vdash \Diamond_n B$.

Theorem 3 (weak normal forms). *Every formula* $A \in \mathbb{F}_n^\nabla$ *is equivalent in* RC^∇ *to a formula of the form*

$$\nabla_n A_n \wedge \nabla_{n+1} A_{n+1} \wedge \cdots \wedge \nabla_{n+k} A_{n+k},$$

for some k, *where* $A_i \in \mathbb{W}_i$ *for all* $i = n, \ldots, n+k$.

Proof. Induction on the build-up of $A \in \mathbb{F}_n^\nabla$. We consider the following cases.

(1) $A = B \wedge C$. The induction hypothesis is applicable to B and C, so it is sufficient to prove: for any $B_i, C_i \in \mathbb{W}_i$ there is a word $A_i \in \mathbb{W}_i$ such that

$$\nabla_i B_i \wedge \nabla_i C_i = \nabla_i A_i.$$

By Lemma 7 we can take one of B_i, C_i as A_i.

(2) $A = \nabla_i B$, for some $i \geqslant n$. Then we obtain

$$\nabla_i B = \nabla_i(\nabla_n B_n \wedge \nabla_{n+1} B_{n+1} \wedge \cdots \wedge \nabla_{n+k} B_{n+k}) =$$
$$= \nabla_n B_n \wedge \cdots \wedge \nabla_{i-1} B_{i-1} \wedge \nabla_i(\nabla_i B_i \wedge \cdots \wedge \nabla_{n+k} B_{n+k}) =$$
$$= \nabla_n B_n \wedge \cdots \wedge \nabla_{i-1} B_{i-1} \wedge \nabla_i B_i',$$

for some $B_i' \in \mathbb{W}_i$, by Theorem 2.

(3) $A = \Diamond_i B$, for some $i \geqslant n$. Then we obtain, using Lemma 3,

$$\Diamond_i B = \Diamond_i(\nabla_n B_n \wedge \nabla_{n+1} B_{n+1} \wedge \cdots \wedge \nabla_{n+k} B_{n+k}) =$$
$$= \Diamond_n B_n \wedge \cdots \wedge \Diamond_{i-1} B_{i-1} \wedge \Diamond_i(\nabla_i B_i \wedge \cdots \wedge \nabla_{n+k} B_{n+k}) =$$
$$= \nabla_n \Diamond_n B_n \wedge \cdots \wedge \nabla_{i-1} \Diamond_{i-1} B_{i-1} \wedge \nabla_i \Diamond_i B_i',$$

for some $B_i' \in \mathbb{W}_i$, by Theorem 2.

Weak normal forms are, in general, not unique. However, the following lemma shows that the "tails" of the weak normal forms are invariant (up to equivalence in RC^{∇}).

Lemma 8. Let $A \overset{\circ}{=} \nabla_n A_n \wedge \nabla_{n+1} A_{n+1} \wedge \cdots \wedge \nabla_k A_k$ and $B \overset{\circ}{=} \nabla_n B_n \wedge \nabla_{n+1} B_{n+1} \wedge \cdots \wedge \nabla_m B_m$ be weak normal forms and $A \vdash B$. Then $k \geqslant m$ and for all i such that $n \leqslant i \leqslant k$ there holds

(i) $\nabla_i A_i \wedge \cdots \wedge \nabla_k A_k \vdash_i \nabla_i B_i \wedge \cdots \wedge \nabla_m B_m$;
(ii) $\nabla_i A_i \wedge \cdots \wedge \nabla_k A_k \vdash \nabla_i B_i \wedge \cdots \wedge \nabla_m B_m$.

Proof. Obviously, Claim (ii) implies Claim (i). We first prove (i) and then strengthen it to (ii). For $i = n$ both claims are vacuous, so we assume $i > n$.

Denote $\overline{A_i} := \nabla_i A_i \wedge \cdots \wedge \nabla_k A_k$ and $\overline{B_i} := \nabla_i B_i \wedge \cdots \wedge \nabla_m B_m$. By Lemma 3 we have either $\overline{A_i} \vdash \Diamond_i \overline{B_i}$ or $\overline{B_i} \vdash \Diamond_i \overline{A_i}$ or $\overline{A_i} \equiv_i \overline{B_i}$. In the first and in the third case we obviously have $\overline{A_i} \vdash_i \overline{B_i}$ as required.

Assume $\overline{B_i} \vdash \Diamond_i \overline{A_i}$. Consider the formula

$$C := \Diamond_n A_n \wedge \cdots \wedge \Diamond_{i-1} A_{i-1} \wedge \overline{B_i}.$$

We show that $C \vdash \Diamond_i C$ contradicting Corollary 1.

Using our assumption and Lemma 3 (i) we obtain

$$C \vdash \Diamond_n A_n \wedge \cdots \wedge \Diamond_{i-1} A_{i-1} \wedge \Diamond_i \overline{A_i}$$
$$\vdash \Diamond_i(\nabla_n A_n \wedge \cdots \wedge \nabla_{i-1} A_{i-1} \wedge \overline{A_i})$$
$$\vdash \Diamond_n A_n \wedge \cdots \wedge \Diamond_{i-1} A_{i-1} \wedge \Diamond_i A$$
$$\vdash \Diamond_n A_n \wedge \cdots \wedge \Diamond_{i-1} A_{i-1} \wedge \Diamond_i B$$
$$\vdash \Diamond_i(\Diamond_n A_n \wedge \cdots \wedge \Diamond_{i-1} A_{i-1} \wedge B)$$
$$\vdash \Diamond_i C.$$

This proves Claim (i).

To prove (ii) assume the contrary and consider the maximal number i such that $\overline{A_i} \nvdash \overline{B_i}$. Such an i exists, since both A and B have finitely many terms. Thus, we have $\overline{A_{i+1}} \vdash \overline{B_{i+1}}$ and

$$\nabla_i A_i \wedge \overline{A_{i+1}} \nvdash \nabla_i B_i \wedge \overline{B_{i+1}}.$$

It follows that $\nabla_i A_i \wedge \overline{A_{i+1}} \nvdash \nabla_i B_i = \nabla_i \nabla_i B_i$, hence $\overline{A_i} \nvdash_i \nabla_i B_i$. Since $\overline{B_i} \vdash \nabla_i B_i$, we obtain $\overline{A_i} \nvdash_i \overline{B_i}$ contradicting Claim (i).

A corollary of Lemma 8 is that in every weak normal form of a given formula each tail $\nabla_i A_i \wedge \cdots \wedge \nabla_k A_k$ is defined uniquely up to equivalence in RC^∇.

There are two formats for graphically unique normal forms. We call them 'fat' and 'thin', because the former consist of larger expressions, whereas the latter are obtained by pruning certain parts of a given formula. Fat normal forms, presented below, have a natural proof-theoretic meaning and are tightly related to collections of proof-theoretic ordinals called *conservativity spectra* or *Turing-Taylor expansions* [20].

Definition 3. A formula $A \in \mathbb{F}_n^\nabla$ is in the *fat normal form* if either $A \overset{\circ}{=} \top$ or it has the form $\nabla_n A_n \wedge \nabla_{n+1} A_{n+1} \wedge \cdots \wedge \nabla_{n+k} A_{n+k}$, where for all $i = n, \ldots, n+k$, $A_i \in \mathbb{W}_i$, $A_{n+k} \not\overset{\circ}{=} \top$ and

$$\nabla_i A_i \vdash \nabla_i (\nabla_i A_i \wedge \cdots \wedge \nabla_{n+k} A_{n+k}). \tag{$*$}$$

Theorem 4.

(i) Every $A \in \mathbb{F}_n^\nabla$ is equivalent to a formula in the fat normal form.
(ii) For any A, the words A_i in the fat normal form of A are unique modulo equivalence in RC.

Proof. (i) First, we apply Theorem 2. Then, by induction on k we show that any formula $\nabla_n A_n \wedge \cdots \wedge \nabla_{n+k} A_{n+k}$ can be transformed into one satisfying $(*)$.

For $k = 0$ the claim is trivial. Otherwise, by the induction hypothesis we can assume that $(*)$ holds for $i = n+1, \ldots, n+k$. Then we argue using Lemma 6 as follows:

$$\nabla_n A_n \wedge \nabla_{n+1} A_{n+1} \wedge \cdots \wedge \nabla_{n+k} A_{n+k} =$$
$$= \nabla_n (\nabla_n A_n \wedge \nabla_{n+1} A_{n+1} \wedge \cdots \wedge \nabla_{n+k} A_{n+k}) \wedge \nabla_{n+1} A_{n+1} \wedge \cdots \wedge \nabla_{n+k} A_{n+k} =$$
$$= \nabla_n A_n' \wedge \nabla_{n+1} A_{n+1} \wedge \cdots \wedge \nabla_{n+k} A_{n+k},$$

where $A' \in \mathbb{W}_n$ is obtained from Theorem 2. Notice that

$$\nabla_n A_n' \vdash \nabla_n (\nabla_n A_n \wedge \nabla_{n+1} A_{n+1} \wedge \cdots \wedge \nabla_{n+k} A_{n+k}) \vdash$$
$$\vdash \nabla_n (\nabla_n A_n' \wedge \nabla_{n+1} A_{n+1} \wedge \cdots \wedge \nabla_{n+k} A_{n+k}),$$

hence $(*)$ holds for $i = n$. This proves Claim (i).

To prove Claim (ii) we apply Lemma 8. Assume $A \vdash B$, $A = \nabla_n A_n \wedge \cdots \wedge \nabla_{n+k} A_{n+k}$ is in the fat normal form and $B = \nabla_n B_n \wedge \cdots \wedge \nabla_{n+m} B_{n+m}$ is in a weak normal form. Then $k \geqslant m$ and, for all $i = n, \ldots, n+m$, $\nabla_i A_i \vdash \nabla_i B_i$.

It follows that, if $A, B \in \mathbb{F}_n^\nabla$ are both in the fat normal form and $A = B$ in RC$^\nabla$, then $m = k$ and $\nabla_i A_i = \nabla_i B_i$, for $i = n, \ldots, n + k$. Since $\mathbb{W}_i$ is linearly preordered by $<_i$ in RC, the latter is only possible if $A_i =_{\mathrm{RC}} B_i$.

Corollary 5. *The set of variable-free sequents $A \vdash B$ provable in* RC$^\nabla$ *is decidable.*

Corollary 6. *Suppose A, B are variable-free and T is a sound Gödelian extension of* EA. *Then $A \vdash_{\mathrm{RC}^\nabla} B$ iff $A^* \leqslant_T B^*$, for any arithmetical interpretation $*$ in* $\overline{\mathfrak{G}}_T$.

A   Iterating Monotone Operators on $\mathfrak{G}_{\mathrm{EA}}$

Transfinite iterations of reflection principles play an important role in proof theory starting from the work of A. Turing on recursive progressions [26]. Here we present a general result on defining iterations of monotone semi-idempotent operators in $\mathfrak{G}_{\mathrm{EA}}$.

An operator $R : \mathfrak{G}_{\mathrm{EA}} \to \mathfrak{G}_{\mathrm{EA}}$ is called *computable* if so is the function $\ulcorner \sigma \urcorner \mapsto \ulcorner R(\sigma) \urcorner$. By extension of terminology we also call computable any operator R' such that $\forall \sigma \in \mathfrak{G}_{\mathrm{EA}} \, R'(\sigma) =_{\mathrm{EA}} R(\sigma)$, for some computable R.

An operator $R : \mathfrak{G}_{\mathrm{EA}} \to \mathfrak{G}_{\mathrm{EA}}$ is called *uniformly definable* if there is an elementary formula $\mathrm{Ax}_R(x, y)$ such that

(i) For each $\sigma \in \mathfrak{G}_{\mathrm{EA}}$ one has $R(\sigma) =_{\mathrm{EA}} \mathrm{Ax}_R(x, \overline{\ulcorner \sigma \urcorner})$,
(ii) EA $\vdash \forall x, y \, (\mathrm{Ax}_R(x, y) \to x \geqslant y)$.

The operators R_n and Π_{n+1} are uniformly definable in a very special way. For example, the formula $\mathrm{R}_n(\sigma)$ is obtained by substituting $\sigma(x)$ for $X(x)$ into a fixed elementary formula containing a single positive occurrence of a predicate variable X. In fact, the following more general proposition holds that we leave here without proof. (Nothing below depends on it.)

Proposition 1. *An operator $R : \mathfrak{G}_{\mathrm{EA}} \to \mathfrak{G}_{\mathrm{EA}}$ is uniformly definable iff R is computable.*

Definition 4. A uniformly definable R is called

- *provably monotone* if EA $\vdash \forall \sigma, \tau \, (\text{``}\tau \leqslant_{\mathrm{EA}} \sigma\text{''} \to \text{``}R(\tau) \leqslant_{\mathrm{EA}} R(\sigma)\text{''})$,
- *reflexively monotone* if EA $\vdash \forall \sigma, \tau \, (\text{``}\tau \leqslant_{\mathrm{EA}} \sigma\text{''} \to \text{``}R(\tau) \leqslant R(\sigma)\text{''})$.

Here, σ, τ range over Gödel numbers of elementary formulas in one free variable, "$\tau \leqslant_{\mathrm{EA}} \sigma$" abbreviates $\Box_{\mathrm{EA}} \forall x \, (\Box_\sigma(x) \to \Box_\tau(x))$ and "$R(\tau) \leqslant R(\sigma)$" stands for $\forall x \, (\Box_{\mathrm{Ax}_R(\cdot, \bar{\sigma})}(x) \to \Box_{\mathrm{Ax}_R(\cdot, \bar{\tau})}(x))$. Reflexivity here refers to the fact that "$R(\tau) \leqslant R(\sigma)$" is the statement of inclusion of theories rather than provable inclusion. Since the formula "$\tau \leqslant_{\mathrm{EA}} \sigma$" implies its own provability in EA, reflexively monotone operators are (provably) monotone but not necessarily vice versa.

It is also easy to see that the operators R_n (along with all the usual reflection principles) are reflexively monotone.

An *elementary well-ordering* is a pair of bounded formulas $D(x)$ and $x \prec y$ and a constant 0 such that in the standard model the relation $\prec$ well-orders the domain D and is provably linear in EA with the least element 0. Given an elementary well-ordering $(D, \prec, 0)$, we will denote its elements by Greek letters and will identify them with an initial segment of the ordinals.

Let R be an uniformly definable monotone operator. The α-th iterate of R along $(D, \prec)$ is a map associating with any numeration σ the Gödelian extension of EA numerated by an elementary formula $\rho(\overline{\alpha}, x)$ such that provably in EA:

$$\rho(\alpha, x) \leftrightarrow ((\alpha = 0 \wedge \sigma(x)) \vee \exists \beta \prec \alpha \; \mathrm{Ax}_R(x, \ulcorner \rho(\overline{\beta}, x) \urcorner)). \qquad (3)$$

We notice that the natural Gödel numbering of formulas and terms should satisfy the inequalities $\ulcorner \rho(\overline{\beta}, x) \urcorner \geqslant \ulcorner \overline{\beta} \urcorner \geqslant \beta$. Hence, the quantifier on β in Eq. (3) can be bounded by x. Thus, some elementary formula $\rho(\alpha, x)$ satisfying (3) can always be constructed by the fixed point lemma.

The parametrized family of theories numerated by $\rho(\alpha, x)$ will be denoted $R^\alpha(\sigma)$ and the formula $\rho(\alpha, x)$ will be more suggestively written as $x \in R^\alpha(\sigma)$. Then, Eq. (3) can be interpreted as saying that $R^0(\sigma) =_{\mathrm{EA}} \sigma$ and, if $\alpha \succ 0$,

$$R^\alpha(\sigma) =_{\mathrm{EA}} \bigcup \{R(R^\beta(\sigma)) : \beta \prec \alpha\}.$$

Lemma 9. *Suppose R is uniformly definable.*

(i) If $0 \prec \alpha \preceq \beta$ then $R^\beta(\sigma) \leqslant_{\mathrm{EA}} R^\alpha(\sigma)$;
(ii) EA $\vdash \forall \alpha, \beta \, (0 \prec \alpha \prec \beta \to$ "$R^\beta(\sigma) \leqslant R^\alpha(\sigma)$").

Proof. Obviously, Claim (i) follows from Claim (ii). For the latter we unwind the definition of $\rho(\alpha, x)$ and prove within EA

$$\forall \alpha, \beta \, (0 \prec \alpha \prec \beta \to \forall x \, (\rho(\alpha, x) \to \rho(\beta, x))).$$

This is sufficient to obtain from the same premise $\forall x \, (\Box_{\rho(\alpha, \cdot)}(x) \to \Box_{\rho(\beta, \cdot)}(x))$.

Reason within EA: If $\rho(\alpha, x)$ and $\alpha \neq 0$ then there is a $\gamma \prec \alpha$ such that $\mathrm{Ax}_R(x, \ulcorner \rho(\overline{\gamma}, x) \urcorner)$. By the provable transitivity of $\prec$ from $\alpha \prec \beta$ we obtain $\gamma \prec \beta$, hence $\rho(\beta, x)$.

Lemma 10. *Suppose R is reflexively monotone. If $\tau \leqslant_{\mathrm{EA}} \sigma$ then $R^\alpha(\tau) \leqslant_{\mathrm{EA}} R^\alpha(\sigma)$ and, moreover, EA $\vdash \forall \alpha$ "$R^\alpha(\tau) \leqslant R^\alpha(\sigma)$".*

Proof. We argue by reflexive induction similarly to [4], that is, we prove in EA that

$$\forall \beta \prec \alpha \, \Box_{\mathrm{EA}} \forall x \, (\Box_{R^{\overline{\beta}}(\sigma)}(x) \to \Box_{R^{\overline{\beta}}(\tau)}(x)) \to \forall x \, (\Box_{R^\alpha(\sigma)}(x) \to \Box_{R^\alpha(\tau)}(x))$$

and then apply Löb's theorem in EA. Assume $\tau \leqslant_{\mathrm{EA}} \sigma$.

Reason within EA: If $\Box_{R^\alpha(\sigma)}(x)$ then either $\alpha = 0 \wedge \Box_\sigma(x)$, or there is a $\beta \prec \alpha$ such that $\Box_{R(R^{\overline{\beta}}(\sigma))}(x)$. In the first case we obtain $\Box_\tau(x)$ by the external assumption $\tau \leqslant_{\mathrm{EA}} \sigma$ and are done. In the second case, by the premise and the reflexive monotonicity of R we obtain $\Box_{R(R^{\overline{\beta}}(\tau))}(x)$ which yields $\Box_{R^\alpha(\tau)}(x)$.

Corollary 7. *The iteration of R along $(D, \prec)$ is uniquely defined, that is, equation (3) has a unique solution modulo $=_{\mathrm{EA}}$.*

The following corollary is most naturally stated for elementary well-orderings equipped with an elementary successor function $\alpha \mapsto \alpha + 1$ such that provably in EA

$$\forall \alpha \, (\alpha \prec \alpha + 1 \wedge \forall \beta \prec \alpha + 1 \, (\alpha \prec \beta \vee \alpha = \beta)).$$

Corollary 8. *Suppose R is provably monotone and semi-idempotent. Then*

   (i) $R^0(\sigma) =_{\mathrm{EA}} \sigma$,
   (ii) $R^{\alpha+1}(\sigma) =_{\mathrm{EA}} R(R^\alpha(\sigma))$,
   (iii) $R^\lambda(\sigma) =_{\mathrm{EA}} \bigcup \{R^\alpha(\sigma) : \alpha < \lambda\}$ *if $\lambda \in \mathrm{Lim}$.*

Proof. For Claim (ii), the implication $R^{\alpha+1}(\sigma) \leqslant_{\mathrm{EA}} R(R^\alpha(\sigma))$ is easy, since provably $\alpha \prec \alpha + 1$. For the opposite implication it is sufficient to prove

$$\mathrm{EA} \vdash \forall x \, (x \in R^{\alpha+1}(\sigma) \rightarrow \Box_{R(R^\alpha(\sigma))}(x)).$$

Then one will be able to conclude (within $\mathrm{EA} + \mathrm{B}\Sigma_1$) that $\forall x \, (\Box_{R^{\alpha+1}(\sigma)}(x) \rightarrow \Box_{R(R^\alpha(\sigma))}(x))$ and then appeal to the Π_2^0-conservativity of $\mathrm{B}\Sigma_1$ over EA.

Reason in EA: Assume $x \in R^{\alpha+1}(\sigma)$ then (since $\alpha + 1 \neq 0$) there is a $\beta \prec \alpha + 1$ such that $x \in R(R^\beta(\sigma))$. If $\beta = \alpha$ then $x \in R(R^\alpha(\sigma))$ and we are done. Otherwise, $\beta \prec \alpha$ and we consider two cases.

If $\beta \succ 0$ then we have $R^\alpha(\sigma) \leqslant_{\mathrm{EA}} R^\beta(\sigma)$ by Lemma 9 (ii). By the provable monotonicity of R we obtain $R(R^\alpha(\sigma)) \leqslant_{\mathrm{EA}} R(R^\beta(\sigma))$, whence $\Box_{R(R^\alpha(\sigma))}(x)$.

If $\beta = 0$ then $x \in R(\sigma)$. We have $R^\alpha(\sigma) \leqslant_{\mathrm{EA}} R(\sigma)$, since $\alpha \succ 0$. Hence, $R(R^\alpha(\sigma)) \leqslant_{\mathrm{EA}} R(R(\sigma)) \leqslant_{\mathrm{EA}} R(\sigma)$ by the semi-idempotence of R. So, from $x \in R(\sigma)$ we infer $\Box_{R(R^\alpha(\sigma))}(x)$.

B   Expressibility of Iterated Reflection

In this section we confuse the arithmetical and reflection calculus notation. We write $\Diamond_n$ for R_n and ∇_n for Π_{n+1}. Our goal is to show that iterated operators $\Diamond_n^\alpha$, for natural ordinal notations $\alpha < \varepsilon_0$, are expressible in the language of RC^∇. We will rely on the so-called *reduction property* (cf. [5], the present version is somewhat more general and follows from [4, Theorem 2], see also [10]).

We write $\Diamond_{n,\sigma}(\tau)$ for $\Diamond_n(\sigma \wedge \tau)$, hence $\Diamond_{n,\sigma}$ is a monotone semi-idempotent operator, for each σ. Let EA^+ denote the theory $R_1(\mathrm{EA})$ which is known to be equivalent to $\mathrm{EA} + \mathrm{Supexp}$.

Theorem 5 (reduction property). *For all $\sigma \in \mathfrak{G}_{\mathrm{EA}}$, $n \in \omega$,*

$$\Diamond_{n,\sigma}^\omega(1) =_{\mathrm{EA}^+} \nabla_n \Diamond_{n+1}(\sigma).$$

We also remark that the theory $\Diamond_{n,\sigma}^\omega(1)$ is equivalent to the one axiomatized over EA by the set $\bigcup \{Q_n^k(\sigma) : k < \omega\}$, where $Q_n^0(\sigma) := \Diamond_n \sigma$ and $Q_n^{k+1}(\sigma) := \Diamond_n(\sigma \wedge Q_n^k(\sigma))$ are formulas in one variable expressible in RC.

Concerning these formulas we note three well-known facts.

Lemma 11. *Provably in* EA,

1. $\forall B \in \mathbb{W}_n \ \forall k \ Q_n^{k+1}(B) \vdash_{\mathrm{RC}} Q_n^k(B) \wedge \Diamond_n Q_n^k(B)$;
2. $\forall B \in \mathbb{W}_n \ \forall k \ Q_n^k(B) <_n \Diamond_{n+1} B$;
3. $\forall B \in \mathbb{W}_n \ \forall k \ \exists A \in \mathbb{W}_n \ Q_n^k(B) =_{\mathrm{RC}} A$.

The first two of these claims are proved by an easy induction on k. The third one is a consequence of a more general theorem that any variable-free formula of RC is equivalent to a word. An explicit rule for calculating such an A is also well-known and related to the so-called *Worm sequence*, see [6, Lemma 5.9].

We consider the set of words $(\mathbb{W}_n, <_n)$ modulo equivalence in RC, together with its natural representation in EA, as an elementary well-ordering. For each $A \in \mathbb{W}_n$, let $o_n(A)$ denote the order type of $\{B <_n A : B \in \mathbb{W}_n\}$ modulo $=_{\mathrm{RC}}$. In a formalized context, the ordinal $o_n(A)$ is represented by its notation, the word A, however we still write $o_n(A)$, as it reminds us that A must be viewed as an ordinal.

From the reduction property we obtain the following theorem that was stated in [5] in a somewhat more restricted way.

Theorem 6. *For all words* $A \in \mathbb{W}_n$, *in* $\overline{\mathfrak{G}}_{\mathrm{EA}^+}$ *there holds*

$$\nabla_n A_\sigma^* =_{\mathrm{EA}^+} \Diamond_{n,\sigma}^{o_n(A)}(1).$$

Theorem 6 of [5] used extensions of σ rather than extensions of EA^+ by iterated reflection principles and as a result involved unnecessary restrictions of the complexity of σ. However, the idea of the present proof is the same.

Proof. We argue by reflexive induction in EA^+ and prove that, for all $\sigma \in \mathfrak{G}_{\mathrm{EA}^+}$ and all $n < \omega$,

$$\mathrm{EA}^+ \vdash \forall B <_n A \ ``\nabla_n B_\sigma^* =_{\mathrm{EA}^+} \Diamond_{n,\sigma}^{o_n(B)}(1)" \ \rightarrow \ ``\nabla_n A_\sigma^* = \Diamond_{n,\sigma}^{o_n(A)}(1)".$$

Arguing inside EA^+, we will omit the quotation marks and read the expressions $\tau \leqslant \nu$ as $\forall x \ (\Box_\nu(x) \leftrightarrow \Box_\tau(x))$ and $\tau = \nu$ as $\forall x \ (\Box_\tau(x) \leftrightarrow \Box_\nu(x))$.

If $A \stackrel{\circ}{=} \top$ the claim amounts to $\nabla_n 1 = 1$.

If $A \stackrel{\circ}{=} \Diamond_n B$ then $o_n(A) = o_n(B) + 1$, hence by the reflexive induction hypothesis $\nabla_n B_\sigma^* =_{\mathrm{EA}^+} \Diamond_{n,\sigma}^{o_n(B)}(1)$. It follows that $\Diamond_{n,\sigma}(\Diamond_{n,\sigma}^{o_n(B)}(1)) = \Diamond_{n,\sigma} \nabla_n B_\sigma^* = \Diamond_{n,\sigma} B_\sigma^*$. Therefore, we obtain

$$\Diamond_{n,\sigma}^{o_n(A)}(1) = \Diamond_{n,\sigma}(\Diamond_{n,\sigma}^{o_n(B)}(1)) = \Diamond_{n,\sigma} B_\sigma^* = A_\sigma^* = \nabla_n A_\sigma^*.$$

If $A \stackrel{\circ}{=} \Diamond_{m+1} B$ with $m \geqslant n$ then $\nabla_n A_\sigma^* = \nabla_n \nabla_m \Diamond_{m+1,\sigma} B_\sigma^*$. By the reduction property $\nabla_m \Diamond_{m+1,\sigma} B_\sigma^* = \bigcup_{k<\omega} (Q_m^k(B))_\sigma^*$. Moreover, by Lemma 11 (i), if a sentence is provable in $\bigcup_{k<\omega} (Q_m^k(B))_\sigma^*$, it must be provable in $(Q_m^k(B))_\sigma^*$, for some $k < \omega$. Hence, we can infer

$$\nabla_n A_\sigma^* = \nabla_n \bigcup_{k<\omega} (Q_m^k(B))_\sigma^* = \bigcup_{k<\omega} \nabla_n (Q_m^k(B))_\sigma^* = \bigcup_{k<\omega} \Diamond_{n,\sigma} (Q_m^k(B))_\sigma^*.$$

By Lemma 11(ii) and (iii), each of $Q_m^k(B)$ is $<_n$-below $A \stackrel{\circ}{=} \diamondsuit_{m+1}B$ and is equivalent to a word in $\mathbb{W}_n$. Hence,

$$\bigcup_{C<_n A} \diamondsuit_{n,\sigma} C_\sigma^* \leqslant \bigcup_{k<\omega} \diamondsuit_{n,\sigma}(Q_m^k(B))_\sigma^*.$$

By the reflexive induction hypothesis, for each $C <_n A$ we have

$$\diamondsuit_{n,\sigma} C_\sigma^* = \diamondsuit_{n,\sigma} \nabla_n C_\sigma^* = \diamondsuit_{n,\sigma} \diamondsuit_{n,\sigma}^{o_n(C)}(1).$$

It follows that

$$\diamondsuit_{n,\sigma}^{o_n(A)}(1) = \bigcup_{C<_n A} \diamondsuit_{n,\sigma} \diamondsuit_{n,\sigma}^{o_n(C)}(1) = \bigcup_{C<_n A} \diamondsuit_{n,\sigma} C_\sigma^* \leqslant$$

$$\leqslant \bigcup_{k<\omega} \diamondsuit_{n,\sigma}(Q_m^k(B))_\sigma^* = \nabla_n A_\sigma^*.$$

On the other hand, if $C <_n A$ then $A_\sigma^* \leqslant \diamondsuit_{n,\sigma} C_\sigma^*$ and $\nabla_n A_\sigma^* \leqslant \nabla_n \diamondsuit_{n,\sigma} C_\sigma^* \leqslant \diamondsuit_{n,\sigma} C_\sigma^*$. Hence,

$$\nabla_n A_\sigma^* \leqslant \bigcup_{C<_n A} \diamondsuit_{n,\sigma} C_\sigma^* = \diamondsuit_{n,\sigma}^{o_n(A)}(1).$$

Thus, we have proved $\nabla_n A_\sigma^* = \diamondsuit_{n,\sigma}^{o_n(A)}(1)$, as required.

For ordinals $\alpha < \varepsilon_0$, let $A_\alpha^n \in \mathbb{W}_n$ denote a canonical notation for α in the system $(\mathbb{W}_n, <_n)$. Thus, $o_n(A_\alpha^n) = \alpha$. We are going to show that the operations $\diamondsuit_n^\alpha$ are expressible in RC^∇ in the following sense.

Theorem 7. *For each $n < \omega$ and $0 < \alpha < \varepsilon_0$ there is an RC-formula $A(p)$ such that $\forall \sigma \in \mathfrak{G}_{\mathrm{EA}^+} \diamondsuit_n^\alpha(\sigma) =_{\mathrm{EA}^+} \nabla_n A(\sigma).$*

Note that $A(\sigma)$ denote the interpretation of the formula $A[p/\top]$ in $\overline{\mathfrak{G}}_{\mathrm{EA}^+}$ sending p to σ. The theorem follows from the following main lemma.

Lemma 12. *For all $n < \omega$ and $\alpha < \varepsilon_0$, for all $\sigma \in \mathfrak{G}_{\mathrm{EA}^+}$,*

$$\diamondsuit_n^\alpha \diamondsuit_n(\sigma) =_{\mathrm{EA}^+} \nabla_n A_\alpha^n \diamondsuit_n(\sigma).$$

The proof relies on a few observations.

Lemma 13. *Suppose $\diamondsuit_n \sigma \leqslant_{\mathrm{EA}^+} \sigma$. Then, for all $A \in \mathbb{W}_n$,*

(i) $A(\sigma) \leqslant_{\mathrm{EA}^+} \sigma$;
(ii) $A_\sigma^* =_{\mathrm{EA}^+} A(\sigma)$;
(iii) $\forall m \geqslant n \, \forall \alpha < \varepsilon_0 \, \diamondsuit_{m,\sigma}^\alpha 1 =_{\mathrm{EA}^+} \diamondsuit_m^\alpha(\sigma).$

Proof. Claim (i) is proved by induction on the build-up of A. Claim (ii) follows from (i), since $\diamondsuit_{m,\sigma}\tau = \diamondsuit_m(\sigma \wedge \tau) = \diamondsuit_m \tau$, if $\tau \leqslant \sigma$. Claim (iii) is proved by reflexive induction using the same observation as in (i) and (ii).

Proof of Theorem 7. We observe that the formula $\diamondsuit_n \sigma$ satisfies the assumption of Lemma 13. Let $B := A_\alpha^n$, then by Theorem 6

$$\nabla_n B(\diamondsuit_n \sigma) = \nabla_n B_{\diamondsuit_n \sigma}^*(1) = \diamondsuit_{n,\diamondsuit_n\sigma}^{o_n(B)}(1) = \diamondsuit_n^{o_n(B)} \diamondsuit_n \sigma.$$

However, $o_n(B) = \alpha$ and the claim follows.

References

1. Beklemishev, L.: A note on strictly positive logics and word rewriting systems. Preprint ArXiv:1509.00666 [math.LO] (2015)
2. Beklemishev, L.D., Onoprienko, A.A.: On some slowly terminating term rewriting systems. Sbornik: Math. **206**, 1173–1190 (2015)
3. Beklemishev, L.D.: Reflection calculus and conservativity spectra. ArXiv:1703.09314 [math.LO], March 2017
4. Beklemishev, L.D.: Proof-theoretic analysis by iterated reflection. Arch. Math. Logic **42**, 515–552 (2003). doi:10.1007/s00153-002-0158-7
5. Beklemishev, L.D.: Provability algebras and proof-theoretic ordinals, I. Ann. Pure Appl. Logic **128**, 103–123 (2004)
6. Beklemishev, L.D.: Reflection principles and provability algebras in formal arithmetic. Russ. Math. Surv. **60**(2), 197–268 (2005). Russian original. Uspekhi Matematicheskikh Nauk **60**(2), 3–78 (2005)
7. Beklemishev, L.D.: The Worm principle. In: Chatzidakis, Z., Koepke, P., Pohlers, W. (eds.) Lecture Notes in Logic 27. Logic Colloquium 2002, pp. 75–95. AK Peters (2006). Preprint: Logic Group Preprint Series 219, Utrecht University, March 2003
8. Beklemishev, L.D.: Calibrating provability logic: from modal logic to reflection calculus. In: Bolander, T., Braüner, T., Ghilardi, S., Moss, L. (eds.) Advances in Modal Logic, vol. 9, pp. 89–94. College Publications, London (2012)
9. Beklemishev, L.D.: Positive provability logic for uniform reflection principles. Ann. Pure Appl. Logic **165**(1), 82–105 (2014)
10. Beklemishev, L.D.: On the reduction property for GLP-algebras. Doklady: Math. **95**(1), 50–54 (2017)
11. Beklemishev, L.D., Joosten, J., Vervoort, M.: A finitary treatment of the closed fragment of Japaridze's provability logic. J. Logic Comput. **15**(4), 447–463 (2005)
12. Boolos, G.: The Logic of Provability. Cambridge University Press, Cambridge (1993)
13. Dashkov, E.V.: On the positive fragment of the polymodal provability logic GLP. Matematicheskie Zametki **91**(3), 331–346 (2012). English translation. Math. Notes **91**(3), 318–333 (2012)
14. de Jongh, D., Japaridze, G.: The Logic of Provability. In: Buss, S.R. (ed.) Handbook of Proof Theory. Studies in Logic and the Foundations of Mathematics, vol. 137, pp. 475–546. Elsevier, Amsterdam (1998)
15. Feferman, S.: Arithmetization of metamathematics in a general setting. Fundamenta Math. **49**, 35–92 (1960)
16. Goris, E., Joosten, J.J.: A new principle in the interpretability logic of all reasonable arithmetical theories. Logic J. IGPL **19**(1), 1–17 (2011)
17. Ignatiev, K.N.: On strong provability predicates and the associated modal logics. J. Symbolic Logic **58**, 249–290 (1993)
18. Japaridze, G.K.: The modal logical means of investigation of provability. Thesis in Philosophy, in Russian, Moscow (1986)
19. Joosten, J., Reyes, E.H.: Turing Schmerl logic for graded Turing progressions. ArXiv:1604.08705v1 [math.LO] (2016)
20. Joosten, J.J.: Turing-Taylor expansions of arithmetical theories. Stud. Log. **104**, 1225–1243 (2015). doi:10.1007/s11225-016-9674-z
21. Kikot, S., Kurucz, A., Tanaka, Y., Wolter, F., Zakharyaschev, M.: On the completeness of EL-equiations: first results. In: 11th International Conference on Advances in Modal Logic, Short Papers (Budapest, 30 August–2 September 2016), pp. 82–87 (2016)

22. Pakhomov, F.: On the complexity of the closed fragment of Japaridze's provability logic. Arch. Math. Logic **53**(7), 949–967 (2014)

23. Pakhomov, F.: On elementary theories of ordinal notation systems based on reflection principles. Proc. Steklov Inst. Math. **289**, 194–212 (2015)

24. Shamkanov, D.: Nested sequents for provability logic GLP. Logic J. IGPL **23**(5), 789–815 (2015)

25. Shapirovsky, I.: PSPACE-decidability of Japaridze's polymodal logic. In: Areces, C., Goldblatt, R. (eds.) Advances in Modal Logic, vol. 7, pp. 289–304. King's College Publications (2008)

26. Turing, A.M.: System of logics based on ordinals. Proc. Lond. Math. Soc. **45**, 161–228 (1939)

27. Visser, A.: An overview of interpretability logic. In: Kracht, M., de Rijke, M., Wansing, H., Zakhariaschev, M. (eds.) Advances in Modal Logic, CSLI Lecture Notes, vol. 1, no. 87, pp. 307–359. CSLI Publications, Stanford (1998)

On the Length of Medial-Switch-Mix Derivations

Paola Bruscoli[1]([✉]) and Lutz Straßburger[2]

[1] University of Bath, Bath, UK
P.Bruscoli@Bath.ac.uk
[2] Inria, Palaiseau, France

Abstract. Switch and medial are two inference rules that play a central role in many deep inference proof systems. In specific proof systems, the mix rule may also be present. In this paper we show that the maximal length of a derivation using only the inference rules for switch, medial, and mix, modulo associativity and commutativity of the two binary connectives involved, is quadratic in the size of the formula at the conclusion of the derivation. This shows, at the same time, the termination of the rewrite system.

1 Introduction

Deep inference is a well established methodology in proof theory; it generalises the more traditional proof theoretical methods, while simultaneously improving our ability in studying proofs from the point of view of normalisation and complexity, addressing therefore the problem of proof identity. For the interested reader, the web site http://alessio.guglielmi.name/res/cos/ provides a detailed overview of the collective developments in deep inference, spanning almost two decades of activities by several research groups and individuals.

For the purposes of this paper, however, we will limit ourselves to recall the essential feature that distinguishes deep inference from the traditional proof formalisms, when describing inference rules of some logical proof system. Deep inference applies logical inference in contexts, i.e. it allows to manipulate formulas at arbitrary depth in any context. In contrast, in the more traditional formalisms, including sequent calculus or natural deduction, the decomposition of a formula around its main connective strictly determines the shape of the inference rules of the proof system, and ultimately the shape of its proofs.

Therefore, on a technical level, deep inference lends itself to be studied also from the perspective of modern term rewriting, and in an easier way than traditional formalisms. This will not come at the expenses of our ability in observing and studying the fundamental proof theoretical properties, including cut-elimination, in a conservative way.

In this paper, we show some upper bounds on the length of deep inference derivations in a (sub)system for classical logic, by taking advantage of both worlds, proof theory and rewriting.

© Springer-Verlag GmbH Germany 2017
J. Kennedy and R.J.G.B. de Queiroz (Eds.): WoLLIC 2017, LNCS 10388, pp. 68–79, 2017.
DOI: 10.1007/978-3-662-55386-2_5

In the standard deep inference proof system KS for classical logic [3], we can find the following two rules

$$s \frac{F\{([A \vee B] \wedge C)\}}{F\{[A \vee (B \wedge C)]\}} \qquad m \frac{F\{[(A \wedge B) \vee (C \wedge D)]\}}{F\{([A \vee C] \wedge [B \vee D])\}} \qquad (1)$$

called *switch* and *medial*, respectively, where $F\{\ \}$ stands for an arbitrary (positive) formula context and A, B, C, and D are formula variables. In the system KS, the two rules switch and medial are applied modulo associativity and commutativity of conjunction and disjunction. The switch rule has been well investigated from the proof-theoretic as well as from the category-theoretic point of view because of its important role in linear logic [2]. The properties of the medial rule have originally been investigated in [20].

Switch and medial in combination have been studied in [18,21] from the perspective of semantics of proofs in deep inference. Moreover, the development of atomic flows in [13], centred on proof normalisation, provides a more abstract view of classical proofs by hiding switch-medial steps. A preliminary account from perspective of proof complexity can be found in [5].

In this paper we look at the two rules of switch and medial as rewriting system. Often these two rules may operate in presence of the mix rule, i.e.

$$mix \frac{F\{[A \vee B]\}}{F\{(A \wedge B)\}} \qquad (2)$$

that may be induced by the interplay of the units under specific conditions in the proof system, or just be made explicitly available in the system. For classical logic, for example, it is a consequence of weakening.

Our main result, in terms of complexity, is that the length of a derivation using only switch, medial, and mix is bounded by a quadratic function in the size of the conclusion of the derivation. Clearly, any such bound for a system with mix also holds for the system without mix. We will however also include the presentation of a more immediate cubic bound that has independent interest.

2   Rewriting with Switch and Medial (and Mix)

Formulas are generated from a countable set $\mathscr{A} = \{a, b, c, \ldots\}$ of atoms via the binary connectives $\wedge$ and $\vee$, called *and* and *or*, respectively. To ease readability of large formulas we will use [ ] for parentheses around disjunctions and ( ) for parentheses around conjunctions.

To simplify the presentation, we do not use the units $\top$ (*truth*) and $\bot$ (*falsum*) in this paper. It follows from the work by Das [9] that our results would also hold under the presence of the units.

The *size* of a formula A, denoted by σA, is the number of atom occurrences in A:

$$\sigma a = 1$$
$$\sigma[A \vee B] = \sigma A + \sigma B$$
$$\sigma(A \wedge B) = \sigma A + \sigma B$$

Example: $\sigma([a \lor b] \land [[a \lor c] \lor b]) = 5.$

The *tree-or-number* of a formula A, denoted by θA, is the number of occurrences of the symbol $\lor$ in A:

$$\theta a = 0$$
$$\theta[A \lor B] = \theta A + \theta B + 1$$
$$\theta(A \land B) = \theta A + \theta B$$

Example: $\theta([a \lor b] \land [[a \lor c] \lor b]) = 3.$

The *relweb-or-number* of a formula A, denoted by $\otimes A$, is the number of $\lor$-edges in the relation web [1] of A:

$$\otimes a = 0$$
$$\otimes[A \lor B] = \otimes A + \otimes B + \sigma A \cdot \sigma B$$
$$\otimes(A \land B) = \otimes A + \otimes B$$

Example: $\otimes([a \lor b] \land [[a \lor c] \lor b]) = 4.$

These values are stable under context application, as shown below.

Lemma 2.1. *Let P and Q be formulas with $\sigma P = \sigma Q$, and let $F\{\ \}$ be a formula context.*

1. If $\theta P = \theta Q$, then $\theta F\{P\} = \theta F\{Q\}$.
2. If $\theta P > \theta Q$, then $\theta F\{P\} > \theta F\{Q\}$.
3. If $\otimes P = \otimes Q$, then $\otimes F\{P\} = \otimes F\{Q\}$.
4. If $\otimes P > \otimes Q$, then $\otimes F\{P\} > \otimes F\{Q\}$.

Proof. By induction on the structure of $F\{\ \}$.   □

Consider the following rewrite rules on formulas:

$$\text{m}\, \frac{[(A \land B) \lor (C \land D)]}{([A \lor C] \land [B \lor D])} \qquad \text{s}\, \frac{([A \lor B] \land C)}{[A \lor (B \land C)]} \qquad \text{mix}\, \frac{(A \land B)}{[A \lor B]} \qquad (3)$$

The rules in (3) are written in the style of inference rules in proof theory (premiss on top implies the conclusion at the bottom), but they behave as rewrite rules in term rewriting, i.e., they can be applied inside any formula context $F\{\ \}$.

The rewriting rules in (3) are applied modulo associativity and commutativity for $\land$ and $\lor$. More precisely, we will do rewriting modulo the equational theory generated by

$$\begin{aligned} (A \land (B \land C)) &= ((A \land B) \land C) & (A \land B) &= (B \land A) \\ [A \lor [B \lor C]] &= [[A \lor B] \lor C] & [A \lor B] &= [B \lor A] \end{aligned} \qquad (4)$$

---

[1] The relation web of a formula provides a graph-based representation of a formula in deep inference contextual rewriting, as in [12]. An equivalent definition of the relweb-or-number is the number of edges in the cograph for $\lor$.

Lemma 2.2. *If $P = Q$, then $\theta P = \theta Q$ and $\otimes P = \otimes Q$.*

Proof. Consider the equation of associativity of disjunction. We have

$$\theta[A \vee [B \vee C]] = \theta A + \theta B + \theta C + 2$$
$$= \theta[[A \vee B] \vee C]$$

and

$$\otimes[A \vee [B \vee C]] = \otimes A + (\otimes B + \otimes C + \sigma B \cdot \sigma C) + \sigma A \cdot (\sigma B + \sigma C)$$
$$= \otimes A + \otimes B + \otimes C + \sigma A \cdot \sigma B + \sigma A \cdot \sigma C + \sigma B \cdot \sigma C$$
$$= \otimes[[A \vee B] \vee C]$$

Now apply Lemma 2.1. The other cases are similar (and simpler).   □

Lemma 2.3. *Let the rule $\rho \dfrac{Q}{P}$ be given.*

1. *If ρ is m, then $\theta Q < \theta P$.*
2. *If ρ is s, then $\theta Q = \theta P$ and $\otimes Q < \otimes P$.*
3. *If ρ is mix, then $\theta Q < \theta P$ and $\otimes Q < \otimes P$.*

Proof. 1. Case of medial:

$$\theta[(A \wedge B) \vee (C \wedge D)] = \theta A + \theta B + 1 + \theta C + \theta D + 1$$
$$< \theta A + \theta B + \theta C + \theta D + 1$$
$$= \theta([A \vee C] \wedge [B \vee D])$$

2. Case of switch:

$$\theta([A \vee B] \wedge C) = \theta A + \theta B + 1 + \theta C$$
$$= \theta[A \vee (B \wedge C)]$$

and

$$\otimes([A \vee B] \wedge C) = \otimes A + \otimes B + \sigma A \cdot \sigma B + \otimes C$$
$$< \otimes A + \otimes B + \otimes C + \sigma A \cdot \sigma B + \sigma A \cdot \sigma C$$
$$= \otimes[A \vee (B \wedge C)]$$

3. Case of mix:

$$\theta(A \wedge B) = \theta A + \theta B \qquad\qquad \otimes(A \wedge B) = \otimes A + \otimes B$$
$$< \theta A + \theta B + 1 \qquad \text{and} \qquad\quad < \otimes A + \otimes B + \sigma A \cdot \sigma B$$
$$= \theta[A \vee B] \qquad\qquad\qquad = \otimes[A \vee B]$$

Now apply Lemma 2.1.   □

3   The Cubic Bound

Before showing the quadratic bound, we present a cubic bound that we consider of independent interest for the following reasons. First, the proof of the cubic bound is rather simple and flexible, so it might be of interest for other logics (not just classical logic) especially in relation to aspects of system implementations [17]. Second, the cubic bound on medial-switch-mix derivations has been generalised to arbitrary (sound) linear systems for classical logic [10,11], in the sense that any derivation of super-cubic length must derive a (semantically) trivial inference.

Separating out the different proofs for the cubic and quadratic bound might help to answer the open question whether the quadratic bound can also be generalized, or whether it is truly specific to medial-switch-mix derivations.

For a formula P, define its *MSM-measure*, denoted by $\mathsf{msm}\,P$, as the pair

$$\mathsf{msm}\,P = \langle \theta P, \otimes P \rangle.$$

We use the lexicographic ordering for this measure on formulae:

$$\mathsf{msm}\,Q < \mathsf{msm}\,P \quad \text{iff} \quad \theta Q < \theta P \quad \text{or} \quad (\,\theta Q = \theta P \quad \text{and} \quad \otimes Q < \otimes P\,).$$

In the sequel, we assume the common notions and terminology on derivations and proof construction, given a (finite) set of inference rules.

The notation $\begin{array}{c} Q \\ \mathscr{S} \,\big\|\, \Delta \\ P \end{array}$ stands for a derivation Δ, with premiss Q and conclusion P, obtained with the rules in $\mathscr{S}$; by $\mathsf{length}(\Delta)$ we intend the number of instances of rules of $\mathscr{S}$ that have been applied in Δ.

Proposition 3.1. *Let Δ be the following derivation* $\begin{array}{c} Q \\ \{\mathsf{m,s,mix}\} \,\big\|\, \Delta \\ P \end{array}$, *where $\sigma P = n$.*

Then, $\mathsf{length}(\Delta) < n^3$.

Proof. By Lemma 2.2 we have that msm is stable under the equivalence of formulas, and from Lemma 2.3 we can conclude that msm strictly decreases at each step when going bottom-up in the derivation. We also have that $\theta P < n$ and $\otimes P \le n^2$. Hence, $\mathsf{length}(\Delta) < n \cdot n^2$. $\qquad\qquad\square$

4   The Quadratic Bound

The analysis that delivers a quadratic bound is performed by treating separately, in a given derivation, those sub-derivations that use only switch rule from those that use both mix and medial. We will keep the presentation slightly more informal to help the intuition.

Lemma 4.1. *Let* Δ *be the following derivation* $\begin{array}{c} Q \\ {}_{\{m,mix\}} \Big\| \Delta, \\ P \end{array}$ *where* $\sigma P = n$.

Then $\mathsf{length}(\Delta) < n$.

Proof. As observed in the proof of Proposition 3.1, we have that $\theta P < n$, and this value strictly decreases with each application of medial and mix. □

Moreover, let γP be the number of $\wedge$-occurrences in a formula P; then, it is trivial to show that

$$\sigma P = \theta P + \gamma P + 1.$$

In the following, we identify a formula P with its formula-tree, and every node of that tree is identified with the subformula occurrence rooted at that node.

Let $P = S\{R\}$, where R is an $\wedge$-node in P, i.e., $R = (A \wedge B)$ for some A and B. We define the following notions:

- the *switch-potential of R in P* is the number of $\vee$-nodes in the context $S\{\ \}$;
- the *switch-potential of P*, denoted by $\mathsf{sp}P$ is the sum of the switch-potentials of all $\wedge$-nodes in P.

Example:
The formula $([a \vee b] \wedge [[a \vee c] \vee b])$ contains only one $\wedge$-node, and its switch-potential is 0. But the formula $A = [(a \wedge b) \vee ((a \wedge c) \wedge b)]$ has 3 $\wedge$-nodes, each of which has switch-potential 1. Hence $\mathsf{sp}A = 3$.

The switch-potential of formulae is preserved through associativity and commutativity of the two operators and under context closure.

Lemma 4.2. *If* P *and* Q *are formulas and* $P \doteq Q$ *then* $\mathsf{sp}P = \mathsf{sp}Q$.

Proof. First note, that whenever $P \doteq Q$ and $\mathsf{sp}P = \mathsf{sp}Q$, then for all contexts $S\{\ \}$ we have $\mathsf{sp}S\{P\} = \mathsf{sp}S\{Q\}$. This can be shown by a straightforward induction on $S\{\ \}$. Hence, it suffices to show that for each equation in (4), the switch-potential of the left-hand side is equal to the switch-potential of the right-hand side. This is straightforward. □

We then consider the switch-potential of premiss and conclusion of the rules switch, medial and mix.

Lemma 4.3. *If* $\mathsf{s}\dfrac{Q}{P}$ *is a correct application of the switch rule, then* $\mathsf{sp}Q < \mathsf{sp}P$.

Proof. We have that $P = S\{A \vee (B \wedge C)\}$ and $Q = S\{[A \vee B] \wedge C\}$ for some context $S\{\ \}$ and some formulas A, B and C. The $\wedge$-nodes in $S\{\ \}$, in A, and in B do not change their switch-potentials. The switch-potentials of the $\wedge$-nodes in C (if present) are reduced by 1. However, note that the $\wedge$-node $([A \vee B] \wedge C)$ in Q has strictly smaller switch-potential than the $\wedge$-node $(B \wedge C)$ in P. Hence $\mathsf{sp}Q < \mathsf{sp}P$. □

Corollary 4.4. *Let* ${}_{\{s\}}\!\!\begin{array}{c} Q \\ \big\| \, \Delta \\ P \end{array}$ *be given. Then* $\mathsf{length}(\Delta) \leq \gamma P \cdot \theta P < \frac{1}{4}n^2$.

Proof. By Lemma 4.3, we immediately get $\mathsf{length}(\Delta) \leq \mathsf{sp}P$. By definition we have $\mathsf{sp}P \leq \gamma P \cdot \theta P$. Since $\gamma P + \theta P = n - 1$, we have $\frac{1}{4}n^2$ as upper bound. □

Observation 4.5. Note that in the derivation Δ in Corollary 4.4 we have $\theta Q = \theta P$ and $\gamma Q = \gamma P$.

Lemma 4.6. *If* $\mathsf{mix}\,\dfrac{Q}{P}$ *is a correct application of the mix rule, then* $\mathsf{sp}Q < \mathsf{sp}P + \theta P$.

Proof. We have that $P = S\{A \vee B\}$ and $Q = S\{A \wedge B\}$ for some context $S\{\ \}$ and some formulas A and B. The switch-potentials of the $\wedge$-nodes in $S\{\ \}$, in A, and in B either remains unchanged by the inference step or is smaller in Q than in P, because one $\vee$-node is removed. However, Q has one $\wedge$-node more than P. Hence, its switch-potential can be increased by the switch-potential of that $\wedge$-node, which is at most θQ. Hence, $\mathsf{sp}Q \leq \mathsf{sp}P + \theta Q$. Since $\theta P = \theta Q + 1$, we get $\mathsf{sp}Q < \mathsf{sp}P + \theta P$. □

Lemma 4.7. *If* $\mathsf{m}\,\dfrac{Q}{P}$ *is a correct application of the medial rule, then* $\mathsf{sp}Q \leq \mathsf{sp}P + \theta P$.

Proof. We have that $P = S\{[A \vee C] \wedge [B \vee D]\}$ and $Q = S\{(A \wedge B) \vee (C \wedge D)\}$ for some context $S\{\ \}$ and some formulas A, B, C and D. The switch-potentials of the $\wedge$-nodes in $S\{\ \}$, and in A, B, C, D could be smaller in Q than in P, because one $\vee$-node is removed. However, one $\wedge$-node in P is replaced by two $\wedge$-nodes in Q, which have both a bigger switch-potential because of the new $\vee$-node as parent. This can be counted as follows: the first new $\wedge$-node in Q increases its switch-potential by 1, compared to the $\wedge$-node in P, whereas the second $\wedge$-node in Q has a switch-potential of at most θQ. Hence $\mathsf{sp}Q \leq \mathsf{sp}P + 1 + \theta Q = \mathsf{sp}P + \theta P$. □

We can now combine these results.

Proposition 4.8. *Let* Δ *be a given derivation* ${}_{\{m,s,mix\}}\!\!\begin{array}{c} Q \\ \big\| \, \Delta \\ P \end{array}$ *where* $\sigma P = n$. *Then* $\mathsf{length}(\Delta) < \frac{1}{2}(n^2 + n)$.

Proof. Without loss of generality, we can assume that Δ has shape

$$
\begin{array}{c}
Q_m \\
\{s\} \,\Big\|\, \Delta_m \\
\rho_m \dfrac{P_m}{Q_{m-1}} \\
\{s\} \,\Big\|\, \Delta_{m-1} \\
\vdots \\
\{s\} \,\Big\|\, \Delta_2 \\
\rho_2 \dfrac{P_2}{Q_1} \\
\{s\} \,\Big\|\, \Delta_1 \\
\rho_1 \dfrac{P_1}{Q_0} \\
\{s\} \,\Big\|\, \Delta_0 \\
P_0
\end{array}
$$

where $P = P_0$, $Q = Q_m$, and where ρ_i is an instance of m or mix, for every $i \in \{1, \dots, m\}$. Let us consider, for every $i \in \{0, \dots, m\}$, the numbers of disjunctions and conjunctions in P_i, respectively denoted by d_i and c_i, as follows:

$$
d_i = \theta P_i = \theta Q_i \qquad \text{and} \qquad c_i = \gamma P_i = \gamma Q_i.
$$

Observe that $d_i + c_i = n - 1$. By the same argument as in the proof of Lemma 4.1, we have that $m = c_m - c_0 = d_0 - d_m$. In particular, we have that

$$
m < d_0 < n, \tag{5}
$$

and for all $i \in \{0, \dots, m-1\}$ we have that

$$
d_{i+1} = d_i - 1 \qquad \text{and} \qquad c_{i+1} = c_i + 1. \tag{6}
$$

We define, for every $i \in \{0, \dots, m\}$, the switch-potentials as

$$
p_i = \mathsf{sp} P_i \qquad \text{and} \qquad q_i = \mathsf{sp} Q_i,
$$

and we have that $p_i \leq d_i c_i$. If we let $l_i = \mathsf{length}(\Delta_i)$, then we obtain

$$
l_i \leq p_i - q_i. \tag{7}
$$

By Lemmas 4.6 and 4.7 we also have

$$
p_{i+1} \leq q_i + d_i. \tag{8}
$$

The remainder of the proof is a simple calculation:

$$
\begin{aligned}
\mathsf{length}(\Delta) &= m + l_0 + l_1 + \cdots + l_{m-1} + l_m \\
&\leq m + (p_0 - q_0) + (p_1 - q_1) + \cdots + (p_{m-1} - q_{m-1}) + (p_m - q_m) \\
&\leq m + (d_0 c_0 - q_0) + (q_0 + d_0 - q_1) + \cdots \\
&\qquad\qquad + (q_{m-2} + d_{m-2} - q_{m-1}) + (q_{m-1} + d_{m-1} - q_m) \\
&\leq m + d_0 c_0 + d_0 + d_1 + \cdots + d_{m-2} + d_{m-1} \\
&= m + d_0(n - 1 - d_0) + d_0 + (d_0 - 1) + \cdots \\
&\qquad\qquad + (d_0 - (m-2)) + (d_0 - (m-1)) \\
&= m + d_0(n - 1 - d_0) + m d_0 - \sum_{i=1}^{m-1} i \\
&= m + d_0 n - d_0 - d_0^2 + m d_0 - \frac{m(m-1)}{2} \\
&= \frac{1}{2}(2 n d_0 + 2 m d_0 - 2 d_0^2 - m^2 + 3m - 2 d_0) \\
&< \frac{1}{2}(n^2 + n),
\end{aligned}
$$

where the last inequality follows by observing that

$$
\begin{aligned}
0 &< (n - m) + (d_0 - m) + (d_0 - m) \\
&= n + 2 d_0 - 3m \\
&= n - (3m - 2 d_0)
\end{aligned}
$$

and

$$
\begin{aligned}
0 &< (n - d_0)^2 + (d_0 - m)^2 \\
&= n^2 - 2 n d_0 + d_0^2 + d_0^2 - 2 d_0 m + m^2 \\
&= n^2 - (2 n d_0 + 2 d_0 m - 2 d_0^2 - m^2),
\end{aligned}
$$

completing, thus, the proof.   □

Some remarks are in order, to comment on the bounds that we have obtained in this study.

Remark 4.9. The linear bound given by Lemma 4.1 cannot be reduced. For derivations containing mix, this is obvious, but even when only medial is allowed, we can form the following derivation

$$
\frac{[(a_{11} \wedge a_{21} \wedge \cdots \wedge a_{m1}) \vee \cdots \vee (a_{1k} \wedge a_{2k} \wedge \cdots \wedge a_{mk})]}{([a_{11} \vee a_{12} \vee \cdots \vee a_{1k}] \wedge \cdots \wedge [a_{m1} \vee a_{m2} \vee \cdots \vee a_{mk}])} \, {}_{\{m\}} \Big\| \, \Delta \tag{9}
$$

that contains $(k-1)(m-1)$ instances of medial. The size of the formulas in Δ is $n = km$. Hence $\mathsf{length}(\Delta) = km - k - m + 1$. If $k = m$, then $\mathsf{length}(\Delta) = n - 2\sqrt{n} + 1$. Lamarche proposes in [18] a matrix notation for denoting the derivation Δ in (9).

Remark 4.10. For derivations that use only the switch rule, the following example shows that the quadratic bound of Lemma 4.4 cannot be pushed further down, and also that the constant factor of $\frac{1}{4}$ is already optimal. Consider the following derivation where only switch is applied:

$$((\cdots(([a_1 \vee [a_2 \vee [a_3 \vee \cdots \vee [a_m \vee b] \cdots]]] \wedge c_1) \wedge c_2) \cdots \wedge c_{k-1}) \wedge c_k)$$
$$\{s\} \,\Big\|\, \Delta_1$$
$$((\cdots([a_1 \vee [a_2 \vee [a_3 \vee \cdots \vee [a_m \vee (b \wedge c_1)] \cdots]]] \wedge c_2) \cdots \wedge c_{k-1}) \wedge c_k)$$
$$\{s\} \,\Big\|\, \Delta_2$$
$$\vdots \tag{10}$$
$$\{s\} \,\Big\|\, \Delta_{k-1}$$
$$([a_1 \vee [a_2 \vee [a_3 \vee \cdots \vee [a_m \vee (\cdots((b \wedge c_1) \wedge c_2) \cdots \wedge c_{k-1})] \cdots]]] \wedge c_k)$$
$$\{s\} \,\Big\|\, \Delta_k$$
$$[a_1 \vee [a_2 \vee [a_3 \vee \cdots \vee [a_m \vee ((\cdots((b \wedge c_1) \wedge c_2) \cdots \wedge c_{k-1}) \wedge c_k)] \cdots]]]$$

For every $i \in \{1, \ldots, k\}$, we have that each Δ_i consists of m switches. Hence $\mathsf{length}(\Delta) = mk$. If we let $m = k$, we have $n = 2k + 1$ and $\mathsf{length}(\Delta) = k^2 = \frac{1}{4}n^2 - \frac{1}{2}n + \frac{1}{4}$.

Remark 4.11. The previous remark also shows that the quadratic bound of Proposition 4.8 cannot be improved. However, it is not known whether the constant factor $\frac{1}{2}$ can be improved (although we know that it must be $\geq \frac{1}{4}$.)

5 Conclusions

Earlier versions of this paper exist since 2008, for research primarily motivated by the need of better understanding the role and shape of the medial rule, especially from the perspective of normalisation. The medial rule is in fact needed in deep inference systems for classical logic to obtain an atomic contraction rule, which, in turns, contributes a form of atomic sharing that influences both normalisation and complexity.

Over time, we noted that variants of the medial rule appear, possibly in disguise, in several different logics of the linear kind, including those with non-commutative operators. It is therefore appropriate studying switch-medial-mix derivations independently from the specific units of the logic and their associated equations. In this sense, the approach based on term rewriting adds an element of generality that could result useful also for aspects of implementations.

The switch-medial-mix fragment is at the core of several investigations from different perspectives, including also the development of atomic flows [13,14] and of the atomic lambda calculus [16]. In particular, atomic flows provide an abstract view of classical derivations by making switch and medial unobservable (hence indirectly related to this topic of study) and enabling the discovery of interesting transformations from the complexity perspective, such as [6–8].

The results of this paper have supported well also the study of the length of derivations consisting only of linear inferences, as developed in [9], and more recently in [10,11]. The two last works show that our cubic bound also holds for non-trivial derivations made of arbitrary sound linear inferences, not just switch and medial. It still remains an open question whether the general case also has a quadratic bound.

As a remark, our medial-switch-mix system differs from the system studied in [1], which contains a rule that is the "inverse" of our medial rule, with premiss and conclusion swapped around.

As a matter of fact, an interesting line of enquiry is what happens when we combine various forms of switch and medial for various connectives. Examples of such a combination is the local system for linear logic [19], and the systems that extend the basic BV that contains a sequential operator inspired by process algebras [4,15]. More recently, a very exciting development that generalises switch and medial through the use of *subatomic logic* is in [22].

For the central role that switch and medial rules have in all deep inference systems, and for the richness of results collected from different perspectives that confirm also our bounds, we hope that this paper proves useful also in relation to implementations as well as further studies in complexity.

Acknowledgements. We are grateful to Alessio Guglielmi and Tom Gundersen, for their encouragements and checks, and to the anonymous referees. Paola Bruscoli was funded by EPSRC Project EP/K018868/1 "Efficient and Natural Proof Systems". This research is also supported by ANR Project FISP – "The Fine Structure of Formal Proof Systems and their Computational Interpretations."

References

1. Bechet, D., Groote, P., Retoré, C.: A complete axiomatisation for the inclusion of series-parallel partial orders. In: Comon, H. (ed.) RTA 1997. LNCS, vol. 1232, pp. 230–240. Springer, Heidelberg (1997). doi:10.1007/3-540-62950-5_74

2. Blute, R., Cockett, R., Seely, R., Trimble, T.: Natural deduction and coherence for weakly distributive categories. J. Pure Appl. Algebra **113**, 229–296 (1996)

3. Brünnler, K., Tiu, A.F.: A local system for classical logic. In: Nieuwenhuis, R., Voronkov, A. (eds.) LPAR 2001. LNCS (LNAI), vol. 2250, pp. 347–361. Springer, Heidelberg (2001). doi:10.1007/3-540-45653-8_24

4. Bruscoli, P.: A purely logical account of sequentiality in proof search. In: Stuckey, P.J. (ed.) ICLP 2002. LNCS, vol. 2401, pp. 302–316. Springer, Heidelberg (2002). doi:10.1007/3-540-45619-8_21

5. Bruscoli, P., Guglielmi, A.: On the proof complexity of deep inference. ACM Trans. Comput. Logic **10**(2), 1–34 (2009). Article 14

6. Bruscoli, P., Guglielmi, A., Gundersen, T., Parigot, M.: A quasipolynomial cut-elimination procedure in deep inference via atomic flows and threshold formulae. In: Clarke, E.M., Voronkov, A. (eds.) LPAR 2010. LNCS (LNAI), vol. 6355, pp. 136–153. Springer, Heidelberg (2010). doi:10.1007/978-3-642-17511-4_9

7. Bruscoli, P., Guglielmi, A., Gundersen, T., Parigot, M.: A quasipolynomial normalisation in deep inference via atomic flows and threshold formulae. Log. Methods Comput. Sci. **12**(1:5), 1–30 (2016)

8. Das, A.: Complexity of deep inference via atomic flows. In: Cooper, S.B., Dawar, A., Löwe, B. (eds.) CiE 2012. LNCS, vol. 7318, pp. 139–150. Springer, Heidelberg (2012). doi:10.1007/978-3-642-30870-3_15

9. Das, A.: Rewriting with linear inferences in propositional logic. In: van Raamsdonk, F. (ed) 24th International Conference on Rewriting Techniques and Applications (RTA). Leibniz International Proceedings in Informatics (LIPIcs), vol. 21, pp. 158–173. Schloss Dagstuhl-Leibniz-Zentrum für Informatik (2013)

10. Das, A., Straßburger, L.: No complete linear term rewriting system for propositional logic. In: Fernandez, M. (ed) 26th International Conference on Rewriting Techniques and Applications, RTA 2015. LIPIcs, Warsaw, Poland, 29 June to 1 July 2015, vol. 36, pp. 127–142 (2015)

11. Das, A., Straßburger, L.: On linear rewriting systems for boolean logic and some applications to proof theory. Log. Methods Comput. Sci. **12**(4) (2016)

12. Guglielmi, A.: A system of interaction and structure. ACM Trans. Comput. Logic **8**(1), 1–64 (2007)

13. Guglielmi, A., Gundersen, T.: Normalisation control in deep inference via atomic flows. Log. Methods Comput. Sci. **4**(1:9), 1–36 (2008)

14. Guglielmi, A., Gundersen, T., Straßburger, L.: Breaking paths in atomic flows for classical logic. In: LICS 2010 (2010)

15. Guglielmi, A., Straßburger, L.: Non-commutativity and MELL in the calculus of structures. In: Fribourg, L. (ed.) CSL 2001. LNCS, vol. 2142, pp. 54–68. Springer, Heidelberg (2001). doi:10.1007/3-540-44802-0_5

16. Gundersen, T., Heijltjes, W., Parigot, M.: Atomic lambda calculus: a typed lambda-calculus with explicit sharing. In: 28th Annual ACM/IEEE Symposium on Logic in Computer Science, LICS 2013, New Orleans, LA, USA, 25–28 June 2013, pp. 311–320. IEEE Computer Society (2013)

17. Kahramanogullari, O.: Interaction and depth against nondeterminism in proof search. Log. Methods Comput. Sci. **10**(2) (2014)

18. Lamarche, F.: Exploring the gap between linear and classical logic. Theory Appl. Categ. **18**(18), 473–535 (2007)

19. Straßburger, L.: A local system for linear logic. In: Baaz, M., Voronkov, A. (eds.) LPAR 2002. LNCS (LNAI), vol. 2514, pp. 388–402. Springer, Heidelberg (2002). doi:10.1007/3-540-36078-6_26

20. Straßburger, L.: A characterization of medial as rewriting rule. In: Baader, F. (ed.) RTA 2007. LNCS, vol. 4533, pp. 344–358. Springer, Heidelberg (2007). doi:10.1007/978-3-540-73449-9_26

21. Straßburger, L.: On the axiomatisation of Boolean categories with and without medial. Theory Appl. Categ. **18**(18), 536–601 (2007)

22. Aler Tubella, A.: A study of normalisation through subatomic logic. PhD thesis, University of Bath (2016)

Proof Theory and Ordered Groups

Almudena Colacito and George Metcalfe[✉]

Mathematical Institute, University of Bern, Bern, Switzerland
{almudena.colacito,george.metcalfe}@math.unibe.ch

Abstract. Ordering theorems, characterizing when partial orders of a group extend to total orders, are used to generate hypersequent calculi for varieties of lattice-ordered groups (ℓ-groups). These calculi are then used to provide new proofs of theorems arising in the theory of ordered groups. More precisely: an analytic calculus for abelian ℓ-groups is generated using an ordering theorem for abelian groups; a calculus is generated for ℓ-groups and new decidability proofs are obtained for the equational theory of this variety and extending finite subsets of free groups to right orders; and a calculus for representable ℓ-groups is generated and a new proof is obtained that free groups are orderable.

1 Introduction

Considerable success has been enjoyed recently in obtaining uniform algebraic completeness proofs for analytic sequent and hypersequent calculi with respect to varieties of residuated lattices [3,4,20]. These methods do not encompass, however, "ordered group-like" structures: algebras with a group reduct such as lattice-ordered groups (ℓ-groups) [1,14] and others admitting representations via ordered groups such as MV-algebras [5], GBL-algebras [13], and varieties of cancellative residuated lattices [17]. Hypersequent calculi have indeed been defined for abelian ℓ-groups, MV-algebras, and related classes in [15,16] and for ℓ-groups in [9], but the completeness proofs in these papers are largely syntactic, proceeding using cut elimination or restricted quantifier elimination.

The first aim of the work reported here is to use ordering theorems for groups, characterizing when a partial (right) order of a group extends to a total (right) order, to generate hypersequent calculi for varieties of lattice-ordered groups, thereby taking a first step towards a general algebraic proof theory for ordered group-like structures. A second aim is to then use these calculi to provide new syntactic proofs of various theorems arising in the theory of ordered groups.

More concretely, this paper makes the following contributions:

(i) A theorem of Fuchs [8] for extending partial orders of abelian groups to total orders is used to generate an analytic (cut-free) hypersequent calculus for the variety of abelian ℓ-groups. This system can be viewed as a one-sided version of the two-sided hypersequent calculus introduced in [15].

Supported by Swiss National Science Foundation grant 200021_146748 and the EU Horizon 2020 research and innovation programme under the Marie Skłodowska-Curie grant agreement No. 689176.

J. Kennedy and R.J.G.B. de Queiroz (Eds.): WoLLIC 2017, LNCS 10388, pp. 80–91, 2017.
DOI: 10.1007/978-3-662-55386-2_6

(ii) A theorem of Kopytov and Medvedev [14] for extending partial right orders of groups to total right orders is used to generate a hypersequent calculus for the variety of ℓ-groups, a variant of a calculus appearing in [9]. The method also provides a correspondence between validity of equations in ℓ-groups and the extension of finite subsets of free groups to total right orders, giving new proofs of decidability for these problems.

(iii) A theorem of Fuchs [8] for extending partial orders of groups to total orders is used to generate a calculus for representable ℓ-groups (equivalently, ordered groups) and to provide a new proof that free groups are orderable.

2   Ordered Groups

In this section, we recall some pertinent definitions and basic facts about ordered groups, referring to [1,14] for further details. Consider a group $\mathbf{G} = \langle G, \cdot, ^{-1}, e \rangle$. A partial order $\leq$ of G is called a *partial right order* of $\mathbf{G}$ if for all $a, b, c \in G$,

$$a \leq b \implies ac \leq bc.$$

Its *positive cone* $P_\leq = \{a \in G : e < a\}$ is a subsemigroup of $\mathbf{G}$ that omits e. Conversely, if P is a subsemigroup of $\mathbf{G}$ omitting e, then

$$a \leq^P b \iff ba^{-1} \in P \cup \{e\}$$

is a partial right order of $\mathbf{G}$ satisfying $P_{\leq^P} = P$. Hence partial right orders of $\mathbf{G}$ can be identified with subsemigroups of $\mathbf{G}$ omitting e. Note also that for $S \subseteq G$, the subsemigroup of $\mathbf{G}$ generated by S, denoted by $\langle S \rangle$, is a partial right order of $\mathbf{G}$ if and only if $e \notin \langle S \rangle$. Partial left orders of $\mathbf{G}$ are defined analogously.

A partial left and right order $\leq$ of $\mathbf{G}$ is called a *partial order* of $\mathbf{G}$. In this case, the positive cone $P_\leq$ is a normal subsemigroup of $\mathbf{G}$ omitting e; that is, whenever $a \in P_\leq$ and $b \in G$, also $bab^{-1} \in P_\leq$. Conversely, if a subset $P \subseteq G$ has these properties, then $\leq^P$ is a partial order of $\mathbf{G}$; hence, partial orders of $\mathbf{G}$ can be identified with normal subsemigroups of $\mathbf{G}$ omitting e. Also, for $S \subseteq G$, the normal subsemigroup of $\mathbf{G}$ generated by S, denoted by $\langle\langle S \rangle\rangle$, is a partial order of $\mathbf{G}$ if and only if $e \notin \langle\langle S \rangle\rangle$.

A partial order or partial right order $\leq$ of $\mathbf{G}$ is called, respectively, a *(total) order* or *(total) right order* of $\mathbf{G}$ if $G = P_\leq \cup P_\leq^{-1} \cup \{e\}$. Note also that if $\leq$ is an order or a right order of $\mathbf{G}$, then the same holds for the *inverse order* defined by $a \leq^\delta b$ if and only if $b \leq a$. In this paper we focus mostly on (right) orders of a finitely generated free (abelian) group $\mathbf{F}$ and address the following problem.

Problem 1. Does a given finite $S \subseteq F$ extend to an order or a right order of $\mathbf{F}$?

We also consider a purely algebraic perspective on ordered groups. That is, a *lattice-ordered group* (or ℓ-group) may be defined as an algebraic structure $\mathbf{L} = \langle L, \wedge, \vee, \cdot, ^{-1}, e \rangle$ satisfying

(i) $\langle L, \cdot, ^{-1}, e \rangle$ is a group;

(ii) $\langle L, \wedge, \vee \rangle$ is a lattice (with $a \leq b \Leftrightarrow a \wedge b = a$, for all $a, b \in L$);

(iii) $a \leq b \implies cad \leq cbd$, for all $a, b, c, d \in L$.

It follows also that $\langle L, \wedge, \vee \rangle$ must be a distributive lattice and that $\mathbf{L}$ satisfies $e \leq a \vee a^{-1}$ for all $a \in L$ (see [1]). If $\leq$ is a total order of the group $\langle L, \cdot, ^{-1}, e \rangle$, then $\mathbf{L}$ is called an *ordered group* (or *o-group*), observing that $\mathbf{L}$ can also be obtained by adding to the group operations the meet and join operations for $\leq$. An ℓ-group whose group operation is commutative is called an *abelian ℓ-group*.

Example 1. Standard examples of abelian ℓ-groups are subgroups of the additive group over the real numbers equipped with the usual order, e.g.,

$$\mathbf{Z} = \langle \mathbb{Z}, \min, \max, +, -, 0 \rangle.$$

Indeed this algebra generates the variety $\mathcal{A}$ of all abelian ℓ-groups [21], which means in particular that an equation is valid in $\mathcal{A}$ if and only if is valid in $\mathbf{Z}$.

Example 2. Fundamental examples of (non-abelian) ℓ-groups are provided by considering the order-preserving bijections of some totally-ordered set $\langle \Omega, \leq \rangle$. These form an ℓ-group $\mathbf{Aut}(\langle \Omega, \leq \rangle)$ under coordinate-wise lattice operations, functional composition, and functional inverse. Indeed, it has been shown by Holland that every ℓ-group embeds into an ℓ-group $\mathbf{Aut}(\langle \Omega, \leq \rangle)$ for some totally-ordered set $\langle \Omega, \leq \rangle$ [10], and that the variety $\mathcal{LG}$ of ℓ-groups is generated by $\mathbf{Aut}(\langle \mathbb{R}, \leq \rangle)$, where $\leq$ is the usual order on $\mathbb{R}$ [11]. This means in particular that an ℓ-group equation is valid in $\mathcal{LG}$ if and only if is valid in $\mathbf{Aut}(\langle \mathbb{R}, \leq \rangle)$.

Let us turn our attention now to the syntax of ℓ-groups. We call a variable x and its inverse x^{-1} *literals*, and consider terms $s, t, \ldots$ built from literals over variables $x_1, x_2, \ldots$, operation symbols e, $\wedge$, $\vee$, and $\cdot$, defining also inductively

$$\overline{x} = x^{-1} \qquad \overline{x^{-1}} = x \qquad \overline{e} = e$$
$$\overline{s \wedge t} = \overline{s} \vee \overline{t} \qquad \overline{s \vee t} = \overline{s} \wedge \overline{t} \qquad \overline{s \cdot t} = \overline{t} \cdot \overline{s}.$$

Using the strong distributivity properties of the ℓ-group operations, it follows that every ℓ-group term is equivalent in $\mathcal{LG}$ to a term of the form $\bigwedge_{i \in I} \bigvee_{j \in J_i} t_{ij_i}$ where each t_{ij_i} is a group term. Hence to check the validity of equations in some class $\mathcal{K}$ of ℓ-groups, it suffices to address the following problem.

Problem 2. Given group terms $t_1, \ldots, t_n$, does it hold that

$$\mathcal{K} \models e \leq t_1 \vee \ldots \vee t_n?$$

Let us therefore define a *sequent* Γ as a finite sequence of literals $\ell_1, \ldots, \ell_n$ with inverse $\overline{\Gamma} = \overline{\ell_n}, \ldots, \overline{\ell_1}$, and a *hypersequent* $\mathcal{G}$ as a finite set of sequents, written

$$\Gamma_1 \mid \ldots \mid \Gamma_n.$$

In what follows, we identify a sequent $\ell_1, \ldots, \ell_n$ with the group term $\ell_1 \cdot \ldots \cdot \ell_n$ for $n > 0$ and e for $n = 0$, and a non-empty hypersequent $\Gamma_1 \mid \ldots \mid \Gamma_n$ with the

ℓ-group term $\Gamma_1 \vee \ldots \vee \Gamma_n$. We will say that a non-empty hypersequent $\mathcal{G}$ is *valid* in a class of ℓ-groups $\mathcal{K}$ and write $\mathcal{K} \models \mathcal{G}$, if $\mathcal{K} \models e \leq \mathcal{G}$. We will also say that a sequent Γ is *group valid* if $\Gamma \approx e$ is valid in all groups.

A *hypersequent rule* is a set of *instances*, each instance consisting of a finite set of hypersequents called the *premises* and a hypersequent called the *conclusion*. Such rules are typically written schematically using $\Gamma, \Pi, \Sigma, \Delta$ and $\mathcal{G}, \mathcal{H}$ to denote arbitrary sequents and hypersequents, respectively. A *hypersequent calculus* GL is a set of hypersequent rules, and a GL-*derivation* of a hypersequent $\mathcal{G}$ is a finite tree of hypersequents with root $\mathcal{G}$ such that each node and its parents form an instance of a rule of GL. In this case, we write $\vdash_{GL} \mathcal{G}$. A hypersequent rule is said to be GL-*admissible* if for each of its instances, whenever the premises are GL-derivable, the conclusion is GL-derivable.

Remark 1. Sequents are often defined (see, e.g., [3,4,15,16]) as ordered pairs of finite sequences (or sets or multisets) of terms, and hypersequents as finite multisets of sequents. Here we exploit the strong duality properties of ℓ-groups to restrict to one-sided sequents and define hypersequents as finite sets of sequents to emphasize the connection with finite sets of group terms.

3   A Hypersequent Calculus for Abelian ℓ-Groups

We use the following ordering theorem for abelian groups to rediscover a single-sided version of the hypersequent calculus for abelian ℓ-groups defined in [15].

Theorem 1 (Fuchs 1963 [8]). *Every partial order of a torsion-free abelian group* **G** *extends to an order of* **G**.

Let $\mathcal{A}b$ be the variety of abelian groups and let $\mathbf{T}(k)$ be the algebra of group terms on $k \in \mathbb{N}$ generators. We may identify the free abelian group $\mathbf{F}_{\mathcal{A}b}(k)$ on k generators with the quotient $\mathbf{T}(k)/\Theta_{\mathcal{A}b}$, where $\Theta_{\mathcal{A}b}$ is the congruence on $\mathbf{T}(k)$ defined by $s\Theta_{\mathcal{A}b}t \Leftrightarrow \mathcal{A}b \models s \approx t$ (see [2] for further details). For convenience, we will use $t \in T(k)$ to denote also $t/\Theta_{\mathcal{A}b}$ in $F_{\mathcal{A}b}(k)$, noting that $\mathcal{A}b \models s \approx t$ if and only if $s = t$ in $\mathbf{F}_{\mathcal{A}b}(k)$. It follows easily that $\mathbf{F}_{\mathcal{A}b}(k)$ is torsion-free.

Theorem 2. *The following are equivalent for* $t_1, \ldots, t_n \in T(k)$:

(1) $\mathcal{A} \models e \leq t_1 \vee \ldots \vee t_n$.
(2) $\{t_1, \ldots, t_n\}$ *does not extend to an order of* $\mathbf{F}_{\mathcal{A}b}(k)$.
(3) $e \in \langle\langle \{t_1, \ldots, t_n\} \rangle\rangle$.
(4) $\mathcal{A}b \models e \approx t_1^{\lambda_1} \cdots t_n^{\lambda_n}$ *for some* $\lambda_1, \ldots, \lambda_n \in \mathbb{N}$ *not all* 0.

Proof. (1) $\Rightarrow$ (2). By contraposition. If $\{t_1, \ldots, t_n\}$ extends to an order of $\mathbf{F}_{\mathcal{A}b}(k)$, then, taking the inverse order, we obtain an ordered abelian group where $t_1, \ldots, t_n$ are negative. But this ordered abelian group may also be viewed as an abelian ℓ-group and taking the evaluation mapping $t \in T(k)$ to $t \in F_{\mathcal{A}b}(k)$, we obtain $\mathcal{A} \not\models e \leq t_1 \vee \ldots \vee t_n$.

$$\frac{}{\mathcal{G} \mid \Delta, \overline{\Delta}} \; (\text{ID}) \qquad \frac{\mathcal{G} \mid \Pi, \Delta, \Gamma}{\mathcal{G} \mid \Pi, \Gamma, \Delta} \; (\text{EX}) \qquad \frac{\mathcal{G} \mid \Gamma, \Delta}{\mathcal{G} \mid \Gamma \mid \Delta} \; (\text{SPLIT})$$

Fig. 1. The hypersequent calculus GA

(2) $\Rightarrow$ (3). Suppose that $\{t_1, \ldots, t_n\}$ does not extend to an order of $\mathbf{F}_{Ab}(k)$. Then, since $\mathbf{F}_{Ab}(k)$ is torsion-free, by Theorem 1, the subsemigroup $\langle\{t_1, \ldots, t_n\}\rangle$ is not a partial order of $\mathbf{F}_{Ab}(k)$. That is, $e \in \langle\{t_1, \ldots, t_n\}\rangle$.

(3) $\Rightarrow$ (4). Suppose that $e \in \langle\{t_1, \ldots, t_n\}\rangle$. Then $e = t_1^{\lambda_1} \cdots t_n^{\lambda_n}$ in $\mathbf{F}_{Ab}(k)$ for some $\lambda_1, \ldots, \lambda_n \in \mathbb{N}$ not all 0, and hence $\mathcal{A}b \models e \approx t_1^{\lambda_1} \cdots t_n^{\lambda_n}$.

(4) $\Rightarrow$ (1). Suppose that $\mathcal{A}b \models e \approx t_1^{\lambda_1} \cdots t_n^{\lambda_n}$ for some $\lambda_1, \ldots, \lambda_n \in \mathbb{N}$ not all 0. Then also $\mathcal{A} \models e \leq t_1^{\lambda_1} \cdots t_n^{\lambda_n}$. It is easily proved that $\mathcal{A} \models e \leq uvt$ implies $\mathcal{A} \models e \leq u \vee v \vee t$ (see, e.g. [9]). Hence, applying this implication repeatedly, we obtain $\mathcal{A} \models e \leq t_1 \vee \ldots \vee t_n$. $\qquad\square$

Remark 2. Theorem 2 may be interpreted geometrically as a variant of Gordan's theorem of the alternative (with integers swapped for real numbers) and close relative of Farkas' lemma (see, e.g., [7]). Namely, given an $m \times n$ integer matrix $A = (a_{ij})$, exactly one of the following systems has a solution:

(a) $y^T A < 0$ for some $y \in \mathbb{Z}^m$.
(b) $Az = 0$ for some $z \in \mathbb{N}^n \setminus \{0\}$.

To prove this, define $t_i = x_1^{a_{1i}} \cdot \ldots \cdot x_m^{a_{mi}}$ for $i = 1, \ldots, n$. Then (a) is equivalent to $\mathbf{Z} \not\models e \leq t_1 \vee \ldots \vee t_n$, which is in turn equivalent to $\mathcal{A} \not\models e \leq t_1 \vee \ldots \vee t_n$ (see Example 1). So, by Theorem 2, (a) fails if and only if $\mathcal{A}b \models e \approx t_1^{\lambda_1} \cdots t_n^{\lambda_n}$ for some $\lambda_1, \ldots, \lambda_n \in \mathbb{N}$ not all 0, which is in turn equivalent to (b).

Theorem 2 can be used to establish soundness and completeness for the hypersequent calculus GA presented in Fig. 1.

Theorem 3. *For any non-empty hypersequent $\mathcal{G}$, $\mathcal{A} \models \mathcal{G}$ if and only if $\vdash_{GA} \mathcal{G}$.*

Proof. By Theorem 2, $\mathcal{A} \models \Gamma_1 \mid \ldots \mid \Gamma_n$ if and only if $\mathcal{A}b \models e \approx \Gamma_1^{\lambda_1} \cdots \Gamma_n^{\lambda_n}$ for some $\lambda_1, \ldots, \lambda_n \in \mathbb{N}$ not all 0. But if this latter condition holds, then the number of occurrences of a variable x in $\Gamma_1^{\lambda_1} \cdots \Gamma_n^{\lambda_n}$ must equal the number of occurrences of x^{-1}, and, using (EX) and (ID), we obtain $\vdash_{GA} \Gamma_1^{\lambda_1} \cdots \Gamma_n^{\lambda_n}$. Hence also, using (SPLIT) repeatedly, $\vdash_{GA} \Gamma_1 \mid \ldots \mid \Gamma_n$. Conversely, we can prove by induction on the height of a derivation that whenever $\vdash_{GA} \Gamma_1 \mid \ldots \mid \Gamma_n$, there exist $\lambda_1, \ldots, \lambda_n \in \mathbb{N}$ not all 0 such that $\mathcal{A}b \models e \approx \Gamma_1^{\lambda_1} \cdots \Gamma_n^{\lambda_n}$. The cases for (ID) and (EX) are immediate, and the case of (SPLIT) follows directly by an application of the induction hypothesis. $\qquad\square$

Remark 3. The calculus for abelian ℓ-groups presented in [15] uses hypersequents defined as finite multisets of two-sided sequents, each consisting of an ordered pair of finite multisets of ℓ-group terms, and therefore requires a quite different set of rules. In particular, this calculus contains rules for operation symbols and external contraction and weakening structural rules, but not the exchange rule (EX). These differences are of an essentially cosmetic nature, however. We can easily add sound and invertible rules for the operation symbols $\cdot$, e, $\wedge$, and $\vee$ to the calculus GA that serve to rewrite hypersequents of arbitrary terms into hypersequents built only from literals, and it remains then simply to translate two-sided sequents $\Gamma \Rightarrow \Delta$ into one-sided sequents $\overline{\Gamma}, \Delta$.

4   Right Orders on Free Groups and Validity in ℓ-Groups

Let $\mathcal{G}$ be the variety of groups and $\mathbf{F}(k)$ the free group over k generators, which, as before, we may identify with $\mathbf{T}(k)/\Theta_{\mathcal{G}}$, where $\Theta_{\mathcal{G}}$ is the congruence on $\mathbf{T}(k)$ defined by $s\Theta_{\mathcal{G}}t \iff \mathcal{G} \models s \approx t$. An element of $F(k)$ can again be represented by a term from $T(k)$: in particular, by a reduced term obtained by cancelling all occurrences of xx^{-1} and $x^{-1}x$. Our first aim in this section will be to show that checking validity of equations in ℓ-groups is equivalent to checking whether finite subsets of $F(k)$ extend to right orders on $\mathbf{F}(k)$.

Theorem 4. *The following are equivalent for* $t_1, \ldots, t_n \in T(k)$:

(1) $\mathcal{LG} \models e \le t_1 \vee \ldots \vee t_n$.
(2) $\{t_1, \ldots, t_n\}$ *does not extend to a right order of* $\mathbf{F}(k)$.
(3) *There exist* $s_1, \ldots, s_m \in F(k) \setminus \{e\}$ *such that*

$$e \in \langle\{t_1, \ldots, t_n, s_1^{\delta_1}, \ldots, s_m^{\delta_m}\}\rangle \text{ for all } \delta_1, \ldots, \delta_m \in \{-1, 1\}.$$

Observe that the equivalence of (2) and (3) is an immediate consequence of the following ordering theorem for groups.

Theorem 5 (Kopytov and Medvedev 1994 [14]). *A subset S of a group* **G** *extends to a right order of* **G** *if and only if for all* $a_1, \ldots, a_m \in G \setminus \{e\}$, *there exist* $\delta_1, \ldots, \delta_m \in \{-1, 1\}$ *such that* $e \notin \langle S \cup \{a_1^{\delta_1}, \ldots, a_m^{\delta_m}\}\rangle$.

Condition (3) corresponds directly to derivability in the hypersequent calculus GLG* presented in Fig. 2. It is not so easy, however, to show directly that the calculus GLG* is sound with respect to ℓ-groups (i.e., to show that $\vdash_{\mathrm{GLG}^*} \mathcal{G}$ implies $\mathcal{LG} \models \mathcal{G}$), since the rule ($*$) is not valid as an implication between premises and conclusion in all ℓ-groups. We therefore consider also a further hypersequent calculus GLG, displayed in Fig. 3, and establish the following relationship between the calculi.

Lemma 1. *For any non-empty hypersequent $\mathcal{G}$, if $\vdash_{\mathrm{GLG}^*} \mathcal{G}$, then $\vdash_{\mathrm{GLG}} \mathcal{G}$.*

Proof. It suffices to show that the rules (SPLIT) and ($*$) of GLG* are GLG-admissible. First, it is easily shown, by an induction on the height of a derivation, that the following rule is GLG-admissible:

$$\frac{\mathcal{G}}{\mathcal{G} \mid \mathcal{H}} \text{ (EW)}$$

$$\frac{}{\mathcal{G} \mid \Gamma} \text{ (GV)} \qquad \frac{\mathcal{G} \mid \Gamma, \Delta}{\mathcal{G} \mid \Gamma \mid \Delta} \text{ (SPLIT)} \qquad \frac{\mathcal{G} \mid \Delta \quad \mathcal{G} \mid \overline{\Delta}}{\mathcal{G}} \text{ (*)}$$

Γ group valid Δ not group valid.

Fig. 2. The hypersequent calculus GLG*

$$\frac{}{\mathcal{G} \mid \Gamma} \text{ (GV)} \qquad \frac{}{\mathcal{G} \mid \Delta \mid \overline{\Delta}} \text{ (EM)} \qquad \frac{\mathcal{G} \mid \Gamma, \Delta \quad \mathcal{G} \mid \overline{\Delta}, \Sigma}{\mathcal{G} \mid \Gamma, \Sigma} \text{ (CUT)}$$

Γ group valid

Fig. 3. The hypersequent calculus GLG

Now for (SPLIT), if $\vdash_{\text{GLG}} \mathcal{G} \mid \Gamma, \Delta$, then, by (EW), we obtain $\vdash_{\text{GLG}} \mathcal{G} \mid \Gamma, \Delta \mid \Delta$. But also, by (EM), $\vdash_{\text{GLG}} \mathcal{G} \mid \overline{\Delta} \mid \Delta$, so, by (CUT), we obtain $\vdash_{\text{GLG}} \mathcal{G} \mid \Gamma \mid \Delta$.

To show that (*) is admissible in GLG, we consider a restricted version of the calculus where (CUT) is never applied to some particular sequent. For a hypersequent $\mathcal{G}$ and a sequent Π, we call the ordered pair $\langle \Pi, \mathcal{G} \rangle$ a *pointed hypersequent* (just a hypersequent with one sequent marked) and transfer the usual definitions for hypersequent calculi to pointed hypersequent calculi. We let the pointed hypersequent calculus GLG^P consist of all pointed hypersequents $\langle \Pi, \mathcal{G} \rangle$ such that either some $\Gamma \in \mathcal{G} \cup \{\Pi\}$ is group valid or there exist Δ and $\overline{\Delta}$ in $\mathcal{G} \cup \{\Pi\}$, together with the restricted cut rule

$$\frac{\langle \Pi, (\mathcal{G} \mid \Gamma, \Delta) \rangle \quad \langle \Pi, (\mathcal{G} \mid \overline{\Delta}, \Sigma) \rangle}{\langle \Pi, (\mathcal{G} \mid \Gamma, \Sigma) \rangle} \text{ (CUT)}$$

Claim. $\vdash_{\text{GLG}} \mathcal{G} \mid \Pi$ if and only if $\vdash_{\text{GLG}^P} \langle \Pi, \mathcal{G} \rangle$.

Proof of Claim. The right-to-left direction is a simple induction on the height of a derivation of $\langle \Pi, \mathcal{G} \rangle$ in GLG^P. For the left-to-right direction, we first note that (by a straightforward induction) whenever $\vdash_{\text{GLG}^P} \langle \Pi, \mathcal{G} \rangle$, also $\vdash_{\text{GLG}^P} \langle \Pi, \mathcal{G} \mid \mathcal{H} \rangle$ and $\vdash_{\text{GLG}^P} \langle \Delta, \mathcal{G} \mid \Pi \rangle$. It suffices now to prove that

$$\vdash_{\text{GLG}^P} \langle (\Gamma, \Delta), \mathcal{G} \rangle \text{ and } \vdash_{\text{GLG}^P} \langle (\overline{\Delta}, \Sigma), \mathcal{H} \rangle \implies \vdash_{\text{GLG}^P} \langle (\Gamma, \Sigma), \mathcal{G} \mid \mathcal{H} \rangle.$$

We proceed by induction on the sum of heights of derivations for $\vdash_{\text{GLG}^P}$ $\langle (\Gamma, \Delta), \mathcal{G} \rangle$ and $\vdash_{\text{GLG}^P} \langle (\overline{\Delta}, \Sigma), \mathcal{H} \rangle$.

For the base case, there are several possibilities. If $\mathcal{G}$ or $\mathcal{H}$ contains a group valid sequent or both Π and $\overline{\Pi}$, then the conclusion follows trivially. If Γ, Δ and $\overline{\Delta}, \Sigma$ are both group valid, then Γ, Σ is group valid and so $\vdash_{\text{GLG}^P} \langle (\Gamma, \Sigma), \mathcal{G} \mid \mathcal{H} \rangle$. Suppose then that $\mathcal{G} = \mathcal{G}' \mid \overline{\Delta}, \overline{\Gamma}$, that is, $\vdash_{\text{GLG}^P} \langle (\Gamma, \Delta), \mathcal{G}' \mid \overline{\Delta}, \overline{\Gamma} \rangle$. Observe that

$$\vdash_{\text{GLG}^P} \langle (\Gamma, \Sigma), \mathcal{G}' \mid \mathcal{H} \mid \overline{\Delta}, \Sigma \rangle \text{ and } \vdash_{\text{GLG}^P} \langle (\Gamma, \Sigma), \mathcal{G}' \mid \mathcal{H} \mid \overline{\Sigma}, \overline{\Gamma} \rangle.$$

Hence, by (CUT), we get $\vdash_{\text{GLG}^P} \langle (\Gamma, \Sigma), \mathcal{G}' \mid \mathcal{H} \mid \overline{\Delta}, \overline{\Gamma} \rangle$; that is, $\vdash_{\text{GLG}^P} \langle (\Gamma, \Sigma), \mathcal{G} \mid \mathcal{H} \rangle$ as required. The case where $\mathcal{H} = \mathcal{H}' \mid \overline{\Sigma}, \Delta$ is symmetrical.

For the induction step, we apply the induction hypothesis twice to the premises of an application of (CUT), and the result follows by applying (CUT). □

Now to prove that $(*)$ is admissible in GLG^P, it suffices by the claim to show that for Δ not group valid,

$$\vdash_{\text{GLG}^\text{P}} \langle \Delta, \mathcal{G} \rangle \text{ and } \vdash_{\text{GLG}^\text{P}} \langle \overline{\Delta}, \mathcal{H} \rangle \implies \vdash_{\text{GLG}} \mathcal{G} \mid \mathcal{H}.$$

We proceed by induction on the height of a GLG^P-derivation of $\langle \Delta, \mathcal{G} \rangle$. For the base case, there are several possibilities. If $\mathcal{G}$ contains a group valid sequent or both Π and $\overline{\Pi}$, then the conclusion follows trivially. Suppose that $\langle \Delta, \mathcal{G} \rangle$ has the form $\langle \Delta, \mathcal{G}' \mid \overline{\Delta} \rangle$. Since $\vdash_{\text{GLG}^\text{P}} \langle \overline{\Delta}, \mathcal{H} \rangle$, also $\vdash_{\text{GLG}} \mathcal{H} \mid \overline{\Delta} \mid \mathcal{G}'$, i.e., $\vdash_{\text{GLG}} \mathcal{G} \mid \mathcal{H}$. For the induction step, suppose that $\mathcal{G} = \mathcal{G}' \mid \Gamma, \Sigma$ and that $\langle \Delta, \mathcal{G} \rangle$ is the conclusion of an application of (CUT) with premises $\langle \Delta, \mathcal{G}' \mid \Gamma, \Pi \rangle$ and $\langle \Delta, \mathcal{G}' \mid \overline{\Pi}, \Sigma \rangle$. By the induction hypothesis twice, $\vdash_{\text{GLG}} \mathcal{G}' \mid \Gamma, \Pi \mid \mathcal{H}$ and $\vdash_{\text{GLG}} \mathcal{G}' \mid \overline{\Pi}, \Sigma \mid \mathcal{H}$. Hence, by (CUT), we obtain $\vdash_{\text{GLG}} \mathcal{G}' \mid \Gamma, \Sigma \mid \mathcal{H}$; that is, $\vdash_{\text{GLG}} \mathcal{G} \mid \mathcal{H}$. □

We now have all the ingredients required to complete the proof of Theorem 4.
Proof of Theorem 4.
(1) $\Rightarrow$ (2). Suppose contrapositively that $\{t_1, \ldots, t_n\}$ extends to a right order of $\mathbf{F}(k)$. Then the inverse order is a right order $\leq$ of $\mathbf{F}(k)$ where $t_1, \ldots, t_n$ are negative. Consider the ℓ-group $\mathbf{Aut}(\langle F(k), \leq \rangle)$ and evaluate each variable x by the map $s \mapsto sx$. Then each group term t is evaluated by the map $s \mapsto st$. In particular, each t_i maps e to $t_i < e$, and hence $t_1 \vee \ldots \vee t_n$ maps e to some $t_j < e$, where $j \in \{1, \ldots, n\}$. That is, $e \not\leq t_1 \vee \ldots \vee t_n$ in $\mathbf{Aut}(\langle F(k), \leq \rangle)$ and we obtain $\mathcal{LG} \not\models e \leq t_1 \vee \ldots \vee t_n$.
(2) $\Rightarrow$ (3). Immediate from Theorem 5.
(3) $\Rightarrow$ (1). Consider $s_1, \ldots, s_m \in F(k) \backslash \{e\}$ where $e \in \langle \{t_1, \ldots, t_n, s_1^{\delta_1}, \ldots, s_m^{\delta_m}\} \rangle$ for all $\delta_1, \ldots, \delta_m \in \{-1, 1\}$. We prove first that $\vdash_{\text{GLG}^*} t_1 \mid \ldots \mid t_n$. For each particular choice of $\delta_1, \ldots, \delta_m \in \{-1, 1\}$, there exist $\lambda_1, \ldots, \lambda_n, \mu_1, \ldots, \mu_m \in \mathbb{N}$ not all 0 such that $e = t_1^{\lambda_1} \cdot \ldots \cdot t_n^{\lambda_n} \cdot (s_1^{\delta_1})^{\mu_1} \cdot \ldots \cdot (s_m^{\delta_m})^{\mu_m}$ in $\mathbf{F}(k)$. Hence $\mathcal{G} \models e \approx t_1^{\lambda_1} \cdot \ldots \cdot t_n^{\lambda_n} \cdot (s_1^{\delta_1})^{\mu_1} \cdot \ldots \cdot (s_m^{\delta_m})^{\mu_m}$ and, by (GV), $\vdash_{\text{GLG}^*} t_1^{\lambda_1} \cdot \ldots \cdot t_n^{\lambda_n} \cdot (s_1^{\delta_1})^{\mu_1} \cdot \ldots \cdot (s_m^{\delta_m})^{\mu_m}$. But using (SPLIT) repeatedly, $\vdash_{\text{GLG}^*} t_1 \mid \ldots \mid t_n \mid s_1^{\delta_1} \mid \ldots \mid s_m^{\delta_m}$. So, by applying $(*)$ iteratively, $\vdash_{\text{GLG}^*} t_1 \mid \ldots \mid t_n$. It follows now by Lemma 1 that $\vdash_{\text{GLG}} t_1 \mid \ldots \mid t_n$. But then a simple induction on the height of a derivation in GLG, shows that $\mathcal{LG} \models e \leq t_1 \vee \ldots \vee t_n$ as required. □

Soundness and completeness results for GLG* and GLG follow directly.

Corollary 1. *The following are equivalent for any hypersequent* $\mathcal{G}$:

$$(1) \; \mathcal{LG} \models \mathcal{G}; \quad (2) \; \vdash_{\text{GLG}} \mathcal{G}; \quad (3) \; \vdash_{\text{GLG}^*} \mathcal{G}.$$

In the last part of this section, we use Theorem 4 to derive new decision procedures for Problems 1 and 2 (see Sect. 2). Let us denote the *length* of a reduced term t in $\mathbf{F}(k)$ by $|t|$, and for $N \in \mathbb{N}$, let $F_N(k)$ denote the set of all elements of $\mathbf{F}(k)$ of length $\leq N$. Given a subset S of $\mathbf{F}(k)$ which omits e, we call S an N-*truncated right order* on $\mathbf{F}(k)$ if $S = \langle S \rangle \cap F_N(k)$ and, for all

$t \in F_{N-1}(k) \setminus \{e\}$, either $t \in S$ or $t^{-1} \in S$. It has been shown that this notion precisely characterizes the finite subsets of $F(k)$ that extend to a right order.

Theorem 6. (Clay and Smith [6,19]). *A finite subset S of $F(k)$ extends to a right order of $\mathbf{F}(k)$ if and only if S extends to an N-truncated right order of $\mathbf{F}(k)$ for some $N \in \mathbb{N}$.*

The condition described in this theorem can be decided as follows. Let N be the maximal length of an element in S. Extend S to the finite set S^* by adding st whenever s, t occur in the set constructed so far and $|st| \leq N$. This ensures that $S^* = \langle S^* \rangle \cap F_N(k)$. If $e \in S^*$, then stop. Otherwise, for every $t \in F_{N-1}(k) \setminus \{e\}$ such that $t \notin S^*$ and $t^{-1} \notin S^*$, add t to S^* to obtain S_1 and t^{-1} to S^* to obtain S_2, and repeat the process with these sets. This procedure terminates because $F_N(k)$ is finite. Hence we obtain a decision procedure for Problem 1.

Corollary 2. *The problem of checking whether a given finite set of elements of a finitely generated free group extends to a right order is decidable.*

Moreover, using Theorem 4, we obtain also a decision procedure for Problem 2.

Corollary 3. *The problem of checking whether an equation is valid in all ℓ-groups is decidable.*

Example 3. Consider $S = \{xx, yy, x^{-1}y^{-1}\} \subseteq F(2)$. By adding all products in $F_2(2)$ of members of S, we obtain

$$S^* = \{xx, yy, x^{-1}y^{-1}, xy^{-1}, x^{-1}y, xy\}.$$

We then consider all possible signs δ for $x, y \in F_1(2)$. If we add x^{-1} or y^{-1} to S^* and take products, then clearly, using xx or yy, we obtain e. Similarly, if we add x and y to S^*, then, taking products, using $x^{-1}y^{-1}$, we obtain e. Hence we may conclude that S does not extend to a right order of $\mathbf{F}(2)$ and obtain

$$\mathcal{LG} \models e \leq xx \vee yy \vee \overline{x}\,\overline{y}.$$

Consider now $T = \{xx, xy, yx^{-1}\} \subseteq F(2)$. By adding all products in $F_2(2)$ of members of T, we obtain

$$T^* = \{xx, xy, yx^{-1}, yx, yy\}.$$

We choose $x, y \in F_1(2)$ to be positive and obtain $\{xx, xy, yx^{-1}, yx, yy, x, y\}$, a 2-truncated right order of $\mathbf{F}(2)$. Hence T extends to a right order of $\mathbf{F}(2)$ and

$$\mathcal{LG} \not\models e \leq xx \vee xy \vee y\overline{x}.$$

The decidability result stated in Corollary 3 was first established by Holland and McCleary in [12] using a quite different decision procedure. Let S be a finite set of reduced terms from $\mathbf{F}(k)$. We denote by $\mathrm{is}(S)$ the set of initial subterms of elements of S, and define $\mathrm{cis}(S)$ to consist of all reduced non-identity terms $s^{-1}t$, where $s, t \in \mathrm{is}(S)$. The following equivalence (expressed quite differently using "diagrams") is proved in [12].

Theorem 7 (Holland and McCleary [12]). *The following are equivalent for* $t_1, \ldots, t_n \in T(k)$:

(1) $\mathcal{LG} \models e \leq t_1 \vee \ldots \vee t_n$.
(2) *There exist* $s_1, \ldots, s_m \in \text{cis}(\{t_1, \ldots, t_n\})$ *such that*

$$e \in \langle\{t_1, \ldots, t_n, s_1^{\delta_1}, \ldots, s_m^{\delta_m}\}\rangle \; for all \; \delta_1, \ldots, \delta_m \in \{-1, 1\}.$$

Since the set $\text{cis}(\{t_1, \ldots, t_n\})$ is finite and checking $e \in \langle S \rangle$ for a finite subset S of $F(k)$ is decidable, we obtain a decision procedure for Problem 2. Moreover, again using Theorem 4, we obtain also a decision procedure for Problem 1.

Remark 4. Variants of the hypersequent calculi GLG* and GLG were defined already in [9], but without the connection to right orders on free groups. They were used to give an alternative proof of Holland's theorem (see [11]) that the algebra $\mathbf{Aut}(\langle\mathbb{R}, \leq\rangle)$ generates the variety $\mathcal{LG}$ of ℓ-groups and also to prove that the equational theory of ℓ-groups is co-NP complete. Let us note here that it follows from the results above that the problem of checking whether a finite subset of $\mathbf{F}(k)$ extends to a right order must also be in co-NP; hardness, however, is still an open problem. Let us also remark that in [9], the following *analytic* (i.e., having the subformula property) hypersequent calculus is shown to be sound and complete for ℓ-groups:

$$\frac{}{\mathcal{G} \mid \Gamma} \;(\text{GV}) \qquad \frac{\mathcal{G} \mid \Gamma \quad \mathcal{G} \mid \Delta}{\mathcal{G} \mid \Gamma, \Delta} \;(\text{MIX}) \qquad \frac{\mathcal{G} \mid \Gamma, \Sigma \quad \mathcal{G} \mid \Pi, \Delta}{\mathcal{G} \mid \Gamma, \Delta \mid \Pi, \Sigma} \;(\text{COM})$$
$$\Gamma \text{ group valid}$$

The proof, however, relies on a rather complicated cut elimination procedure and it is not yet clear how this calculus might relate to right orders on free groups.

5 Ordering Free Groups and Validity in Ordered Groups

In this section, we consider the variety $\mathcal{RG}$ of *representable ℓ-groups* generated by the class of o-groups. Similarly to the previous section, we establish the following theorem relating validity of equations in this variety (equivalently, the class of o-groups) to extending finite subsets of free groups to (total) orders.

Theorem 8. *The following are equivalent for* $t_1, \ldots, t_n \in T(k)$:

(1) $\mathcal{RG} \models e \leq t_1 \vee \ldots \vee t_n$.
(2) $\{t_1, \ldots, t_n\}$ *does not extend to an order of* $\mathbf{F}(k)$.
(3) *There exist* $s_1, \ldots, s_m \in F(k) \setminus \{e\}$ *such that*

$$e \in \langle\langle\{t_1, \ldots, t_n, s_1^{\delta_1}, \ldots, s_m^{\delta_m}\}\rangle\rangle \; for \; all \; \delta_1, \ldots, \delta_m \in \{-1, 1\}.$$

In this case, we will not be able to obtain any decision procedure for checking these equivalent conditions. However, we do obtain a new syntactic proof of the orderability of finitely generated free groups [18].

Corollary 4. *Every finitely generated free group is orderable.*

Proof. The equation $e \leq x$ is not valid in the o-group $\mathbf{Z}$, so $\mathcal{RG} \not\models e \leq x$. But then, by Theorem 8, there must exist an order of $\mathbf{F}(k)$ where x is positive. $\square$

The proof of Theorem 8 makes use of the following ordering theorem for groups.

Theorem 9 (Fuchs 1963 [8]). *A subset S of a group $\mathbf{G}$ extends to an order of $\mathbf{G}$ if and only if for all $a_1, \ldots, a_m \in G \setminus \{e\}$, there exist $\delta_1, \ldots, \delta_m \in \{-1, 1\}$ such that $e \notin \langle\langle S \cup \{a_1^{\delta_1}, \ldots, a_m^{\delta_m}\}\rangle\rangle$.*

Similarly to the previous section, we introduce hypersequent calculi GRG* and GRG as extensions of, respectively, GLG* and GLG with the rule

$$\frac{\mathcal{G} \mid \Delta, \Gamma}{\mathcal{G} \mid \Gamma, \Delta} \text{ (CYCLE)}$$

and establish the following relationship between these calculi.

Lemma 2. *For any non-empty hypersequent $\mathcal{G}$, if $\vdash_{\text{GRG*}} \mathcal{G}$, then $\vdash_{\text{GRG}} \mathcal{G}$.*

Proof. The proof is almost exactly the same as that of Lemma 1 except that we must take account also of the extra rule (CYCLE). That is, we define the pointed hypersequent calculus GRG^{P} as the extension of GLG^{P} with the restricted rule

$$\frac{\langle \Pi, (\mathcal{G} \mid \Delta, \Gamma) \rangle}{\langle \Pi, (\mathcal{G} \mid \Gamma, \Delta) \rangle} \text{ (CYCLE)}$$

and prove that $\vdash_{\text{GRG}} \mathcal{G} \mid \Pi$ if and only if $\vdash_{\text{GRG}^{\text{P}}} \langle \Pi, \mathcal{G} \rangle$. In this case, we also prove by a straightforward induction on the height of a derivation in GRG^{P} that

$$\vdash_{\text{GRG}^{\text{P}}} \langle (\Gamma, \Delta), \mathcal{G} \rangle \implies \vdash_{\text{GRG}^{\text{P}}} \langle (\Delta, \Gamma), \mathcal{G} \rangle.$$

Finally, the proof that $(*)$ is admissible in GRG^{P} proceeds in exactly the same way as in the proof of Lemma 1. $\square$

Proof of Theorem 8.
(1) $\Rightarrow$ (2). Suppose contrapositively that $\{t_1, \ldots, t_n\}$ extends to an order of $\mathbf{F}(k)$. Then the inverse order is an order $\leq$ of $\mathbf{F}(k)$ where $t_1, \ldots, t_n$ are negative. But this ordered group may also be viewed as a representable ℓ-group and taking the evaluation mapping $t \in T(k)$ to $t \in F(k)$, we obtain $\mathcal{RG} \not\models e \leq t_1 \vee \ldots \vee t_n$.
(2) $\Rightarrow$ (3). Immediate from Theorem 9.
(3) $\Rightarrow$ (1). Consider $s_1, \ldots, s_m \in F(k) \setminus \{e\}$ where $e \in \langle\langle \{t_1, \ldots, t_n, s_1^{\delta_1}, \ldots, s_m^{\delta_m}\} \rangle\rangle$ for all $\delta_1, \ldots, \delta_m \in \{-1, 1\}$. We prove first that $\vdash_{\text{GRG*}} t_1 \mid \ldots \mid t_n$. For each choice of $\delta_1, \ldots, \delta_m \in \{-1, 1\}$, there exist $l > 0$ and conjugates $r_1, \ldots, r_l$ of $t_1, \ldots, t_n, s_1^{\delta_1}, \ldots, s_m^{\delta_m}$ such that $e = r_1 \cdot \ldots \cdot r_l$ in $\mathbf{F}(k)$. So $\mathcal{G} \models e \approx r_1 \cdot \ldots \cdot r_l$ and, by (GV), $\vdash_{\text{GRG*}} r_1 \cdot \ldots \cdot r_l$. But then, by (SPLIT) and (CYCLE), also $\vdash_{\text{GRG*}} t_1 \mid \ldots \mid t_n \mid s_1^{\delta_1} \mid \ldots \mid s_m^{\delta_m}$. Hence, by repeated applications of $(*)$, we get

$\vdash_{\text{GRG}^*} t_1 \mid \ldots \mid t_n$. It follows now also by Lemma 2 that $\vdash_{\text{GRG}} t_1 \mid \ldots \mid t_n$. Finally, a simple induction on the height of a derivation in GLG shows that $\mathcal{RG} \models e \le t_1 \vee \ldots \vee t_n$ as required. $\qquad\square$

Soundness and completeness for GRG* and GRG follow directly.

Corollary 5. *The following are equivalent for any hypersequent $\mathcal{G}$:*

$$(1)\ \mathcal{RG} \models \mathcal{G}; \quad (2)\ \vdash_{\text{GRG}} \mathcal{G}; \quad (3)\ \vdash_{\text{GRG}^*} \mathcal{G}.$$

References

1. Anderson, M.E., Feil, T.H.: Lattice-Ordered Groups: An Introduction. Springer, Heidelberg (1988)
2. Burris, S., Sankappanavar, H.P.: A Course in Universal Algebra. Springer, Heidelberg (1981)
3. Ciabattoni, A., Galatos, N., Terui, K.: Algebraic proof theory for substructural logics: cut-elimination and completions. Ann. Pure Appl. Logic **163**(3), 266–290 (2012)
4. Ciabattoni, A., Galatos, N., Terui, K.: Algebraic proof theory: hypersequents and hypercompletions. Ann. Pure Appl. Logic **168**(3), 693–737 (2017)
5. Cignoli, R., D'Ottaviano, I.M.L., Mundici, D.: Algebraic foundations of many-valued reasoning. Kluwer, Berlin (1999)
6. Clay, A., Smith, L.H.: Corrigendum to [19]. J. Symb. Comput. **44**(10), 1529–1532 (2009)
7. Dantzig, G.B.: Linear Programming and Extensions. Princeton University, Press (1963)
8. Fuchs, L.: Partially Ordered Algebraic Systems. Pergamon Press, Oxford (1963)
9. Galatos, N., Metcalfe, G.: Proof theory for lattice-ordered groups. Ann. Pure Appl. Logic **8**(167), 707–724 (2016)
10. Holland, W.C.: The lattice-ordered group of automorphisms of an ordered set. Mich. Math. J. **10**, 399–408 (1963)
11. Holland, W.C.: The largest proper variety of lattice-ordered groups. Proc. Am. Math. Soc. **57**, 25–28 (1976)
12. Holland, W.C., McCleary, S.H.: Solvability of the word problem in free lattice-ordered groups. Houston J. Math. **5**(1), 99–105 (1979)
13. Jipsen, P., Montagna, F.: Embedding theorems for classes of GBL-algebras. J. Pure Appl. Algebra **214**(9), 1559–1575 (2010)
14. Kopytov, V.M., Medvedev, N.Y.: The Theory of Lattice-Ordered Groups. Kluwer, Alphen aan den Rijn (1994)
15. Metcalfe, G., Olivetti, N., Gabbay, D.: Sequent and hypersequent calculi for abelian and Łukasiewicz logics. ACM Trans. Comput. Log. **6**(3), 578–613 (2005)
16. Metcalfe, G., Olivetti, N., Gabbay, D.: Proof Theory for Fuzzy Logics. Springer, Heidelberg (2008)
17. Montagna, F., Tsinakis, C.: Ordered groups with a conucleus. J. Pure Appl. Algebra **214**(1), 71–88 (2010)
18. Neumann, B.H.: On ordered groups. Am. J. Math. **71**(1), 1–18 (1949)
19. Smith, L.H.: On ordering free groups. J. Symb. Comput. **40**(6), 1285–1290 (2005)
20. Terui, K.: Which structural rules admit cut elimination? – An algebraic criterion. J. Symbolic Logic **72**(3), 738–754 (2007)
21. Weinberg, E.C.: Free abelian lattice-ordered groups. Math. Ann. **151**, 187–199 (1963)

Constructive Canonicity for Lattice-Based Fixed Point Logics

Willem Conradie[1], Andrew Craig[1(✉)], Alessandra Palmigiano[1,2], and Zhiguang Zhao[2]

[1] Department of Pure and Applied Mathematics,
University of Johannesburg, Johannesburg, South Africa
`acraig@uj.ac.za`
[2] Faculty of Technology, Policy and Management,
Delft University of Technology, Delft, The Netherlands

Abstract. In the present paper, we prove canonicity results for lattice-based fixed point logics in a constructive meta-theory. Specifically, we prove two types of canonicity results, depending on how the fixed-point binders are interpreted. These results smoothly unify the constructive canonicity results for inductive inequalities, proved in a general lattice setting, with the canonicity results for fixed point logics on a bi-intuitionistic base, proven in a non-constructive setting.

Keywords: Canonicity · Lattice-based fixed point logics · Logics for categorization · Unified correspondence

1 Introduction

Recently, lattice- and distributive lattice-based fixed point logics have received increasing attention. In [4], substructural based epistemic logics are studied, in which the knowledge is interpreted as the information confirmed by a reliable source, formally defined as a backward-looking diamond operation; an ongoing direction studies multiagent interaction in this framework, especially adding the common knowledge operator to the language, which is defined in terms of fixed point operators. In [2,20], fixed point operators are added to linear logic, making it possible to study and define processes that are characterized by infinite and iterative behavior. In [9], an epistemic interpretation of lattice-based modal logics is given in terms of categorization systems; in this context, fixed points can be used to model several forms of group knowledge, common grounds or social agreement shared among agents, and are a promising tool towards a proper mathematical formalization of the notion of default categories.

Canonicity is an important notion in logic, since it is closely related to the completeness of logical systems. We say that a formula is canonical if its validity is preserved

The research of the first author has been funded by the National Research Foundation of South Africa, Grant number 81309. The research of the third and fourth author has been funded by the NWO Vidi grant 016.138.314, the NWO Aspasia grant 015.008.054, and a Delft Technology Fellowship awarded in 2013.

© Springer-Verlag GmbH Germany 2017
J. Kennedy and R.J.G.B. de Queiroz (Eds.): WoLLIC 2017, LNCS 10388, pp. 92–109, 2017.
DOI: 10.1007/978-3-662-55386-2_7

under taking canonical extensions (cf. Definition 2). Sahlqvist [28] identified a class of modal formulas (which are nowadays called *Sahlqvist formulas*) such that these formulas have first-order correspondents and they are canonical. Recently, in *unified correspondence theory* [11, 15], the algebraic and order-theoretic principles of Sahlqvist correspondence and canonicity are identified, which makes it possible to uniformly extend the state-of-the-art in Sahlqvist correspondence and canonicity to large families of nonclassical logics which include lattice-based (modal) logics [13, 14], regular modal logics [27], monotone modal logic [19], hybrid logics [17], and very recently, many-valued logics [5].

The present paper contributes to the mathematical background of lattice-based fixed point logics by proving the canonicity (cf. Sect. 3) of two classes of μ-inequalities in a constructive meta-theory of normal lattice expansions (cf. Definition 1) in an algorithmic way. The results of the present paper simultaneously generalize and unify extant canonicity results for fixed-point languages based on a bi-intuitionistic bi-modal logic [6], and constructive canonicity results for inductive inequalities [12] (restricted to normal lattice expansions). Besides the greater generality, the unification of these strands refines and simplifies the existing proofs for the canonicity of μ-formulas and inequalities. Indeed, remarkably, the two canonicity results[1] of [6] are fully generalized to the constructive setting and to normal lattice expansions (LEs, see Definition 1), but the rules of the algorithm ALBA used for this result have a *simpler* formulation than the rules of [6], and are analogous to those of [12], with no additional rules added specifically to handle the fixed point binders. Rather, fixed points are accounted for by certain restrictions on the application of the rules, concerning the order-theoretic properties of the term functions associated with the formulas to which the rules are applied.

Further research directions are opened up by the results of the present paper, in connection with the recent applications of unified correspondence to the algebraic methods for structural proof theory. Specifically, in [23], a methodology is introduced which uses perfect algebras as the semantic environment of proper display calculi, and provides a general and uniform semantic argument for proving that a given calculus is conservative with respect to the Hilbert-style axiomatization it intends to capture.[2] This semantic argument crucially uses the algebraic canonicity of the axioms and rules of the given Hilbert-style axiomatization. In the light of this connection, the canonicity results established in the present paper pave the way to designing analytic (proper display) calculi (i.e. calculi in which rules are such that formulas in the premises are subformulas of formulas in the conclusion) for large classes of axiomatic extensions of lattice-based fixed point logics.

Structure of the Paper. In Sect. 2, we collect preliminaries on the syntax and algebraic semantics of lattice-based mu-calculi. In Sect. 3, we expand on two notions of canonicity based on different interpretations of the fixed point operators. In Sect. 4, we define the classes of mu-inequalities to which the canonicity results apply. In Sects. 5 and 7, we outline the adapted version of the algorithm ALBA, the main tool for the canonicity results of the present paper. In Sect. 6, we state our main results. Due to space con-

---

[1] Namely, the tame and proper canonicity, cf. Sect. 3.

[2] The same methodology can be used also to define *Gentzen* calculi, as is witnessed by [25] in the setting of strict implication logics.

straints we do not include proofs, but these may be found online in an expanded version of the present paper [7].

2   Preliminaries

In the present section, we collect preliminaries on the language and algebraic semantics of fixed point expansions of LE-logics (cf. [6,14] for more details). We keep our notation as uniform as possible and intersperse comments to highlight the adaptations needed for the specific setting of the present paper.

2.1   Lattice-Based Logics and Their Semantics

In the present subsection, we introduce the propositional fragments of the lattice-based mu-languages we consider in this paper, as well as their semantics.

Before introducing the language, we will introduce some auxiliary definitions. an *order-type* over $n \in \mathbb{N}$ is a tuple $\varepsilon \in \{1, \partial\}^n$, and its *opposite* order-type ε^∂ is defined such that $\varepsilon_i^\partial = 1$ iff $\varepsilon_i = \partial$ for every $1 \leq i \leq n$.

The LE-language $\mathcal{L}(\mathcal{F}, \mathcal{G})$ (we omit the parameters when they are clear from the context) consists of a set of propositional variables PROP and disjoint sets of connectives $\mathcal{F}$ and $\mathcal{G}$. Each $f \in \mathcal{F}$ (resp. $g \in \mathcal{G}$) has arity $n_f \in \mathbb{N}$ (resp. $n_g \in \mathbb{N}$), and has order-type ε_f over n_f (resp. ε_g over n_g). The formulas of $\mathcal{L}$ are defined as follows:

$$\varphi ::= p \mid \bot \mid \top \mid \varphi \wedge \varphi \mid \varphi \vee \varphi \mid f(\overline{\varphi}) \mid g(\overline{\varphi})$$

where $p \in$ PROP, $f \in \mathcal{F}, g \in \mathcal{G}$.

The algebras interpreting LE-logics are defined as follows:

Definition 1. *For any LE-language $\mathcal{L} = \mathcal{L}(\mathcal{F}, \mathcal{G})$, a normal lattice expansion (LE, or $\mathcal{L}$-algebra) is a tuple $\mathbb{A} = (A, \mathcal{F}^{\mathbb{A}}, \mathcal{G}^{\mathbb{A}})$ such that A is a bounded lattice, $\mathcal{F}^{\mathbb{A}} = \{f^{\mathbb{A}} \mid f \in \mathcal{F}\}$ and $\mathcal{G}^{\mathbb{A}} = \{g^{\mathbb{A}} \mid g \in \mathcal{G}\}$, such that every $f^{\mathbb{A}} \in \mathcal{F}^{\mathbb{A}}$ (resp. $g^{\mathbb{A}} \in \mathcal{G}^{\mathbb{A}}$) is an n_f-ary (resp. n_g-ary) operation on $\mathbb{A}$ and preserves finite joins (resp. meets) in the i-th coordinate with $\varepsilon_f(i) = 1$ (resp. $\varepsilon_g(i) = 1$) and reverses finite meets (resp. joins) in j-th coordinate with $\varepsilon_f(j) = \partial$ (resp. $\varepsilon_g(j) = \partial$).*

For any lattice $\mathbb{A}$, we let $\mathbb{A}^1 := \mathbb{A}$ and $\mathbb{A}^\partial$ be the dual lattice. For any order-type ε, we let $\mathbb{A}^\varepsilon := \prod_{i=1}^n \mathbb{A}^{\varepsilon_i}$.

Example 1. For a given set Ag of agents, consider the language $\mathcal{L}(\mathcal{F}, \mathcal{G})$ where $\mathcal{F} = \varnothing$, and $\mathcal{G} = \{\Box_a \mid a \in \text{Ag}\}$ with $n_{\Box_a} = 1$ and $\varepsilon_{\Box_a} = 1$ for every $a \in$ Ag. This language is used in [9] as an epistemic logic of *categories* (or *formal concepts*). For each formal context, categories are completely specified by their *extension* (the set of members of the category), and their *intension* (the set of properties shared by the members of the category). Terms φ in this language denote categories, and inequalities such as $\varphi \leq \psi$ encode that category φ is a subcategory of category ψ. In this context, $\Box_a \varphi$ denotes category φ as perceived/known/believed to be by agent a.

2.2 The Expanded Language $\mathcal{L}^t$

Any LE-language $\mathcal{L} = \mathcal{L}(\mathcal{F}, \mathcal{G})$ can be associated with the expanded language $\mathcal{L}^t = \mathcal{L}(\mathcal{F}^t, \mathcal{G}^t)$, where $\mathcal{F}^t \supseteq \mathcal{F}$ and $\mathcal{G}^t \supseteq \mathcal{G}$ are obtained by adding:

1. for $f \in \mathcal{F}$ and $1 \leq i \leq n_f$, the n_f-ary connective $f_i^\sharp$, the intended interpretation of which is the right residual of f in its i-th coordinate[3] if $\varepsilon_f(i) = 1$ (resp. its Galois-adjoint[4] if $\varepsilon_f(i) = \partial$);
2. for $g \in \mathcal{G}$ and $1 \leq i \leq n_g$, the n_g-ary connective $g_i^\flat$, the intended interpretation of which is the left residual of g in its i-th coordinate if $\varepsilon_g(i) = 1$ (resp. its Galois-adjoint if $\varepsilon_g(i) = \partial$).

We stipulate that $f_i^\sharp \in \mathcal{G}^t$ if $\varepsilon_f(i) = 1$, and $f_i^\sharp \in \mathcal{F}^t$ if $\varepsilon_f(i) = \partial$. Dually, $g_i^\flat \in \mathcal{F}^t$ if $\varepsilon_g(i) = 1$, and $g_i^\flat \in \mathcal{G}^t$ if $\varepsilon_g(i) = \partial$. Regarding order-type, for any $f \in \mathcal{F}$ and $g \in \mathcal{G}$,

1. if $\varepsilon_f(i) = 1$, then $\varepsilon_{f_i^\sharp}(i) = 1$ and $\varepsilon_{f_i^\sharp}(j) = (\varepsilon_f(j))^\partial$ for any $j \neq i$.
2. if $\varepsilon_f(i) = \partial$, then $\varepsilon_{f_i^\sharp} = \varepsilon_f$.
3. if $\varepsilon_g(i) = 1$, then $\varepsilon_{g_i^\flat}(i) = 1$ and $\varepsilon_{g_i^\flat}(j) = (\varepsilon_g(j))^\partial$ for any $j \neq i$.
4. if $\varepsilon_g(i) = \partial$, then $\varepsilon_{g_i^\flat} = \varepsilon_g$.

The expanded language will be used in the execution of the algorithm ALBA defined later in this paper.

Example 2. If $\mathcal{L}(\mathcal{F}, \mathcal{G})$ is the language of Example 1, the expanded language $\mathcal{L}^t(\mathcal{F}^t, \mathcal{G}^t)$ is given by $\mathcal{F}^t = \{\blacklozenge_a \mid a \in Ag\}$, with $n_{\blacklozenge_a} = 1$ and $\varepsilon_{\blacklozenge_a} = 1$ for every $a \in Ag$, and $\mathcal{G}^t = \{\Box_a \mid a \in Ag\}$. The term $\blacklozenge_a \varphi$ denotes a different (e.g. lax, where $\Box_a$ is strict) epistemic approximation of the category φ according to the agent a.

2.3 Constructive Canonical Extensions

Canonical extensions provide a purely algebraic encoding of Stone-type dualities. However, purely algebraic constructions are also available, such as those of [18,21], which do not rely on principles equivalent to some forms of the axiom of choice.

Definition 2. *Let $\mathbb{A}$ be a (bounded) sublattice of a complete lattice $\mathbb{A}'$.*

1. *$\mathbb{A}$ is dense in $\mathbb{A}'$ if every element of $\mathbb{A}'$ can be expressed both as a join of meets and as a meet of joins of elements from $\mathbb{A}$.*
2. *$\mathbb{A}$ is compact in $\mathbb{A}'$ if, for all $S, T \subseteq \mathbb{A}'$, if $\bigvee S \leq \bigwedge T$ then $\bigvee S' \leq \bigwedge T'$ for some finite $S' \subseteq S$ and $T' \subseteq T$.*

---

[3] We say that $g : \mathbb{A}^n \to \mathbb{A}$ is the right residual of $f : \mathbb{A}^n \to \mathbb{A}$ in its i-th coordinate if for any $a_1, \ldots, a_n, b \in \mathbb{A}$, $f(a_1, \ldots, a_i, \ldots, a_n) \leq b$ iff $a_i \leq g(a_1, \ldots, b, \ldots, a_n)$. We also say that f is the left residual of g in its j-th coordinate.

[4] We say that $g : \mathbb{A}^n \to \mathbb{A}$ is the left Galois adjoint of $f : \mathbb{A}^n \to \mathbb{A}$ in its i-th coordinate if for any $a_1, \ldots, a_n, b \in \mathbb{A}$, $f(a_1, \ldots, a_i, \ldots, a_n) \leq b$ iff $g(a_1, \ldots, b, \ldots, a_n) \leq a_i$. We say that g is the right Galois adjoint of f in its i-th coordinate if for any $a_1, \ldots, a_n, b \in \mathbb{A}$, $b \leq f(a_1, \ldots, a_i, \ldots, a_n)$ iff $a_i \leq g(a_1, \ldots, b, \ldots, a_n)$.

3. *The canonical extension of a lattice* A *is a complete lattice* A^δ *containing* A *as a dense and compact sublattice.*

It can be shown that the canonical extension of a bounded lattice is unique up to iso-morphism (for a proof, see e.g. Proposition 2.7 in [21]). In meta-theoretic settings in which Zorn's lemma is available, the canonical extension of a lattice A is a *perfect* lat-tice, which is complete, completely join-generated by the set $J^\infty(A)$ of the completely join-irreducible elements of A, and completely meet-generated by the set $M^\infty(A)$ of the completely meet-irreducible elements of A. In the constructive setting, canonical extensions might not be perfect.

Let $K(A^\delta)$ and $O(A^\delta)$ denote the meet-closure and the join-closure of A in A^δ, i.e. the collection of elements in A^δ which are meets/joins of elements in A, respectively. The elements of $K(A^\delta)$ are referred to as *closed* elements, and elements of $O(A^\delta)$ as *open* elements.

The canonical extension of an LE A will be defined as a suitable expansion of the canonical extension of the underlying lattice of A.

Definition 3. *For every unary, order-preserving operation* $f : A \to B$, *the* σ- *and* π-*extension of* f *are defined as follows:*

$$f^\sigma(u) = \bigvee\{\bigwedge\{f(a) : k \le a \in A\} : u \ge k \in K(A^\delta)\}$$
$$f^\pi(u) = \bigwedge\{\bigvee\{f(a) : o \ge a \in A\} : u \le o \in O(A^\delta)\}.$$

Definition 4. *The canonical extension of an* $\mathcal{L}$-*algebra* $A = (L, \mathcal{F}^A, \mathcal{G}^A)$ *is the* $\mathcal{L}$-*algebra* $A^\delta := (L^\delta, \mathcal{F}^{A^\delta}, \mathcal{G}^{A^\delta})$ *such that* f^{A^δ} *and* g^{A^δ} *are defined as the* σ-*extension of* f^A *and as the* π-*extension of* g^A *respectively, for all* $f \in \mathcal{F}$ *and* $g \in \mathcal{G}$.

2.4 Adding the Fixed Point Operators

In this subsection, we extend LE-languages by adding two kinds of fixed point opera-tors. Let FVAR be the set of *fixed point variables*.

For any LE-language $\mathcal{L}$, let $\mathcal{L}_1$ be the set of terms which extends $\mathcal{L}$ by allowing terms $\mu X.t(X)$ and $\nu X.t(X)$ where $t \in \mathcal{L}_1$, $X \in$ FVAR and $t(X)$ is positive in X. Terms in $\mathcal{L}$ are interpreted in LEs as described above. If $t(X_1, X_2, \ldots, X_n) \in \mathcal{L}_1$ and $a_1, \ldots, a_{n-1} \in A$, then

$$\mu X.t(X, a_1, \ldots, a_{n-1}) := \bigwedge\{a \in A \mid t(a, a_1, \ldots, a_{n-1}) \le a\}$$

if this meet exists, otherwise $\mu X.t(X, a_1, \ldots, a_{n-1})$ is undefined. Similarly,

$$\nu X.t(X, a_1, \ldots, a_{n-1}) := \bigvee\{a \in A \mid a \le t(a, a_1, \ldots, a_{n-1})\}$$

if this join exists, otherwise $\nu X.t(X, a_1, \ldots, a_{n-1})$ is undefined.

The second extension is denoted $\mathcal{L}_2$ and extends $\mathcal{L}$ by allowing construction of the terms $\mu_2 X.t(X)$ and $\nu_2 X.t(X)$ where $t \in \mathcal{L}_2$, $X \in$ FVAR and $t(X)$ is positive in X. The interpretation of terms from $\mathcal{L}_2$ is defined below. For each ordinal α we define $t^\alpha(\bot, a_1, \ldots, a_{n-1})$ as follows:

$$t^0(\bot, a_1, \ldots, a_{n-1}) = \bot, \qquad t^{\alpha+1}(\bot, a_1, \ldots, a_{n-1}) = t(t^\alpha(\bot, a_1, \ldots, a_{n-1}), a_1, \ldots, a_{n-1}),$$
$$t^\lambda(\bot, a_1, \ldots, a_{n-1}) = \bigvee\nolimits_{\alpha<\lambda} t^\alpha(\bot, a_1, \ldots, a_{n-1}) \qquad \text{for limit ordinals } \lambda;$$
$$t_0(\top, a_1, \ldots, a_{n-1}) = \top, \qquad t_{\alpha+1}(\top, a_1, \ldots, a_{n-1}) = t(t_\alpha(\top, a_1, \ldots, a_{n-1}), a_1, \ldots, a_{n-1}),$$
$$t_\lambda(\top, a_1, \ldots, a_{n-1}) = \bigwedge\nolimits_{\alpha<\lambda} t_\alpha(\top, a_1, \ldots, a_{n-1}) \qquad \text{for limit ordinals } \lambda$$

If $t(X_1, \ldots, X_n) \in \mathcal{L}_2$, then we let

$$\mu_2 X.t(X, a_1, \ldots, a_{n-1}) := \bigvee\nolimits_{\alpha\geq0} t^\alpha(\bot, a_1, \ldots, a_{n-1})$$

$$\nu_2 X.t(X, a_1, \ldots, a_{n-1}) := \bigwedge\nolimits_{\alpha\geq0} t_\alpha(\top, a_1, \ldots, a_{n-1})$$

if this join and meet exist, otherwise are undefined.

A lattice expansion $\mathbb{A}$ is *of the first* (resp. *second*) *kind* if $t^{\mathbb{A}}(a_1, \ldots, a_n)$ is defined for all $a_1, \ldots, a_n \in \mathbb{A}$ and all $t \in \mathcal{L}_1$ ($t \in \mathcal{L}_2$). Henceforth we will refer to these algebras as *T1-algebras* (resp. *T2-algebras*). When restricted to the Boolean case, our $T1$-algebras are essentially the modal mu-algebras defined in [3, Definition 2.2] and [1, Definition 5.1]. Every $T2$-algebra is a $T1$-algebra. (cf. [1, Proposition 2.4] and [6, Lemma 2.2]. These proofs straightforwardly extend to the setting of general LEs). Hence, the interpretation of the two types of fixed point binders on $T2$-algebras will agree, i.e. $\mu X.\varphi(X) = \mu_2 X.\varphi(X)$ and $\nu X.\psi(X) = \nu_2 X.\psi(X)$ in $T2$-algebras.

The final sets of terms, $\mathcal{L}_*$ (resp. $\mathcal{L}_*^t$), are obtained as extensions of $\mathcal{L}$ (resp. $\mathcal{L}^t$) by allowing $\mu^* X.s(X)$ and $\nu^* X.s(X)$ whenever $s \in \mathcal{L}_*$ (resp. $s \in \mathcal{L}_*^t$) and is positive in X. Terms in $\mathcal{L}_*$ and $\mathcal{L}_*^t$ are only interpreted in the constructive canonical extensions $\mathbb{A}^\delta$ of lattice expansions $\mathbb{A}$. If $s(X_1, X_2, \ldots, X_n) \in \mathcal{L}_* \cup \mathcal{L}_*^t$ and $a_1, \ldots, a_{n-1} \in \mathbb{A}^\delta$, then $\mu^* X_1.s(X_1, a_1, \ldots, a_{n-1}) := \bigwedge\{a \in \mathbb{A} \mid s(a, a_1, \ldots, a_{n-1}) \leq a\}$ and $\nu^* X_1.s(X_1, a_1, \ldots, a_{n-1}) := \bigvee\{a \in \mathbb{A} \mid a \leq s(a, a_1, \ldots, a_{n-1})\}$. As the canonical extension $\mathbb{A}^\delta$ is a complete lattice, the interpretation of $\mu^* X.t(X)$ or $\nu^* X.t(X)$ is always defined. For any term $\varphi \in \mathcal{L}_1^t$ we let φ^* denote the $\mathcal{L}_*^t$ term obtained from φ by replacing all occurrences of μ and ν with μ^* and ν^*, respectively. The main feature of the μ^* and ν^* binders is that their interpretation does not change from $\mathbb{A}$ to $\mathbb{A}^\delta$.

Example 3. Consider the inequality $p \leq \nu X.(p \wedge (\Box_1 X \wedge \Box_2 X))$ in the fixed-point expansion of the language $\mathcal{L}$ of Example 1. It can be easily verified that the unfolding of the term $\nu X.(p \wedge \Box_1 X \wedge \Box_2 X)$ coincides with the common knowledge-type connective $C(p)$ introduced in [9, Sect. 4]. As discussed in [10, Sect. 4.2], $C(p)$ can be understood as the category identified by the objects that a given group of agents (in this case a group of two agents) consider *typical members* of p. In this context, the inequality above expresses that (relative to this group of agents) all members of p are considered typical members of p, i.e. p has maximal contrast.

Example 4. The inequality $p \leq \mu X.\blacklozenge(X \vee p)$, where $\blacklozenge$ is an $\mathcal{F}$-operator with $n_{\blacklozenge} = 1$ and $\varepsilon_{\blacklozenge} = 1$, is well known from the Boolean setting (cf. [3, Example 6.7]). As observed in [8, Sect. 4.2][5], this inequality is equivalent to $\nu X.\Box(X \wedge p) \leq p$. In the setting of [9,10], the intuitive content of the latter inequality is that every object that the agent regards as a typical member of p is an actual member of p.

---

[5] Although that argument is given in the distributive setting, it can be easily verified that it does not make use of any specific feature of that setting.

2.5   The Language of Constructive ALBA for LEs

The expanded language used in the algorithm μ^*-ALBA contains, in addition to the connectives in $\mathcal{L}^t$, a set of special variables NOM called *nominals*, a set of special variables CO-NOM, called *co-nominals*, and fixed point binders. In the constructive setting, there are not enough completely join-irreducible and the completely meet-irreducible elements, therefore nominals and co-nominals are interpreted as elements of $K(\mathbb{A}^\delta)$ and $O(\mathbb{A}^\delta)$, respectively.

Formulas in the extended language $\mathcal{L}_1^+$ are defined by the following recursion:

$$\varphi ::= \bot \mid \top \mid p \mid X \mid \mathbf{j} \mid \mathbf{m} \mid \varphi \wedge \psi \mid \varphi \vee \psi \mid f(\overline{\varphi}) \mid g(\overline{\varphi}) \mid \mu X.\varphi(X) \mid \nu X.\varphi(X)$$

where $p \in$ PROP, $X \in$ FVAR, $\mathbf{j} \in$ NOM, $\mathbf{m} \in$ CNOM, $f \in \mathcal{F}^t$, $g \in \mathcal{G}^t$, and φ is positive in X in $\mu X.\varphi(X)$ and $\nu X.\varphi(X)$. Formulas in $\mathcal{L}_*^+$ are defined by replacing the fixed point operators μ, ν with μ^*, ν^*.

A formula is *pure* if it contains no ordinary propositional variables but only nominals and co-nominals, and is a *sentence* if it contains no free fixed point variables. A *quasi-inequality* is an expression of the form $\varphi_1 \le \psi_1 \& \cdots \& \varphi_n \le \psi_n \Rightarrow \varphi \le \psi$ where the $\varphi_i, \psi_i, \varphi$ and ψ are formulas of the corresponding languages.

Semantics. Constructive canonical extensions of $\mathcal{L}$-algebras provide natural semantics for the language $\mathcal{L}^+$. For any $\mathcal{L}$-algebra $\mathbb{A}$, an *assignment* on $\mathbb{A}$ sends propositional variables to elements of $\mathbb{A}$. An *assignment* on $\mathbb{A}^\delta$ is a map $V :$ PROP$\cup$NOM$\cup$CO-NOM $\rightarrow$ $\mathbb{A}^\delta$ sending propositional variables to elements of $\mathbb{A}^\delta$, nominals to $K(\mathbb{A}^\delta)^6$ and co-nominals to $O(\mathbb{A}^\delta)$. An *admissible assignment* on $\mathbb{A}^\delta$ is an assignment which sends propositional variables to elements of $\mathbb{A}$. An $\mathcal{L}$-inequality $\alpha \le \beta$ is *valid* on $\mathbb{A}$, denoted $\mathbb{A} \models \alpha \le \beta$, if it holds under all assignments. An $\mathcal{L}^+$-inequality $\alpha \le \beta$ is *admissibly valid* on $\mathbb{A}$, denoted $\mathbb{A}^\delta \models_\mathbb{A} \alpha \le \beta$, if it holds under all admissible assignments. A quasi-inequality $\varphi_1 \le \psi_1 \& \cdots \& \varphi_n \le \psi_n \Rightarrow \varphi \le \psi$ is satisfied under an assignment V in an algebra $\mathbb{A}$, written $\mathbb{A}, V \models \varphi_1 \le \psi_1 \& \cdots \& \varphi_n \le \psi_n \Rightarrow \varphi \le \psi$ if $\mathbb{A}, V \models \varphi_i \le \psi_i$ for all $1 \le i \le n$ imply $\mathbb{A}, \nu \models \varphi \le \psi$. A quasi-inequality is (admissibly) valid in an algebra if it is satisfied by every (admissible) assignment.

3   Two Kinds of Canonicity, Constructively

In this section we give a brief conceptual and methodological overview of the main results of this paper. Generally, algorithmic canonicity proofs go in the "U-shaped" argument[7] described in Fig. 1 (see [11] for a more detailed discussion):

---

[6] In other settings, nominals are interpreted as completely join-irreducible elements in $\mathbb{A}^\delta$, while in the constructive setting, the constructive canonical extensions might not be perfect, therefore there might not be enough completely join-irreducible elements, therefore nominals are interpreted as closed elements instead in the constructive setting.

[7] Indeed, the U-shaped argument is the algorithmic version of the Sambin–Vaccaro canonicity proof [29]. In [26], this method has been unified with Jónsson's canonicity proof [24]; in [12], it has been unified with constructive canonicity introduced in Ghilardi and Meloni [22]; in [16], the canonicity via pseudo-correspondence has been presented as an instance of this method.

$$\mathbb{A} \models \alpha \leq \beta \qquad\qquad\qquad \mathbb{A}^\delta \models \alpha \leq \beta$$

$$\Updownarrow$$

$$\mathbb{A}^\delta \models_\mathbb{A} \alpha \leq \beta \qquad\qquad\qquad \Updownarrow$$

$$\Updownarrow$$

$$\mathbb{A}^\delta \models_\mathbb{A} \mathrm{ALBA}(\alpha \leq \beta) \qquad \Longleftrightarrow \qquad \mathbb{A}^\delta \models \mathrm{ALBA}(\alpha \leq \beta)$$

Fig. 1. The U-shaped argument

The U-shaped argument starts from the validity of the inequality $\alpha \leq \beta$ in the algebra $\mathbb{A}$. This validity is equivalent to the *admissible validity* in $\mathbb{A}^\delta$, i.e. the validity of $\alpha \leq \beta$ in $\mathbb{A}^\delta$ while restricting the assignments of propositional variables to $\mathbb{A}$. Then by the algorithm ALBA, the inequality $\alpha \leq \beta$ can be equivalently transformed into a set of pure quasi-inequalities $\mathrm{ALBA}(\alpha \leq \beta)$. Then by the fact that admissible validity and validity are the same for pure quasi-inequalities, the bottom line of the argument goes through. Then by the soundness of the rules with respect to the constructive canonical extension $\mathbb{A}^\delta$, we proceed up the right-hand arm of the U-shaped argument.

However, the argument stated above need to be adapted in the fixed point setting. Indeed, for the step from $\mathbb{A} \models \alpha \leq \beta$ to $\mathbb{A}^\delta \models_\mathbb{A} \alpha \leq \beta$, it depends on the property that $\alpha^{\mathbb{A}^\delta}$ and $\alpha^\mathbb{A}$ agree on arguments from $\mathbb{A}$. However, for formulas with fixed point operators, this property fails, since $(\alpha(X))^{\mathbb{A}^\delta}$ can have more pre-fixed points in $\mathbb{A}^\delta$ than $(\alpha(X))^\mathbb{A}$ has in $\mathbb{A}$, and so $(\mu X.\alpha(X))^{\mathbb{A}^\delta}$ would generally be smaller than $(\mu X.\alpha(X))^\mathbb{A}$, which creates additional difficulties for the argument.

To overcome this difficulty, one solution is to restrict the pre-fixed points used in calculating $(\mu X.\varphi(X))^{\mathbb{A}^\delta}$ to be in $\mathbb{A}$, which is indeed $(\mu^* X.\varphi(X))^{\mathbb{A}^\delta}$.

Thus, in [6], two different notions of canonicity for the mu-calculus were considered, and their two ensuing canonicity results were shown for certain classes of mu-inequalities. We prove that the counterparts of these results hold in a constructive general lattice environment. Specifically, following [6], we call an inequality $\varphi \leq \psi$ *tame canonical* when $\mathbb{A} \models \varphi \leq \psi$ if and only if $\mathbb{A}^\delta \models \varphi^* \leq \psi^*$ for all mu-algebras $\mathbb{A}$. We generalize the *tame inductive* mu-inequalities of [6] to the LE-setting, and prove that they are tame canonical in a constructive meta-theory.

Of course, the usual notion of canonicity may also be applied to formulas with fixed-point binders, i.e., that $\mathbb{A} \models \varphi \leq \psi$ implies $\mathbb{A}^\delta \models \varphi \leq \psi$, where fixed points are interpreted in the standard way, e.g. least fixed points in $\mathbb{A}^\delta$ are calculated as the meet of all pre-fixed points *in* $\mathbb{A}^\delta$. A canonicity result of this kind can be proved by generalizing the class of *restricted inductive mu-inequalities* of [6] to the LE-setting, and showing that they are preserved under constructive canonical extensions of $T2$-algebras. Whenever a *tame* run (cf. Sect. 7, p. 17) of μ^*-ALBA succeeds on a mu-inequality $\varphi \leq \psi$, we have that $\varphi \leq \psi$ is *tame canonical*. Moreover, whenever a *proper* run succeeds on a mu-inequality $\alpha \leq \beta$, then $\alpha \leq \beta$ will be *canonical*. Finally, for every tame inductive mu-inequality (respectively, a restricted inductive mu-inequality), there exists a tame (respectively, proper) run of μ^*-ALBA which succeeds on that inequality.

4   Recursive, Restricted Inductive, and Tame-Inductive Inequalities

In this section, we introduce three syntactically defined classes of mu-inequalities: the 'general lattice' counterpart of the recursive mu-inequalities introduced in [8], and two subclasses of it, for which the two canonicity results hold. Our presentation follows the usual notational conventions of unified correspondence (cf. [23] for discussion).

Signed Generation Trees. For any formula/term φ in $\mathcal{L}_1^+$ and $\mathcal{L}_*^+$, we assign two *signed generation trees* $+\varphi$ and $-\varphi$. Each node is signed as follows:

- the root node of $+\varphi$ is signed $+$ and the root node of $-\varphi$ is signed $-$;
- if a node is $\vee$, $\wedge$ assign the same sign to its children nodes;
- if a node is $h \in \mathcal{F}^t \cup \mathcal{G}^t$, assign the same (resp. the opposite) sign to every node corresponding to a coordinate i such that $\varepsilon_h(i) = 1$ (resp. $\varepsilon_h(i) = \partial$);
- if a node is $\mu X.\varphi(X)$, $\mu^* X.\varphi(X)$, $\nu X.\varphi(X)$ or $\nu^* X.\varphi(X)$ then assign the same sign to the child node.

A node in a signed generation tree is *positive* (resp. *negative*) if it is signed "$+$" (resp. "$-$").[8] For any $\mathcal{L}_1$-sentence $\varphi(p_1, \ldots p_n)$, order-type ε over n, and $1 \leq i \leq n$, an ε-*critical node* in a signed generation tree of φ is a (leaf) node $+p_i$ with $\varepsilon_i = 1$, or $-p_i$ with $\varepsilon_i = \partial$. An ε-*critical branch* is a branch terminating in an ε-critical node. Variable occurrences corresponding to ε-critical nodes are those which ALBA will *solve for*. A *live branch* is a branch ending in a propositional variable. All critical branches are live. A branch is not live iff it ends in a propositional constant ($\top$ or $\bot$) or in a fixed point variable. For every $\mathcal{L}_1$-sentence $\varphi(p_1, \ldots p_n)$ and order-type ε, we say that $+\varphi$ (resp. $-\varphi$) *agrees with* ε, and write $\varepsilon(+\varphi)$ (resp. $\varepsilon(-\varphi)$), if every leaf node in the signed generation tree $+\varphi$ (resp. $-\varphi$) which is labelled with a propositional variable is ε-critical. We use the *sub-tree relation* $\gamma \prec \varphi$, which extends to signed generation trees, and write $\varepsilon(\gamma) \prec *\varphi$ to indicate that γ, regarded as a sub- (signed generation) tree of $*\varphi$, agrees with ε.

The following definition generalizes the definition of recursive inequalities introduced in [8] to the setting of general lattices. The key difference is in the order-theoretic properties of $\vee$ and $\wedge$, which is reflected in their syntactic classification in Table 1. See [11,14] for further discussion on methodology and nomenclature.

Definition 5. *Nodes in signed generation trees are classified as* Skeleton *nodes and* PIA *nodes and further classified as Δ-adjoint, SLR, Binders, SLA, SRA or SRR, according to the specification given in Table 1. Let $\varphi(p_1, \ldots, p_n)$ be a formula, and ε be an order-type on $\{1, \ldots, n\}$. A branch in a signed generation tree $*\varphi$, for $* \in \{+, -\}$, ending in a propositional variable is an ε-good branch if, apart from the leaf, it is the concatenation of three paths P_1, P_2, and P_3, each of which may possibly be of length 0, such that P_1 is a path from the leaf consisting only of PIA-nodes, P_2 consists only of inner skeleton-nodes, and P_3 consists only of outer skeleton-nodes and, moreover, it satisfies conditions (GB1), (GB2) and (GB3) below.*

---

[8] A term φ is *positive* (*negative*) in a variable p if in the signed generation tree $+\varphi$ all p-nodes are signed $+$ ($-$). An inequality $\varphi \leq \psi$ is *positive* (*negative*) in p if φ is negative (positive) in p and ψ is positive (negative) in p.

Table 1. Skeleton and PIA nodes.

Outer Skeleton (P_3)	Inner Skeleton (P_2)	PIA (P_1)
Δ-adjoints	Binders	Binders
$+ \vee$	$+ \mu$	$+ \nu$
$- \wedge$	$- \nu$	$- \mu$
SLR	SLA	SRA
$+ f$	$+ \vee f \ (n_f = 1)$	$+ \wedge g \ (n_g = 1)$
$- g$	$- \wedge g \ (n_g = 1)$	$- \vee f \ (n_f = 1)$
	SLR	SRR
	$+ \quad f \ (n_f \geq 2)$	$+ \quad g \ (n_g \geq 2)$
	$- \quad g \ (n_g \geq 2)$	$- \quad f \ (n_f \geq 2)$

(GB1). *The formula corresponding to the uppermost node on P_1 is a sentence.*
(GB2). *For every SRR-node in P_1 of the form $h(\overline{\gamma}, \beta)$, where β is the coordinate where the branch lies, every γ in $\overline{\gamma}$ is a mu-sentence and $\varepsilon^\partial(\gamma) \prec *\varphi$ (i.e., each γ contains no ε-critical node).*
(GB3). *For every SLR-node in P_2 of the form $h(\overline{\gamma}, \beta)$, where β is the coordinate where the branch lies, every γ in $\overline{\gamma}$ is a mu-sentence and $\varepsilon^\partial(\gamma) \prec *\varphi$.*

The definition above and similar definitions (cf. [6,8,14]) are modular and independent of the specific signature thanks to the fact that they adapt the same order-theoretic principles in different semantic settings. Our main interest is in ε-good branches satisfying some of the additional properties, reported in the following definition.

Definition 6. *For a formula $\varphi(p_1, \ldots, p_n)$, order-type ε on $\{1, \ldots, n\}$, and strict partial order $<_\Omega$ on $p_1, \ldots p_n$, an ε-good branch may satisfy one or more of the following:*

(NB-PIA). *P_1 contains no fixed point binders.*
(NL). *For every SLR-node in P_2 of the form $h(\overline{\gamma}, \beta)$, where β is the coordinate where the branch lies, the signed generation tree of each γ contains no live branches.*
(Ω-CONF). *For every SRR-node in P_1 of the form $h(\overline{\gamma}, \beta)$, where β is the coordinate where the branch lies, $p_j <_\Omega p_i$ for every p_j occurring in γ, where p_i is the propositional variable labelling the leaf of the branch.*

Definition 7. *For any order-type ε and strict partial order $<_\Omega$ on the variables $p_1, \ldots p_n$, the signed generation tree $*\varphi$, $* \in \{-, +\}$, of a term $\varphi(p_1, \ldots p_n)$ is*

1. *ε-recursive if every ε-critical branch is ε-good.*
2. *(Ω, ε)-inductive if it is ε-recursive and every ε-critical branch satisfies (Ω-CONF).*
3. *restricted (Ω, ε)-inductive if it is (Ω, ε)-inductive and*
   (a) every ε-critical branch satisfies (NB-PIA) and (NL),
   (b) every occurrence of a binder is on an ε-critical branch.
4. *tame (Ω, ε)-inductive if it is (Ω, ε)-inductive and*
   (a) no binder occurs on any ε-critical branch,
   (b) the only nodes involving binders which are allowed to occur are $+\nu$ and $-\mu$.

An inequality $\varphi \leq \psi$ is *restricted ε-recursive* (resp. *tame (Ω, ε)-inductive*) if $+\varphi$ and $-\psi$ are both restricted ε-recursive (resp. tame (Ω, ε)-inductive). An inequality $\varphi \leq \psi$ is *restricted recursive* (resp. *tame inductive*) if $\varphi \leq \psi$ is restricted (Ω, ε)-inductive (resp. tame (Ω, ε)-inductive) for some strict partial order Ω and order-type ε.

Example 5. The inequality $p \leq vX.(p \wedge (\square_1 X \wedge \square_2 X))$ from Example 3 is restricted (Ω, ε)-inductive for the order-type $\varepsilon(p) = \partial$ (and $\Omega = \varnothing$). Indeed, the only ε-critical branch in the signed generation tree below is $-vX, -\wedge, -p$, where $-vX, -\wedge$ are Inner Skeleton nodes, and P_1 is empty. Thus, (GB1) and (NB-PIA) are vacuously satisfied. Since this branch does not contain SRR or SLR nodes, (GB2), (GB3), (NL), (Ω-CONF) are also vacuously satisfied. With a symmetric argument, one shows that the $\mathcal{L}^t$-inequality $\mu X.(p \vee \blacklozenge_1 X \vee \blacklozenge_2 X) \leq p$ is restricted (Ω, ε)-inductive for the order-type $\varepsilon(p) = 1$ (and $\Omega = \varnothing$).

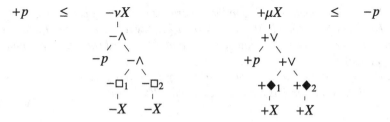

Example 6. The inequality $p \leq \mu X.\blacklozenge(X \vee p)$ from Example 4 is tame (Ω, ε)-inductive for $\varepsilon(p) = 1$ (and $\Omega = \varnothing$). Indeed, the only critical branch in the signed generation tree of $p \leq \mu X.\blacklozenge(X \vee p)$ is $+p$, therefore (GB1), (GB2), (GB3), (Ω-CONF) are vacuously satisfied. It is also easy to see that no binders occur on any ε-critical branch, and the only fixed point binder node is $-\mu$. By an analogous argument, it can be verified that $vX.\square(X \wedge p) \leq p$ is tame (Ω, ε)-inductive for $\varepsilon(p) = \partial$ (and $\Omega = \varnothing$).

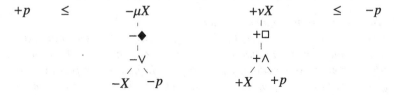

5   Constructive μ^*-ALBA

In this section, we introduce the fundamental tool to prove the constructive canonicity results via the argument discussed in Sect. 3. *Constructive μ^*-ALBA* is a calculus for eliminating propositional variables from lattice-based fixed point inequalities, while maintaining admissible validity (cf. page 7) in a constructive metatheory. The purpose of this is to make the transition from *admissible validity* to *validity* in the argument for canonicity. Below, we outline its general strategy.

Constructive μ^*-ALBA, from now on abbreviated as ALBA, takes an inequality $\varphi \leq \psi$ in $\mathcal{L}_1$ (cf. Sect. 2.4) as input, and executes in three stages.

The first stage, the *preprocessing* stage, aims to eliminate all propositional variables occurring only positively or only negatively, applies distribution rules for $f \in \mathcal{F}$ and $g \in \mathcal{G}$ and splitting rules exhaustively, and converts all occurrences of $\mu X.\varphi(X)$ to $\mu^* X.\varphi(X)$ and $\nu X.\psi(X)$ to $\nu^* X.\psi(X)$. We emphasize that this step is required in both tame *and* proper runs of ALBA (see Sect. 7). The preprocessing produces a finite set of inequalities, $\{\varphi_i' \leq \psi_i'\}_{i=1}^n$, on each of which it proceeds separately. Now ALBA forms the *initial quasi-inequalities* $\& S_i \Rightarrow \mathsf{Ineq}_i$, compactly represented as tuples (S_i, Ineq_i) referred to as *systems*, with each S_i initialized to $\varnothing$ and Ineq_i initialized to $\varphi_i' \leq \psi_i'$.

The second stage, the *reduction* stage, transforms S_i and Ineq_i through the application of transformation rules (see Sect. 7). This stage aims at eliminating propositional variables and reach *pure* or *purified* systems, i.e. systems containing no propositional variables but only nominals and co-nominals. The Ackermann rules (cf. Appendix) are the ones which eliminate the propositional variables, and the other rules transform a given system into one on which the Ackermann rules are applicable. Once all propositional variables are eliminated, this phase terminates and returns the pure quasi-inequalities $\& S_i \Rightarrow \mathsf{Ineq}_i$.

The third stage, the *output* stage, either reports failure if some system could not be purified, or else returns the conjunction of the pure quasi-inequalities $\& S_i \Rightarrow \mathsf{Ineq}_i$, which we denote by $\mathsf{ALBA}(\varphi \leq \psi)$. A more complete outline of each of the three stages will be given in Sect. 7.

6   Main Results

This section collects the main results of the paper: all restricted inductive inequalities are constructively canonical, and all tame inductive inequalities are constructively tame canonical. The proof strategy is the same in both cases, and has been discussed in Sect. 3: one first proves, by means of a 'U-shaped' argument, that successful runs of constructive ALBA satisfying certain conditions (cf. proper, pivotal, tame runs defined on page 17) guarantee these types of canonicity. Then one shows that ALBA is successful on all members of the two classes of inequalities, by means of runs respectively satisfying these conditions. The main canonicity results follow as corollaries of these facts. Due to space limitations we do not include the proofs, which can be found in [7].

Proposition 1

1. *Let* $\mathbb{A}$ *be a T2-algebra (cf. Sect. 2.4) and* $\varphi \leq \psi$ *be an* $\mathcal{L}_1$-*inequality on which a proper and pivotal run of ALBA succeeds. If* $\mathbb{A} \models \varphi \leq \psi$ *then* $\mathbb{A}^\delta \models \varphi \leq \psi$.
2. *Let* $\mathbb{A}$ *be a T1-algebra and* $\varphi \leq \psi$ *be an* $\mathcal{L}_1$-*inequality on which a tame and pivotal run of ALBA succeeds. If* $\mathbb{A} \models \varphi \leq \psi$ *then* $\mathbb{A}^\delta \models \varphi^* \leq \psi^*$.

Proposition 2

1. *ALBA succeeds on all restricted inductive* $\mathcal{L}_1$-*inequalities (cf. Definition 7) by means of proper and pivotal runs.*
2. *ALBA succeeds on all tame inductive* $\mathcal{L}_1$-*inequalities (cf. Definition 7) by means of tame and pivotal runs.*

The canonicity of restricted inductive $\mathcal{L}_1$-inequalities follows from Propositions 1.1 and 2.1; the tame canonicity of tame inductive $\mathcal{L}_1$-inequalities follows from Propositions 1.2 and 2.2.

Theorem 1

1. *All restricted inductive $\mathcal{L}_1$-inequalities are constructively canonical over T2-algebras.*
2. *All tame inductive $\mathcal{L}_1$-inequalities are constructively canonical over T1-algebras.*

7 Appendix: Stages and Rules of Constructive μ^*-ALBA

As remarked in the introduction, the version of ALBA presented below is a significant simplification of the one given in [6], and is rather similar to that of [12]. In particular, there are no rules specific to fixed points, which are accounted for by restrictions on the shape of formulas.

Stage 1: Preprocessing and Initialization. ALBA receives an $\mathcal{L}_1$-inequality $\varphi \leq \psi$ as input. It applies the following rules exhaustively:

Monotone and Antitone Variable-Elimination Rules.

$$\frac{\alpha(p) \leq \beta(p)}{\alpha(\bot) \leq \beta(\bot)} \qquad \frac{\gamma(p) \leq \delta(p)}{\gamma(\top) \leq \delta(\top)}$$

for $\alpha(p) \leq \beta(p)$ positive and $\gamma(p) \leq \delta(p)$ negative in p, respectively.

Distribution Rules. Push down $+f \in \mathcal{F}$ (resp. $-g \in \mathcal{G}$) by distributing them over Skeleton nodes labelled $+\vee$ (resp. $-\wedge$).

Splitting-Rules.

$$\frac{\alpha \leq \beta \wedge \gamma}{\alpha \leq \beta \quad \alpha \leq \gamma} \qquad \frac{\alpha \vee \beta \leq \gamma}{\alpha \leq \gamma \quad \beta \leq \gamma}$$

This gives rise to a set of inequalities $\{\varphi'_i \leq \psi'_i \mid 1 \leq i \leq n\}$. For each of them, ALBA converts all occurrences of $\mu X.\varphi(X)$ to $\mu^* X.\varphi(X)$ and all occurrences of $\nu X.\psi(X)$ to $\nu^* X.\psi(X)$, and forms the *initial quasi-inequality* $\&\, S_i \Rightarrow \mathsf{Ineq}_i$, compactly represented as a tuple (S_i, Ineq_i) referred as *initial system*, with each S_i initialized to the empty set and Ineq_i initialized to $\varphi'_i \leq \psi'_i$. Each initial system is processed separately in stage 2, described below, where we will suppress indices i.

Stage 2: Reduction and Elimination. This stage aims at eliminating all propositional variables from a given system (S, Ineq). To this effect, the following rules (collectively called *reduction rules*) are applied: *approximation rules*, *residuation rules*, *splitting rules*, and *Ackermann rules*. The terms and inequalities in this subsection are from $\mathcal{L}^+_*$.

Approximation Rules. There are four approximation rules, which simplify Ineq and add an inequality to S. We write $\alpha(!x)$ to indicate that the placeholder variable x has a unique occurrence in formula α.

Left-positive approximation rule.

$$\frac{(S, \ \varphi'(\gamma/!x) \leq \psi)}{(S \cup \{\mathbf{j} \leq \gamma\}, \ \varphi'(\mathbf{j}/!x) \leq \psi)} \ (L^+A)$$

with $+x \prec +\varphi'(!x)$, the branch of $+\varphi'(!x)$ starting at $+x$ subject to the restrictions detailed below, γ belonging to the smaller language $\mathcal{L}_*$ and $\mathbf{j}$ being the first nominal variable not occurring in S or $\varphi'(\gamma/!x) \leq \psi$.

Left-negative approximation rule.

$$\frac{(S, \ \varphi'(\gamma/!x) \leq \psi)}{(S \cup \{\gamma \leq \mathbf{m}\}, \ \varphi'(\mathbf{m}/!x) \leq \psi)} \ (L^-A)$$

with $-x \prec +\varphi'(!x)$, the branch of $+\varphi'(!x)$ starting at $-x$ subject to the restrictions detailed below, γ belonging to the smaller language $\mathcal{L}_*$ and $\mathbf{m}$ being the first co-nominal not occurring in S or $\varphi'(\gamma/!x) \leq \psi$.

Right-positive approximation rule.

$$\frac{(S, \ \varphi \leq \psi'(\gamma/!x))}{(S \cup \{\mathbf{j} \leq \gamma\}, \ \varphi \leq \psi'(\mathbf{j}/!x))} \ (R^+A)$$

with $+x \prec -\psi'(!x)$, the branch of $-\psi'(!x)$ starting at $+x$ subject to the restrictions detailed below, γ belonging to the smaller language $\mathcal{L}_*$ and $\mathbf{j}$ being the first nominal not occurring in S or $\varphi \leq \psi'(\gamma/!x)$.

Right-negative approximation rule.

$$\frac{(S, \ \varphi \leq \psi'(\gamma/!x))}{(S \cup \{\gamma \leq \mathbf{m}\}, \ \varphi \leq \psi'(\mathbf{m}/!x))} \ (R^-A)$$

with $-x \prec -\psi'(!x)$, the branch of $-\psi'(!x)$ starting at $-x$ subject to the restrictions detailed below, γ belonging to the smaller language $\mathcal{L}_*$ and $\mathbf{m}$ being the first co-nominal not occurring in S or $\varphi \leq \psi'(\gamma/!x))$.

The restrictions on φ' and ψ' in the approximation rules above are formulated in terms of the following:

Definition 8. *For any mu-algebra $\mathbb{C}$ and order-type τ, a join $\bigvee S$ in $(\mathbb{C}^\delta)^\tau$ is called $\mathbb{C}^\tau$-targeted if $\bigvee S \in \mathbb{C}^\tau$. A map $f : (\mathbb{C}^\delta)^\tau \to \mathbb{C}^\delta$ preserves $\mathbb{C}^\tau$-targeted joins if $f(\bigvee S) = \bigvee_{s \in S} f(s)$ for every $S \subseteq (\mathbb{C}^\delta)^\tau$ such that $\bigvee S$ is $\mathbb{C}^\tau$-targeted. Targeted meets and their preservation are defined order-dually.*

Let us now list the requirements on φ' and ψ':

1. $\varphi', \psi' \in \mathcal{L}_*$;

2. the branches of φ' and ψ' starting at x going up to the root consist only of Skeleton nodes.[9]
3. for every node of the form $+\mu^*X.\theta(\overline{x}, X)$ or of the form $-\nu^*X.\delta(\overline{x}, X)$ in such branches, which is not in the scope of another binder, all propositional variables and free fixed point variables in $\theta(\overline{x}, X)$ and $\delta(\overline{x}, X)$ must be among $\overline{x}$ and X; moreover,
   (a) the associated term function $\theta(\overline{x}, X)$: $(\mathbb{C}^\delta)^\tau \times \mathbb{C}^\delta \rightarrow \mathbb{C}^\delta$ preserves $(\mathbb{C}^\tau \times \mathbb{C})$-targeted joins for all μ-algebras $\mathbb{C}$; moreover $\theta(\overline{x}, X)$ is required to be positive (negative) in x_i if $\tau_i = 1$ ($\tau_i = \partial$), i.e. $\theta(\overline{x}, X)$ must be τ-positive in $\overline{x}$;
   (b) the associated term function $\delta(\overline{x}, X)$: $(\mathbb{C}^\delta)^\tau \times \mathbb{C}^\delta \rightarrow \mathbb{C}^\delta$ preserves $(\mathbb{C}^\tau \times \mathbb{C})$-targeted meets for all μ-algebras $\mathbb{C}$; moreover $\delta(\overline{x}, X)$ is required to be positive (negative) in x_i if $\tau_i = 1$ ($\tau_i = \partial$), i.e. $\delta(\overline{x}, X)$ must be τ-positive in $\overline{x}$.

Remark 1

1. We will restrict the applications of approximation rules to nodes $!x$ giving rise to *maximal* skeleton branches in order to simplify the proof of canonicity. Such applications will be called *pivotal*. An execution of ALBA in which approximation rules are applied only pivotally will be referred to as *pivotal*.
2. In [6], approximation rules were formulated specifically for formulas having a fixed point binder as main connective. These rules had a substantially more cumbersome formulation than the one given above, which, modulo the restrictions about the preservation of targeted joins and meets, follows verbatim the approximation rules of [14]. Moreover, the approximation rules [6] could give rise to the splitting of the quasi-inequality into a set of quasi-inequalities, which is not the case of the present setting. This is thanks to the fact that nominals and co-nominals are not interpreted as completely join-primes (resp. meet-primes), but as closed and open elements.

Residuation Rules. The residuation rule rewrite a given inequality in S. For every $f \in \mathcal{F}$ and $g \in \mathcal{G}$, and any $1 \le i \le n_f$ and $1 \le j \le n_g$, the following rules are defined:

$$\frac{f(\varphi_1, \ldots, \varphi_i, \ldots, \varphi_{n_f}) \le \psi}{\varphi_i \le f_i^\#(\varphi_1, \ldots, \psi, \ldots, \varphi_{n_f})} \varepsilon_f(i) = 1 \qquad \frac{f(\varphi_1, \ldots, \varphi_i, \ldots, \varphi_{n_f}) \le \psi}{f_i^\#(\varphi_1, \ldots, \psi, \ldots, \varphi_{n_f}) \le \varphi_i} \varepsilon_f(i) = \partial$$

$$\frac{\psi \le g(\varphi_1, \ldots, \varphi_i, \ldots, \varphi_{n_g})}{g_i^\flat(\varphi_1, \ldots, \psi, \ldots, \varphi_{n_g}) \le \varphi_i} \varepsilon_g(i) = 1 \qquad \frac{\psi \le g(\varphi_1, \ldots, \varphi_i, \ldots, \varphi_{n_g})}{\varphi_i \le g_i^\flat(\varphi_1, \ldots, \psi, \ldots, \varphi_{n_g})} \varepsilon_g(i) = \partial$$

---

[9] The purpose of this restriction is to enforce preservation of non-empty joins by the term function $\varphi'^\mathbb{C}$. The soundness of the rule is founded upon this and approximation of the argument γ as the join of all closed elements below it. In the non-constructive setting of [14] the same strategy is followed, except that the approximation is done by means of completely join-irreducibles. Since this can give rise to empty sets of approximants and hence empty joins, $+\vee$ is excluded in the analogous approximation rule in [14], as the join does not preserve empty joins coordinate-wise. In the present setting, the set of closed approximants is never empty, and hence this restriction may be dropped. Similar considerations apply to $-\wedge$.

Ackermann Rules. The Ackermann rules operate on the whole system of inequalities and aim at eliminating propositional variables.

Right Ackermann-Rule.

$$\frac{(\{\alpha_i \leq p \mid 1 \leq i \leq n\} \cup \{\beta_j(p) \leq \gamma_j(p) \mid 1 \leq j \leq m\}, \ \ \mathsf{Ineq})}{(\{\beta_j(\bigvee_{i=1}^n \alpha_i) \leq \gamma_j(\bigvee_{i=1}^n \alpha_i) \mid 1 \leq j \leq m\}, \ \ \mathsf{Ineq})} \ (RAR)$$

where the α_i are p-free and syntactically closed, the β_j are positive in p and syntactically closed, while the γ_j are negative in p and syntactically open (cf. Definition 9).

Left Ackermann-Rule.

$$\frac{(\{p \leq \alpha_i \mid 1 \leq i \leq n\} \cup \{\gamma_j(p) \leq \beta_j(p) \mid 1 \leq j \leq m\}, \ \ \mathsf{Ineq})}{(\{\gamma_j(\bigwedge_{i=1}^n \alpha_i) \leq \beta_j(\bigwedge_{i=1}^n \alpha_i) \mid 1 \leq j \leq m\}, \ \ \mathsf{Ineq})} \ (LAR)$$

where the α_i are p-free and syntactically open, the β_j are positive in p and syntactically open, while the γ_j are negative in p and syntactically closed.

Syntactically Open and Closed Formulas. In the following definition we will use $f \in \mathcal{F}$ and $g \in \mathcal{G}$ to denote connectives of the original signature, and $h \in \mathcal{F}^+ \setminus \mathcal{F}$ and $k \in \mathcal{G}^+ \setminus \mathcal{G}$ to denote connectives of the expanded language. To simplify notation, we will disregard the actual order of the coordinates, but keep track of their polarity. So, for instance we will write $f(\overline{\psi}, \overline{\varphi})$ and $k(\overline{\varphi}, \overline{\psi})$, where in both cases the coordinates are divided in two possibly empty arrays, the first (resp. second) of which contains the positive (resp. negative) coordinates.

Definition 9. *The syntactically open formulas φ and syntactically closed formulas ψ are defined by simultaneous mutual recursion as follows:*

$$\varphi ::= \bot \mid \top \mid p \mid \mathbf{m} \mid \varphi_1 \wedge \varphi_2 \mid \varphi_1 \vee \varphi_2 \mid g(\overline{\varphi}, \overline{\psi}) \mid f(\overline{\varphi}, \overline{\psi}) \mid k(\overline{\varphi}, \overline{\psi}) \mid \nu^* X.\varphi$$

$$\psi ::= \bot \mid \top \mid p \mid \mathbf{i} \mid \psi_1 \wedge \psi_2 \mid \psi_1 \vee \psi_2 \mid f(\overline{\psi}, \overline{\varphi}) \mid g(\overline{\psi}, \overline{\varphi}) \mid h(\overline{\psi}, \overline{\varphi}) \mid \mu^* X.\psi$$

where $p \in \mathsf{PROP}$, $\mathbf{i} \in \mathsf{NOM}$, and $\mathbf{m} \in \mathsf{CNOM}$.

Stage 3: Success, Failure and Output. If stage 2 succeeded in eliminating all propositional variables from each system, the algorithm returns the conjunction of these purified quasi-inequalities, denoted by $\mathrm{ALBA}(\varphi \leq \psi)$. Otherwise, the algorithm reports failure and terminates.

Special Runs of ALBA. A *tame run* of ALBA is one during which the approximation rules are applied only to formulas $\varphi'(\gamma/!x)$ and $\psi'(\gamma/!x)$ such that no fixed point binder occurs in the branch from x to the root of φ' and ψ'. By contrast, a *proper run* of ALBA is one during which *all* occurrences of fixed point binders lie along some branch ending with a subterm γ which the application of an approximation rule extracts. We say that a run of the algorithm ALBA *succeeds* if all propositional variables are eliminated from the input inequality, $\varphi \leq \psi$, and denote the resulting set of pure quasi-inequalities by $\mathrm{ALBA}(\varphi^* \leq \psi^*)$. An inequality on which some run of ALBA succeeds is called a μ^*-*ALBA inequality.*

References

1. Ambler, S., Kwiatkowska, M., Measor, N.: Duality and the completeness of the modal μ-calculus. Theoret. Comput. Sci. **151**, 3–27 (1995)
2. Baelde, D.: Least and greatest fixed points in linear logic. ACM Trans. Comput. Log. (TOCL) **13**, 2 (2012)
3. Bezhanishvili, N., Hodkinson, I.: Sahlqvist theorem for modal fixed point logic. Theoret. Comput. Sci. **424**, 1–19 (2012)
4. Bilkova, M., Majer, O., Pelis, M.: Epistemic logics for sceptical agents. J. Log. Comput. **26**, 1815–1841 (2016)
5. Britz, C.: Correspondence theory in many-valued modal logics. Master's thesis, University of Johannesburg, South Africa (2016)
6. Conradie, W., Craig, A.: Canonicity results for mu-calculi: an algorithmic approach. J. Log. Comput. **27**, 705–748 (2017)
7. Conradie, W., Craig, A., Palmigiano, A., Zhao, Z.: Constructive canonicity for lattice-based fixed point logics. ArXiv preprint arXiv:1603.06547
8. Conradie, W., Fomatati, Y., Palmigiano, A., Sourabh, S.: Algorithmic correspondence for intuitionistic modal mu-calculus. Theoret. Comput. Sci. **564**, 30–62 (2015)
9. Conradie, W., Frittella, S., Palmigiano, A., Piazzai, M., Tzimoulis, A., Wijnberg, N.M.: Categories: how I learned to stop worrying and love two sorts. In: Väänänen, J., Hirvonen, Å., de Queiroz, R. (eds.) WoLLIC 2016. LNCS, vol. 9803, pp. 145–164. Springer, Heidelberg (2016). doi:10.1007/978-3-662-52921-8_10
10. Conradie, W., Frittella, S., Palmigiano, A., Piazzai, M., Tzimoulis, A., Wijnberg, N.M.: Towards an epistemic-logical theory of categorization (2017, submitted). https://sites.google.com/site/willemconradie/files/EpLogicalViewCategorization.pdf
11. Conradie, W., Ghilardi, S., Palmigiano, A.: Unified correspondence. In: Baltag, A., Smets, S. (eds.) Johan van Benthem on Logic and Information Dynamics, Outstanding Contributions to Logic, vol. 5, pp. 933–975. Springer International Publishing, Cham (2014)
12. Conradie W., Palmigiano, A.: Constructive canonicity of inductive inequalities (submitted). ArXiv preprint arXiv:1603.08341
13. Conradie, W., Palmigiano, A.: Algorithmic correspondence and canonicity for distributive modal logic. Ann. Pure Appl. Log. **163**, 338–376 (2012)
14. Conradie, W., Palmigiano, A.: Algorithmic correspondence and canonicity for non-distributive logics (submitted). ArXiv preprint arXiv:1603.08515
15. Conradie, W., Palmigiano, A., Sourabh, S.: Algebraic modal correspondence: Sahlqvist and beyond. J. Log. Algebr. Methods Program. (2016). doi:10.1016/j.jlamp.2016.10.006
16. Conradie, W., Palmigiano, A., Sourabh, S., Zhao, Z.: Canonicity and relativized canon-icity via pseudo-correspondence: an application of ALBA (submitted). ArXiv preprint arXiv:1511.04271
17. Conradie, W., Robinson, C.: On Sahlqvist theory for hybrid logic. J. Log. Comput. **27**, 867–900 (2017)
18. Dunn, J.M., Gehrke, M., Palmigiano, A.: Canonical extensions and relational completeness of some substructural logics. J. Symb. Log. **70**, 713–740 (2005)
19. Frittella, S., Palmigiano, A., Santocanale, L.: Dual characterizations for finite lattices via correspondence theory for monotone modal logic. J. Log. Comput. **27**, 639–678 (2017)
20. Gavazzo, F.: Investigations into linear logic with fixed-point operators. ILLC MoL thesis (2015)
21. Gehrke, M., Harding, J.: Bounded lattice expansions. J. Algebra **238**, 345–371 (2001)
22. Ghilardi, S., Meloni, G.: Constructive canonicity in non-classical logics. Ann. Pure Appl. Log. **86**, 1–32 (1997)

23. Greco, G., Ma, M., Palmigiano, A., Tzimoulis, A., Zhao, Z.: Unified correspondence as a proof-theoretic tool. J. Log. Comput. (2016). doi:10.1093/logcom/exw022
24. Jónsson, B.: On the canonicity of Sahlqvist identities. Stud. Logica **53**, 473–491 (1994)
25. Ma, M., Zhao, Z.: Unified correspondence and proof theory for strict implication. J. Log. Comput. **27**, 921–960 (2017)
26. Palmigiano, A., Sourabh, S., Zhao, Z.: Jónsson-style canonicity for ALBA-inequalities. J. Log. Comput. **27**, 817–865 (2017)
27. Palmigiano, A., Sourabh, S., Zhao, Z.: Sahlqvist theory for impossible worlds. J. Log. Comput. **27**, 775–816 (2017)
28. Sahlqvist, H.: Completeness and correspondence in the first and second order semantics for modal logic. Stud. Log. Found. Math. **82**, 110–143 (1975)
29. Sambin, G., Vaccaro, V.: A new proof of Sahlqvist's theorem on modal definability and completeness. J. Symb. Log. **54**(3), 992–999 (1989)

Non-commutative Logic for Compositional Distributional Semantics

Karin Cvetko-Vah[1]([⊠]), Mehrnoosh Sadrzadeh[2], Dimitri Kartsaklis[2], and Benjamin Blundell[3]

[1] Faculty of Mathematics and Physics, University of Ljubljana, Ljubljana, Slovenia
`karin.cvetko@fmf.uni-lj.si`
[2] School of Electronic Engineering and Computer Science,
Queen Mary University of London, London, UK
`{m.sadrzadeh,d.kartsaklis}@qmul.ac.uk`
[3] ITS Research, Queen Mary University of London, London, UK
`b.blundell@qmul.ac.uk`

Abstract. Distributional models of natural language use vectors to provide a contextual foundation for meaning representation. These models rely on large quantities of real data, such as corpora of documents, and have found applications in natural language tasks, such as word similarity, disambiguation, indexing, and search. Compositional distributional models extend the distributional ones from words to phrases and sentences. Logical operators are usually treated as noise by these models and no systematic treatment is provided so far. In this paper, we show how skew lattices and their encoding in upper triangular matrices provide a logical foundation for compositional distributional models. In this setting, one can model commutative as well as non-commutative logical operations of conjunction and disjunction. We provide theoretical foundations, a case study, and experimental results for an entailment task on real data.

Keywords: Non-commutative logic · Compositional semantics · Distributional semantics · Vector semantics · Skew lattices · Meaning · Entailment

1 Introduction

Distributional semantics is a model of natural language that works with vector representations of words embedded in a vector space of features. The vector representations are formalisations of insights of Firth and Harris [8,10] that words that occur in similar contexts have similar meanings. These models are

K. Cvetko-Vah acknowledges the financial support from the Slovenian Research Agency (research core funding No. P1-0222). M. Sadrzadeh, D. Kartsaklis and B. Blundell acknowledge financial support from AFOSR International Scientific Collaboration Grant FA9550-14-1-0079.

J. Kennedy and R.J.G.B. de Queiroz (Eds.): WoLLIC 2017, LNCS 10388, pp. 110–124, 2017.
DOI: 10.1007/978-3-662-55386-2_8

contrasted with traditional approaches to formal semantics where words are treated as indices in a dictionary or vocabulary list, a string of letters, or the set of their denotations. The vector representations of words are built from co-occurrence matrices [23], the columns of which are features of a text, the rows of which are words, and the entries of which contain degrees of co-occurrences of the two. Vector space models provide different ways of modelling similarity relations between words [21,24,25], a concept that has found applications in areas such as question answering, summarisation and classification.

In [4] a mathematical framework for a unification of the distributional method and a compositional theory of grammatical types was introduced. This unification is important because the insights on which the distributional method is built mostly make sense for words. In order to obtain vector representations for phrases and sentences and to reason about their degrees of similarity, one needs to extend the distributional method from words to phrases and sentences. The unified model combines the formal grammar models of language [3,15,16] with the distributional theories of meaning. The result is a vector space model where the meaning of a sentence is represented by a vector computed from the vectors corresponding to the meanings of the words therein and the grammatical structure of the sentence. In [4], the two approaches are connected by the use of compact closed categories, which admit purely diagrammatic computations. These computations are related to the work by Abramsky and Coecke on the flow of information in the context of quantum information protocols [1].

Several questions were posed in [4], including extending the fragment covered there by adding their natural language coordination words to it and find proper operators corresponding to the logical connectives "and", "or", "but", "unless". This question turned out to be a challenge, since the category does not have separate products and coproducts, neither do the usual vector product and sum are fully distributive. Hence they will not correspond to logical conjunction and disjunction and their variants.

In the present paper, we connect the vector model of words and sentences to skew lattice theory. Skew lattices present a non-commutative generalization of lattices, they were introduced in 1949 by Jordan [11], the author of the quantum field theory. The idea to study algebras of non-commutative idempotents arised from the realization that a pair of observables A, B corresponding to properties studied in quantum mechanics is compatible (i.e. they can be simultaneously observed) if and only if any projection corresponding to A commutes with any projection corresponding to B. The theory of skew lattices was later developed mostly by Leech, cf. [17,20]. The idea that the conjunction and disjunction in the natural language are sometimes non-commutative is not new. For instance, in [9] the authors argue that: "*A candle was burning on the table and the room was brightly lit* is not the same as *The room was brightly lit and the candle was burning on the table*" ([9], p. 76–77).

This paper is structured as follows. In Sect. 2, we briefly recall the case of Boolean vectors when the connectives "and", "or" are commutative. In Sect. 3, we generalize this setting to the non-Boolean non-commutative case.

We represent our data by vectors, or more generally by matrices, so that they form a skew lattice. Then we use the skew lattice operations to represent the non-commutative connectives "and", "or". We present a case study from real data and show how entailment can be used to distinguish the non-commutative conjunction from the commutative one. We use this case study as a pilot and perform an experiment on real data. The experiment is an entailment task on a dataset of verb-object conjuncts with the two types of conjunctions. The results show that the non-commutative conjunction operator of skew lattices recognises the non-commutative conjunctive entailments better than the commutative ones.

2 Commutative Connectives *and, or*

We assume that pieces of information are encoded as vectors. Let n be a natural number and consider the vector space $\mathbb{Z}_2^n$. The connectives *NOT, AND, OR* are encoded by the following operations:

NOT $\neg(x_1, x_2, \ldots x_n) = (1 - x_1, 1 - x_2, \ldots, 1 - x_n)$,
AND : $(x_1, x_2, \ldots, x_n) \wedge (y_1, y_2, \ldots, y_n) = (x_1 y_1, x_2 y_2, \ldots, x_n y_n)$,
OR : $(x_1, x_2, \ldots, x_n) \wedge (y_1, y_2, \ldots, y_n) = (x_1 \circ y_1, x_2 \circ y_2, \ldots, x_n \circ y_n)$,

where $u \circ v = u + v - uv$.

It is an easy exercise to verify that $(\mathbb{Z}_2^n; \wedge, \vee)$ is a distributive lattice. In fact, it is a bounded distributive lattice with bottom $b = (0, 0, \ldots, 0)$ and top $t = (1, 1, \ldots, 1)$ in which every element is complemented ($\neg x$ being the complement of x). Hence $(\mathbb{Z}_2^n; \wedge, \vee, b, t)$ is a Boolean algebra.

Example 1. Assume that we measure the properties "love" and "see", like in John loves Mary *or* John sees Mary. *The information about the pair (John, Mary) is encoded by a 4-dimensional vector (i, j, k, l), where each property is encoded by a two-dimensional vector. More precisely, $(i, j) = (1, 0)$ if John loves Mary and $(i, j) = (0, 1)$ if he hates her (which we consider to be the negation of loving her), and $(k, l) = (1, 0)$ if John can see Mary and $(k, l) = (0, 1)$ otherwise. Consider the statements:*

s_1 : John loves Mary and sees her. s_2 : John doesn't love Mary and he sees her.

What is the conjunction of statements s_1, s_2 with respect to the above definition?

$$s_1 \text{ AND } s_2 : (1, 0, 1, 0) \wedge (0, 1, 1, 0) = (0, 0, 1, 0).$$

How are we to interpret this result? Denoting $\perp = (0, 0)$ and likewise $\perp = (\perp, j, k) = (i, j, \perp)$ which is interpreted by false, we obtain that s_1 AND $s_2 = \perp$ which sounds reasonable. Similarly, the disjunction of s_1, s_2 is obtained by:

$$s_1 \text{ OR } s_2 : (1, 0, 1, 0) \vee (0, 1, 1, 0) = (1, 1, 1, 0).$$

Denoting $\top = (1, 1)$ and $(\top, j, k) = (j, k), (i, j, \top) = (i, j)$ we obtain $s_1 \vee s_2 = (1, 0)$ which corresponds to the statement John sees Mary.

3   Non-commutative Connectives *and, or*

There are examples in everyday life where the meaning of the connective *and* is essentially non-commutative. Consider the following sentences:

> Sentence 1: *Alice found gold and ran away.*
> Sentence 2: *Alice ran away and found gold.*

Although the above two sentences are both composed from the same pair of simple sentences, i.e. *Alice found gold.* and *Alice ran away.* which are connected by the connective *and*, their meaning is not the same. In the case of Sentence 1, Alice most probably found gold and *then* ran away in order not to get caught or get the gold stolen from her, while in the case of Sentence 2, she first ran away from something that we are not aware of and for the reason that we don't know, and then while running away she ran into gold and found it. In the above example we saw that the connective AND as used in the natural language can have a time component implicit in it. The first action might be implicitly assumed to come before the second one, and that can effect the meaning of the sentence. There are cases in the natural language where given two actions that are connected by an AND it is natural to assume that the second one has a deeper impact. For instance, consider the following sentences:

> Sentence 1: *I drank the wine and filled the glass.*
> Sentence 2: *I filled the glass and drank the wine.*

While I drank a glass of wine in both cases, when the action of sentence s_1 was completed I still had a full glass, while I ended up with an empty glass when the action of sentence s_2 was completed. As we shall see below there are other instances when it is natural to glue together pieces of information by a non-commutative connective AND.

There are also situations where a non-commutative version of the connective OR is used in the natural language. To see this, consider the connective *unless* in the sentence: *Buy a blue car unless you can get a red car.* If we want to follow the above instruction, we are going to end up with either a red or a blue car, however if both colours are available, we are going to choose the red one. However, if the instruction was: *Buy a red car unless you can get a blue car*, in the case that cars of both colours were available, we would choose a blue car.

What is the right mathematical frame to encapture the above situations where the connectives AND, OR can be non-commutative? We adopt the definition below from [17].

Definition 1. *A skew lattice is an algebra* $(S; \wedge, \vee)$ *satisfying the following:*

- associativity *of* $\wedge$: $(x \wedge y) \wedge z = x \wedge (y \wedge z)$,
- associativity *of* $\vee$: $(x \vee y) \vee z = x \vee (y \vee z)$,
- idempotency *of* $\wedge$: $x \wedge x = x$,
- idempotency *of* $\vee$: $x \vee x = x$,
- absorptions: $x \vee (x \wedge y) = x = x \wedge (x \vee y)$ *and* $(x \wedge y) \vee y = y = (x \vee y) \wedge y$.

A skew lattice is called *strongly distributive* if it satisfies the identities

$$x \wedge (y \vee z) = (x \wedge y) \vee (x \wedge z) \text{ and } (x \vee y) \wedge z = (x \wedge z) \vee (y \wedge z).$$

The following result is due to Leech [18, 19].

Proposition 1. *Let $\mathcal{P}(A, B)$ be the set of all partial functions from A to B, for A and B non-empty sets. Given partial functions $f, g \in \mathcal{P}(A, B)$ we set:*

Restriction: $f \wedge g = g|_{\text{dom}f \cap \text{dom}g}$,   Override: $f \vee g = f \cup g|_{\text{dom}g \setminus \text{dom}f}$.

Then $(\mathcal{P}(A, B); \wedge, \vee)$ is a strongly distributive skew lattice.

Notice that given $f, g \in \mathcal{P}(A, B)$ as above then if $\text{dom}f = \text{dom}$ then $f \wedge g = g$ and $f \vee g = f$. The algebras with operations *restriction* and *override* were first studied in [2], while their connection to skew lattices was established in [7].

One more example of non-commutative conjunction, this time between adverbs: *to paint the fence white and (then) brown* vs. *to paint the fence brown and (then) white*. In the first case we end up with a brown fence, while in the second case the fence is white after we finish painting it. Again, the right one won, just like in our definition of the operation *restriction* above.

Although the connective *unless* can be seen as a non-commutative OR it is not consistent with the definition of the *override* operation above. That is because *unless* prefers the second statement, while *override* prefers the first. An interpretation of non-commutative OR that is consistent with our setting was established in [6] where the *override* operation was interpreted as $x \vee y = q(x, x, y)$, where $q(x, y, z)$ is the *Church algebra* operation satisfying the fundamental properties of the *if-then-else* connective: $q(1, x, y) = x$ and $q(0, x, y) = y$. Thus the *override* $q(x, x, y)$ can be interpreted as *if x then x else y*. In everyday language we may encounter an instance of *override* in a situation like: *Buy a blue car or else buy a red car*. Our interpretation of this sentence is: *Buy a blue car if there is one; if not, then buy a red one*. Hence, the first option is preferred.

Example 2. Assume that we observe our information by measuring the colour (or the wavelength) of an object and its size. We assign to each measurement a pair (x, y), where x is either 3 (blue), 2 (red), 1 (green) or 0 (not seen); and y is a positive real number (in meters, for example) or 0 (not seen). Applying AND (restriction), OR (override) and interpreting 0 (not seen) as not-defined we get:

$$(3, 2) \wedge (0, 1) = (0, 1), (3, 2) \vee (0, 1) = (3, 2)$$
$$(0, 1) \wedge (3, 2) = (0, 2), (0, 1) \vee (3, 2) = (3, 1)$$

The first conjunction corresponds to (blue, 2 m) AND (not seen, 1 m) = (not seen, 1 m), others are similarly unfolded. Notice that "blue and two meters high" is denoted by (blue, 2 m), while "blue and 2 m high, and not seen and 1 m high" is denoted by (blue, 2 m) AND (not seen, 1 m). The connective AND is used for the connection between vectors, i.e. pieces of full information (although some of it might be partial in that it may contain "not seen").

Another way to encode the information of Example 2 is by use of upper-triangular matrices (over the reals, for example) with 0–1 diagonals and possibly non-zero elements in the last column (0 elsewhere). Each 1 on the diagonal denotes that the property was observed (0: not observed), the element that lies in the far right column and in the row of the particular 1 denotes the value of the observed property. When we wish to encode more information we can also allow non-zero elements in the first row of the matrices.

Given the matrix ring $M_n(\mathbb{R})$ of all $n \times n$ real matrices, a subset $S \subseteq M_n(\mathbb{R})$ is called a *band* if it is closed under multiplication and $A^2 = A$ holds for all $A \in S$. Let S be such a band and let $A, B \in S$. We denote:

$$A \circ B = A + B - AB$$
$$A \nabla B = (A \circ B)^2 = A + B + BA - ABA - BAB.$$

Note that if S is closed under $\circ$ then $A \circ B = A \nabla B$ holds for all $A, B \in S$.

Proposition 2 [5]. *Let S consist of all $(k+2) \times (k+2)$ real matrices of following form, then $(S; \cdot, \nabla)$ is a skew lattice.*

$$\begin{bmatrix} 0 & a_1 & \dots & a_k & c \\ 0 & e_1 & \dots & 0 & b_1 \\ \vdots & \vdots & \vdots & \vdots & \vdots \\ 0 & 0 & \dots & e_k & b_k \\ 0 & 0 & \dots & 0 & 0 \end{bmatrix} \quad \text{where:} \quad \begin{array}{l} \text{(i) each } e_i = 0 \text{ or } 1, \\ \text{(ii) } b_i = 0 \text{ for all } i \text{ s.t. } e_i = 0, \\ \text{(iii) } a_i = 0 \text{ for all } i \text{ s.t. } e_i = 0, \\ \text{(iv) } c = a_1 b_1 + \dots + a_k b_k. \end{array}$$

Corollary 1. *Let S consist of all $(k+1) \times (k+1)$ real matrices of the following form, then $(S; \cdot, \circ)$ is a skew lattice.*

$$\begin{bmatrix} e_1 & 0 & \dots & 0 & b_1 \\ 0 & e_2 & \dots & 0 & b_2 \\ \vdots & \vdots & \vdots & \vdots & \vdots \\ 0 & 0 & \dots & e_k & b_k \\ 0 & 0 & \dots & 0 & 0 \end{bmatrix} \quad \text{where:} \quad \begin{array}{l} \text{(i) each } e_i = 0 \text{ or } 1, \\ \text{(ii) } b_i = 0 \text{ for all } i \text{ s.t. } e_i = 0, \end{array}$$

(blue, $2m$) AND (not seen, $1m$)   (blue, $2m$) OR (not seen, $1m$)

$$u = \begin{bmatrix} 1 & 0 & 3 \\ 0 & 1 & 2 \\ 0 & 0 & 0 \end{bmatrix} \quad \begin{bmatrix} 1 & 0 & 3 \\ 0 & 1 & 2 \\ 0 & 0 & 0 \end{bmatrix} \cdot \begin{bmatrix} 0 & 0 & 0 \\ 0 & 1 & 1 \\ 0 & 0 & 0 \end{bmatrix} = \begin{bmatrix} 0 & 0 & 0 \\ 0 & 1 & 1 \\ 0 & 0 & 0 \end{bmatrix} \quad \begin{bmatrix} 1 & 0 & 3 \\ 0 & 1 & 2 \\ 0 & 0 & 0 \end{bmatrix} \circ \begin{bmatrix} 0 & 0 & 0 \\ 0 & 1 & 1 \\ 0 & 0 & 0 \end{bmatrix} = \begin{bmatrix} 1 & 0 & 3 \\ 0 & 1 & 2 \\ 0 & 0 & 0 \end{bmatrix}$$

(not seen, $1m$) AND (blue, $2m$)   (not seen, $1m$) OR (blue, $2m$)

$$v = \begin{bmatrix} 0 & 0 & 0 \\ 0 & 1 & 1 \\ 0 & 0 & 0 \end{bmatrix} \quad \begin{bmatrix} 0 & 0 & 0 \\ 0 & 1 & 1 \\ 0 & 0 & 0 \end{bmatrix} \cdot \begin{bmatrix} 1 & 0 & 3 \\ 0 & 1 & 2 \\ 0 & 0 & 0 \end{bmatrix} = \begin{bmatrix} 0 & 0 & 0 \\ 0 & 1 & 2 \\ 0 & 0 & 0 \end{bmatrix} \quad \begin{bmatrix} 0 & 0 & 0 \\ 0 & 1 & 1 \\ 0 & 0 & 0 \end{bmatrix} \circ \begin{bmatrix} 1 & 0 & 3 \\ 0 & 1 & 2 \\ 0 & 0 & 0 \end{bmatrix} = \begin{bmatrix} 1 & 0 & 3 \\ 0 & 1 & 1 \\ 0 & 0 & 0 \end{bmatrix}$$

Fig. 1. Examples of commutative versus non-commutative conjunction and disjunction

Example 3. We interpret the data from Example 2 in upper triangular 3×3 matrices from Corollary 1. The first row corresponds to colour, the second one to height, and the third is zero (always, we need it so that the usual multiplication of matrices works). We assign to the vectors (blue, 2 m) and (not seen, 1 m) matrices from which we then obtain the correspondence between vectors and matrices. These are shown in Fig. 1.

4   Boolean and Skew Semantics for Natural Language

For demonstration purposes, consider a simple fragment of English generated by the context free grammar of Fig. 2.

S	→	NP VP	NP →	John, Mary, $\cdots$
VP →		V NP	Adj →	lucky, tall, red, $\cdots$
S	→	S and/or S	Adv →	deeply, slowly, quickly, $\cdots$
NP →		Adj NP	VP →	sneeze, sleep, $\cdots$
VP →		VP Adv	V →	love, kiss, $\cdots$

Fig. 2. An exemplary context free grammar for a simple fragment of English

A model for the language generated by this grammar minus the logical rule S → S and/or S, is a pair $(U, [\![\,]\!])$, where U is universal reference set and $[\![\,]\!]$ is an interpretation function defined by induction as follows. For terminals we have:

- The interpretation of a terminal $y \in \{np, adj, adv, vp\}$ generated by either NP → np, Adj → adj, VP → vp, Adv → adv, is $[\![y]\!] \subseteq U$. That is, noun phrases, adjectives, verb phrases, and adverbs are interpreted as unary predicates over the reference set.
- The interpretation of a terminal y generated by V → y is $[\![y]\!] \subseteq U \times U$; verbs are interpreted as binary predicates over the reference set.

For non-terminals, for all rules except for S → S and/or S, we have:

$$[\![\text{V NP}]\!] = [\![v]\!]([\![np]\!]) \quad [\![\text{NP VP}]\!] = [\![vp]\!]([\![np]\!])$$
$$[\![\text{Adj NP}]\!] = [\![adj]\!]([\![np]\!]) \quad [\![\text{VP Adv}]\!] = [\![adv]\!]([\![vp]\!])$$

Here, for $R \subseteq U \times U$ and $A \subseteq U$, by $R(A)$ we mean the forward image of R on A, that is $R(A) = \{y \mid (x, y) \in R, \text{for } x \in A\}$. To keep the notation unified, for R a unary relation $R \subseteq U$, we use the same notation and define $R(A) = \{y \mid y \in R, \text{for } x \in A\}$, i.e. $R \cap A$.

In order to interpret the logical rule S → S and/or S, we have to move to a lattice over U. If our connectives are Boolean, this lattice is $\mathcal{P}(U)$ and we have:

$$[\![\text{S and/or S}]\!] = [\![\text{S}]\!] \wedge / \vee [\![\text{S}]\!]$$

for $\wedge/\vee$ the Boolean lattice operations. In this case, we are working in a Boolean model $(\mathcal{P}(U), [\![\,]\!])$. For non-Boolean non-commutative logical operations, we work with a skew lattice over U.

Definition 2. *A skew lattice semantics for the language generated by the grammar of Fig. 2 is $(S(U); \wedge, \vee; [\![\ ]\!])$, where U is a universal reference set, $S(U)$ consists of the real matrices defined in Proposition 2 and $[\![\ ]\!]$ is an interpretation function defined by induction as follows. To terminals we assign:*

- *to each np, vp a skew matrix $[\![np]\!] := u_{np}, [\![vp]\!] := u_{vp}$ satisfying $e_1 = \cdots = e_k = 1$ and all $a_i = b_i \neq 0$,*
- *to each v a diagonal matrix $[\![v]\!] := u_v$ of the form (1) with at least one 1 on the diagonal (and all other entries 0),*
- *to each adj a skew matrix $[\![adj]\!] := u_{adj}$ of the form $e_1 = \cdots = e_k = 1$, $a_1 = \cdots = a_k \neq 0$ and $b_1 = \cdots = b_k = 0$,*
- *to each adv a skew matrix $[\![adv]\!] := u_{adv}$ of the form $e_1 = \cdots = e_k = 1$, $a_1 = \cdots = a_k = 0$ and $b_1 = \cdots = b_k \neq 0$.*

To non-terminals we assign:

- *to each $x \to y\ z$, the skew matrix $[\![x]\!] = u_x := [\![y]\!] \times [\![z]\!]$,*
- *to $S \to S$ and/or S, the skew matrix $[\![S]\!] := [\![S]\!] \wedge / \vee [\![S]\!]$, for $\wedge, \vee$.*

Note that each e_i-position in the verb item corresponds to a particular verb. We call an index i a *defining index* for a skew matrix A if $a_{ii} \neq 0$. The product $A \cdot B$ can only be nonzero if A and B have at least one common defining index.

Consider the set of terminals "John, loves, sees, Mary, sleeps, lucky, deeply". We encode our data by the 5×5 matrices, presented in Fig. 3. The $*$ element in u_{Mary} equals $3m^2$ and the $*$ element in u_{John} equals $3j^2$. All our matrices are idempotent, i.e. they satisfy $A^2 = A$. So we have $[\![\text{Mary}]\!] = u_{\text{Mary}}$, $[\![\text{John}]\!] = u_{\text{John}}$, $[\![\text{loves}]\!] = u_{\text{loves}}$, and $[\![\text{John loves Mary}]\!] := u_{\text{John}} \times u_{\text{loves}} \times u_{\text{Mary}}$.

5 A Case Study from Real Data

In this section, we use real data to build our matrices and detail the computations for one set of the example sentences of the paper. We perform a case study with 2 verbs, two nouns, and one adjective. We work with 5×5 matrices, where one dimension of each verb matrix is reserved such that the two verbs of the example have a common defining index.

Consider the lemmatised versions of the sentences "filled glass and drank wine" and "drank wine and filled glass" of Sect. 3. We fill the matrices of words of these sentences with real data and compute their conjunction, in the two presented orders. After the first conjunct, the glass will be full and after the second conjunct, the glass will be empty. We verify if this fact indeed follows from real data, by computing each entailment and observing which one of the conjuncts entails "full glass" with a larger degree.

The terminals of our sentences are "filled, drank, glass, wine, full, empty". We build skew matrices for these. The a_i, b_i entries for nouns and adjective matrices are obtained from the PPMI-normalised version (see Appendix for the PPMI formula and its explanation) of the degree of co-occurrence of each word with the two features of "full" and "empty". The entries $a_2 = b_2$ correspond to

$$u_{\text{Mary}} = \begin{bmatrix} 0 & m & m & m & * \\ 0 & 1 & 0 & 0 & m \\ 0 & 0 & 1 & 0 & m \\ 0 & 0 & 0 & 1 & m \\ 0 & 0 & 0 & 0 & 0 \end{bmatrix}, u_{\text{John}} = \begin{bmatrix} 0 & j & j & j & * \\ 0 & 1 & 0 & 0 & j \\ 0 & 0 & 1 & 0 & j \\ 0 & 0 & 0 & 1 & j \\ 0 & 0 & 0 & 0 & 0 \end{bmatrix}, u_{\text{loves}} = \begin{bmatrix} 0 & 0 & 0 & 0 & 0 \\ 0 & 1 & 0 & 0 & 0 \\ 0 & 0 & 0 & 0 & 0 \\ 0 & 0 & 0 & 0 & 0 \\ 0 & 0 & 0 & 0 & 0 \end{bmatrix},$$

$$u_{\text{sleeps}} = \begin{bmatrix} 0 & 0 & 0 & 0 & 0 \\ 0 & 0 & 0 & 0 & 0 \\ 0 & 0 & 0 & 0 & 0 \\ 0 & 0 & 0 & 1 & 0 \\ 0 & 0 & 0 & 0 & 0 \end{bmatrix}, u_{\text{lucky}} = \begin{bmatrix} 0 & l & l & l & 0 \\ 0 & 1 & 0 & 0 & 0 \\ 0 & 0 & 1 & 0 & 0 \\ 0 & 0 & 0 & 1 & 0 \\ 0 & 0 & 0 & 0 & 0 \end{bmatrix}, u_{\text{deeply}} = \begin{bmatrix} 0 & 0 & 0 & 0 & 0 \\ 0 & 1 & 0 & 0 & d \\ 0 & 0 & 1 & 0 & d \\ 0 & 0 & 0 & 1 & d \\ 0 & 0 & 0 & 0 & 0 \end{bmatrix},$$

$$[\![\text{Mary loves John}]\!] = \begin{bmatrix} 0 & m & 0 & 0 & mj \\ 0 & 1 & 0 & 0 & j \\ 0 & 0 & 0 & 0 & 0 \\ 0 & 0 & 0 & 0 & 0 \\ 0 & 0 & 0 & 0 & 0 \end{bmatrix}, [\![\text{Mary sleeps}]\!] = \begin{bmatrix} 0 & 0 & 0 & m & 0 \\ 0 & 0 & 0 & 0 & 0 \\ 0 & 0 & 0 & 0 & 0 \\ 0 & 0 & 0 & 1 & 0 \\ 0 & 0 & 0 & 0 & 0 \end{bmatrix},$$

$$[\![\text{lucky Mary}]\!] = \begin{bmatrix} 0 & l & l & l & 3lm \\ 0 & 1 & 0 & 0 & m \\ 0 & 0 & 1 & 0 & m \\ 0 & 0 & 0 & 1 & m \\ 0 & 0 & 0 & 0 & 0 \end{bmatrix}, \quad [\![\text{Mary sleeps deeply}]\!] = \begin{bmatrix} 0 & 0 & 0 & m & 0 \\ 0 & 0 & 0 & 0 & 0 \\ 0 & 0 & 0 & 0 & 0 \\ 0 & 0 & 0 & 1 & d \\ 0 & 0 & 0 & 0 & 0 \end{bmatrix}.$$

Fig. 3. Example skew matrices for words

the feature "empty" and the entries $a_3 = b_3$ to the feature "full". Dimension 4 records the common defining index. We copy the information corresponding to the feature "full" in this dimension. The reason we are copying this dimension and not dimension 2 is because we are verifying the degree of entailment with the phrase "full glass" (and not with "empty glass"), thus entries $a_4 = b_4$ should be the same as $a_3 = b_3$. This is important for matrices of the verbs, where cell c_{33} records the common defining index of the "drank" and "filled" matrix.

Note that, in a Boolean setting the two properties of being full and being empty are opposites of each other: if one of them is 1, the other will be 0. In real scenarios, however, this is not necessarily the case. For instance, i the data is presented in Fig. 4, in the matrix of "glass", the PPMI-normalised versions

$$\text{glass} = \begin{bmatrix} 0 & 10.2 & 11 & 11 & 346 \\ 0 & 1 & 0 & 0 & 10.2 \\ 0 & 0 & 1 & 0 & 11 \\ 0 & 0 & 0 & 1 & 11 \\ 0 & 0 & 0 & 0 & 0 \end{bmatrix} \text{wine} = \begin{bmatrix} 0 & 8.7 & 10.9 & 10.9 & 313.3 \\ 0 & 1 & 0 & 0 & 8.7 \\ 0 & 0 & 1 & 0 & 10.9 \\ 0 & 0 & 0 & 1 & 10.9 \\ 0 & 0 & 0 & 0 & 0 \end{bmatrix} \text{full glass} = \begin{bmatrix} 0 & 1 & 1 & 1 & 32.2 \\ 0 & 1 & 0 & 0 & 10.2 \\ 0 & 0 & 1 & 0 & 11 \\ 0 & 0 & 0 & 1 & 11 \\ 0 & 0 & 0 & 0 & 0 \end{bmatrix}$$

$$\text{drank} = \begin{bmatrix} 0 & 0 & 0 & 0 & 0 \\ 0 & 1 & 0 & 0 & 0 \\ 0 & 0 & 1 & 0 & 0 \\ 0 & 0 & 0 & 0 & 0 \\ 0 & 0 & 0 & 0 & 0 \end{bmatrix} \text{filled} = \begin{bmatrix} 0 & 0 & 0 & 0 & 0 \\ 0 & 0 & 0 & 0 & 0 \\ 0 & 0 & 1 & 0 & 0 \\ 0 & 0 & 0 & 1 & 0 \\ 0 & 0 & 0 & 0 & 0 \end{bmatrix} \text{full} = \begin{bmatrix} 0 & 1 & 1 & 1 & 0 \\ 0 & 1 & 0 & 0 & 0 \\ 0 & 0 & 1 & 0 & 0 \\ 0 & 0 & 0 & 1 & 0 \\ 0 & 0 & 0 & 0 & 0 \end{bmatrix}.$$

Fig. 4. A set of word matrices derived from data

of the number of times "glass" occurred 5 words close to features "full" and "empty" are 11 and 10.2, respectively. This is because a glass can be empty and it can be full in different contexts.

We compose these word matrices an obtain matrices for phrases, we then form their two possible conjunctions, resulting in matrices of Fig. 5. We see that "drank wine and filled glass" is closer (although very slightly so) to "full glass" than "filled glass and drank wine". This is because the entry b_3 of the first conjunct is 11, this is closer to the same entry in "full glass" that is, 11, than the b_3 of the second conjunct, which is 10.9. All the other entries are of equal distance of the entries of "full glass". The difference is small due to the fact that we took the features "full" and "empty" to be the same as the words "full" and "empty" and that these words are examples of words that do often occur in similar contexts, thus we get very similar numbers for them, i.e. 11 and 10.2. A more refined analysis on features that are not similar will reflect better on data.

$$\text{drank wine and filled glass} = \begin{bmatrix} 0 & 0 & 0 & 0 & 0 \\ 0 & 1 & 0 & 0 & 8.7 \\ 0 & 0 & 1 & 0 & 10.9 \\ 0 & 0 & 0 & 0 & 0 \\ 0 & 0 & 0 & 0 & 0 \end{bmatrix} \cdot \begin{bmatrix} 0 & 0 & 0 & 0 & 0 \\ 0 & 0 & 0 & 0 & 0 \\ 0 & 0 & 1 & 0 & 11 \\ 0 & 0 & 0 & 1 & 11 \\ 0 & 0 & 0 & 0 & 0 \end{bmatrix} = \begin{bmatrix} 0 & 0 & 0 & 0 & 0 \\ 0 & 0 & 0 & 0 & 0 \\ 0 & 0 & 1 & 0 & 11 \\ 0 & 0 & 0 & 0 & 0 \\ 0 & 0 & 0 & 0 & 0 \end{bmatrix}$$

$$\text{filled glass and drank wine} = \begin{bmatrix} 0 & 0 & 0 & 0 & 0 \\ 0 & 0 & 0 & 0 & 0 \\ 0 & 0 & 1 & 0 & 11 \\ 0 & 0 & 0 & 1 & 11 \\ 0 & 0 & 0 & 0 & 0 \end{bmatrix} \cdot \begin{bmatrix} 0 & 0 & 0 & 0 & 0 \\ 0 & 1 & 0 & 0 & 8.7 \\ 0 & 0 & 1 & 0 & 10.9 \\ 0 & 0 & 0 & 0 & 0 \\ 0 & 0 & 0 & 0 & 0 \end{bmatrix} = \begin{bmatrix} 0 & 0 & 0 & 0 & 0 \\ 0 & 0 & 0 & 0 & 0 \\ 0 & 0 & 1 & 0 & 10.9 \\ 0 & 0 & 0 & 0 & 0 \\ 0 & 0 & 0 & 0 & 0 \end{bmatrix}$$

Fig. 5. The set of conjunctive phrase matrices built from word matrices Fig. 4.

6   Large Scale Entailment Experiment

In previous work [12,13], we developed theory for and experimented with entailment in compositional distributional semantics. We built three entailment datasets from real data by using linguistic resources such as WordNet. These datasets consist of subject-verb and verb-object phrases and subject-verb-object sentences. We worked with different degrees of feature inclusion on the vectors and matrices of these phrases and sentences and measured the entailments thereof based on these degrees. In this section, we repeat the experiment of the previous section on a logical extension of the verb-object part of this dataset.

Our skew matrices are 300×300 and their entries are normalised using probabilistic versions of raw co-occurrences and their non-negative logarithms, a measure known as Positive Pointwise Mutual Information (PPMI); the formulae and explanations for these are given in the Appendix. The raw co-occurrence counts (before normalisation) were collected in the context of a 5-word window around the words. The verb-object dataset has 436 verb-object pairs, 218 of which stand in a positive entailment relationship with each other and 218 in a negative one. A snapshot of the positive entailments is presented in Fig. 6. The negative entries are the reverses of the positive ones. For an explanation on how these datasets are built, please see [12].

We extended the above dataset with commutative and non-commutative conjunctions in the following way. From each two entries of the dataset $vo_1 \vdash vo_2$ and $vo'_1 \vdash vo'_2$, we form two conjunctive entries, of the following forms

$$vo_1 \wedge vo'_1 \vdash vo_2 \wedge vo'_2 \quad \text{and} \quad vo'_1 \wedge vo_1 \vdash vo_2 \wedge vo'_2$$

We then compute a matrix for each of the vo's (i.e. for vo_1, vo_2, vo'_1, vo'_2). We compute their skew conjunctions, with the goal of verifying whether this non-commutative conjunction does perform better on recognising the conjunctive entailments of the first case above. In the first case, vo_1 entails vo_2 and vo'_1 entails vo'_2, hence $vo_1 \wedge vo'_1$ should entail $vo_2 \wedge vo'_2$. This entailment fails for second case, because the conjunction is non-commutative, that is vo'_1 does entail vo_2, similarly, vo_1 does not entail vo'_2, thus the entailment between their non-commutative conjunctions fails.

	Inclusion	APinc	BAPinc	SAPinc	SBAPinc
PPMI					
non-comm.	0.52	0.67	0.60	0.82	0.82
comm.	0.51	0.63	0.58	0.81	0.80
Probability					
non-comm.	0.58	0.65	0.61	0.80	0.79
comm.	0.56	0.60	0.57	0.79	0.77

$VO \vdash V'O'$
sign contract ⊢ write agreement
publish book ⊢ produce publication
sing song ⊢ perform music
reduce number ⊢ decrease amount
promote development ⊢ support event

Fig. 6. Examples from the verb-object entailment dataset and results of the non-commutative conjunction experiment with the PPMI and probability ratio on the 1st sample of dataset.

The results are evaluated by a binary classification of the existing entailment measures: APinc, BAPinc, SAPinc, SBAPinc. These are from the distributional literature on degrees of entailment between words and sentences, the formulae for computing them and explanations thereof are presented in the Appendix. As we are not working with Boolean models, we will have degrees of entailment and report Area Under Curve; this returns an evaluation of the entailment at every possible non-zero threshold. The baseline is labelled "Inclusion": the binary entailment between the features. Since our sample size is large (about 6000, obtained by recasting all of the conjuncts against each other), we performed the experiments on random subsets of the dataset, each with size 1000. The results of the first sample are in right hand table of Fig. 6. The results of the second sample are in the Appendix.

With all of the measures and in both normalisation schemes, the non-commutative conjunction comes out as a more appropriate operation for judging the non-commutative entailments. These results are preliminary, they are based on word-word matrices. A more appropriate empirical evaluation will be obtained by working on word-feature matrices, where the columns are not just word, but a set of words clustered together using feature induction techniques such as Single Value Decomposition (SVD).

7 Conclusions and Future Work

We reviewed the theory of skew lattices, which formalise a logic with non-commutative conjunction and disjunction. We motivated the existence of these operations in natural language. We presented an account of compositional distri-butional semantics where meanings of words, phrases, and sentences are vectors. We then showed how the data represented by skew lattices is encoded in matrices and developed a skew lattice semantics for compositional distributional models. Treating logical operators has been a challenge to these models and this paper provides a solution. We related our work to real data by first recasting one of the examples of the paper against co-occurrence matrices and then performed an experiment on a conjunctive entailment task. A similar experiment can be per-formed for the non-commutative disjunction, this is left to future work. On the theoretical side, with the current definitions, the matrices of verbs have to have 0–1 entries. We aim to generalise the setting, either by changing the matrices of nouns and adjectives, or the definition of the non-commutative conjunction, so we can populate all the matrices with real co-occurrence counts.

A Appendix

A.1 Normalisation Schemes

The raw co-occurrence counts are normalised using two measures:

- Probability Ratio

$$\frac{P(w,f)}{P(w)P(f)}$$

  where $P(w,c)$ is the probability that words w and feature f have occurred together, and $P(w)$ and $P(f)$ are probabilities of occurrences of w and f. This measure tells us how often w and f were observed together in comparison to how often they would have occurred were they independent.
- Positive Pointwise Mutual Information (PPMI)

$$\max(log(\frac{P(w,f)}{P(w)P(f)}),0)$$

  This is the positive version of the logarithm of probability ratio, where the negative logarithmic values are sent to 0.

A.2   Formulae for Computing Entailment

APinc is the average precision applied to feature inclusion. It measures a ranked version of feature inclusion on vectors $\overrightarrow{u}$ and $\overrightarrow{v}$, from highest to lowest:

$$APinc(u,v) = \frac{\sum_r \left[P(r) \cdot rel'(f_r) \right]}{|F(\overrightarrow{u})|} \tag{1}$$

In the above, f_r is the feature in $\overrightarrow{u}$, denoted by $F(\overrightarrow{u})$, with rank r; $P(r)$ is the precision at rank r, which measures how many of $\overrightarrow{v}$'s features are included at rank r in the features of $\overrightarrow{u}$, and $rel'(f_r)$ is a relevance measure reflecting how important f_r is in $\overrightarrow{v}$. It is computed as follows:

$$rel'(f) = \begin{cases} 1 - \frac{rank(f, F(\overrightarrow{v}))}{|F(\overrightarrow{v})|+1} & f \in F(\overrightarrow{v}) \\ 0 & o.w. \end{cases} \tag{2}$$

BAPinc balances *APinc* with the *LIN* degree of similarity between the vectors. *BAPinc* was developed in [14] after realising that *APinc* returns poor results when the vectors had a radically different number of non-zero features; the *LIN* measure was included to balance out the extra dimensions of the longer vector.

$$BAPinc(u,v) = \sqrt{LIN(u,v) \cdot APinc(u,v)} \tag{3}$$

LIN is a similarity measure between vectors and was defined in [22]. It can be replaced with any other similarity measure, such as the cosine measure.

SAPinc is a measure developed in [12], based on *BAPinc*, but for dense vectors. Whereas *APinc* and *BAPinc* were developed to compute the degree of entailment between word vectors, which are usually sparse since word vectors live in high dimensional spaces (e.g. 5000), *SAPinc* was developed to deal with phrase and sentence vectors. These are obtained by composing the vectors of words in lower dimension (e.g. 300), where the compositional operators accumulate the information and return dense results.

$$SAPinc(u,v) = \frac{\sum_r \left[P(r) \cdot rel'(f_r) \right]}{|\overrightarrow{u}|} \tag{4}$$

Here, $P(r)$ and $rel'(f_r)$ are defined differently, as shown below:

$$P(r) = \frac{\left| \{ f_r^{(u)} | f_r^{(u)} \le f_r^{(v)}, 0 < r \le |\overrightarrow{u}| \} \right|}{r} \tag{5}$$

$$rel'(f_r) = \begin{cases} 1 & f_r^{(u)} \le f_r^{(v)} \\ 0 & o.w. \end{cases} \tag{6}$$

For more explanations on these measures please see [12,13].

A.3   Experimental Results for a Second Sample

The results of the experiment of Sect. 6, with PPMI and probability ratio matrices on the second 1000 sample of the dataset are presented in Fig. 7.

Similar to the results presented in the paper, the non-commutative operation performs better on recognising the non-commutative conjunctive entailments.

	inclusion	APinc	BAPinc	SAPinc	SBAPinc
PPMI non-comm.	0.58	0.64	0.60	0.81	0.80
PPMI: comm.	0.57	0.61	0.58	0.79	0.78
Prob: non-comm.	0.57	0.63	0.60	0.82	0.80
Prob: comm.	0.56	0.60	0.58	0.80	0.78

Fig. 7. Results of the non-commutative conjunction experiment with the PPMI and probability ratio on the 2nd sample of dataset.

References

1. Abramsky, S., Coecke, B.: A categorical semantics of quantum protocols. In: Proceedings of the 19th Annual IEEE Symposium on Logic in Computer Science (LiCS 2004). IEEE Computer Science Press (2004). arXiv:quant-ph/0402130
2. Berendsen, J., Jansen, D.N., Schmaltz, J., Vaandrager, F.W.: The axiomatization of override and update. J. Appl. Log. **8**, 141–150 (2010)
3. Chomsky, N.: Three models for the description of language. IRE Trans. Inf. Theory **2**, 113–124 (1956)
4. Coecke, B., Sadrzadeh, M., Clark, S.: Mathematical foundations for distributed compositional model of meaning. Lambek Festschr. Linguist. Anal. **36**, 345–384 (2010)
5. Cvetko-Vah, K.: Skew lattices of matrices in rings. Algebra Univers. **53**, 471–479 (2005)
6. Cvetko-Vah, K., Salibra, A.: The connection of skew Boolean algebras and discriminator varieties to church algebras. Algebra Univers. **73**, 369–390 (2015)
7. Cvetko-Vah, K., Leech, J., Spinks, M.: Skew lattices and binary operations on functions. J. Appl. Log. **11**, 253–265 (2013)
8. Firth, J.: A synopsis of linguistic theory 1930–1955. In: Studies in Linguistic Analysis (1957)
9. Galatos, N., Jipsen, P., Kowalski, T., Ono, H.: Residuated Lattices: An Algebraic Glimpse at Substructural Logics. Studies in Logic and the Foundations of Mathematics, vol. 151. Elsevier, Amsterdam (2007)
10. Harris, Z.: Distributional structure. Word **10**, 146–162 (1954)
11. Jordan, P.: Über nichtkommutative verbände. Arch. Math. **2**, 56–59 (1949)
12. Kartsaklis, D., Sadrzadeh, M.: A compositional distributional inclusion hypothesis. In: Amblard, M., de Groote, P., Pogodalla, S., Retoré, C. (eds.) LACL 2016. LNCS, vol. 10054, pp. 116–133. Springer, Heidelberg (2016). doi:10.1007/978-3-662-53826-5_8
13. Kartsaklis, D., Sadrzadeh, M.: Distributional inclusion hypothesis for tensor-based composition. In: COLING 2016, 26th International Conference on Computational Linguistics, Proceedings of the Conference: Technical Papers, Osaka, Japan, 11–16 December 2016, pp. 2849–2860. ACL (2016)
14. Kotlerman, L., Dagan, I., Szpektor, I., Zhitomirsky-Geffet, M.: Directional distributional similarity for lexical inference. Nat. Lang. Eng. **16**(4), 359–389 (2010)

15. Lambek, J.: Type grammar revisited. In: Lecomte, A., Lamarche, F., Perrier, G. (eds.) LACL 1997. LNCS, vol. 1582, pp. 1–27. Springer, Heidelberg (1999). doi:10. 1007/3-540-48975-4_1

16. Lambek, J.: The mathematics of sentence structure. Am. Math. Mon. **65**, 154–170 (1958)

17. Leech, J.: Skew lattices in rings. Algebra Univers. **26**, 48–72 (1989)

18. Leech, J.: Skew Boolean algebras. Algebra Univers. **27**, 497–506 (1990)

19. Leech, J.: Normal skew lattices. Semigroup Forum **44**, 1–8 (1992)

20. Leech, J.: Recent developments in the theory of skew lattices. Algebra Univers. **52**, 7–24 (1996)

21. Lin, D.: Automatic retrieval and clustering of similar words. In: Proceedings of the 17th International Conference on Computational Linguistics, vol. 2, pp. 768–774. Association for Computational Linguistics (1998)

22. Lin, D.: An information-theoretic definition of similarity. In: Proceedings of the International Conference on Machine Learning, pp. 296–304 (1998)

23. Rubenstein, H., Goodenough, J.: Contextual correlates of synonymy. Commun. ACM **8**(10), 627–633 (1965)

24. Schuetze, H.: Automatic word sense discrimination. Comput. Linguist. **24**(1), 97–123 (1998)

25. Weeds, J., Weir, D., McCarthy, D.: Characterising measures of lexical distributional similarity. In: Proceedings of the 20th International Conference on Computational Linguistics, COLING 2004. Association for Computational Linguistics (2004)

On Fragments of Higher Order Logics that on Finite Structures Collapse to Second Order

Flavio Ferrarotti[1]([✉]), Senén González[1], and José María Turull-Torres[2,3]

[1] Software Competence Center Hagenberg, Hagenberg im Mühlkreis, Austria
{Flavio.Ferrarotti,Senen.Gonzalez}@scch.at
[2] Depto. de Ingeniería e Investigaciones Tecnológicas,
Universidad Nacional de La Matanza, San Justo, Argentina
J.M.Turull@massey.ac.nz
[3] Massey University, Palmerston North, New Zealand

Abstract. We define new fragments of higher-order logics of order three and above, and investigate their expressive power over finite models. The key unifying property of these fragments is that they all admit inexpensive algorithmic translations of their formulae to equivalent second-order logic formulae. That is, within these fragments we can make use of third- and higher-order quantification without paying the extremely high complexity price associated with them. Although theoretical in nature, the results reported here are more significant from a practical perspective. It turns out that there are many examples of properties of finite models (queries from the perspective of relational databases) which can be simply and elegantly defined by formulae of the higher-order fragments studied in this work. For many of those properties, the equivalent second-order formulae can be very complicated and unintuitive. In particular when they concern properties of complex objects, such as hyper-graphs, and the equivalent second-order expressions require the encoding of those objects into plain relations.

1 Introduction

There are many examples of properties of finite models (queries from the perspective of relational databases) that can be defined by simple and elegant sentences of higher-order logics of order three and above. Take for instance the property of a graph of being an n-hypercube graph $\mathbf{Q}_n$, i.e., an undirected graph whose vertices are binary n-tuples and such that two vertices are adjacent iff they differ in exactly one bit. We can build an $(n+1)$-hypercube $\mathbf{Q}_{n+1}$ by simply taking

Work supported by **Austrian Science Fund (FWF): [I2420-N31]**. Project: *Higher-Order Logics and Structures*. Initiated during a project sponsored visit of Prof. José María Turull-Torres. The research reported in this paper has been partly supported by the Austrian Ministry for Transport, Innovation and Technology, the Federal Ministry of Science, Research and Economy, and the Province of Upper Austria in the frame of the COMET center SCCH.

J. Kennedy and R.J.G.B. de Queiroz (Eds.): WoLLIC 2017, LNCS 10388, pp. 125–139, 2017.
DOI: 10.1007/978-3-662-55386-2_9

two isomorphic copies of an n-hypercube $\mathbf{Q}_n$ and adding edges between the corresponding vertices. This strategy can be formally expressed by means of a clear and elegant third-order logic sentence which expresses that $\mathbf{G}$ is an n-hypercube graph for some n iff the following holds:

- There is a sequence $\mathcal{S}$ of graphs, i.e., a *third-order linear digraph* whose nodes are undirected (second-order) graphs.
- The sequence $\mathcal{S}$ starts with a K_2-graph and ends with $\mathbf{G}$.
- For every graph $\mathbf{G}_{succ}$ and its immediate predecessor $\mathbf{G}_{pred}$ in the sequence $\mathcal{S}$, there is a pair of injective functions f_1, f_2 from $\mathbf{G}_{pred}$ to $\mathbf{G}_{succ}$ such that
  - f_1 and f_2 induce in $\mathbf{G}_{succ}$ two isomorphic copies of $\mathbf{G}_{pred}$,
  - f_1 and f_2 define a partition in the vertex set of $\mathbf{G}_{succ}$, and
  - for every edge (x, y) of $\mathbf{G}_{succ}$, $f_1^{-1}(x) = f_2^{-1}(y)$ or either the edge $(f_1^{-1}(x), f_1^{-1}(y))$ or $(f_2^{-1}(x), f_2^{-1}(y))$ belongs to $\mathbf{G}_{pred}$.

Yet another example of a property that can be expressed by a simple and elegant third-order sentence is given by the *formula-value query*, consisting on determining whether a propositional formula φ with constants in $\{F, T\}$ evaluates to true. We can express it in third-order logic by writing that there is a sequence of propositional formulae (represented as finite structures) which starts with φ, ends with the formula T, and such that every formula φ_{suc} in the sequence results from applying to exactly one sub-formula of its immediate predecessor φ_{pred} an operations of conjunction, disjunction, or negation which is "ready" to be evaluated (e.g., the conjunction in "$(T \wedge F)$"), or the elimination of a pair of redundant parenthesis (e.g., the parenthesis in "(T)").

The high expressive power of third-order logic is *not* really necessary to characterize hypercube graphs, since they can be recognized in non-deterministic polynomial time (NP) and by Fagin's theorem [6] existential second-order logic is then powerful enough to define this property. Nevertheless, to define the class of hypercube graphs in second-order logic is certainly more challenging than to define it in third-order logic (see the two strategies for hypercube graphs in [8]). Likewise, we do *not* really need third-order logic to express the formula-value query, since it is in DLOGSPACE [2].

It is then relevant to *distinguish* formulae of order three or higher which *do have* a second-order equivalent formula, from those which (most likely) do not. Beyond the significance of this questions to advance the theory of descriptive complexity, such a development can clearly empower us to write simpler and more intuitive queries, *although still formal*, by taking advantage of the higher level of expressivity of higher-order logics. Provided that those queries can be translated into formal languages with lower complexity of evaluation, this can be done without paying the extremely high complexity price which is associated to higher-order logics. Note that by the results in [6,13], existential second-order logic captures NP while existential third-order logic already captures $NTIME(2^{n^{O(1)}})$.

Outline of Contributions. We define new fragments of higher-order logics of order three and above, and investigate their expressive power over finite models.

The key unifying property of these fragments is that they all admit inexpensive algorithmic translations of their formulae to equivalent second-order logic formulae.

We start by defining in Sect. 3 a general *schema* of existential third-order formulae. The schema generalizes the approach described in our previous examples for hypercube graph and formula-value query. It essentially allows us to express an iteration of polynomial length. This iteration is represented by an unfolded sequence of relational structures which can be seen as a computation or derivation. Transitions are then specified by explicitly stating the operations which can be involved in the construction of a given structure in the sequence, when applied to the previous one. As further discussed in Sect. 3, this is a very usual, intuitive, and convenient schema in the expression of properties.

In Sect. 4, we characterize a broader fragment of third-order logic which is no longer restricted to formulae of a fixed schema as in the previous section. We call this fragment TO^P, for *polynomial third-order*, and give a constructive proof of the fact that it collapses to second-order logic. Although the schema of existential third-order formulae proposed in Sect. 3 turned out to be a special case of TO^P, it is still relevant. The translation of the formulae of that schema yields second-order formulae which are more intuitive and clearer than the general translation of TO^P formulae. Moreover, taking into consideration the examples in this paper, the translation proposed in Sect. 3 always results in second-order formulae which use relation variables of considerable smaller arity than the equivalent, but also more general translation proposed in this section. Since the maximum arity of the relation variables in the second-order formulae is *relevant* for the complexity of their evaluation (see [13] among others), it makes sense to study specific schemas of third-order formulae with the aim of finding more efficient translations.

In Sect. 5 we generalize the result in Sect. 4 by characterizing, for each order $i \geq 4$, a fragment $HO^{i,P}$ of the i-th order logic which collapses to second-order. Again this result has interesting practical applications. As an example, consider a multilevel PERT chart such as those commonly used in engineering for planning and scheduling tasks of complex projects. The encoding of higher-order relations of order ≥ 3 into second-order relations can be exploited as a normal form to store such type of complex multilevel PERT charts into a standard relational database. Under certain conditions, higher-order queries of order ≥ 3 could then be synthesised into efficient SQL queries over such normalized relational database. Notice that, a related approach with synthesisation to efficient algorithms was already taken in [14].

We conclude the paper in Sect. 6 where we discuss in detail the expressive power of different fragments of the $HO^{i,P}$ logics and their relationship with known fragments of second-order logic. In particular, adapting Makowsky and Pnueli [16] approach to prove hierarchies of arity and alternation of second-order formulae, we are able to prove interesting strict hierarchies of $HO^{i,P}$ formulae.

Due to space limitations, in most cases we only present sketches of the proofs. Nevertheless, a technical report with the omitted details in Sects. 3–5 is accessible as a CoRR abstract in [9].

2 Preliminaries

We assume familiarity with the basic concepts of finite model theory [5,15]. We only consider signatures, or vocabularies, which are purely *relational*. We use the classical Tarski's semantics, except that in the context of finite model theory, *only finite* structures or interpretations are considered. If **A** is a structure of vocabulary σ, we denote its finite domain by $dom(\mathbf{A})$ or A. By $\varphi(x_1, \ldots, x_r)$ we denote a formula of some logic whose free variables are exactly $\{x_1, \ldots, x_r\}$. We write $\mathbf{A} \models \varphi(x_1, \ldots, x_r)[\bar{a}]$ to denote that φ is satisfied by the structure **A** under *all* valuations v such that $v(x_i) = a_i$ for $1 \leq i \leq r$.

With HO^i we denote the i-th order logic which extends first-order logic with quantifiers of any order $2 \leq j \leq i$, which in turn bind j-th order relation variables. In particular, HO^2 denotes second-order logic as usually studied in the context of finite model theory [5,15], and HO^3 denotes third-order logic. A *third-order relation type of width w* is a w-tuple $\tau = (r_1, \ldots, r_w)$ where $w, r_1, \ldots, r_w \geq 1$, and $r_1, \ldots, r_w$ are arities of (second-order) relations. For $i \geq 4$, an *i-th order relation type of width w* is a w-tuple $\tau = (\rho_1, \ldots, \rho_w)$ where $w \geq 1$ and $\rho_1, \ldots, \rho_w$ are $(i-1)$-th order relation types. A *second-order relation* is a relation in the usual sense. A *third-order relation* of type $\tau = (r_1, \ldots, r_w)$ is a set of tuples of (second-order) relations of arities $r_1, \ldots, r_w$, respectively. For $i \geq 4$, an *i-th order relation* of type $\tau = (\rho_1, \ldots, \rho_w)$ is a set of tuples of $(i-1)$-th order relations, of types $\rho_1, \ldots, \rho_w$, respectively. A^τ denotes the set of all higher-order relations of type τ over the domain of individuals A. We use uppercase calligraphic letters $\mathcal{X}^i, \mathcal{Y}^i, \mathcal{Z}^i, \ldots$ to denote i-th order variables of order $i \geq 3$, uppercase letters $X, Y, Z, \ldots$ to denote second-order variables, and lower case letters $x, y, z, \ldots$ to denote first-order variables. With $\mathcal{X}^{i,\tau}$ we denote an i-th order variable of type τ. If $\mathcal{X}^i$ is a third-order variable, we tend to omit the superscript. We sometimes use X^r to denote that X is a second-order variable or arity r. Second-order variables of arity r are valuated with r-ary relations. For $i \geq 3$, i-th order relation variables are valuated with sets of tuples of $(i-1)$-th order relations according to their relation types. Thus, if v is a valuation, **A** is a structure and $\mathcal{X}^{i,\tau}$ is a higher-order variable, then $v(\mathcal{X}^{i,\tau})$ is a i-th order-relation of type τ in A^τ. Independently of the order and type of the variables, we say that two valuations v and v' are $\mathcal{X}$-equivalent if $v(\mathcal{Y}) = v'(\mathcal{Y})$ for every variable $\mathcal{Y}$ other than $\mathcal{X}$. For any $i \geq 3$, we define the notion of *satisfaction* in HO^i by extending the usual notion of satisfaction of second-order logic formula as follows: $\mathbf{A}, v \models \exists \mathcal{X}^{i,(\rho_1, \ldots, \rho_w)}(\varphi(\mathcal{X}))$, where $\mathcal{X}$ is an i-th order relation variable and φ is a well-formed formula, iff there is a i-th order relation $\mathcal{R}$ of type $\tau = (\rho_1, \ldots, \rho_w)$ in A^τ, such that $\mathbf{A}, v' \models \varphi(\mathcal{X})$ whenever v' is $\mathcal{X}$-equivalent to v and $v'(\mathcal{X}) = \mathcal{R}$. Likewise, $\mathbf{A}, v \models \forall \mathcal{X}^{i,(\rho_1, \ldots, \rho_w)}(\varphi(\mathcal{X}))$ iff for all i-th order relation $\mathcal{R}$ in A^τ, it holds that $\mathbf{A}, v' \models \varphi(\mathcal{X})$ whenever v' is $\mathcal{X}$-equivalent to v and $v'(\mathcal{X}) = \mathcal{R}$.

3   A General Schema of Existential Third-Order Formulae

We define next a *general schema* of ∃TO formulae which consists of existentially quantifying a third-order linear digraph of polynomial length (i.e., a sequence of structures that represents a computation) by explicitly stating which operations are the ones which can be involved in the construction of a given structure in the sequence, when applied to the previous one. The schema is as follows:

$$\exists \mathcal{C}^{\bar{s}} \mathcal{O}^{\bar{s}\bar{s}} \big(\text{TotalOrder}(\mathcal{C}, \mathcal{O}) \wedge$$
$$\forall G \big(\text{First}(G) \to \alpha_{\text{First}}(G) \wedge \text{Last}(G) \to \alpha_{\text{Last}}(G) \big) \wedge$$
$$\forall G_{pred} G_{succ} \big(\mathcal{C}(G_{pred}) \wedge \mathcal{C}(G_{succ}) \wedge \text{Pred}(G_{pred}, G_{succ}) \tag{1}$$
$$\to \varphi(G_{pred}, G_{succ}) \big) \big),$$

where

- $\mathcal{C}$ ranges over third-order relations of type $\bar{s} = (i_1, \ldots, i_s)$, i.e., over sets of s-tuples of relations of arities $i_1, \ldots, i_s \geq 1$.
- $\text{TotalOrder}(\mathcal{C}, \mathcal{O})$, $\text{First}(G)$, $\text{Last}(G)$ and $\text{Pred}(G_{pred}, G_{succ})$ denote fixed second-order formulae which express that $\mathcal{O}$ is a total order over $\mathcal{C}$, G is the first relational structure in $\mathcal{O}$, G is the last relational structure in $\mathcal{O}$, and G_{pred} is the immediate predecessor of G_{succ} in $\mathcal{O}$, respectively.
- $\alpha_{\text{First}}(G)$ and $\alpha_{\text{Last}}(G)$ denote arbitrary second-order formulae which define, respectively, the properties that the first and last structure in $\mathcal{O}$ should satisfy.
- $\varphi(G_{pred}, G_{succ})$ denotes an arbitrary second-order formula that expresses the transition from G_{pred} to G_{succ}, i.e., which operations can be used to obtain G_{succ} from G_{pred}.

This is a very usual, intuitive, and convenient schema in the expression of natural properties of finite models. For a start, it can clearly be used to express the hypercube and formula-value query as described in the introduction. Significant additional examples are provided by the different relationships between pairs of undirected graphs (G, H) that can be defined as orderings of special sorts. Using schema (1) these relationships can be expressed by defining a set of possible operations that can be applied repeatedly to H, until a graph which is isomorphic to G is obtained. In particular, the following relationships fall into this category: (a) $G \leq_{immersion} H$: G is an *immersion* in H (see [1,4,12]); (b) $G \leq_{top} H$: G is *topologically embedded* or *topologically contained* in H (see [1,4,12]); (c) $G \leq_{minor} H$: G is a *minor* of H (see [4,11]); (d) $G \leq_{induced-minor} H$: G is an *induced minor* of H (see [4]). Interestingly, in all these cases the length of the sequence is at most linear. The operations on graphs needed to define those orderings are: (E) delete an edge, (V) delete a vertex, (C) contract an edge, (T) degree 2 contraction, or subdivision removal, and (L) lift an edge. In particular the set of allowable operations for each of those orderings are: $\{E, V, L\}$ for $\leq_{immersion}$, $\{E, V, C\}$ for $\leq_{minor}$, $\{E, V, T\}$ for $\leq_{top}$, and $\{V, C\}$ for $\leq_{induced-minor}$ (see [4]).

The classical Kuratowski definition of *planarity*, provides yet another example of a property that can be defined using our schema (1) and also results in a polynomially bounded sequence of structures. By Wagner's characterization [3]: a graph is planar if and only if it contains neither K_5 nor $K_{3,3}$ as a minor.

Provided some simple conditions are met, every third-order formula of the schema (1) can be translated into an equivalent second-order formula. Note that the proof of this results directly implies that the translation can be done by means of a simple and inexpensive algorithm.

Theorem 1. *Every third-order formulae* $\Psi \equiv \exists \mathcal{C}^{\bar{s}} \mathcal{O}^{\bar{s}\bar{s}} \psi(\mathcal{C}, \mathcal{O})$ *of the schema (1) can be translated into an equivalent second-order formula* Ψ' *whenever the following conditions hold.*

i. The sub formulae α_{First}, α_{Last} *and* φ *of* Ψ *are second-order formulae.*
ii. There is a $d \geq 0$ *such that for every valuation* v *with* $v(\mathcal{C}) = \mathcal{R}$, *if* $\mathbf{A}, v \models \exists \mathcal{O}^{\bar{s}\bar{s}} \psi(\mathcal{C}, \mathcal{O})$, *then* $|\mathcal{R}| \leq |dom(\mathbf{A})|^d$.

Proof (Sketch). Let us first consider the case in which $\mathcal{C}$ is valuated with sets of non-empty graphs. Let t be the degree of a polynomial bounding the size of those graphs. Our strategy consists on encoding $\mathcal{C}$ as a pair of second-order variables C and E_C of arities $d + t$ and $2(d + t)$, respectively. Notice that every formula that complies with schema (1) stipulates that $\mathcal{O}$ is a linear order of the graphs in $\mathcal{C}$ which represents the stages (or steps) of a computation. Consequently the number of stages needed is bounded by n^d, where n is the size of the structure. Since in turn each stage has a bound on the number of elements it adds or changes (at most n^t), we have to consider a set of $(d + t)$-tuples.

The encoding into second-order is completed by a total relation $R \subseteq ST \times C$, where C is the union of the domains of all the structures in the sequence. Every node in ST represents one stage, and through the forest R defines a subset of nodes, which is the vertex set of a sub graph (not necessarily connected) of the whole graph (C, E_C). We use $C|_{R(\bar{x})}$, $E_C|_{R(\bar{x})}$ to denote the restriction of C and E_C, respectively, to $R(\bar{x})$, i.e., $C|_{R(\bar{x})} = \{\bar{y} \mid C(\bar{y}) \wedge R(\bar{x}, \bar{y})\}$ and $E_C|_{R(\bar{x})} = \{(\bar{v}, \bar{w}) \in E_C \mid R(\bar{x}, \bar{v}) \wedge R(\bar{x}, \bar{w})\}$. The sub graph of (C, E_C) which corresponds to the stage $ST(\bar{x})$ is denoted as $(C|_{R(\bar{x})}, E_C|_{R(\bar{x})})$. In this way, Ψ can then be translated into an equivalent second-order formula Ψ' as follows:

$$\exists C^{d+t} E_C^{2(d+t)} ST^d E_{ST}^{2d} R^{2d+t} \big(\text{Linear}(ST, E_{ST}) \wedge R \subseteq ST \times C \wedge \text{Total}(R) \wedge$$
$$\forall \bar{x} \forall \bar{y} \big((\text{First}(\bar{x}) \to \hat{\alpha}_{\text{First}}) \wedge (\text{Last}(\bar{x}) \to \hat{\alpha}_{\text{Last}}) \wedge$$
$$((ST(\bar{x}) \wedge ST(\bar{y}) \wedge \text{Pred}(\bar{x}, \bar{y})) \to \tag{2}$$
$$\hat{\varphi}((C|_{R(\bar{x})}, E_C|_{R(\bar{x})}), (C|_{R(\bar{y})}, E_C|_{R(\bar{y})})))))\big),$$

where

- Linear(ST, E_{ST}), First$(\bar{x})$, Last$(\bar{x})$ and Pred$(\bar{x}, \bar{y})$ denote second-order formulae which express that (ST, E_{ST}) is a linear digraph, $\bar{x}$ is the first node in (ST, E_{ST}), $\bar{x}$ is the last node in (ST, E_{ST}), and $\bar{x}$ is the immediate predecessor of $\bar{y}$ in (ST, E_{ST}), respectively.
- $R \subseteq ST \times C$ and Total(R) are shorthands for $\forall \bar{x} \bar{y}(R(\bar{x}, \bar{y}) \to (ST(\bar{x}) \wedge C(\bar{y})))$ and $\forall \bar{x}(ST(\bar{x}) \to \exists \bar{y}(R(\bar{x}, \bar{y})))$, respectively.

– $\hat{\alpha}_{\text{First}}$ and $\hat{\alpha}_{\text{Last}}$ are second-order formulae built from α_{First} and α_{Last}, respectively, by modifying them to talk about the graph described by $\bar{x}$ through $ST(\bar{x})$, E_{ST} and R.

– $\hat{\varphi}$ is an second-order formula built from φ by modifying it to talk about the graphs described by $\bar{x}$ and $\bar{y}$ through $ST(\bar{x})$, $ST(\bar{y})$, E_{ST} and R.

For the case of relations of arbitrary arity, say S of arity $r \geq 1$, we simply need to consider E_C as an r-ary relation (denoted E_C^S). Thus $E_C^S|_{R(\bar{x})} = \{(\bar{v}_1, \ldots, \bar{v}_r) \in E_C^S : R(\bar{x}, \bar{v}_1) \wedge \ldots \wedge R(\bar{x}, \bar{v}_r)\}$. If we have a tuple of relations, say $l \geq 1$ relations of arities $r_1, \ldots, r_l \geq 1$, respectively, then we have to consider similarly $E_{C_1}^{S_1}, \ldots, E_{C_l}^{S_l}$. □

Remark 1. Every property definable by a third-order formula of the schema (1), where α_{First}, α_{Last} and φ are *existential* second-order formulae and condition (ii) in Theorem 1 also holds, can be checked in NP exactly as it happens for every property definable in existential second-order. It suffices to additionally guess polynomial-sized valuations for the existentially quantified third-order variables. Then, by Fagin's theorem, we get that every property definable by such kind of third-order formulae can also be defined in existential second-order. Our approach however is fundamentally different. Instead of producing a non deterministic Turing machine, we produce a clear and intuitive second-order formula.

4   TO^P: A Restricted Third Order Logic

We define the logic TO^P as third-order logic restricted to third-order quantification ranging over third-order relations of cardinality bounded by a polynomial in the size of the structure. By contrast, the cardinality of an arbitrary third-order relation $\mathcal{R}$ over a structure $\mathbf{A}$ is exponentially bounded by $2^{|dom(\mathbf{A})|^{O(1)}}$.

Beyond the usual symbols, the alphabet of TO^P includes a third-order quantifier $\exists^{P,d}$ and countably many third order variable symbols $\mathcal{X}^{d,\bar{r}}$ for every $d \geq 0$ and third-order type $\bar{r}$. Whenever it is clear from the context, we avoid the superscript d in the TO^P variables. A *valuation* in a structure $\mathbf{A}$ assigns to each variable $\mathcal{X}^{d,\bar{r}}$ a third-order relation $\mathcal{R}$ in $A^{\bar{r}}$, such that $|\mathcal{R}| \leq |dom(\mathbf{A})|^d$. The quantifier $\exists^{P,d}$ has the following semantics: $\mathbf{A} \models \exists^{P,d}\mathcal{X}^{d,\bar{r}}\varphi(\mathcal{X})$ iff there is TO relation $\mathcal{R}^{\bar{r}}$ of type $\bar{r}$, such that $\mathbf{A} \models \varphi(\mathcal{X})[\mathcal{R}]$ and $|\mathcal{R}| \leq |dom(\mathbf{A})|^d$.

The following result shows that the expressive power of TO^P collapses to second-order logic. Same as in the previous section, the proof is constructive and directly implies that the translation can be done algorithmically.

Theorem 2. *Every* TO^P *formula* α *can be translated into an equivalent second-order formula* α'.

Proof (Sketch). Let $\mathbf{A}$ be a structure and $\mathcal{R}^{(r_1, \ldots, r_s)}$ be a TO^P relation of type $(r_1, \ldots, r_s)$ in $A^{(r_1, \ldots, r_s)}$ which is bounded by a polynomial of degree $d \geq 0$, i.e., such that $|\mathcal{R}| \leq |dom(\mathbf{A})|^d$. Assuming that all relations which appear in the tuples of $\mathcal{R}$ are non-empty, we can use a second-order relation $R_{\mathcal{R}}^{d+r_1+\ldots+r_s}$ of arity $(d + r_1 + \ldots + r_s)$ to encode $\mathcal{R}^{(r_1,\ldots,r_s)}$. More precisely, we can use d-tuples from $dom(\mathbf{A})^d$ as identifiers of tuples of second-order relations in $\mathcal{R}$, so that whenever a tuple $(a_1,\ldots,a_d,a_{d+1},\ldots,a_{d+r_1},\ldots,$ $a_{d+r_1+\ldots+r_{s-1}+1},\ldots,a_{d+r_1+\ldots+r_s}) \in R_{\mathcal{R}}$, then there is a tuple in $\mathcal{R}$ which can be identified by $(a_1,\ldots,a_d)$, which consists of s second-order relations $S_1^{r_1},\ldots,S_s^{r_s}$, of arities $r_1,\ldots,r_s$, respectively, such that $(a_{d+1},\ldots,a_{d+r_1}) \in$ $S_1,\ldots,(a_{d+r_1+\ldots+r_{s-1}+1},\ldots,a_{d+r_1+\ldots+r_s}) \in S_s$.

The actual translation can be done by structural induction on the TO^P-formula α. We present next the two non-trivial cases.

Atomic Formulae. Let α be of the form $\mathcal{X}^{d,(r_1,\ldots,r_s)}(X_1^{r_1},\ldots,X_s^{r_s})$, where $s, r_1,\ldots,r_s \geq 1$ and $\mathcal{X}$ is a TO^P variable.

Note that there are 2^s possible patterns of empty and non-empty relations in an s-tuple of second-order relations. We denote by $\omega = (i_1,\ldots,i_{|\omega|})$ the pattern of *empty* relations, with $1 \leq i_1 < i_2 < \ldots < i_{|\omega|} \leq s$ being the indices of the components which are empty. Correspondingly, we denote by $\bar{\omega} = (j_1,\ldots,j_{|\bar{\omega}|})$ the pattern of *non-empty* relations. By abuse of notation, we denote as $\{\omega\}$ and $\{\bar{\omega}\}$ the sets of indices in ω and $\bar{\omega}$, respectively. In particular if $\{\omega\} = \emptyset$ and $\bar{\omega} = (1,\ldots,s)$, then all the components of the s-tuple of second-order relations are non-empty.

The idea for the translation is to replace $\mathcal{X}$ with 2^s second-order variables, one for each pattern ω of empty second-order relations. We use $X_{\mathcal{X},e,\omega}$ to denote the second-order variable that encodes those tuples of second-order relations (in the TO^P relation that valuates $\mathcal{X}$) which follow ω. The arity of the second-order relation $X_{\mathcal{X},e,\omega}$ is $d + r_{j_1} + \cdots + r_{j_{|\omega|}}$ where $\bar{\omega} = (j_1,\ldots,j_{|\omega|})$. In what follows we use $\bar{f}_{\bar{\omega}} = \bar{f}_{j_1}\ldots\bar{f}_{j_{|\bar{\omega}|}}$ to denote a tuple of first-order variables formed by the concatenation of the tuples of first-order variables $\bar{f}_{j_1} = (f_{j_1 1},\ldots,f_{j_1 r_{j_1}}),\ldots,$ $\bar{f}_{j_{|\bar{\omega}|}} = (f_{j_{|\bar{\omega}|}1},\ldots,f_{j_{|\bar{\omega}|}r_{j_{|\bar{\omega}|}}})$.

Let $\Omega = \{\omega \mid \omega = (i_1,\ldots,i_{|\omega|}); 1 \leq i_1 < i_2 < \ldots < i_{|\omega|} \leq s; 0 \leq |\omega| \leq s; \bar{\omega} = (j_1,\ldots,j_{|\bar{\omega}|}); \{\bar{\omega}\} \cup \{\omega\} = \{1,\ldots,s\}; \{\bar{\omega}\} \cap \{\omega\} = \emptyset\}$. The translation to second-order of $\mathcal{X}^{d,(r_1,\ldots,r_s)}(X_1^{r_1},\ldots,X_s^{r_s})$ is as follows:

$$\bigvee_{\omega \in \Omega} \Big(\text{``}(X_{i_1} = \emptyset \wedge \ldots \wedge X_{i_{|\omega|}} = \emptyset)\text{''} \wedge \text{``}(X_{j_1} \neq \emptyset \wedge \ldots \wedge X_{j_{|\bar{\omega}|}} \neq \emptyset)\text{''} \wedge$$

$$\exists v_1 \ldots v_d \big(X_{\mathcal{X},e,\omega}(v_1,\ldots,v_d,\bar{f}_{\bar{\omega}}) \vee$$

$$\forall \bar{f}_{\bar{\omega}} \big(X_{\mathcal{X},e,\omega}(v_1,\ldots,v_d,\bar{f}_{\bar{\omega}}) \leftrightarrow \bigwedge_{l \in \{j_1,\ldots,j_{|\bar{\omega}|}\}} X_l(f'_{l1},\ldots,f'_{lr_l}) \big) \big) \Big)$$

Existential Case. Let α be of the form $\exists^{P,d}\mathcal{X}^{d,(r_1,\ldots,r_s)}(\varphi)$. In the translation we simply replace the existentially quantified $\mathcal{X}$ by its corresponding 2^s second-order variables and state that no d-tuple can be in more than one of the different second-order relations that encode the value of $\mathcal{X}$. The formula is as follows:

$$\hat{\alpha} \equiv \{\exists X_{\mathcal{X},e,\omega}^{d+|\bar{f}_{\bar{\omega}}|}\}_{\omega \in \Omega} (\forall z_1 \dots z_d [\bigwedge_{\substack{\omega=(i_1,\dots,i_{|\omega|}) \\ 0 \le i_1 < i_2 \dots < i_{|\omega|} \le s \\ 1 \le |\omega| \le s}} \forall \bar{f}_{\bar{\omega}}[X_{\mathcal{X},e,\omega}(z_1,\dots,z_d,\bar{f}_{\bar{\omega}}) \rightarrow$$

$$(\bigwedge_{\substack{\omega'=(i'_1,\dots,i'_{|\omega'|}) \\ 1 \le i'_1 < i'_2 \dots < i'_{|\omega'|} \le s \\ 0 \le |\omega'| \le s;\ \omega' \ne \omega}} \forall \bar{f}'_{\bar{\omega}'}(\neg X_{\mathcal{X},e,\omega'}(z_1,\dots,z_d,\bar{f}'_{\bar{\omega}'})))]]) \wedge \hat{\varphi},$$

where Ω is as before and φ' is the second-order formula equivalent to the TO^P formula φ, obtained by applying inductively the described translations.   □

5   HO^{i,P}: Restricted Higher Order Logics

We say that a third-order relation $\mathcal{R}$ is *downward polynomially bounded* by d in a structure $\mathbf{A}$ if $|\mathcal{R}| \le |dom(\mathbf{A})|^d$. Likewise, we say that a relation $\mathcal{R}$ of order $i > 3$ is *downward polynomially bounded* by d in $\mathbf{A}$ if $|\mathcal{R}| \le |dom(\mathbf{A})|^d$ and further every relation $\mathcal{R}_i^j$ of order $3 \le j < i$ which appears in a tuple of $\mathcal{R}$ is downward polynomially bounded by d.

For $i = 4$, we define HO^{4,P} as the extension of TO^P with quantifiers that range over downward polynomially bounded relations of order 4. More general, for $i \ge 5$ we define HO^{i,P} as the extension of HO$^{i-1,P}$ with quantifiers that range over downward polynomially bounded relations of order i.

Beyond the symbols in the alphabet of TO^P, the alphabet of HO^{i,P} includes a j-th order quantifier $\exists^{j,P,d}$ for every $i \ge j \ge 4$, as well as countably many variable symbols $\mathcal{X}^{j,d,\tau}$ for every j-th order type τ. We sometimes avoid the superscripts d and τ for clarity.

A *valuation* in a structure $\mathbf{A}$ assigns to each variable $\mathcal{X}^{j,d,\tau}$ a j-th order relation $\mathcal{R}$ in A^τ which is downward polynomially bounded by d. The quantifier $\exists^{j,P,d}$ has the following semantics: $\mathbf{A} \models \exists^{j,P,d}\mathcal{X}^{j,d,\tau}\varphi(\mathcal{X})$ iff there is a j-th order relation $\mathcal{R}$ of type τ, such that $\mathbf{A} \models \varphi(\mathcal{X})[\mathcal{R}]$ and $\mathcal{R}$ is downward polynomially bounded by d in $\mathbf{A}$.

Same as with TO^P, for every order $i \ge 4$ the expressive power of HO^{i,P} collapses to second-order logic.

Theorem 3. *For every order $i \ge 3$, every HO^{i,P} formula α can be translated into an equivalent second-order formula α'.*

The actual proof of this theorem is quite long and cumbersome. Due to space limitations we omit it here. The details can nevertheless be consulted in the technical report in [9]. To gain some intuition on how this translations works, let us consider the case of $HO^{4,P}$. The general idea is to represent the fourth-order relations as a *normalized relational database*. Assume w.l.o.g. that the type of every relations of order 3 and 4 has width $s \ge 1$, and that every such relation is downward polynomially bounded by $d \ge 1$. In the case of a formula of the form $\mathcal{X}^{4,d,\tau}(\mathcal{Y}_1^3,\dots,\mathcal{Y}_s^3)$ of HO^{4,P} we can represent the fourth-order variable $\mathcal{X}^{4,d,\tau}$

using 2^s second-order variables $X_{\chi^4, \omega_{3,\chi}}$ of arities between d and $d + (s \cdot (d+s))$, depending on the pattern $\omega_{3,\chi}$ of non-empty relations. Thus, each $X_{\chi^4, \omega_{3,\chi}}$ can encode the tuples of (non-empty) third-order relations whose pattern is $\omega_{3,\chi}$. In turns, each $\mathcal{Y}_j^3$ can be represented by 2^s second-order variables as explained in Sect. 4 for the case of TO^P.

6   Fragments of $HO^{i,P}$ Formulae

The aim of this section is to gain a better understanding of the syntactic restrictions relevant as to the expressive power of the $HO^{i,P}$ logics.

In [17] we showed that for any $i \geq 3$ the deterministic inflationary fixed-point quantifier (IFP) in HO^i (i.e., where the variable which is bound by the IFP quantifier is an $(i+1)$-th order variable) is expressible in $\exists HO^{i+1}$. Let $IFP|_P$ denote the restriction of IFP where there is a positive integer d such that in every structure $\mathbf{A}$, the number of stages of the fixed-point is bounded by $|dom(\mathbf{A})|^d$. And let $(SO + IFP)$ denote second-order logic extended with the deterministic inflationary fixed-point quantifier, where the variable which is bound by the IFP quantifier is a third order variable. Note that the addition of such IFP quantifier to second-order means that we can express iterations of length exponential in $|dom(\mathbf{A})|$, so that it is strongly conjectured that $(SO + IFP)$ strictly includes second-order logic as to expressive power. However, as a consequence of Theorem 2 this is not the case with $IFP|_P$.

Corollary 1. *For every formula in $(SO+IFP|_P)$ there is an equivalent second-order formula.*

Let us define ΣTO_n^P as the restriction of TO^P to prenex formulae of the form $Q_1 V_1 \ldots Q_k V_k (\varphi)$ such that:

- $Q_1, \ldots, Q_k \in \{\forall^{P,d}, \exists^{P,d}, \forall, \exists\}$.
- V_i for $1 \leq i \leq k$ is either a second-order or TO^P variable (depending on Q_i).
- φ is a TO^P formula free of TO^P as well as second-order quantifiers (first-order quantifiers as well as free TO^P and second-order variables are allowed).
- The prefix $Q_1 V_1 \ldots Q_k V_k$ starts with an existential block of quantifiers and has at most n alternating (between universal and existential) blocks.

By the well known Fagin-Stockmeyer characterization [18] of the polynomial-time hierarchy, for every $n \geq 1$ the prenex fragment Σ_n of second-order logic captures the level Σ_n^{poly} of the polynomial-time hierarchy. From the proof of Theorem 2 is immediate that every formula in ΣTO_n^P can be translated into an equivalent second-order formula in Σ_n.

Corollary 2. ΣTO_n^P *captures* Σ_n^{poly}.

Consider the $AA(r, m)$ classes of second-order logic formulae where all quantifiers of *whichever* order are grouped together at the beginning of the formula, forming up to m alternating blocks of consecutive existential and universal quantifiers, and such that the arity of the second-order variables is bounded by r. Note that, the order of the quantifiers in the prefix may be mixed. As shown by Makowsky and Pnueli [16], the $AA(r,m)$ classes constitute a strict hierarchy of arity and alternation. Their strategy to prove this result consisted in considering the set $AUTOSAT(AA(r,m))$ of formulae of $AA(r,m)$ which, encoded as finite structures, satisfy themselves. As the well known diagonalization argument applies, it follows that $AUTOSAT(AA(r,m))$ is not definable by any formulae of $AA(r,m)$, but it is definable in a higher level of the same hierarchy. Similarly to Makowsky and Pnueli arity and alternation hierarchy of second-order formulae, we can we define hierarchies of HO^i and $HO^{i,P}$ formulae as follows.

Definition 1 (AA^i- and AAD^i-hierarchies). *The* maximum-width *of a type* $\tau = (\rho_1, \ldots, \rho_s)$ *of order* $i \geq 3$ *(denoted as* max-width(τ)*) is defined as follows:*

- max-width$(\tau) = \max(\{s, \rho_1, \ldots, \rho_s\})$ *if* $i = 3$.
- max-width$(\tau) = \max(\{s, \text{max-width}(\rho_1), \ldots, \text{max-width}(\rho_s)\})$ *if* $i > 3$.

For $r, m, i \geq 1$, *we define the level* $AA^i(r,m)$ *of the* AA^i*-hierarchy as the class of formulae* $\varphi \in HO^{i+1}$ *of the form* $Q_1 V_1 \ldots Q_k V_k(\psi)$ *such that:*

i *ψ is a quantifier-free HO^i-formula.*
ii *For $j = 1, \ldots, k$, each Q_j is either an existential or universal quantifier and each V_j is a variable of order $\leq i + 1$.*
iii *The prefix $Q_1 V_1 \ldots Q_k V_k$ has at most m alternating blocks of quantifiers.*
iv *If V_j is a second-order variable, then its arity is bounded by r.*
v *If V_j is of order ≥ 3, then the maximum-width of the type τ of V_j is $\leq r$.*

For $r, m, i, d \geq 1$, *the level* $AAD^i(r, m, d)$ *of the* AAD^i*-hierarchy is obtained by adding the following condition to the definition of* $AA^i(r,m)$.

- *If V_j is a variable of order ≥ 3, then the quantifier Q_j has a superscript $d_j \leq d$, which denotes the degree of the polynomial bounding the size of the valuations of V_j (recall definition of $HO^{i,P}$).*

Note that in the formulae of the AA^i and AAD^i hierarchies the quantifiers of the highest order do not necessarily precede all the remaining quantifiers in the prefix, as it is the case in the Σ_m^i hierarchies of higher-order logics.

As proven in [7,10], it is possible to gerenarlize Makowsky and Pnueli result regarding the AA-hierarchy [16] to every higher-order logic of order $i \geq 2$.

Theorem 4 (Theorem 4.28 in [7]). *For every $i, r, m \geq 1$, there are Boolean queries not expressible in $AA^i(r,m)$ but expressible in $AA^i(r+c(r), m+6)$, where $c(r) = 1$ for $r > 1$ and $c(r) = 2$ for $r = 1$.*

The proof of the previous result in [7] also follows the strategy introduced in [16]. That is, it uses the diagonalization argument to prove the lower bound for the definability of $AUTOSAT(AA^i(r,m))$, and shows a formula in $AA^i(r + c(r), m + 6)$ that defines $AUTOSAT(AA^i(r,m))$ to prove the upper bound.

Interestingly, the formula $\psi_A \in AA^i(r + c(r), m + 6)$ used in the proof of Proposition 4.27 in [7] to define $AUTOSAT(AA^i(r,m))$ can be straightforwadly translated to a formula ψ_A^d that defines $AUTOSAT(AAD^i(r, m, d))$. We simply need to qualify with an appropriate superscript $d_j \leq d$ the existential and universal quantifiers associated to each higher-order variable V_j of order ≥ 3 which appears in ψ_A, so that V_j becomes restricted to range over higher-order relations which are downward polynomially bounded by d_j. Since d as well as the order and the maximum-width of the higher-order types are bounded, a finite set of variables is still sufficient to encode the valuations for the different variables that might appear in any arbitrary sentence in $AAD^i(r, m, d)$. The resulting ψ_A^d formula is clearly in $AAD^i(r + c(r), m + 6, d)$. It no longer defines $AUTOSAT(AA^i(r, m))$ when $i \geq 2$, since the size of the higher-order relations that interpret the higher-order variables of order ≥ 3 in ψ_A^d are downward polynomially bounded by d and thus insufficient to encode every possible valuation for variables of order ≥ 3. This is clearly not a problem if we only consider sentences in $AAD^i(r, m, d)$. In fact, in this latter case we need higher-order variables that encode only those valuations which are downward polynomially bounded by some positive integer $\leq d$.

The previous observation together with the fact that by using the same diagonalization argument we can prove that $AUTOSAT(AAD^i(r, m, d))$ is not definable in $AAD^i(r, m, d)$, gives us the following strict hierarchies of $HO^{i,P}$ formulae for every order i.

Theorem 5. *For every $i \geq 2$ and $r, m, d \geq 1$, there are Boolean queries not expressible in $AAD^i(r, m, d)$ but expressible in $AAD^i(r + c(r), m + 6, d)$, where $c(r) = 1$ for $r > 1$ and $c(r) = 2$ for $r = 1$.*

Lemma 1 is a direct consequence of the translations in Theorems 2 and 3.

Lemma 1. *Let $r, m, d \geq 1$ and $i \geq 2$. For every sentence φ in $AAD^i(r, m, d)$, there is an equivalent sentence φ' in $AA^1(r^i + d \cdot (i - 1), m + 2)$.*

The following result suggest that the exact converse of the previous lemma is unlikely to hold. It further shows the relationship between the levels of the arity and alternation hierarchy of Makowsky and Pnueli and the AAD^i hierarchies.

Lemma 2. *Let $r, m \geq 1$. For every second-order sentence φ in $AA^1(r, m)$, there are three sentences φ', φ'' and φ''', each of them equivalent to φ, such that:*

i. $\varphi' \in AAD^{\lceil \frac{r}{2} \rceil + 1}(2, m, 2)$.
ii $\varphi'' \in AAD^2(2, m, r)$.
iii $\varphi''' \in AAD^2(\lceil \sqrt[3]{r} \rceil, m, \lceil \sqrt[2]{r} \rceil (\lceil \sqrt[3]{r} \rceil - 1))$.

A Sketch of the proof of Lemma 2 is included in Appendix A.

Our final result gives a fine grained picture of the effect, as to expressive power, of simultaneously bounding the arity, alternation and maximum degree of the $HO^{i,P}$-sentences.

Theorem 6. *For every $r, m, d \geq 1$, there are Boolean queries not expressible in $AAD^2(r, m, d)$ but expressible in $AAD^c(2, m + 8, 2)$ as well as in $AAD^2(2, m + 8, (r + 1)^2 + d)$ and in $AAD^2(q, m + 8, q(q - 1))$, where $c = \left\lceil \frac{((r+1)^2+d)}{2} \right\rceil + 1$ and $q = \left\lceil \sqrt[2]{((r + 1)^2 + d)} \right\rceil$.*

Proof. By Lemma 1, we get that $AAD^2(r, m, d) \subseteq AA^1(r^2 + d, m + 2)$, i.e., the class of Boolean queries definable by TOP-sentences in $AAD^2(r, m, d)$ is included in those definable by second-order sentences in $AA^1(r^2 + d, m + 2)$. In turns, by Makowsky and Pnueli [16] result (first level $(i = 1)$ in Theorem 4), we get that $AA^1(r^2 + d, m + 2) \subset AA^1(r^2 + d + 1, m + 8)$. Finally, by Lemma 2 we get that the class of Boolean queries definable in $AA^1(r^2 + d + 1, m + 8)$ is included in $AAD^c(2, m + 8, 2)$ as well as in $AAD^2(2, m + 8, (r + 1)^2 + d)$ and in $AAD^2(q, m + 8, q(q - 1))$. $\qquad \square$

Appendix A Proof Sketch of Lemma 2

All three sentences φ', φ'' and φ''' can be defined by structural induction on φ. We show only the non trivial cases.

If φ is an atomic formula of the form $X(x_1, \ldots, x_s)$ where $s \leq r$. Then

- φ' is $\mathcal{X}^c(\mathcal{X}_1^{c-1}, \mathcal{X}_2^{c-1}) \wedge \mathcal{X}_1^{c-1}(\mathcal{X}_1^{c-2}) \wedge \mathcal{X}_2^{c-1}(\mathcal{X}_2^{c-2}, \mathcal{X}_3^{c-2}) \wedge \cdots \wedge$
  $\mathcal{X}_1^3(X_1) \wedge \mathcal{X}_2^3(X_2) \wedge \cdots \wedge \mathcal{X}_{c-2}^3(X_{c-2}, X_{c-1}) \wedge$
  $X_{c-1}(x_{s-1}, x_s) \wedge X_{c-2}(x_{s-3}, x_{s-2}) \wedge \cdots \wedge X_1(\bar{x})$,
  where $\bar{x} = (x_1)$ or $\bar{x} = (x_1, x_2)$ depending on whether s is odd or even, respectively, and $c = \lceil \frac{r}{2} \rceil + 1$.

- φ'' is $\bigwedge_{\bar{w} \in W} \left(\left(\bigwedge_{1 \leq j < l \leq s} \alpha_{i,j} \right) \to \psi_{\bar{w}} \right)$, where
  - $W = \{(i_1, i_2, \cdots, i_s) \mid 1 \leq j \leq s \text{ and } 1 \leq i_j \leq j\}$
  - $\alpha_{i,j} = \{^{x_i = x_j \ if \ i = j}_{x_i \neq x_j \ if \ i \neq j}$
  - $\psi_{\bar{w}}$ is $X_{1_{\bar{w}}}(x_1, x_1)$ if $i_u = i_v$ for every $i_u, i_v \in \bar{w}$.
    Otherwise $\psi_{\bar{w}}$ is $\bigwedge_{(u,v) \in A_{\bar{w}}} X_{1_{\bar{w}}}(x_u, x_v)$, where
    $A_{\bar{w}} = \{(u, v) \mid 1 \leq u < v \leq s, \ i_u, i_v \text{ are the } u\text{-th and } v\text{-th elements of } \bar{w},$
    respectively, $i_u = u$, $i_v = v$, and $i_t < t$ for all $v < t < u\}$.

- φ''' is $\mathcal{X}(X_1, \cdots, X_t) \bigwedge_{1 \leq i \leq t-1} X_i(x_{k_i+1}, \cdots, x_{k_i+u}) \wedge X_t(x_{s-t}, \cdots, x_s)$
  Where $t = \lceil \sqrt[2]{r} \rceil$, $k_i = ((i - 1)t) + min((i - 1), (s - t) \mod (t - 1))$, $u = |\frac{s-t}{t-1}| + c$, and $c = 1$ if $i \leq (s - t) \mod (t - 1)$, $c = 0$ otherwise.

If φ is a formula of the form $\exists X(\psi)$ where X is a second-order variable of arity $s \leq r$. Then

- φ' is $\exists^{c,P,2} \mathcal{X}^c \exists^{c-1,P,2} \mathcal{X}_1^{c-1} \mathcal{X}_2^{c-1} \exists^{c-2,P,2} \mathcal{X}_1^{c-2} \mathcal{X}_2^{c-2} \mathcal{X}_3^{c-2} \cdots$
  $\exists^{P,2} \mathcal{X}_1^3 \mathcal{X}_2^3 \cdots \mathcal{X}_{c-2}^3 \exists X_1 X_2 \cdots X_{c-1}(\psi')$,
  where again $c = \lceil \frac{s}{2} \rceil + 1$, and ψ' is the formula in $AAD^c(2, m, 2)$ equivalent to ψ, obtained by applying the translation inductively.

- φ'' is $\exists^{P,s} \mathcal{X}_{\bar{w}_1} \mathcal{X}_{\bar{w}_2} \cdots \mathcal{X}_{\bar{w}_{|W|}} \exists X_{1_{\bar{w}_1}} X_{2_{\bar{w}_1}} X_{1_{\bar{w}_2}} X_{2_{\bar{w}_2}} \cdots X_{1_{\bar{w}_{|W|}}} X_{2_{\bar{w}_{|W|}}} (\psi'')$,
  where $\bar{w}_1, \bar{w}_2, \cdots, \bar{w}_{|W|}$ is an arbitrary lexicographic order of W and ψ'' is the formula in $AAD^2(2, m, s)$ equivalent to ψ, obtained by applying the translation inductively.

– φ''' is $\exists^{P,h}\mathcal{X}\exists X_1 X_2 \cdots X_t(\psi''')$ where $t = \lceil \sqrt[2]{s} \rceil$, $h = t(t-1)$ and ψ''' is the formula in $AAD^2(t, m, t(t-1))$ equivalent to ψ, obtained by applying the translation inductively.

It is not difficult to show by structural induction that φ', φ'' and φ''' are equivalent to φ. We only need to see that every second-order relation of arity $s \leq r$ can be encoded as a higher-order relation of the type used in the translation and with the required polynomial bound.

References

1. Abu-Khzam, F.N., Langston, M.A.: Graph coloring and the immersion order. In: Warnow, T., Zhu, B. (eds.) COCOON 2003. LNCS, vol. 2697, pp. 394–403. Springer, Heidelberg (2003). doi:10.1007/3-540-45071-8_40

2. Beaudry, M., McKenzie, P.: Cicuits, matrices, and nonassociative computation. In: Proceedings of the Seventh Annual Structure in Complexity Theory Conference, Boston, Massachusetts, USA, 22–25 June 1992, pp. 94–106 (1992)

3. Bollobás, B.: Modern Graph Theory. Graduate Texts in Mathematics, vol. 184. Springer, Heidelberg (2002). Corrected edition

4. Downey, R.G., Fellows, M.R.: Parameterized Complexity. Monographs in Computer Science. Springer, Heidelberg (1999)

5. Ebbinghaus, H.D., Flum, J.: Finite Model Theory. Perspectives in Mathematical Logic, 2nd edn. Springer, Heidelberg (1999)

6. Fagin, R.: Generalized first-order spectra and polynomial-time recognizable sets. In: Karp, R. (ed.) Complexity of Computations. SIAM-AMS Proceedings, vol. 7, pp. 27–41. American Mathematical Society (1974)

7. Ferrarotti, F.: Expressibility of higher-order logics on relational databases: proper hierarchies. Ph.D. thesis, Department of Information Systems, Massey University, Wellington, New Zealand (2008). http://hdl.handle.net/10179/799

8. Ferrarotti, F., Ren, W., Turull-Torres, J.M.: Expressing properties in second- and third-order logic: hypercube graphs and SATQBF. Logic J. IGPL **22**(2), 355–386 (2014)

9. Ferrarotti, F., Tec, L., Torres, J.M.T.: On higher order query languages which on relational databases collapse to second order logic. CoRR abs/1612.03155 (2016). http://arxiv.org/abs/1612.03155

10. Ferrarotti, F., Turull-Torres, J.M.: Arity and alternation: a proper hierarchy in higher order logics. Ann. Math. Artif. Intell. **50**(1–2), 111–141 (2007)

11. Flum, J., Grohe, M.: Parameterized Complexity Theory. Texts in Theoretical Computer Science. An EATCS Series. Springer-Verlag New York, Inc., Secaucus (2006)

12. Grohe, M., Kawarabayashi, K., Marx, D., Wollan, P.: Finding topological subgraphs is fixed-parameter tractable. In: Proceedings of the Forty-third Annual ACM Symposium on Theory of Computing, STOC 2011, pp. 479–488. ACM, New York (2011)

13. Hella, L., Turull-Torres, J.M.: Computing queries with higher-order logics. Theor. Comput. Sci. **355**(2), 197–214 (2006)

14. Itzhaky, S., Gulwani, S., Immerman, N., Sagiv, M.: A simple inductive synthesis methodology and its applications. SIGPLAN Not. **45**(10), 36–46 (2010)

15. Libkin, L.: Elements of Finite Model Theory. Texts in Theoretical Computer Science, EATCS. Springer, Heidelberg (2004)

16. Makowsky, J.A., Pnueli, Y.B.: Arity and alternation in second-order logic. Ann. Pure Appl. Logic **78**(1–3), 189–202 (1996)
17. Schewe, K.D., Turull-Torres, J.M.: Fixed-point quantifiers in higher order logics. In: Proceedings of the 2006 Conference on Information Modelling and Knowledge Bases XVII, pp. 237–244. IOS Press, Amsterdam (2006)
18. Stockmeyer, L.J.: The polynomial-time hierarchy. Theoret. Comput. Sci. **3**(1), 1–22 (1976)

Computable Quotient Presentations of Models of Arithmetic and Set Theory

Michał Tomasz Godziszewski[1]($\boxtimes$) and Joel David Hamkins[2,3]

[1] Logic Department, Institute of Philosophy, University of Warsaw,
Krakowskie Przedmiescie 3, 00-927 Warszawa, Poland
mtgodziszewski@gmail.com

[2] Mathematics, Philosophy, Computer Science, The Graduate Center of The City
University of New York, 365 Fifth Avenue, New York 10016, USA
jhamkins@gc.cuny.edu

[3] Mathematics, College of Staten Island of CUNY, Staten Island 10314, USA
https://uw.academia.edu/MichalGodziszewski, http://jdh.hamkins.org

Abstract. We prove various extensions of the Tennenbaum phenomenon to the case of computable quotient presentations of models of arithmetic and set theory. Specifically, no nonstandard model of arithmetic has a computable quotient presentation by a c.e. equivalence relation. No Σ_1-sound nonstandard model of arithmetic has a computable quotient presentation by a co-c.e. equivalence relation. No nonstandard model of arithmetic in the language $\{+, \cdot, \leq\}$ has a computably enumerable quotient presentation by any equivalence relation of any complexity. No model of ZFC or even much weaker set theories has a computable quotient presentation by any equivalence relation of any complexity. And similarly no nonstandard model of finite set theory has a computable quotient presentation.

A *computable quotient presentation* of a mathematical structure $\mathcal{A}$ consists of a computable structure on the natural numbers $\langle \mathbb{N}, \star, \ast, \ldots \rangle$, meaning that the operations and relations of the structure are computable, and an equivalence relation E on $\mathbb{N}$, not necessarily computable but which is a congruence with respect

This article is a preliminary report of results following up research initiated at the conference Mathematical Logic and its Applications, held in memory of Professor Yuzuru Kakuda of Kobe University in September 2016 at the Research Institute for Mathematical Sciences (RIMS) in Kyoto. The second author is grateful for the chance twenty years ago to be a part of Kakuda-sensei's logic group in Kobe, a deeply formative experience that he is pleased to see growing into a lifelong connection with Japan. He is grateful to the organizer Makoto Kikuchi and his other Japanese hosts for supporting this particular research visit, as well as to Bakhadyr Khoussainov for insightful conversations. The first author has been supported by the National Science Centre (Poland) research grant NCN PRELUDIUM UMO-2014/13/N/HS1/02058. He also thanks the Mathematics Program of the CUNY Graduate Center in New York for his research visit as a Fulbright Visiting Scholar between September 2016 and April 2017. Commentary concerning this paper can be made at http://jdh.hamkins.org/computable-quotient-presentations.

J. Kennedy and R.J.G.B. de Queiroz (Eds.): WoLLIC 2017, LNCS 10388, pp. 140–152, 2017.
DOI: 10.1007/978-3-662-55386-2_10

to this structure, such that the quotient $\langle \mathbb{N}, \star, *, \dots \rangle / E$ is isomorphic to the given structure $\mathcal{A}$. Thus, one may consider computable quotient presentations of graphs, groups, orders, rings and so on, for any kind of mathematical structure. In a language with relations, it is also natural to relax the concept somewhat by considering the *computably enumerable* quotient presentations, which allow the pre-quotient relations to be merely computably enumerable, rather than insisting that they must be computable.

At the 2016 conference Mathematical Logic and its Applications at the Research Institute for Mathematical Sciences (RIMS) in Kyoto, Khoussainov [Kho16] outlined a sweeping vision for the use of computable quotient presentations as a fruitful alternative approach to the subject of computable model theory. In his talk, he outlined a program of guiding questions and results in this emerging area. Part of this program concerns the investigation, for a fixed equivalence relation E or type of equivalence relation, which kind of computable quotient presentations are possible with respect to quotients modulo E.

In this article, we should like to engage specifically with two conjectures that Khoussainov had made in Kyoto.

Conjecture (Khoussainov)

(1) *No nonstandard model of arithmetic admits a computable quotient presentation by a computably enumerable equivalence relation on the natural numbers.*

(2) *Some nonstandard model of arithmetic admits a computable quotient presentation by a co-c.e. equivalence relation.*

We shall prove the first conjecture and refute several natural variations of the second conjecture, although a further natural variation, perhaps the central case, remains open. In addition, we consider and settle the natural analogues of the conjectures for models of set theory.

Perhaps it will be helpful to mention as background the following observation, amounting to a version of the computable completeness theorem[1], which identifies a general method of producing computable quotient presentations.

Observation 1. *Every consistent c.e. axiomatizable theory T in a functional language admits a computable quotient presentation by an equivalence relation E of low Turing degree.*

Proof. Consider any computably enumerable theory T in a functional language (no relation symbols). Let τ be the computable tree of attempts to build a complete consistent Henkin theory extending T, in the style of the usual computable completeness theorem. To form the tree τ, we first give ourselves sufficient Henkin constants, and then add to T all the Henkin assertions $\exists x \, \varphi(x) \rightarrow \varphi(c_\varphi)$. Next, we enumerate all sentences in this expanded language, and then build the tree τ by adding to T at successive nodes either the next sentence or its negation,

---

[1] This analogy comes with the obvious difference that the computable completeness theorem assumes the theory is decidable.

provided that no contradiction has yet been realized from that theory by that stage. This tree is computable, infinite and at most binary branching. And so by the low basis theorem, it has a branch of low Turing complexity. Fix such a branch. The assertions made on it provide a complete consistent Henkin theory T^+ extending T. Let A be the term algebra generated by the Henkin constants in the language of T. Thus, the elements of A consist of formal terms in this language with the Henkin constants, and we may code the elements of A with natural numbers. The natural operations on this term algebra are computable: to apply an operation to some terms is simply to produce another term. We may define an equivalence relation E on A, by saying that two terms are equivalent $s \: E \: t$, just in case the assertion $s = t$ is in the Henkin theory T^+, and this will be a congruence with respect to the operations in the term algebra, precisely because T^+ proves the equality axioms. Finally, the usual Henkin analysis shows that the quotient A/E is a model of T^+, and in particular, it provides a computable quotient presentation of T^2.   □

The previous observation is closely connected with a fundamental fact of universal algebra, namely, the fact that every algebraic structure is a quotient of the term algebra on a sufficient number of generators. Every countable group, for example, is a quotient of the free group on countably many generators, and more generally, every countable algebra (a structure in a language with no relations) arises as the quotient of the term algebra on a countable number of generators. Since the term algebra of a computable language is a computable structure, it follows that every countable algebra in a computable language admits a computable quotient presentation.

One of the guiding ideas of the theory of computable quotients is to take from this observation the perspective that the complexity of an algebraic structure is contained not in its atomic diagram, often studied in computable model theory, but rather solely in its equality relation. The algebraic structure on the term algebra, after all, is computable; what is difficult is knowing when two terms represent the same object. Thus, the program is to investigate which equivalence relations E or classes of equivalence relations can give rise to a domain $\mathbb{N}/E$ for a given type of mathematical structure. There are many open questions and the theory is just emerging.

We should like to call particular attention to the fact that the proof method of Observation 1 and the related observation of universal algebra breaks down when the language has relation symbols, because the corresponding relation for the resulting Henkin model will not generally be computable on the term algebra or even just on the constants. The complexity of the relation in the quotient structure arises from the particular branch that was chosen through the Henkin tree or equivalently from the Henkin theory itself. So it seems difficult to use the Henkin theory idea to produce computable quotient presentations of relational theories. We shall see later how this relational obstacle plays out in the case of arithmetic, whose usual language $\{+, \cdot, 0, 1, <\}$ includes a relation symbol, and especially in the case of set theory, whose language $\{\in\}$ is purely relational.

---

[2] Obviously, a low degree does not have to be c.e. and cannot be in this construction.

Let us now prove that Khoussainov's first conjecture is true.

Theorem 2. *No nonstandard model of arithmetic has a computable quotient presentation by a c.e. equivalence relation. Indeed, this is true even in the restricted (but fully expressive) language $\{+, \cdot\}$ with only addition and multiplication: there is no computable structure $\langle \mathbb{N}, \oplus, \odot \rangle$ and a c.e. equivalence relation E, which is a congruence with respect to this structure, such that the quotient $\langle \mathbb{N}, \oplus, \odot \rangle / E$ is a nonstandard model of arithmetic.*

Proof. Suppose toward contradiction that E is a computably enumerable equivalence relation on the natural numbers, that $\langle \mathbb{N}, \oplus, \odot \rangle$ is a computable structure with computable binary operations $\oplus$ and $\odot$, that E is a congruence with respect to these operations and that the quotient structure $\langle \mathbb{N}, \oplus, \odot \rangle / E$ is a nonstandard model of arithmetic. A very weak theory of arithmetic suffices for this argument.

Let $\bar{0}$ be a number representing zero in $\langle \mathbb{N}, \oplus, \odot \rangle / E$ and let $\bar{1}$ be a number representing one. Since $\oplus$ is computable, we can computably find numbers $\bar{n}$ representing the standard number n in $\langle \mathbb{N}, \oplus, \odot \rangle / E$ simply by computing $\bar{n} = \bar{1} \oplus \cdots \oplus \bar{1}$.

Let A and B be computably inseparable c.e. sets in the standard natural numbers. So they are disjoint c.e. sets for which there is no computable set containing A and disjoint from B. Fix Turing machine programs p_A and p_B that enumerate A and B, respectively. We shall run these programs inside the nonstandard model $\langle \mathbb{N}, \oplus, \odot \rangle / E$. Although every actual element of A will be enumerated by p_A inside the model at some standard stage, and similarly for B and p_B, the programs p_A will also enumerate nonstandard numbers into the sets, and it is conceivable that at nonstandard stages of computation, the program p_A might place standard numbers into its set, even when those numbers are not in A. In particular, there is no guarantee in general that the sets enumerated by p_A and p_B in $\langle \mathbb{N}, \oplus, \odot \rangle / E$ will be disjoint.

Nevertheless, we proceed as follows. In the quotient structure, fix any non-standard number c, and let $\tilde{A}$ be the set of elements below c that in the quotient structure $\langle \mathbb{N}, \oplus, \odot \rangle / E$ are thought to be enumerated by p_A before they are enumerated by p_B. Since every actual element of A is enumerated by p_A at a standard stage, and not by p_B by that stage, it follows that the elements of A are all in $\tilde{A}$, in the sense that whenever $n \in A$, then $\bar{n}$ is in $\tilde{A}$. Similarly, since the actual elements of B are enumerated by p_B at a standard stage and not by p_A by that stage, it follows that none of the actual elements of B will enter $\tilde{A}$.

$$n \in A \quad \rightarrow \quad \bar{n} \in \tilde{A}$$
$$n \in B \quad \rightarrow \quad \bar{n} \notin \tilde{A}$$

Thus, the set $C = \{ n \mid \bar{n} \in \tilde{A} \}$ contains A and is disjoint from B. We shall prove that C is computable.

Since $\tilde{A}$ is definable inside $\langle \mathbb{N}, \oplus, \odot \rangle / E$, it is coded by an element of this structure. Let us use the prime-product coding method. Namely, inside the non-standard model let p_k be the kth prime number, and let a be the product of the p_k for which $k < c$ and $k \in \tilde{A}$.

Next, the key idea of the proof, we let b be the corresponding code for the complement of $\tilde{A}$ below c. That is, b is the product of the p_k for which $k < c$ and $k \notin \tilde{A}$. We shall use both a and b to decode the set.

Given any number n, we can compute $\bar{p}_n$ and then search for a number x for which $(x \odot \bar{p}_n)\ E\ a$. In other words, we are searching for a witness that $\bar{p}_n$ divides a, from which we could conclude that $\bar{n} \in \tilde{A}$ and so $n \in C$. At the same time, we search for a number y for which $(y \odot \bar{p}_n)\ E\ b$. Such a y would witness that $\bar{p}_n$ divides b and therefore that $\bar{n} \notin \tilde{A}$ and hence $n \notin C$. The main point is that one or the other of these things will happen, since a and b code complementary sets, and so in this way we can compute whether $n \in C$ or not. So C is a computable separation of A and B, contrary to our assumption that they were computably inseparable.   □

By replacing $x \odot \bar{p}_n$ in the proof with $x \oplus x \oplus \cdots \oplus x$, using p_n many factors, we may deduce the Tennenbaum-style result that if $\langle \mathbb{N}, \oplus, \odot \rangle / E$ is a nonstandard model of arithmetic and E is c.e., then $\oplus$ is not computable. That is, we don't need both operations in the pre-quotient structure to be computable. Similar remarks will apply to many of the other theorems in this article, and we shall explore this one-operation-at-a-time issue more fully in our follow-up article.

An alternative proof of Theorem 2 proceeds as follows. Consider the *standard system* of any nonstandard model of arithmetic, which is the collection of traces on the standard $\mathbb{N}$ of the sets that are coded inside the model. Using the prime-product coding, for example, these can be seen as sets of the form $\{ n \mid \bar{p}_n \text{ divides } a \}$, where a is an arbitrary element of the model, p_n means the nth prime number and $\bar{p}_n$ means the object inside the model that represents that prime number. It is a theorem of Scott that the standard systems of the countable nonstandard models of PA are precisely the countable *Scott* sets, which are sets of subsets of $\mathbb{N}$ that form a Boolean algebra, are closed downward under relative computability, and contain paths through any infinite binary tree coded in them. Because there is a computable tree with no computable path, every standard system must have noncomputable sets and therefore non-c.e. sets, since it is closed under complements.

For the alternative proof of Theorem 2, the main point is that the assumptions of the theorem ensure that every set in the standard system of the quotient model $\langle \mathbb{N}, \oplus, \odot \rangle / E$ is c.e., contradicting the fact we just mentioned. The reason is that for any object a, the number n is in the set coded by a just in case $\bar{p}_n$ divides a, and this occurs just in case there is a number x for which $(x \odot \bar{p}_n)\ E\ a$, which is a c.e. property since E is c.e. and $\odot$ is computable. So every set in the standard system would be c.e., contrary to the fact we mentioned earlier.

Another alternative proof of a version of Theorem 2 handles the case of nonstandard models in the full language of arithmetic $\{ +, \cdot, 0, 1, < \}$. Namely, if E is c.e. and $\langle \mathbb{N}, \oplus, \odot, \bar{0}, \bar{1}, \lhd \rangle$ is a computably enumerable structure whose quotient by E is a nonstandard model of arithmetic, then it follows from the next lemma that E must also be co-c.e., and hence computable. And once we know that E is computable, we may construct a computable nonstandard model of arithmetic, by using least representatives in each equivalence class, and this would contradict Tennenbaum's theorem, which says that there is no computable nonstandard model of arithmetic.

Lemma 3. *Suppose that E is an equivalence relation on the natural numbers.*

(1) If E is a congruence with respect to a computable relation $\lhd$ and the quotient $\langle \mathbb{N}, \lhd \rangle / E$ is a strict linear order, then E is computable.

(2) If E is a congruence with respect to a c.e. relation $\lhd$ and the quotient $\langle \mathbb{N}, \lhd \rangle / E$ is a strict linear order, then E is co-c.e.

(3) If E is a congruence with respect to a computable relation $\unlhd$ and the quotient $\langle \mathbb{N}, \unlhd \rangle / E$ is a reflexive linear order or merely an anti-symmetric relation, then E is computable.

(4) If E is a congruence with respect to a c.e. relation $\unlhd$ and the quotient $\langle \mathbb{N}, \unlhd \rangle / E$ is a reflexive linear order $\unlhd$ or merely anti-symmetric, then E is c.e.

Proof. For statement (1), suppose that E is a congruence with respect to a computable relation $\lhd$ and the quotient is a strict linear order. Since the quotient relation obeys

$$x \neq y \quad \leftrightarrow \quad x < y \text{ or } y < x,$$

it follows that

$$\neg(x \mathrel{E} y) \quad \leftrightarrow \quad x \lhd y \text{ or } y \lhd x.$$

Since this latter property is computable, it follows that E is computable. For statement (2), similarly, the latter property is c.e., and so E is co-c.e.

For statement (3), suppose that E is a congruence with respect to a computable relation $\unlhd$, whose quotient is anti-symmetric. Since the quotient relation satisfies

$$x = y \quad \leftrightarrow \quad x \leq y \text{ and } y \leq x,$$

it follows that

$$x \mathrel{E} y \quad \leftrightarrow \quad x \unlhd y \text{ and } y \unlhd x.$$

If $\unlhd$ is computable, as in statement (3), then E will be computable. And if $\unlhd$ is computably enumerable, as in statement (4), then E must be c.e. □

In particular, including $<$ or $\leq$ in the language of arithmetic and asking for a computable or computably enumerable quotient presentation with respect to E will impose certain complexity requirements on E, simply in order that E is a congruence with respect to the order relation.

Using this idea, the following corollary to Theorem 2 settles the version of Khoussainov's second conjecture for the language $\{+, \cdot, \leq\}$. By referring to the language of arithmetic with $\leq$, we intend the theory of arithmetic expressed in terms of the natural reflexive order relation, rather than the usual strict order relation $<$.

Corollary 4. *No nonstandard model of arithmetic in the language $\{+, \cdot, \leq\}$ has a computably enumerable quotient presentation by any equivalence relation, of any complexity. That is, there is no computably enumerable structure $\langle \mathbb{N}, \oplus, \odot, \trianglelefteq \rangle$, where $\oplus$ and $\odot$ are computable binary operations and $\trianglelefteq$ is a computably enumerable relation, and an equivalence relation E that is a congruence with respect to that structure, such that the quotient $\langle \mathbb{N}, \oplus, \odot, \trianglelefteq \rangle / E$ is a nonstandard model of arithmetic in the language $\{+, \cdot, \leq\}$.*

Proof. Suppose toward contradiction that E is an equivalence relation that is a congruence with respect to computable functions $\oplus$ and $\odot$ and c.e. relation $\trianglelefteq$ for which the quotient structure $\langle \mathbb{N}, \oplus, \odot, \trianglelefteq \rangle / E$ is a nonstandard model of arithmetic. Because the quotient of $\trianglelefteq$ by E is a reflexive linear order, it follows by Lemma 3 that E must be c.e., and so the corollary follows directly from Theorem 2. □

Let's now consider another version of the second conjecture and the case of co-c.e. equivalence relations. We shall refute the versions of the second conjecture for which the quotient model is to exhibit a certain degree of soundness.

Let's begin with an extreme version of this phenomenon, where we ask for far too much: models of true arithmetic. A model of *true arithmetic* is a model with the same theory as the standard model of arithmetic. Equivalently, it is an elementary extension of the standard model inside it. After ruling out this extreme case, we shall than sharpen the result to the case of Σ_1-soundness and much less. Recall that a theory T is Σ_1-sound, if for any Σ_1-sentence φ, if φ is provable in T, then φ is true in the standard model $\mathbb{N}$.

Theorem 5. *There is no computable structure $\langle \mathbb{N}, \oplus, \odot \rangle$ and a co-c.e. equivalence relation E, which is a congruence with respect to this structure, such that the quotient $\langle \mathbb{N}, \oplus, \odot \rangle / E$ is a nonstandard model of true arithmetic.*

Proof. Suppose that $\langle \mathbb{N}, \oplus, \odot \rangle$ is a computable structure and E is a co-c.e. equivalence relation, a congruence with respect to this structure, whose quotient $\langle \mathbb{N}, \oplus, \odot \rangle / E$ is a nonstandard model of true arithmetic. As in the earlier proof, let $\bar{1}$ be a representative of the number 1 inside this model and let $\bar{n}$ be the result of adding $\bar{1}$ to itself n times with $\oplus$ inside the model, so that $\bar{n}$ is a representative for what the quotient model thinks is the standard number n.

Since the quotient model satisfies true arithmetic, it follows that it is correct about the halting problem on standard numbers. So there is a number h that codes the halting problem up to some nonstandard length c of computations. In particular, for standard n we shall have that $n \in 0'$ if and only if $\bar{n}$ is in the set coded by h. Another way to say this is that $0'$ is in the standard system of the quotient model, and this is all we actually require of true arithmetic here.

Let A and B be $0'$-computably inseparable sets, that is, sets that are computably enumerable relative to an oracle for the halting problem $0'$, but there is no $0'$-decidable separating set. Let p_A and p_B be the programs that enumerate A and B from an oracle for $0'$. Inside the nonstandard model $\langle \mathbb{N}, \oplus, \odot \rangle / E$, we may run p_A and p_B with the oracle determined by h, which happens to agree with $0'$

on the standard numbers. In particular, on standard input n, the computation with oracle h inside the model will agree at the standard stages of computation with the actual computation using the real oracle $0'$.

Let $\tilde{A}$ be the elements $k < c$ that are enumerated by p_A^h before they are enumerated by p_B^h. As before, our assumptions ensure that every actual element of A is in $\tilde{A}$, and no element of B is in $\tilde{A}$.

$$n \in A \quad \rightarrow \quad \bar{n} \in \tilde{A}$$
$$n \in B \quad \rightarrow \quad \bar{n} \notin \tilde{A}$$

Thus, the set C of standard n for which $\bar{n} \in \tilde{A}$ is a set that contains A and is disjoint from B.

It remains for us to show for the contradiction that C is computable from $0'$. As before, inside the quotient model, let a be the product of p_k for k in $\tilde{A}$, and let b be the product of p_k for k not in $\tilde{A}$. Given n, we want to determine whether $n \in C$ or not, which is equivalent to $\bar{n} \in \tilde{A}$. We can compute $\bar{p}_n$, and then we can try to discover if $\bar{p}_n$ divides a or $\bar{p}_n$ divides b. Note that $\bar{p}_n$ divides a just in case $\exists x \, (x \odot \bar{p}_n) \, E \, a$, which has complexity Σ_2, since E is Π_1. Similarly, the relation $\bar{p}_n$ divides b is also Σ_2. But since these answers are opposite, it follows that both of these relations are Δ_2, and hence computable from $0'$. So the relation $n \in C$ is computable from $0'$, and we have therefore found a $0'$-computable separating set C, contradiction our assumption that A and B were $0'$-computably inseparable. □

We could alternatively have argued as in the alternative proof of Theorem 2 that every element of the standard system of the model is computable from $0'$, which is a contradiction if one knows that $0'$ is in the standard system.

Of course, true arithmetic was clearly much too strong in this theorem, and we could also have given a more direct alternative proof just by extracting higher-order arithmetic truths from this model in a Σ_2 or even Δ_2-manner, since the pre-quotient model is computable and the relation is co-c.e. So a better theorem will eliminate or significantly weaken the true-arithmetic hypothesis, as we do in the following sharper result.

Theorem 6. *There is no computable structure $\langle \mathbb{N}, \oplus, \odot \rangle$ and a co-c.e. equivalence relation E, which is a congruence with respect to this structure, such that the quotient $\langle \mathbb{N}, \oplus, \odot \rangle /E$ is a Σ_1-sound nonstandard model of arithmetic, or even merely a nonstandard model of arithmetic with $0'$ in the standard system of the model.*

Proof. If the model is Σ_1-sound, then it computes the halting problem correctly, and so $0'$ will be in the standard system of the model, which means that it has a code h as in the proof above. That was all that was required in the previous argument, and so the same contradiction is achieved. □

Corollary 7. *No nonstandard model of arithmetic in the language $\{ +, \cdot, 0, 1, < \}$ and with $0'$ in its standard system has a computably enumerable quotient presentation by any equivalence relation, of any complexity.*

Proof. If $\langle \mathbb{N}, \oplus, \odot, \bar{0}, \bar{1}, \lhd \rangle / E$ is such a computably enumerable quotient presentation, then Lemma 3 shows that E must be co-c.e., and so the situation is ruled out by Theorem 6. □

Note that containing $0'$ in the standard system is a strictly weaker property than being Σ_1-sound, since a simple compactness argument allows us to insert any particular set into the standard system of an elementary extension of any particular model of arithmetic.

Our results do not settle what might be considered the central case of the second conjecture, which remains open. We are inclined to expect a negative answer, whereas Khoussainov has conjectured a positive answer.

Question 8. *Is there a nonstandard model of* PA *in the usual language of arithmetic* $\{+, \cdot, 0, 1, <\}$ *that has a computably enumerable quotient presentation by some co-c.e. equivalence relation? Equivalently, is there a nonstandard model of* PA *in that language with a computably enumerable quotient presentation by any equivalence relation, of any complexity?*

The two versions of the question are equivalent by Lemma 3, which shows that in the language with the strict order, the equivalence relation must in any case be co-c.e.

Let us now consider the analogous ideas for the models of set theory, rather than for the models of arithmetic. We take this next theorem to indicate how the program of computable quotient presentations has difficulties with purely relational structures.

Theorem 9. *No model of* ZFC *has a computable quotient presentation. That is, there is no computable relation ϵ and equivalence relation E, a congruence with respect to ϵ, for which the quotient $\langle \mathbb{N}, \epsilon \rangle / E$ is a model of* ZFC. *Indeed, no such computable quotient is a model of* KP *or even considerably weaker set theories.*

Just to emphasize, we do not assume anything about the complexity of the equivalence relation E, which can be arbitrary, or about whether the quotient model of set theory $\langle \mathbb{N}, \epsilon \rangle / E$ is well-founded or ill-founded, standard or nonstandard. Note also the typographic distinction between the relation ϵ, which is the computable relation of the pre-quotient structure $\langle \mathbb{N}, \epsilon \rangle$, and the ordinary set membership relation $\in$ of set theory.

Proof. Suppose toward contradiction that ϵ is a computable relation on $\mathbb{N}$ and that E is an equivalence relation, a congruence with respect to ϵ, for which the quotient $\langle \mathbb{N}, \epsilon \rangle / E$ is a model of set theory. We need very little strength in the set theory, and even an extremely weak set theory suffices for the argument. We shall use the Kuratowski definition of ordered pair in set theory, for which $\langle x, y \rangle = \{\{x\}, \{x, y\}\}$.

Since set theory proves that the set of natural numbers exists, there is some $N \in \mathbb{N}$ that the quotient model thinks represents the set of all natural numbers.

Also, this model thinks that various kinds of sets involving natural numbers exist, such as the set coding the successor relation

$$S = \{ \, \langle n, n+1 \rangle \mid n \in \mathbb{N} \, \}.$$

To be clear, we mean that S is a number in $\mathbb{N}$ that the quotient model $\langle \mathbb{N}, \epsilon \rangle / E$ thinks is the set of the successor relation we identify above. So the ϵ-elements of S will all be thought to be Kuratowski pairs of natural numbers in the model, and this could include nonstandard numbers if there are any.

Similarly, we have sets consisting of the natural number singletons and doubletons.

$$\text{Sing} = \{ \, \{n\} \mid n \in \mathbb{N} \, \},$$
$$\text{Doub} = \{ \, \{n, m\} \mid n \neq m \text{ in } \mathbb{N} \, \}.$$

To be clear, we mean that Sing and Doub are particular elements of $\mathbb{N}$ that in the quotient model $\langle \mathbb{N}, \epsilon \rangle / E$ are thought to be the sets defined by those set-theoretic expressions. We assume that our set theory proves that these sets exist.

Next, we claim that there is a computable function $n \mapsto \bar{n}$, such that $\bar{n}$ represents what the quotient model $\langle \mathbb{N}, \epsilon \rangle / E$ thinks is the standard natural number n.[3] To see this, we may fix a number $\bar{0}$ that represents the number 0. Next, given $\bar{n}$, we search for an element $d \,\epsilon\, S$ that will represent the pair $\langle \bar{n}, m \rangle$, and when found, we set $\overline{n+1} = m$. How shall we recognize this d and m using only ϵ? Well, the d we want has the form $\{ \, \{\bar{n}\}, \, \{\bar{n}, m\} \, \}$ inside the model, and so we search for an element $d \,\epsilon\, S$ that has an element $x \,\epsilon\, d$ with $x \,\epsilon\, \text{Sing}$ and $\bar{n} \,\epsilon\, x$. This x must represent the set $\{\bar{n}\}$, since x is thought to have only one element, since it is in Sing. Having found d, we search for $y \,\epsilon\, d$ with $y \in \text{Doub}$ and an element m with $m \,\epsilon\, y$, but $\neg(m \,\epsilon\, x)$. In this case, it must be that y represents $\{\bar{n}, m\}$, and so we may let $\overline{n+1} = m$ and proceed. So the map $n \mapsto \bar{n}$ is computable.

It follows that every set in the standard system of the model $\langle \mathbb{N}, \epsilon \rangle / E$ is computable. Specifically, if a is any element of the model, then the trace of this object on the natural numbers is the set $\{ \, n \in \mathbb{N} \mid \bar{n} \,\epsilon\, a \, \}$, which would be a computable set, since both ϵ and the map $n \mapsto \bar{n}$ is computable.

But we mentioned earlier that every model of set theory and indeed of arithmetic must have non-computable sets in its standard system, so this is a contradiction.   □

We could have argued a little differently in the proof. Namely, if ϵ is a computable relation with a congruence E and $\langle \mathbb{N}, \epsilon \rangle / E$ is a model of set theory, then by the axiom of extensionality, we have

$$x \neq y \quad \leftrightarrow \quad \exists z \, \neg(z \in x \leftrightarrow z \in y).$$

In the pre-quotient model, this amounts to:

$$\neg(x \, E \, y) \quad \leftrightarrow \quad \exists z \neg(z \,\epsilon\, x \leftrightarrow z \,\epsilon\, y).$$

_____

[3] Apparently, a similar argument appears in [Rab58].

Thus, in the case that ϵ is computable, in analogy with Lemma 3 we may deduce from this that E must be co-c.e., even though we had originally made no assumption on the complexity of E. And in this case, the theorem follows from the next result.

Theorem 9 shows that it is too much to ask for computable quotient presentations of models of set theory. So let us relax the computability requirement on the pre-quotient membership relation ϵ by considering the case of computably enumerable quotient presentations, where ϵ is merely c.e. rather than computable. In this case, we can still settle the second conjecture by ruling out quotient presentations by co-c.e. equivalence relations.

Theorem 10. *There is no c.e. relation ϵ with a co-c.e. equivalence relation E respecting it for which $\langle \mathbb{N}, \epsilon \rangle / E$ is a model of set theory.*

Proof. In the proof of Theorem 9, we had used the computability of ϵ, as opposed to the computable enumerability of ϵ, in the step where we needed to know $\neg(m \, \epsilon \, x)$. At that step of the proof, really what we needed to know was that m and $\bar{n}$ were not representing the same object. But if E is co-c.e., then we can learn that $\bar{n} \neq m$ simply by waiting to see that $\bar{n} \, E \, m$ fails, which if true will happen at some finite stage since E is co-c.e. Indeed, it is precisely with the co-c.e. equivalence relations E that one is entitled to know by some finite stage that two numbers represent different objects in the quotient. Therefore, if E is co-c.e., we still get a computable map $n \mapsto \bar{n}$. And then, in the latter part of the proof, we would conclude that every set in the standard system is c.e., since the trace of any object a in the model on the natural numbers is the set $\{ n \in \mathbb{N} \mid \bar{n} \, \epsilon \, a \}$, which would be c.e. But every standard system must contain non-c.e. sets, by the paths-through-trees argument, since it contains non-computable sets and it is a closed under complements. So again we achieve a contradiction. $\square$

Let us now explore the analogues of the earlier theorems for nonstandard models of finite set theory. Let $ZF^{\neg\infty}$ denote the usual theory of finite set theory, which includes all the usual axioms of ZFC, but without the axiom of infinity, plus the negation of the axiom of infinity and plus the $\in$-induction scheme formulation of the foundation axiom. This theory is true in the structure $\langle HF, \in \rangle$ of hereditarily finite sets, and it is bi-interpretable with PA via the Ackermann relation on natural numbers.[4]

Theorem 11. *There is no computable relation ϵ and equivalence relation E, a congruence with respect to ϵ, of any complexity, such that the quotient $\langle \mathbb{N}, \epsilon \rangle / E$ is a nonstandard model of finite set theory $ZF^{\neg\infty}$.*

---

[4] Some researchers have also considered another strictly weaker version of this theory, omitting the $\in$-induction scheme. But it turns out that this version of the theory is flawed for various reasons: it cannot prove that every set has a transitive closure; it is not bi-interpretable with PA; it does not support the Tennenbaum phenomenon (see [ESV11]). Meanwhile, since all these issues are addressed by the more attractive and fruitful theory $ZF^{\neg\infty}$, we prefer to take this theory as the meaning of 'finite set theory'.

Proof. Assume that ϵ is a computable relation for which $\langle \mathbb{N}, \epsilon \rangle / E$ is a nonstandard model of $ZF^{\neg\infty}$. The ordinals of this model with their usual arithmetic form a nonstandard model of PA, which we may view as the natural numbers of the model. Let N be a number representing a nonstandard such natural number in $\langle \mathbb{N}, \epsilon \rangle / E$. There is a set S representing the set $\{ \langle n, n+1 \rangle \mid n < N \}$ as defined inside the model, and similarly we have sets representing the natural number singletons and doubletons up to N.

$$\text{Sing} = \{ \{n\} \mid n \in \mathbb{N}, \ n < N \},$$
$$\text{Doub} = \{ \{n, m\} \mid n \neq m \text{ in } \mathbb{N}, \ n, m < N \}.$$

So Sing and Doub are particular numbers in $\mathbb{N}$ that in the quotient $\langle \mathbb{N}, \epsilon \rangle / E$ represent the sets we have just defined by those expressions.

We may now run essentially the same argument as in the proof of Theorem 9. Namely, we may define a computable function $n \mapsto \bar{n}$, where $\bar{n}$ represents the natural number n in the model $\langle \mathbb{N}, \epsilon \rangle / E$, by using the parameters S, Sing and Doub and decoding via the Kuratowski pair function as before. This argument uses the computability of ϵ as before in order to produce $\overline{n+1}$ from $\bar{n}$. Finally, we use this function to show that every set in the standard system of the model is computable, since for any object a, the trace of a on the natural numbers is the set of n for which $\bar{n} \epsilon a$, which is a computable property. This contradicts the fact that the standard system of any nonstandard model of $ZF^{\neg\infty}$ must include non-computable sets. $\qquad\square$

Finally, we have the analogue of Theorem 10 for the case of finite set theory.

Theorem 12. *There is no c.e. relation ϵ with a co-c.e. equivalence relation E respecting it for which $\langle \mathbb{N}, \epsilon \rangle / E$ is a nonstandard model of finite set theory $ZF^{\neg\infty}$.*

Proof. This theorem is related to Theorem 11 the same way that Theorem 10 is related to Theorem 9. Namely, in the proof of Theorem 11, we used the computability of the membership relation ϵ in the step computing the function $n \mapsto \bar{n}$. If ϵ is merely computably enumerable, as here, then we can nevertheless still find a computable function $n \mapsto \bar{n}$, provided that the equivalence relation E is co-c.e., since in the details of the proof as explained in Theorem 10, we needed to know that we had found the right value for $\overline{n+1}$ by knowing that a certain number m was actually representing a different number than $\bar{n}$, and it is precisely with a co-c.e. equivalence relation E that one can know such a thing at some finite stage.

If $n \mapsto \bar{n}$ is computable, then with a c.e. relation ϵ, we can deduce that every set in the standard system is c.e., since a codes the set of n for which $\bar{n} \epsilon a$, a c.e. property, and this contradicts the fact that every standard system of a nonstandard model of $ZF^{\neg\infty}$ must contain non-c.e. sets. $\qquad\square$

We expect to follow up this article with a second article containing several more refined results.

References

[ESV11] Enayat, A., Schmerl, J., Visser, A.: ω-models of finite set theory. In: Kennedy, J., Kossak, R. (eds.) Set Theory, Arithmetic, and Foundations of Mathematics: Theorems, Philosophies. Lecture Notes in Logic, no. 36. Cambridge University Press, Cambridge (2011)

[Kho16] Khoussainov, B.: Computably enumerable structures: domain dependence, September 2016. Slides for Conference Talk at Mathematical Logic and Its Applications, Research Institute for Mathematical Sciences (RIMS), Kyoto University. http://www2.kobe-u.ac.jp/~mkikuchi/mla2016khoussainov.pdf

[Rab58] Rabin, M.O.: On recursively enumerable and arithmetic models of set theory. J. Symb. Log. **23**(4), 408–416 (1958)

Lattice Logic Properly Displayed

Giuseppe Greco[1(✉)] and Alessandra Palmigiano[1,2]

[1] Delft University of Technology, Delft, The Netherlands
{G.Greco,A.Palmigiano}@tudelft.nl
[2] University of Johannesburg, Johannesburg, South Africa

Abstract. We introduce a *proper* display calculus for (non-distributive) Lattice Logic which is sound, complete, conservative, and enjoys cut-elimination and subformula property. Properness (i.e. closure under uniform substitution of all parametric parts in rules) is the main interest and added value of the present proposal, and allows for the smoothest Belnap-style proof of cut-elimination, and for the most comprehensive account of axiomatic extensions and expansions of Lattice Logic in a single overarching framework. Our proposal builds on an algebraic and order-theoretic analysis of the semantic environment of lattice logic, and applies the guidelines of the multi-type methodology in the design of display calculi.

Keywords: Lattice logic · Substructural logics · Algebraic proof theory · Sequent calculi · Cut elimination · Display calculi · Multi-type calculi

2010 Math. Subj. Class. 03F52 · 03F05 · 03G10 · 06A15 · 06B15 · 08A68 · 18A40

1 Introduction

Lattice logic (i.e. the restriction of classical propositional logic to the $\{\wedge, \vee, \top, \bot\}$-fragment without distributivity) is the propositional base of many well known 'lattice-based' logics (e.g. the full Lambek calculus [31], bilattice logic [1], orthologic [35], linear logic [33]), and as such is hardly ever studied in isolation. An important question in structural proof theory concerns how to smoothly account for the transition between a given (lattice-based) logic and its axiomatic extensions and expansions [50, Chap. 1], [23, p. 352], [24,27]. In the present paper, we introduce a calculus for lattice logic aimed at supporting this smooth transition for a class of lattice-based logics which is the widest so far.

Toward this goal, research in structural proof theory [24,46,50] has identified two general criteria: (a) all introduction rules for logical connectives are to have one and the same form; (b) the information on the distinctive features of each logical connective and on the interaction between connectives is to be encoded in structural rules satisfying certain requirements, captured by the notion of *analiticity*. These criteria are

This research has been funded by the NWO Vidi grant 016.138.314, the NWO Aspasia grant 015.008.054, and a Delft Technology Fellowship awarded to the second author in 2013.

© Springer-Verlag GmbH Germany 2017
J. Kennedy and R.J.G.B. de Queiroz (Eds.): WoLLIC 2017, LNCS 10388, pp. 153–169, 2017.
DOI: 10.1007/978-3-662-55386-2_11

fulfilled, among others, by *proper display calculi*, a refinement of Belnap's display calculi [2] introduced by Wansing [50]. However, in most calculi for lattice-based logics (cf. e.g. [47,48]), including display calculi [3], the introduction rules for conjunction and disjunction have the so-called *additive* form, while those of the other connectives typically are in *multiplicative form* (see Sect. 2.2). More fundamentally, conjunction and disjunction do not have structural counterparts in these calculi. This non-standard treatment can be explained, in the setting of display calculi, by the following trade-off: introducing the structural counterparts of these connectives would require the addition of the *display postulates* in order to enforce the *display property*, which is key to the Belnap-style cut elimination metatheorem [2,50]; however, the addition of display postulates would make it possible for the resulting calculus to derive the unwanted distributivity axioms as theorems. So, the need to block the derivation of distributivity is at the root of the non-standard design choice of having logical connectives without their structural counterpart (cf. [4]).

In the present paper, we introduce the proper display calculus D.LL for lattice logic which enjoys the *full display* property, and is such that all introduction rules have the same form (namely the *multiplicative* form). We succeed in circumventing the trade-off described above by introducing a richer language, with terms of different types. This solution applies the principles of a design for proof calculi (the *multi-type display calculi*) introduced in [25,26,28,36] with the aim of displaying dynamic epistemic logic and propositional dynamic logic, then successfully applied to several other logics (such as linear logic with exponentials [39], inquisitive logic [29], semi De Morgan logic [34]) which are not properly displayable[1] in their single-type formulation, and has also served as a platform for the design of novel logics [5].

The main feature of D.LL is that it makes it possible to express the interactions between conjunction and disjunction and between them and other connectives at the structural level, by means of *analytic* structural rules. The remarkable property of these rules is that adding them to a given proper multi-type calculus preserves the package of basic properties of that calculus (soundness, completeness, conservativity, cut elimination and subformula property). This is all the more an advantage, because a uniform theory of analytic extensions of proper multi-type calculi is being developed thanks to the systematic connections established in [37] between proper display calculi and unified correspondence theory. These connections have made it possible to characterize the syntactic shape of axioms (the so-called *analytic inductive* axioms) which can be equivalently translated into analytic rules of a proper display calculus. Thus, the main feature of this calculus paves the way to the creation of the most comprehensive and modular proof theory of analytic extensions of lattice-based logics. The specific solution for lattice logic is justified semantically by Birkhoff's representation theorem for complete lattices.

---

[1] *Properly displayable* logics (cf. [50]) are those amenable to be presented in the form of a proper display calculus. The notion of properly displayable logic has been characterized in a purely proof-theoretic way in [9]. In [37], an alternative characterization of properly displayable logics has been proposed which builds on the algebraic theory of unified correspondence [10,12,14,15,18–21,30,42–45]. The techniques and insights of unified correspondence are also available for lattice-based logics, cf. [11,13,16,17]).

Structure of the Paper. In Sect. 2, we briefly report on a Hilbert-style presentation of lattice logic and its algebraic semantics, and discuss the issue of a modular account of its axiomatic extensions and expansions. In Sect. 3, we report on well known order-theoretic facts related with the representation of complete lattices, which help to introduce an equivalent multi-type semantic environment for lattice logic. In Sect. 4, we introduce the multi-type language naturally associated with the semantic environment of the previous section. In Sect. 5, we introduce the multi-type calculus D.LL for lattice logic which constitutes the core contribution of the present paper. In Sect. 6, we discuss the basic properties of D.LL (soundness, completeness, cut-elimination, subformula property, and conservativity). In Sect. B, we collect some derivations, and prove that (the translation of) the distributivity axiom is not derivable in D.LL.

2 Lattice Logic and Its Single-Type Proof Theory

2.1 Hilbert-Style Presentation of Lattice Logic and Its Algebraic Semantics

The language $\mathcal{L}$ of lattice logic over a set AtProp of atomic propositions is so defined:

$$A ::= p \mid \top \mid \bot \mid A \wedge A \mid A \vee A.$$

Lattice logic has the following Hilbert-style presentation:

$$A \vdash A, \quad \bot \vdash A, \quad A \vdash \top, \quad A \vdash A \vee B, \quad B \vdash A \vee B, \quad A \wedge B \vdash A, \quad A \wedge B \vdash B$$

$$\frac{A \vdash B \quad B \vdash C}{A \vdash C} \qquad \frac{A \vdash B}{A[C/p] \vdash B[C/p]} \qquad \frac{A \vdash B \quad A \vdash C}{A \vdash B \wedge C} \qquad \frac{A \vdash C \quad B \vdash C}{A \vee B \vdash C}$$

where $A[C/p]$ indicates that all occurrences of $p \in$ AtProp in A are replaced by C.

The algebraic semantics of lattice logic is given by the class of *bounded lattices* (cf. [6,8]), i.e. $(2,2,0,0)$-algebras $\mathbb{A} = (X, \wedge, \vee, \top, \bot)$ validating the following identities:

Commutative laws	Associative laws
cC. $a \wedge b = b \wedge a$	cA. $a \wedge (b \wedge c) = (a \wedge b) \wedge c$
dC. $a \vee b = b \vee a$	dA. $a \vee (b \vee c) = (a \vee b) \vee c$
Identity laws	Absorption laws
cI. $a \wedge \top = a$	cAb. $a \wedge (a \vee b) = a$
dI. $a \vee \bot = a$	dAb. $a \vee (a \wedge b) = a$

A bounded lattice is *distributive* if it validates the distributivity laws below. A bounded lattice is *residuated* (resp. *dually residuated*) if it validates the residuation law cR (resp. dR). If a lattice is (dually) residuated then is distributive (cf. [22,31]).

Distributivity laws	Residuation laws
cD. $a \wedge (b \vee c) = (a \wedge b) \vee (a \vee c)$	cR. $a \wedge b \leq c$ iff $b \leq a \rightarrow c$
dD. $a \vee (b \wedge c) = (a \vee b) \wedge (a \vee c)$	dR. $a \leq b \vee c$ iff $b \succ a \leq c$

2.2 Towards a Modular Proof Theory for Lattice Logic

To motivate the calculus introduced in Sect. 5, it is useful to discuss preliminarily the following Gentzen-style sequent calculus for lattice logic (cf. e.g. [49]):

- Identity and Cut rules

$$\frac{}{p \vdash p}\ Id \qquad \frac{X \vdash A \qquad A \vdash Y}{X \vdash Y}\ Cut$$

- Operational rules (where $i \in \{1, 2\}$)

$$\frac{}{\bot \vdash I}\ \bot \qquad \frac{X \vdash I}{X \vdash \bot}\ \bot \qquad \wedge_i \frac{A_i \vdash X}{A_1 \wedge A_2 \vdash X} \qquad \frac{X \vdash A \qquad X \vdash B}{X \vdash A \wedge B}\ \wedge$$

$$\top \frac{I \vdash X}{\top \vdash X} \qquad \frac{}{I \vdash \top}\ \top \qquad \vee \frac{A \vdash X \qquad B \vdash X}{A \vee B \vdash X} \qquad \frac{X \vdash A_i}{X \vdash A_1 \vee A_2}\ \vee_i$$

The calculus above, which we refer to as L0, is sound w.r.t. the class of lattices, complete w.r.t. the Hilbert-style presentation of lattice logic, and enjoys cut-elimination. Hence, L0 is perfectly adequate as a calculus for lattice logic, when this logic is regarded in isolation. However, as discussed in the introduction, the main interest of lattice logic lays in its serving as a base for its axiomatic extensions (cf. e.g. [40]) and language-expansions. Axiomatic extensions of lattice logic can be supported by L0 by adding suitable axioms. For instance, modular and distributive lattice logic can be respectively captured by adding the following axioms to L0:

$$((C \wedge B) \vee A) \wedge B \vdash (C \wedge B) \vee (A \wedge B) \quad \text{and} \quad A \wedge (B \vee C) \vdash (A \wedge B) \vee (A \vee C).$$

However, cut elimination for the resulting calculi needs to be proved from scratch. More in general, we lack uniform principles or proof strategies aimed at identifying axioms which can be added to L0 so that cut elimination transfers to the resulting calculus. Another source of nonmodularity arises from the fact that L0 lacks structural rules. Indeed, the additive formulation of the introduction rules of L0 encodes the information which is stored in standard structural rules such as weakening, contraction, associativity, and exchange. Hence, L0 cannot be used as a base to capture logics aimed at 'negotiating' these rules, such as the Lambek calculus [41] and other substructural logics [31]. To remedy this, one can move to the following calculus L1, which fulfills the *visibility* property,[2] isolated by Sambin et al. in [47] to formulate a general strategy for cut elimination. Visibility generalizes Gentzen's idea, realized in his calculus LJ, that intuitionistic logic could be captured by restricting the shape of sequents and admitting at most one formula in succedent position [32]. The calculus L1 has a structural language consisting of one structural constant 'I', interpreted as $\top$ (resp. $\bot$) when occurring in precedent (resp. succedent) position, and one binary connective ',', interpreted as conjunction (resp. disjunction) in precedent (resp. succedent) position. The rules Exchange, Associativity, Weakening and Contraction are the usual ones and are not reported here.

---

[2] A sequent calculus verifies the *visibility* property if both the auxiliary formulas and the principal formula of each operational rule of the calculus occur in an *empty* context. Hence, by design, L1 verifies the visibility property.

- Identity and Cut rules

$$\frac{}{p \vdash p}\, Id \qquad \text{L-Cut}\, \frac{X \vdash A \qquad (Y \vdash Z)[A]^{pre}}{(X \vdash Y)[Z/A]^{pre}} \qquad \frac{(X \vdash Y)[A]^{succ} \qquad A \vdash Z}{(X \vdash Y)[Z/A]^{suc}}\, \text{R-Cut}$$

where $(Y \vdash Z)[A]^{pre}$ (resp. $(Y \vdash Z)[A]^{succ}$) indicates that the A occurs in precedent (resp. succedent) position in the sequent $Y \vdash Z$.

- Operational rules

$$\frac{}{\bot \vdash I}\, \bot \qquad \frac{X \vdash I}{X \vdash \bot}\, \bot \qquad \wedge\, \frac{A, B \vdash X}{A \wedge B \vdash X} \qquad \frac{X \vdash A \qquad Y \vdash B}{X, Y \vdash A \wedge B}\, \wedge$$

$$\top\, \frac{I \vdash X}{\top \vdash X} \qquad \frac{}{I \vdash \top}\, \top \qquad \vee\, \frac{A \vdash X \qquad B \vdash Y}{A \vee B \vdash X, Y} \qquad \frac{X \vdash A, B}{X \vdash A \vee B}\, \vee$$

Unlike the operational rules of L0 which are *additive*, the operational rules of L1 are *multiplicative*.[3] The latter formulation is more general, and implies that weakening, exchange, associativity, and contraction are not anymore subsumed by the introduction rules.

The visibility of L1 blocks the derivation of the distributivity axiom. Hence, to be able to derive distributivity, one option is to relax the visibility constraint both in precedent and in succedent position. This solution is not entirely satisfactory, and suffers from the same lack of modularity which prevents Gentzen's move from LJ to LK to capture intermediate logics. Specifically, relaxing visibility captures the logics of Sambin's cube, but many other logics are left out. Moreover, without visibility, we do not have a uniform strategy for cut elimination.

To conclude, a proof theory for axiomatic extensions and expansions of general lattice logic is comparably not as modular as that of the axiomatic extensions and expansions of the logic of *distributive* lattices, which can rely on the theory of proper display calculi [37,50]. The idea guiding the approach of the present paper, which we will elaborate upon in the next sections, is that, rather than trying to work our way up starting from a calculus for lattice logic, we will obtain a calculus for lattice logic from the standard proper display calculus for the logic of distributive lattices, by endowing it with a suitable mechanism to block the derivation of distributivity.

3   Multi-type Semantic Environment for Lattice Logic

In the present section, we introduce a class of *heterogeneous algebras* [7] which equivalently encodes complete lattices, and which will be useful to motivate the design of the calculus for lattice logic from a semantic viewpoint, as well as to establish its properties. This presentation takes its move from very well known facts in the representation theory of complete lattices, which can be found e.g. in [6,22], formulated—however—in terms of *covariant* (rather than contravariant) adjunction. For every partial order $\mathbb{Q} = (Q, \leq)$,

---

[3] The multiplicative form of the introduction rules is the most important aspect in which L1 departs from the calculus of [47], which adopts the additive formulation for the introduction rules for conjunction and disjunction.

we let $\mathbb{Q}^{op} := (Q, \leq^{op})$, where $\leq^{op}$ denotes the converse ordering. If $\mathbb{Q} = (Q, \wedge, \vee, \bot, \top)$ is a lattice, we let $\mathbb{Q}^{op} := (Q, \wedge^{op}, \vee^{op}, \bot^{op}, \top^{op})$ denote the lattice induced by $\leq^{op}$. Moreover, for any $b \in Q$, we let $b\!\uparrow := \{c \mid c \in Q$ and $b \leq c\}$ and $b\!\downarrow := \{a \mid a \in Q$ and $a \leq b\}$.

A *polarity* is a structure $\mathbb{P} = (X, Y, R)$ such that X and Y are sets and $R \subseteq X \times Y$. Every polarity induces a pair of maps $\rho : \mathcal{P}(Y)^{op} \to \mathcal{P}(X)$, $\lambda : \mathcal{P}(X) \to \mathcal{P}(Y)^{op}$, respectively defined by $Y' \mapsto \{x \in X \mid \forall y(y \in Y' \to xRy)\}$ and $X' \mapsto \{y \in Y \mid \forall x(x \in X' \to xRy)\}$. It is well known (cf. [22]) and easy to verify that these maps form an *adjunction pair*, that is, for any $X' \subseteq X$ and $Y' \subseteq Y$,

$$\lambda(X') \subseteq^{op} Y' \quad \text{iff} \quad X' \subseteq \rho(Y').$$

The map λ is the left adjoint, and ρ is the right adjoint of the pair. By general order-theoretic facts, this implies that λ preserves arbitrary joins and ρ arbitrary meets: that is, for any $S \subseteq \mathcal{P}(X)$ and any $T \subseteq \mathcal{P}(Y)$,

$$\lambda\left(\bigcup S\right) = \bigcup_{s \in S}^{op} \lambda(s) \quad \text{and} \quad \rho\left(\bigcap T\right) = \bigcap_{t \in T}^{op} \rho(t). \tag{1}$$

Other well known facts about adjoint pairs are that $\rho\lambda : \mathcal{P}(X) \to \mathcal{P}(X)$ is a closure operator and $\lambda\rho : \mathcal{P}(Y)^{op} \to \mathcal{P}(Y)^{op}$ an interior operator (cf. [22]). Moreover, $\lambda\rho\lambda = \lambda$, and $\rho\lambda\rho = \rho$ (cf. [22]). That is, $\lambda\rho$ restricted to $\mathsf{Range}(\lambda)$ is the identity map, and likewise, $\rho\lambda$ restricted to $\mathsf{Range}(\rho)$ is the identity map. Hence, $\mathsf{Range}(\rho) = \mathsf{Range}(\rho\lambda)$, $\mathsf{Range}(\lambda) = \mathsf{Range}(\lambda\rho)$ and

$$\mathcal{P}(X) \supseteq \mathsf{Range}(\rho) \cong \mathsf{Range}(\lambda) \subseteq \mathcal{P}(X)^{op}.$$

Furthermore, $\rho\lambda$ being a closure operator on $\mathcal{P}(X)$ implies that $\mathsf{Range}(\rho) = \mathsf{Range}(\rho\lambda)$ is a complete sub $\bigcap$-semilattice of $\mathcal{P}(X)$ (cf. [22]), and hence $\mathbb{L} = \mathsf{Range}(\rho)$ is endowed with a structure of complete lattice, by setting for every $S \subseteq \mathbb{L}$,

$$\bigwedge_{\mathbb{L}} S := \bigcap S \quad \text{and} \quad \bigvee_{\mathbb{L}} S := \rho\lambda\left(\bigcup S\right) \tag{2}$$

Likewise, $\lambda\rho$ being an interior operator on $\mathcal{P}(Y)^{op}$ implies that $\mathsf{Range}(\lambda)$ is a complete sub $\bigcup$-semilattice of $\mathcal{P}(Y)^{op}$, and hence $\mathbb{L} = \mathsf{Range}(\lambda)$ is endowed with a structure of complete lattice, by setting

$$\bigvee_{\mathbb{L}} T := \bigcup^{op} T \quad \text{and} \quad \bigwedge_{\mathbb{L}} T := \lambda\rho\left(\bigcap^{op} T\right) \tag{3}$$

for every $T \subseteq \mathbb{L}$. Finally, for any $S \subseteq \mathsf{Range}(\rho)$,

$$\begin{aligned}
\lambda(\bigvee S) &= \lambda(\rho\lambda(\bigcup S)) \quad (2) \\
&= \lambda(\bigcup S) \qquad \lambda\rho\lambda = \lambda \\
&= \bigcup_{s \in S}^{op} \lambda(s) \quad (1) \\
&= \bigvee_{s \in S} \lambda(s), \quad (3)
\end{aligned}$$

and

$$\bigwedge_{s\in S} \lambda(s) = \lambda\rho(\bigcap_{s\in S}^{op} \lambda(s)) \ (3)$$
$$= \lambda(\bigcap_{s\in S} \rho\lambda(s)) \ (1)$$
$$= \lambda(\bigcap S) \qquad S \subseteq \mathsf{Range}(\rho) \text{ and } \rho\lambda\rho = \rho$$
$$= \lambda(\bigwedge S), \qquad (2)$$

which shows that the restriction of λ to $\mathsf{Range}(\rho)$ is a complete lattice homomorphism. Likewise, one can show that the restriction of ρ to $\mathsf{Range}(\lambda)$ is a complete lattice homomorphism, which completes the proof that the bijection

$$\mathcal{P}(X) \supseteq \mathsf{Range}(\rho) \cong \mathsf{Range}(\lambda) \subseteq \mathcal{P}(X)^{op}$$

is an isomorphism of complete lattices, and justifies the abuse of notation which we made by denoting both the lattice $\mathsf{Range}(\rho)$ and the lattice $\mathsf{Range}(\lambda)$ by $\mathbb{L}$.

Conversely, for every complete lattice $\mathbb{L}$, consider the polarity $\mathbb{P}_{\mathbb{L}} := (L, L, \leq)$ where L is the universe of $\mathbb{L}$ and $\leq$ is the lattice order. Then the maps $\lambda : \mathcal{P}(L) \to \mathcal{P}(L)^{op}$ and $\rho : \mathcal{P}(L)^{op} \to \mathcal{P}(L)$ are respectively defined by the assignments $S \mapsto \{a \in L \mid \forall b(b \in S \to b \leq a)\} = (\bigvee S)\uparrow$ and $T \mapsto \{a \in L \mid \forall b(b \in T \to a \leq b)\} = (\bigwedge T)\downarrow$ for all $S, T \subseteq L$. Since $\bigwedge((\bigvee S)\uparrow) = \bigvee S$ and $\bigvee((\bigwedge T)\downarrow) = \bigwedge T$, the closure operator $\rho\lambda : \mathcal{P}(L) \to \mathcal{P}(L)$ and the interior operator $\lambda\rho : \mathcal{P}(L)^{op} \to \mathcal{P}(L)^{op}$ are respectively defined by

$$S \mapsto (\bigvee S)\downarrow \quad \text{and} \quad T \mapsto (\bigwedge T)\uparrow. \qquad (4)$$

The lattice $\mathbb{L}$ can be mapped injectively both into $\mathsf{Range}(\rho) = \mathsf{Range}(\rho\lambda)$ and into $\mathsf{Range}(\lambda) = \mathsf{Range}(\lambda\rho)$ by the assignments $a \mapsto a\downarrow$ and $a \mapsto a\uparrow$ respectively. Moreover, since $\mathbb{L}$ is complete, the maps defined by these assignments are also *onto* $\mathsf{Range}(\rho\lambda)$ and $\mathsf{Range}(\lambda\rho)$. Finally, for any $S \subseteq \mathbb{L}$,

$$\bigwedge_{\mathsf{Range}(\rho)}\{a\downarrow \mid a \in S\} = \bigcap\{a\downarrow \mid a \in S\} \qquad (2)$$
$$= (\bigwedge S)\downarrow$$

$$\bigvee_{\mathsf{Range}(\rho)}\{a\downarrow \mid a \in S\} = \rho\lambda(\bigcup\{a\downarrow \mid a \in S\}) \quad (2)$$
$$= (\bigvee\bigcup\{a\downarrow \mid a \in S\})\downarrow \ (4)$$
$$= (\bigvee S)\downarrow,$$

which completes the verification that the map $\mathbb{L} \to \mathsf{Range}(\rho)$ defined by the assignment $a \mapsto a\downarrow$ is a complete lattice isomorphism. Similarly, one verifies that the map $\mathbb{L} \to \mathsf{Range}(\lambda)$ defined by the assignment $a \mapsto a\uparrow$ is a complete lattice isomorphism. The discussion so far can be summarized by the following.

Proposition 1. *Any complete lattice $\mathbb{L}$ can be identified both with the lattice of closed sets of some closure operator $c : \mathbb{D} \to \mathbb{D}$ on a complete and completely distributive lattice $\mathbb{D} = (D, \cap, \cup, \wp, \varnothing)$, and with the lattice of open sets of some interior operator $i : \mathbb{E} \to \mathbb{E}$ on a complete and completely distributive lattice $\mathbb{E} = (E, \sqcap, \sqcup, \mathfrak{I}, \emptyset)$.*

Hence, in what follows, $\mathbb{L}$ will be identified both with $\mathsf{Range}(c)$ endowed with its structure of complete lattice defined as in (2) (replacing $\rho\lambda$ by c), and with $\mathsf{Range}(i)$ endowed with its structure of complete lattice defined as in (3) (replacing $\lambda\rho$ by i). Taking these identifications into account, general order-theoretic facts (cf. [22, Chap. 7]) imply that $c = e_\ell \circ \gamma$, where $\gamma : \mathbb{D} \twoheadrightarrow \mathbb{L}$ is defined by $\alpha \mapsto c(\alpha)$ and $e_\ell : \mathbb{L} \hookrightarrow \mathbb{D}$ is the natural embedding, and moreover, these maps form an adjunction pair as follows: for any $a \in \mathbb{L}$ and any $\alpha \in \mathbb{D}$,

$$\gamma(\alpha) \le a \quad \text{iff} \quad \alpha \le e_\ell(a),$$

with the additional property that $\gamma \circ e_\ell = Id_\mathbb{L}$. Likewise, $i = e_r \circ \iota$, where $\iota : \mathbb{E} \twoheadrightarrow \mathbb{L}$ is defined by $\xi \mapsto i(\xi)$ and $e_r : \mathbb{L} \hookrightarrow \mathbb{E}$ is the natural embedding, and moreover, these maps form an adjunction pair as follows: for any $a \in \mathbb{L}$ and any $\xi \in \mathbb{E}$,

$$e_r(a) \le \xi \quad \text{iff} \quad a \le \iota(\xi),$$

with the additional property that $\iota \circ e_r = Id_\mathbb{L}$.

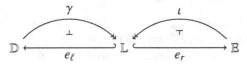

Summing up, any complete lattice $\mathbb{L}$ can be associated with a *heterogeneous LL-algebra*, i.e. a tuple $(\mathbb{L}, \mathbb{D}, \mathbb{E}, e_\ell, \gamma, e_r, \iota)$ such that:

H1. $\mathbb{L} = (L, \le)$ is a bounded poset;[4]
H2. $\mathbb{D}$ and $\mathbb{E}$ are complete and completely distributive lattices;
H3. $\gamma : \mathbb{D} \to \mathbb{L}$ and $e_\ell : \mathbb{L} \to \mathbb{D}$ are such that $\gamma \dashv e_\ell$ and $\gamma \circ e_\ell = Id_\mathbb{L}$;
H4. $\iota : \mathbb{E} \to \mathbb{L}$ and $e_r : \mathbb{L} \to \mathbb{E}$ are such that $e_r \dashv \iota$ and $\iota \circ e_r = Id_\mathbb{L}$.

Conversely, for any heterogeneous LL-algebra as above, the poset $\mathbb{L}$ can be endowed with the structure of a complete lattice inherited by being order-isomorphic both to the poset of closed sets of the closure operator $c := \gamma \circ e_\ell$ on $\mathbb{D}$ and to the poset of open sets of the interior operator $i := \iota \circ e_r$ on $\mathbb{E}$. Finally, no algebraic information is lost when presenting a complete lattice $\mathbb{L}$ as its associated heterogeneous LL-algebra. Indeed, the identification of $\mathbb{L}$ with $\mathsf{Range}(c)$, endowed with the structure of complete lattice defined as in (2), implies that for all $a, b \in \mathbb{L}$,

$$a \vee b = \gamma(e_\ell(a) \cup e_\ell(b)).$$

As discussed above, e_ℓ being a right adjoint and γ a left adjoint imply that e_ℓ is completely meet-preserving and γ completely join-preserving. Therefore, $e_\ell(\top) = \wp$ and $\bot = \gamma(\varnothing)$. Moreover, γ being both surjective and order-preserving implies that $\top = \gamma(\wp)$. Furthermore, for all $a, b \in \mathbb{L}$,

$$a \wedge b = \gamma \circ e_\ell(a \wedge b) = \gamma(e_\ell(a) \cap e_\ell(b)).$$

Thus, the whole algebraic structure of $\mathbb{L}$ can be captured in terms of the algebraic structure of $\mathbb{D}$ and the adjoint maps γ and e_ℓ as follows: for all $a, b \in \mathbb{L}$,

$$\bot = \gamma(\varnothing) \quad \top = \gamma(\wp) \quad a \vee b = \gamma(e_\ell(a) \cup e_\ell(b)) \quad a \wedge b = \gamma(e_\ell(a) \cap e_\ell(b)). \tag{5}$$

Reasoning analogously, one can also capture the algebraic structure of $\mathbb{L}$ in terms of the algebraic structure of $\mathbb{E}$ and the adjoint maps ι and e_r as follows: for all $a, b \in \mathbb{L}$,

$$\top = \iota(\mathfrak{I}) \quad \bot = \iota(\emptyset) \quad a \wedge b = \iota(e_r(a) \sqcap e_r(b)) \quad a \vee b = \iota(e_r(a) \sqcup e_r(b)). \tag{6}$$

---

[4] We overload the symbol $\mathbb{L}$ and use it both to denote the complete lattice and its underlying poset.

4   Multi-type Language for Lattice Logic

In Sect. 3, heterogeneous LL-algebras have been introduced and shown to be equivalent presentations of complete lattices. The toggle between these mathematical structures is reflected in the toggle between the logical languages which are naturally interpreted in the two types of structures. Indeed, the heterogeneous LL-algebras of Sect. 3 provide a natural interpretation for the following multi-type language $\mathcal{L}_{MT}$ over a set AtProp of Lattice-type atomic propositions:

$$\mathsf{Left} \ni \alpha ::= e_\ell(A) \mid \wp \mid \varnothing \mid \alpha \cup \alpha \mid \alpha \cap \alpha$$
$$\mathsf{Right} \ni \xi ::= e_r(A) \mid \Im \mid \emptyset \mid \xi \sqcup \xi \mid \xi \sqcap \xi$$
$$\mathsf{Lattice} \ni A ::= p \mid \gamma(\alpha) \mid \iota(\xi) \mid \top \mid \bot$$

where $p \in$ AtProp. The interpretation of $\mathcal{L}_{MT}$-terms into heterogeneous LL-algebras is defined as the straightforward generalization of the interpretation of propositional languages in algebras of compatible signature. At the end of the previous section, we observed that the algebraic structure of the complete lattice $\mathbb{L}$ can be captured in terms of the algebraic structure of its associated heterogeneous LL-algebra. This observation serves as a base for the definition of the translations $(\cdot)^\ell, (\cdot)^r : \mathcal{L} \to \mathcal{L}_{MT}$ between the original language $\mathcal{L}$ of lattice logic and $\mathcal{L}_{MT}$:

$$
\begin{array}{ll}
p^\ell = \gamma e_\ell(p) & p^r = \iota e_r(p)^\perp \\
\top^\ell = \gamma e_\ell(\top) & \top^r = \iota e_r(\top) \\
\bot^\ell = \gamma e_\ell(\bot) & \bot^r = \iota e_r(\bot) \\
(A \wedge B)^\ell = \gamma(e_\ell(A^\ell) \cap e_\ell(B^\ell)) & (A \wedge B)^r = \iota(e_r(A^r) \sqcap e_r(B^r)) \\
(A \vee B)^\ell = \gamma(e_\ell(A^\ell) \cup e_\ell(B^\ell)) & (A \vee B)^r = \iota(e_r(A^r) \sqcup e_r(B^r))
\end{array}
$$

For every complete lattice $\mathbb{L}$, let $\mathbb{L}^*$ denote its associated heterogeneous LL-algebra as defined in Sect. 3. The proof of the following proposition relies on the observations made at the end of Sect. 3.

Proposition 2. *For all $\mathcal{L}$-formulas A and B and every complete lattice $\mathbb{L}$,*

$$\mathbb{L} \models A \leq B \quad \text{iff} \quad \mathbb{L}^* \models A^\ell \leq B^r.$$

5   Proper Display Calculus for Lattice Logic

5.1   Language

The language of the calculus D.LL includes the types Lattice, Left, and Right, sometimes abbreviated as L, P, and P^{op} respectively.

$$
\mathsf{L}\begin{cases} A ::= p \mid \blacklozenge \alpha \mid \blacksquare \xi \\[4pt] X ::= p \mid I \mid \bullet \Gamma \mid \bullet^{op} \Pi \end{cases}
\qquad
\mathsf{P}\begin{cases} \alpha ::= \square A \\[4pt] \Gamma ::= \circ X \mid \circledS \mid \Gamma . \Gamma \mid \Gamma \supset \Gamma \end{cases}
$$

$$
\mathsf{P}^{op}\begin{cases} \xi ::= \lozenge^{op} A \\[4pt] \Pi ::= \circ^{op} X \mid \circledS^{op} \mid \Pi .^{op} \Pi \mid \Pi \supset^{op} \Pi \end{cases}
$$

Our notational conventions assign different variables to different types. This allows us to drop the subscripts ^{op}, since the parsing of expressions such as $\bullet\Gamma$ and $\bullet\Pi$ is inherently unambiguous, and the parsing of e.g. $\circ X$ is contextually unambiguous.

– Structural and operational pure L-type connectives:[5]

L connectives

I	
$\top$	$\bot$

– Structural and operational pure P-type and P^{op}-type connectives:

P connectives

$\text{\textcircled{S}}$			$\supset$	
$(\wp)$	$(\varnothing)$	$\cap$	$\cup$	$(\supset\!\!-)$	$(-\!\!\supset)$

P^{op} connectives

$\text{\textcircled{S}}^{\mathrm{op}}$		$\cdot^{\mathrm{op}}$	$\supset^{\mathrm{op}}$	
$(\wp^{\mathrm{op}})$	$(\varnothing^{\mathrm{op}})$	$\cap^{\mathrm{op}}$	$\cup^{\mathrm{op}}$	$(\supset\!\!-^{\mathrm{op}})$	$(-\!\!\supset^{\mathrm{op}})$

– Structural and operational multi-type connectives:

$\mathsf{L} \to \mathsf{P}$	$\mathsf{L} \to \mathsf{P}^{\mathrm{op}}$	$\mathsf{P} \to \mathsf{L}$	$\mathsf{P}^{\mathrm{op}} \to \mathsf{L}$
$\circ$	$\circ^{\mathrm{op}}$	$\bullet$	$\bullet$
$\square$	$\lozenge^{\mathrm{op}}$	$\blacklozenge$	$\blacksquare^{\mathrm{op}}$

The connectives $\square$, $\lozenge^{\mathrm{op}}$, $\blacklozenge$ and $\blacksquare^{\mathrm{op}}$ are interpreted in heterogeneous LL-algebras as the maps e_ℓ, e_r, γ, and ι, respectively.

5.2 Rules

In what follows, structures of type L are denoted by the variables X, Y, Z, and W; structures of type P are denoted by the variables Γ, Δ, Θ, and Λ; structures of type P^{op} are denoted by the variables Π, Σ, Ψ, and Ω. Given the semantic environment introduced in Sect. 3, it will come as no surprise that there is a perfect match between the pure P-type rules and the pure P^{op}-type rules. In order to achieve a more compact presentation of the calculus, in what follows we will also reserve the variables S, T, U, and V to denote either P-type structures or P^{op}-type structures, and s, t, u and v to denote operational terms of either P-type or P^{op}-type, with the proviso that they should be interpreted in the *same* type in the same pure type-rule.

– Multi-type display rules

$$\frac{\Gamma \vdash \circ X}{\bullet\Gamma \vdash X}\, D_{P\text{-}L} \qquad \frac{\circ X \vdash \Pi}{X \vdash \bullet\Pi}\, D_{P\text{-}L}$$

– Pure P-type and P^{op}-type display rules

$$D_P\, \frac{S . T \vdash U}{T \vdash S \supset U} \qquad \frac{S \vdash T . U}{T \supset S \vdash U}\, D_P$$

---

[5] We follow the notational conventions introduced in [36]: Each structural connective in the upper row of the synoptic tables is interpreted as the logical connective in the left (resp. right) slot in the lower row when occurring in precedent (resp. succedent) position.

– Structural and operational pure P-type and P^{op}-type rules

$$\dfrac{S \vdash s \quad s \vdash T}{S \vdash T}\;Cut \qquad\qquad \cap\,\dfrac{s\,.\,t \vdash S}{s \cap t \vdash S} \quad \dfrac{S \vdash s \quad T \vdash t}{S\,.\,T \vdash s \cap t}\,\cap$$

$$\text{\textcircled{S}}\,\dfrac{S \vdash T}{S\,.\,\text{\textcircled{S}} \vdash T} \quad \dfrac{S \vdash T}{S \vdash T\,.\,\text{\textcircled{S}}}\,\text{\textcircled{S}} \qquad \cup\,\dfrac{s \vdash S \quad t \vdash T}{s \cup t \vdash S\,.\,T} \quad \dfrac{S \vdash s\,.\,t}{S \vdash s \cup t}\,\cup$$

$$A\,\dfrac{(S\,.\,T)\,.\,U \vdash V}{S\,.\,(T\,.\,U) \vdash V} \quad \dfrac{S \vdash (T\,.\,U)\,.\,V}{S \vdash T\,.\,(U\,.\,V)}\,A \qquad E\,\dfrac{S\,.\,T \vdash U}{T\,.\,S \vdash U} \quad \dfrac{S \vdash T\,.\,U}{S \vdash U\,.\,T}\,E$$

$$W\,\dfrac{S \vdash T}{S\,.\,U \vdash T} \quad \dfrac{S \vdash T}{S \vdash T\,.\,U}\,W$$

$$C\,\dfrac{S\,.\,S \vdash T}{S \vdash T} \quad \dfrac{S \vdash T\,.\,T}{S \vdash T}\,C$$

– Structural and operational pure L-type rules

$$Id\,\dfrac{}{p \vdash p} \qquad \dfrac{X \vdash A \quad A \vdash Y}{X \vdash Y}\;Cut \qquad \top\,\dfrac{\mathrm{I} \vdash X}{\top \vdash X} \quad \dfrac{}{\mathrm{I} \vdash \top}\,\top$$

$$\text{I-W}\,\dfrac{\mathrm{I} \vdash X}{Y \vdash X} \qquad\qquad \bot\,\dfrac{}{\bot \vdash \mathrm{I}} \quad \dfrac{X \vdash \mathrm{I}}{X \vdash \bot}\,\bot$$

– Operational rules for multi-type connectives:

$$\overset{\textstyle \mathsf{L} \to \mathsf{P}^{\mathrm{op}}}{\diamond\,\dfrac{\circ A \vdash \Pi}{\diamond A \vdash \Pi} \quad \dfrac{X \vdash A}{\circ X \vdash \diamond A}\,\diamond} \qquad \overset{\textstyle \mathsf{P}^{\mathrm{op}} \to \mathsf{L}}{\blacksquare\,\dfrac{X \vdash \bullet\xi}{X \vdash \blacksquare\xi} \quad \dfrac{\xi \vdash \Pi}{\blacksquare\xi \vdash \bullet\Pi}\,\blacksquare}$$

$$\overset{\textstyle \mathsf{P} \to \mathsf{L}}{\blacklozenge\,\dfrac{\bullet\alpha \vdash X}{\blacklozenge\alpha \vdash X} \quad \dfrac{\Gamma \vdash \alpha}{\bullet\Gamma \vdash \blacklozenge\alpha}\,\blacklozenge} \qquad \overset{\textstyle \mathsf{L} \to \mathsf{P}}{\Box\,\dfrac{\Gamma \vdash \circ A}{\Gamma \vdash \Box A} \quad \dfrac{A \vdash X}{\Box A \vdash \circ X}\,\Box}$$

6   Properties

Soundness. First, structural symbols are interpreted as logical symbols according to their (precedent or succedent) position, as indicated in the tables of Sect. 5.1. Then, sequents are interpreted as inequalities, and rules as quasi-inequalities in heterogeneous LL-algebras. Rules of D.LL are sound if their corresponding quasi-inequalities are valid in heterogeneous LL-algebras. This verification is routine.

Conservativity. We need to show that, for all formulas A and B of the original language of lattice logic, if $A^\tau \vdash B_\tau$ is a D.LL-derivable sequent, then $A \vdash B$ is a theorem of the Hilbert-style presentation of lattice logic. The argument follows the standard proof strategy discussed in [36,37], using the following facts: (a) the rules of D.LL are sound w.r.t. heterogeneous LL-algebras (cf. Sect. 6); (b) lattice logic is strongly complete w.r.t. the class of complete lattices, and (c) complete lattices are equivalently presented as heterogeneous LL-algebras (cf. Sect. 3), so that Proposition 2 holds.

Cut Elimination and Subformula Property. These can be inferred from the meta-theorem, in [26]; in [39, Theorem A.2] a restricted version of it is stated which specifically applies to *proper multi-type display calculi* (cf. [39, Definition A.1]). The verification is straightforward, and is omitted.

Completeness. First, we translate $\mathcal{L}$-sequents $A \vdash B$ into D.LL-sequents $A^\tau \vdash B_\tau$, using the following translations:

$$\top^\tau ::= \blacklozenge \Box \top \qquad\qquad \top_\tau ::= \blacksquare^{op} \Diamond^{op} \top$$
$$\bot^\tau ::= \blacklozenge \Box \bot \qquad\qquad \bot_\tau ::= \blacksquare^{op} \Diamond^{op} \bot$$
$$p^\tau ::= \blacklozenge \Box p \qquad\qquad p_\tau ::= \blacksquare^{op} \Diamond^{op} p$$
$$(A \wedge B)^\tau ::= \blacklozenge(\Box A^\tau \cap \Box B^\tau) \qquad (A \wedge B)_\tau ::= \blacksquare^{op}(\Diamond^{op} A_\tau \cap^{op} \Diamond^{op} B_\tau)$$
$$(A \vee B)^\tau ::= \blacklozenge(\Box A^\tau \cup \Box B^\tau) \qquad (A \vee B)_\tau ::= \blacksquare^{op}(\Diamond^{op} A_\tau \cup^{op} \Diamond^{op} B_\tau)$$

Proposition 3. *For every $A \in \mathcal{L}$, the multi-type sequent $A^\tau \vdash A_\tau$ is derivable in D.LL.*

Proof. By simultaneous induction on $A \in \mathsf{L}$, $\alpha \in \mathsf{P}$, and $\xi \in \mathsf{P}^{op}$.

In what follows, we collect all the translations of the axioms involving conjunction,[6] and derive some of them in D.LL (cf. Sect. A). The full set of derivations can be found in the extended version of the present paper [38]. All derivations are standard and make use only of Weakening, Contraction and Exchange as structural rules.

Commutative laws	translation
cC1 $(A \wedge B)^\tau \vdash (B \wedge A)_\tau$	$\rightsquigarrow \blacklozenge(\Box A^\tau \cap \Box B^\tau) \vdash \blacksquare(\Diamond B_\tau \cap \Diamond A_\tau)$
cC2 $(B \wedge A)^\tau \vdash (A \wedge B)_\tau$	$\rightsquigarrow \blacklozenge(\Box B^\tau \cap \Box A^\tau) \vdash \blacksquare(\Diamond A_\tau \cap \Diamond B_\tau)$

Associative laws	translation
cA1 $(A \wedge (B \wedge C))^\tau \vdash ((A \wedge B) \wedge C)_\tau$	$\rightsquigarrow \blacklozenge(\Box A^\tau \cap \Box \blacklozenge(\Box B^\tau \cap \Box C^\tau)) \vdash \blacksquare(\Diamond \blacksquare(\Diamond A_\tau \cap \Diamond B_\tau) \cap \Diamond C_\tau)$
cA2 $((A \wedge B) \wedge C)^\tau \vdash (A \wedge (B \wedge C))_\tau$	$\rightsquigarrow \blacklozenge(\Box \blacklozenge(\Box A^\tau \cap \Box B^\tau) \cap \Box C^\tau) \vdash \blacksquare(\Diamond A_\tau \cap \Diamond \blacksquare(\Diamond B_\tau \cap \Diamond C_\tau))$

Identity laws	translation where $A = C \wedge D$
cI1 $(A \wedge \top)^\tau \vdash A_\tau$	$\rightsquigarrow \blacklozenge(\Box \blacklozenge(\Box C^\tau \cap \Box D^\tau) \cap \Box \blacklozenge \top) \vdash \blacksquare(\Diamond C_\tau \cap \Diamond D_\tau)$
cI2 $A^\tau \vdash (A \wedge \top)_\tau$	$\rightsquigarrow \blacklozenge(\Box C^\tau \cap \Box D^\tau) \vdash \blacksquare(\Diamond \blacksquare(\Diamond C_\tau \cap \Diamond D_\tau) \cap \Diamond \blacksquare \Diamond \top)$

Absorption laws	translation where $A = \bot$
cAb1 $(A \wedge (A \vee B))^\tau \vdash A_\tau$	$\rightsquigarrow \blacklozenge(\Box \blacklozenge \Box \bot \cap \Box \blacklozenge(\Box \blacklozenge \Box \bot \cup \Box B^\tau)) \vdash \blacksquare \Diamond \bot$
cAb2 $A^\tau \vdash (A \wedge (A \vee B))_\tau$	$\rightsquigarrow \blacklozenge \Box \bot \vdash \blacksquare(\Diamond \blacksquare \Diamond \bot \cap \Diamond \blacksquare(\Diamond \blacksquare \Diamond \bot \cup \Diamond B_\tau))$

A   Completeness

In the cases of the Identity and Absorption laws a formula occurs in isolation on one side of the turnstile, therefore we need to proceed by cases according to the shape of A (we just show $A = C \wedge D$ and $A = \bot$, respectively).

---

[6] The translations of the axioms involving disjunction are perfectly symmetric, and are omitted.

```
                                                                    D ⊢ D
                                                                 ─────────
              C ⊢ C                                               oD ⊢ ◇D
            ─────────                                            ─────────
            oC ⊢ ◇C                                              D ⊢ •◇D
            ─────────                                          ───────────
            C ⊢ •◇C                                            □D ⊢ o•◇D
          ───────────                                       W ─────────────
       W  □C ⊢ o•◇C                                         E  □D.□C ⊢ o•◇D
         ─────────────                                        ─────────────
         □C.□D ⊢ o•◇C                                         □C.□D ⊢ o•◇D
         ─────────────                                       ───────────────
         □C∩□D ⊢ o•◇C                                        □C∩□D ⊢ o•◇D
        ───────────────                                     ───────────────
        •□C∩□D ⊢ •◇C                                        •□C∩□D ⊢ •◇D
       ─────────────────                                   ─────────────────
       ◆(□C∩□D) ⊢ •◇C                                      ◆(□C∩□D) ⊢ •◇D
      ───────────────────                                 ───────────────────
      □◆(□C∩□D) ⊢ o•◇C                                    □◆(□C∩□D) ⊢ o•◇D
   W ─────────────────────────                          W ─────────────────────────
     □◆(□C∩□D).□◆□⊤ ⊢ o•◇C                                □◆(□C∩□D).□◆□⊤ ⊢ o•◇D
    ─────────────────────────                            ─────────────────────────
    □◆(□C∩□D)∩□◆□⊤ ⊢ o•◇C                                □◆(□C∩□D)∩□◆□⊤ ⊢ o•◇D
   ───────────────────────────                         ───────────────────────────
   •□◆(□C∩□D)∩□◆□⊤ ⊢ •◇C                                •□◆(□C∩□D)∩□◆□⊤ ⊢ •◇D
  ─────────────────────────────                       ─────────────────────────────
  ◆(□◆(□C∩□D)∩□◆□⊤) ⊢ •◇C                              ◆(□◆(□C∩□D)∩□◆□⊤) ⊢ •◇D
 ───────────────────────────────                     ───────────────────────────────
 o◆(□◆(□C∩□D)∩□◆□⊤) ⊢ ◇C                              o◆(□◆(□C∩□D)∩□◆□⊤) ⊢ ◇D

      o◆(□◆(□C∩□D)∩□◆□⊤).o◆(□◆(□C∩□D)∩□◆□⊤) ⊢ ◇C∩◇D
   C ─────────────────────────────────────────────────
        o◆(□◆(□C∩□D)∩□◆□⊤) ⊢ ◇C∩◇D
       ─────────────────────────────────
        ◆(□◆(□C∩□D)∩□◆□⊤) ⊢ •◇C∩◇D
       ─────────────────────────────────
        ◆(□◆(□C∩□D)∩□◆□⊤) ⊢ ■(◇C∩◇D)
```

```
                                                           ⊥ ⊢ I
                                                         ─────────
                                                          ⊥ ⊢ ⊥
                                                         ─────────
                                                          □⊥ ⊢ o⊥
                                                         ─────────
                                                          •□⊥ ⊢ ⊥
                                                         ─────────
                                                          ◆□⊥ ⊢ ⊥
                                                         ───────────
                                                          o◆□⊥ ⊢ ◇⊥
                                                         ───────────
                                                          ◆□⊥ ⊢ •◇⊥
                                                         ───────────
                                                          ◆□⊥ ⊢ ■◇⊥
                                                         ─────────────
                                                          o◆□⊥ ⊢ ◇■◇⊥
                                                         ───────────────
                                                          o◆□⊥ ⊢ ◇■◇⊥.◇B
                                                        W ─────────────────
                                                          o◆□⊥ ⊢ ◇■◇⊥∪◇B
                                                         ─────────────────
                                                          ◆□⊥ ⊢ •◇■◇⊥∪◇B
         ⊥ ⊢ I                                           ─────────────────
       ─────────                   ⊥ ⊢ ⊥                  ◆□⊥ ⊢ ■(◇■◇⊥∪◇B)
        ⊥ ⊢ ⊥                    ─────────               ───────────────────
       ─────────                  □⊥ ⊢ o⊥                 o◆□⊥ ⊢ ◇■(◇■◇⊥∪◇B)
       o⊥ ⊢ ◇⊥                  ─────────
       ─────────                  •□⊥ ⊢ ⊥         C ─────────────────────────────────────────
       ⊥ ⊢ •◇⊥                  ─────────           o◆□⊥.o◆□⊥ ⊢ ◇■◇p∩◇■(◇■◇⊥∪◇B)
       ─────────                  ◆□⊥ ⊢ ⊥          ───────────────────────────────────
       ⊥ ⊢ ■◇⊥                 ───────────          o◆□⊥ ⊢ ◇■◇⊥∩◇■(◇■◇⊥∪◇B)
      ───────────                o◆□⊥ ⊢ ◇⊥         ───────────────────────────────────
       □⊥ ⊢ o■◇⊥               ───────────          ◆□⊥ ⊢ •◇■◇⊥∩◇■(◇■◇⊥∪◇B)
      ───────────                ◆□⊥ ⊢ •◇⊥         ───────────────────────────────────
       •□⊥ ⊢ ■◇⊥               ───────────          ◆□⊥ ⊢ ■(◇■◇⊥∩◇■(◇■◇⊥∪◇B))
      ───────────                ◆□⊥ ⊢ ■◇⊥
       ◆□⊥ ⊢ ■◇⊥              ─────────────
      ─────────────            o◆□⊥ ⊢ ◇■◇⊥
       □◆□⊥ ⊢ o■◇⊥
  W ─────────────────────────────────────────
     □◆□⊥.□◆(□◆□⊥∪□B) ⊢ o■◇⊥
    ───────────────────────────────────────
     □◆□⊥∩□◆(□◆□⊥∪□B) ⊢ o■◇⊥
   ─────────────────────────────────────────
    •□◆□⊥∩□◆(□◆□⊥∪□B) ⊢ ■◇⊥
   ─────────────────────────────────────────
    ◆(□◆□⊥∩□◆(□◆□⊥∪□B)) ⊢ ■◇⊥
```

B   Distributivity Fails

Distributivity law	translation

cD1 $(A\cap(B\cup C))^\tau \vdash ((A\cap B)\cup(A\cup C))_\tau \rightsquigarrow$

$$\blacklozenge\big(\Box A^\tau\cap\Box\blacklozenge(\Box A^\tau\cup\Box B^\tau)\big)\vdash\blacksquare\big(\lozenge\blacksquare(\lozenge A_\tau\cap\lozenge B_\tau)\cup\lozenge\blacksquare(\lozenge A_\tau\cap\lozenge C_\tau)\big)$$

We will show that all the paths in the backward proof-search of the translation of the distributivity axiom end in deadlocks. First, we apply exhaustively all invertible operational rules (modulo display rules):

$$\frac{???}{\bullet\big(\Box A.\Box\blacklozenge(\Box B\cup\Box C)\big)\vdash\bullet\big(\Diamond\blacksquare(\Diamond A\cap\Diamond B).\Diamond\blacksquare(\Diamond A\cap\Diamond C)\big)}$$

$$\frac{}{\Box A.\Box\blacklozenge(\Box B\cup\Box C)\vdash\circ\bullet\big(\Diamond\blacksquare(\Diamond A\cap\Diamond B).\Diamond\blacksquare(\Diamond A\cap\Diamond C)\big)}$$

$$\frac{}{\Box A\cap\Box\blacklozenge(\Box B\cup\Box C)\vdash\circ\bullet\big(\Diamond\blacksquare(\Diamond A\cap\Diamond B).\Diamond\blacksquare(\Diamond A\cap\Diamond C)\big)}$$

$$\frac{}{\bullet\big(\Box A\cap\Box\blacklozenge(\Box B\cup\Box C)\big)\vdash\bullet\big(\Diamond\blacksquare(\Diamond A\cap\Diamond B).\Diamond\blacksquare(\Diamond A\cap\Diamond C)\big)}$$

$$\frac{}{\circ\bullet\big(\Box A\cap\Box\blacklozenge(\Box B\cup\Box C)\big)\vdash\Diamond\blacksquare(\Diamond A\cap\Diamond B).\Diamond\blacksquare(\Diamond A\cap\Diamond C)}$$

$$\frac{}{\circ\bullet\big(\Box A\cap\Box\blacklozenge(\Box B\cup\Box C)\big)\vdash\Diamond\blacksquare(\Diamond A\cap\Diamond B)\cup\Diamond\blacksquare(\Diamond A\cap\Diamond C)}$$

$$\frac{}{\bullet\big(\Box A\cap\Box\blacklozenge(\Box B\cup\Box C)\big)\vdash\bullet\big(\Diamond\blacksquare(\Diamond A\cap\Diamond B)\cup\Diamond\blacksquare(\Diamond A\cap\Diamond C)\big)}$$

$$\frac{}{\bullet\big(\Box A\cap\Box\blacklozenge(\Box B\cup\Box C)\big)\vdash\blacksquare\big(\Diamond\blacksquare(\Diamond A\cap\Diamond B)\cup\Diamond\blacksquare(\Diamond A\cap\Diamond C)\big)}$$

$$\frac{}{\blacklozenge\big(\Box A\cap\Box\blacklozenge(\Box B\cup\Box C)\big)\vdash\blacksquare\big(\Diamond\blacksquare(\Diamond A\cap\Diamond B)\cup\Diamond\blacksquare(\Diamond A\cap\Diamond C)\big)}$$

There are no rules in which $\bullet$ and . interact, hence we are reduced to either isolate

$$X = \Box A.\Box\blacklozenge(\Box B\cup\Box C)$$

in precedent position by the backward application of a display rule, or isolate the following structure in succedent position:

$$Y = \Diamond\blacksquare(\Diamond A\cap\Diamond B).\Diamond\blacksquare(\Diamond A\cap\Diamond C)$$

We only treat the first case, the second being analogous. Once in isolation, we can act on X only via Exchange, Weakening or Residuation. In each case we reach a dead end:

– Case 1: (Exchange or) Residuation.

As an intermediate step, we can isolate any of the substructures of X via Residuation, or via Exchange and Residuation, as shown below. In each case we reach a dead end.

$$\frac{???}{\Box\blacklozenge(\Box B\cup\Box C)\vdash\Box A\supset\circ\bullet\big(\Diamond\blacksquare(\Diamond A\cap\Diamond B).\Diamond\blacksquare(\Diamond A\cap\Diamond C)\big)}{\Box A.\Box\blacklozenge(\Box B\cup\Box C)\vdash\circ\bullet\big(\Diamond\blacksquare(\Diamond A\cap\Diamond B).\Diamond\blacksquare(\Diamond A\cap\Diamond C)\big)}$$

$$\frac{???}{\Box A\vdash\Box\blacklozenge(\Box B\cup\Box C)\supset\circ\bullet\big(\Diamond\blacksquare(\Diamond A\cap\Diamond B).\Diamond\blacksquare(\Diamond A\cap\Diamond C)\big)}$$

$$\frac{\Box\blacklozenge(\Box B\cup\Box C).\Box A\vdash\circ\bullet\big(\Diamond\blacksquare(\Diamond A\cap\Diamond B).\Diamond\blacksquare(\Diamond A\cap\Diamond C)\big)}{\Box A.\Box\blacklozenge(\Box B\cup\Box C)\vdash\circ\bullet\big(\Diamond\blacksquare(\Diamond A\cap\Diamond B).\Diamond\blacksquare(\Diamond A\cap\Diamond C)\big)}$$

– Case 2: (Exchange or) Weakening.

As an intermediate step, we can try to isolate an immediate substructure of X by applying backward Weakening. By directly applying Weakening, we obtain

$$\Box A\vdash\circ\bullet\big(\Diamond\blacksquare(\Diamond A\cap\Diamond B).\Diamond\blacksquare(\Diamond A\cap\Diamond C)\big),$$

and by applying Exchange and Weakening, we obtain

$$\Box\blacklozenge(\Box B\cup\Box C)\vdash\circ\bullet\big(\Diamond\blacksquare(\Diamond A\cap\Diamond B).\Diamond\blacksquare(\Diamond A\cap\Diamond C)\big).$$

Notice that the second subcase can be reduced to the first one as follows:

$$
\cfrac{
\cfrac{
\cfrac{
\cfrac{
\cfrac{\quad\cfrac{??}{\Box B\vdash\circ\bullet\big(\Diamond\blacksquare(\Diamond A\cap\Diamond B).\Diamond\blacksquare(\Diamond A\cap\Diamond C)\big)}\qquad\cfrac{??}{\Box C\vdash\circ\bullet\big(\Diamond\blacksquare(\Diamond A\cap\Diamond B).\Diamond\blacksquare(\Diamond A\cap\Diamond C)\big)}}
{\Box B\cup\Box C\vdash\circ\bullet\big(\Diamond\blacksquare(\Diamond A\cap\Diamond B).\Diamond\blacksquare(\Diamond A\cap\Diamond C)\big).\circ\bullet\big(\Diamond\blacksquare(\Diamond A\cap\Diamond B).\Diamond\blacksquare(\Diamond A\cap\Diamond C)\big)}}
{\Box B\cup\Box C\vdash\circ\bullet\big(\Diamond\blacksquare(\Diamond A\cap\Diamond B).\Diamond\blacksquare(\Diamond A\cap\Diamond C)\big)}}
{\bullet(\Box B\cup\Box C)\big)\vdash\bullet\big(\Diamond\blacksquare(\Diamond A\cap\Diamond B).\Diamond\blacksquare(\Diamond A\cap\Diamond C)\big)}}
{\blacklozenge(\Box B\cup\Box C)\big)\vdash\bullet\big(\Diamond\blacksquare(\Diamond A\cap\Diamond B).\Diamond\blacksquare(\Diamond A\cap\Diamond C)\big)}}
{\Box\blacklozenge(\Box B\cup\Box C)\big)\vdash\circ\bullet\big(\Diamond\blacksquare(\Diamond A\cap\Diamond B).\Diamond\blacksquare(\Diamond A\cap\Diamond C)\big)}
$$

As to the proof of the first subcase, let us preliminarily perform the following steps:

$$
\cfrac{
\cfrac{
\cfrac{??}{\circ A\vdash\Diamond\blacksquare(\Diamond A\cap\Diamond B).\Diamond\blacksquare(\Diamond A\cap\Diamond C)}}
{A\vdash\bullet\big(\Diamond\blacksquare(\Diamond A\cap\Diamond B).\Diamond\blacksquare(\Diamond A\cap\Diamond C)\big)}}
{\Box A\vdash\circ\bullet\big(\Diamond\blacksquare(\Diamond A\cap\Diamond B).\Diamond\blacksquare(\Diamond A\cap\Diamond C)\big)}
$$

Again, we are in a situation in which we can act on Y only via Exchange, Weakening or Residuation, and also in this case any option leads to a dead end. Indeed:

- Case 2.1: Exchange or Weakening. We can delete one of the immediate substructures of Y via Weakening or, respectively, Exchange and Weakening, obtaining

$$\circ A\vdash\Diamond\blacksquare(\Diamond A\cap\Diamond B)\quad\text{and}\quad\circ A\vdash\Diamond\blacksquare(\Diamond A\cap\Diamond C).$$

In each case, we reach a dead end, as shown below:

$$
\cfrac{
\cfrac{
\cfrac{\cfrac{?}{\circ A\vdash\Diamond A\cap\Diamond B}}
{A\vdash\bullet(\Diamond A\cap\Diamond B)}}
{A\vdash\blacksquare(\Diamond A\cap\Diamond B)}}
{\circ A\vdash\Diamond\blacksquare(\Diamond A\cap\Diamond B)}
\qquad\qquad
\cfrac{
\cfrac{
\cfrac{\cfrac{?}{\circ A\vdash\Diamond A\cap\Diamond C}}
{A\vdash\bullet(\Diamond A\cap\Diamond C)}}
{A\vdash\blacksquare(\Diamond A\cap\Diamond C)}}
{\circ A\vdash\Diamond\blacksquare(\Diamond A\cap\Diamond C)}
$$

- Case 2.2: Residuation. We can isolate any of the substructures of Y via Residuation, or via Exchange and Residuation. In each case we reach a dead end:

$$
\cfrac{\cfrac{?}{\Diamond\blacksquare(\Diamond A\cap\Diamond B)\supset\circ A\vdash\Diamond\blacksquare(\Diamond A\cap\Diamond C)}}
{\circ A\vdash\Diamond\blacksquare(\Diamond A\cap\Diamond B).\Diamond\blacksquare(\Diamond A\cap\Diamond C)}
$$

$$
\cfrac{
\cfrac{\cfrac{?}{\Diamond\blacksquare(\Diamond A\cap\Diamond C)\supset\circ A\vdash\Diamond\blacksquare(\Diamond A\cap\Diamond B)}}
{\circ A\vdash\Diamond\blacksquare(\Diamond A\cap\Diamond C).\Diamond\blacksquare(\Diamond A\cap\Diamond B)}}
{\circ A\vdash\Diamond\blacksquare(\Diamond A\cap\Diamond B).\Diamond\blacksquare(\Diamond A\cap\Diamond C)}
$$

References

1. Belnap, N.: A useful four-valued logic. In: Dunn, J.M., Epstein, G. (eds.) Moder Uses of Multiple-Valued Logic, pp. 5–37. D. Reidel Springer Netherlands Edition, Dordrecht (1977)
2. Belnap, N.: Display logic. J. Philos. Logic **11**, 375–417 (1982)
3. Belnap, N.: Linear logic displayed. Notre Dame J. Formal Log. **31**(1), 14–25 (1990)
4. Belnap, N.: Life in the undistributed middle. In: Došen, K., Schroeder-Heister, P. (eds.) Substructural Logics, pp. 31–41. Oxford University Press, Oxford (1993)
5. Bílková, M., Greco, G., Palmigiano, A., Tzimoulis, A., Wijnberg, N.: The logic of resources and capabilities (submitted). arXiv preprint arXiv:1608.02222
6. Birkhoff, G.: Lattice Theory, vol. 25. American Mathematical Society, Providence (1967)
7. Birkhoff, G., Lipson, J.D.: Heterogeneous algebras. J. Comb. Theory **8**(1), 115–133 (1970)
8. Burris, S., Sankappanavar, H.P.: A Course in Universal Algebra. Springer, Heidelberg (2006)
9. Agata, C., Revantha, R.: Power and limits of structural display rules. ACM Trans. Comput. Logic (TOCL) **17**(3), 17 (2016)
10. Conradie, W., Craig, A.: Canonicity results for mu-calculi: an algorithmic approach. J. Logic Comput. (forthcoming). arXiv preprint arXiv:1408.6367
11. Conradie, W., Craig, A., Palmigiano, A., Zhao, Z.: Constructive canonicity for lattice-based fixed point logics (submitted). arXiv preprint arXiv:1603.06547
12. Conradie, W., Fomatati, Y., Palmigiano, A., Sourabh, S.: Algorithmic correspondence for intuitionistic modal mu-calculus. Theoret. Comput. Sci. **564**, 30–62 (2015)
13. Conradie, W., Frittella, S., Palmigiano, A., Piazzai, M., Tzimoulis, A., Wijnberg, N.M.: Categories: how i learned to stop worrying and love two sorts. In: Väänänen, J., Hirvonen, Å., de Queiroz, R. (eds.) WoLLIC 2016. LNCS, vol. 9803, pp. 145–164. Springer, Heidelberg (2016). doi:10.1007/978-3-662-52921-8_10. arXiv preprint arXiv:1604.00777
14. Conradie, W., Ghilardi, S., Palmigiano, A.: Unified correspondence. In: Baltag, A., Smets, S. (eds.) Johan van Benthem on Logic and Information Dynamics. OCL, vol. 5, pp. 933–975. Springer, Cham (2014). doi:10.1007/978-3-319-06025-5_36
15. Conradie, W., Palmigiano, A.: Algorithmic correspondence and canonicity for distributive modal logic. Ann. Pure Appl. Logic **163**(3), 338–376 (2012)
16. Conradie, W., Palmigiano, A.: Constructive canonicity of inductive inequalities (submitted). arXiv preprint arXiv:1603.08341
17. Conradie, W., Palmigiano, A.: Algorithmic correspondence and canonicity for non-distributive logics (submitted). arXiv preprint arXiv:1603.08515
18. Conradie, W., Palmigiano, A., Sourabh, S.: Algebraic modal correspondence: Sahlqvist and beyond. J. Log. Algebr. Methods Programm. (2016). arXiv preprint arXiv:1606.06881
19. Conradie, W., Palmigiano, A., Sourabh, S., Zhao, Z.: Canonicity and relativized canonicity via pseudo-correspondence: an application of ALBA (submitted). arXiv preprint arXiv:1511.04271
20. Conradie, W., Palmigiano, A., Zhao, Z.: Sahlqvist via translation (submitted). arXiv preprint arXiv:1603.08220
21. Conradie, W., Robinson, C.: On Sahlqvist theory for hybrid logic. J. Log. Comput. (2015). doi:10.1093/logcom/exv045
22. Davey, B.A., Priestley, H.A.: Introduction to Lattices and Order. Cambridge University Press, Cambridge (2002)
23. Došen, K.: Sequent systems and groupoid models i. Stud. Logica **47**, 353–389 (1988)
24. Došen, K.: Logical constants as punctuation marks. Notre Dame J. Formal Log. **30**(3), 362–381 (1989)
25. Frittella, S., Greco, G., Kurz, A., Palmigiano, A.: Multi-type display calculus for propositional dynamic logic. J. Log. Comput. **26**(6), 2067–2104 (2016). doi:10.1093/logcom/exu064

26. Frittella, S., Greco, G., Kurz, A., Palmigiano, A., Sikimić, V.: Multi-type sequent calculi. In: Indrzejczak, A., Kaczmarek, J., Zawidski, M. (eds.) Proceedings Trends in Logic XIII, vol. 13, pp. 81–93 (2014)

27. Frittella, S., Greco, G., Kurz, A., Palmigiano, A., Sikimić, V.: A proof-theoretic semantic analysis of dynamic epistemic logic. J. Log. Comput. **26**(6), 1961–2015 (2016). doi:10.1093/logcom/exu063

28. Frittella, S., Greco, G., Kurz, A., Palmigiano, A., Sikimić, V.: A multi-type display calculus for dynamic epistemic logic. J. Log. Comput. **26**(6), 2017–2065 (2016). doi:10.1093/logcom/exu068

29. Frittella, S., Greco, G., Palmigiano, A., Yang, F.: A multi-type calculus for inquisitive logic. In: Väänänen, J., Hirvonen, Å., de Queiroz, R. (eds.) WoLLIC 2016. LNCS, vol. 9803, pp. 215–233. Springer, Heidelberg (2016). doi:10.1007/978-3-662-52921-8_14. arXiv preprint arXiv:1604.00936

30. Frittella, S., Palmigiano, A., Santocanale, L.: Dual characterizations for finite lattices via correspondence theory for monotone modal logic. J. Log. Comput. (2016). doi:10.1093/logcom/exw011

31. Galatos, N., Jipsen, P., Kowalski, T., Ono, H.: Residuated Lattices: An Algebraic Glimpse at Substructural Logics, vol. 151. Elsevier, Amsterdam (2007)

32. Gentzen, G.: The Collected Papers of Gerhard Gentzen/Edited by M.E. Szabo. North-Holland Publishing Company, Amsterdam (1969)

33. Girard, J.-Y.: Linear logic. Theoret. Comput. Sci. **50**(1), 1–101 (1987)

34. Greco, G., Liang, F., Moshier, A., Palmigiano, A.: Multi-type display calculus for semi De Morgan logic. In: Proceedings WoLLIC 2017 (forthcoming)

35. Goldblatt, R.I.: Semantic analysis of orthologic. J. Philos. Log. **3**(1–2), 19–35 (1974)

36. Greco, G., Kurz, A., Palmigiano, A.: Dynamic epistemic logic displayed. In: Grossi, D., Roy, O., Huang, H. (eds.) LORI 2013. LNCS, vol. 8196, pp. 135–148. Springer, Heidelberg (2013). doi:10.1007/978-3-642-40948-6_11

37. Greco, G., Ma, M., Palmigiano, A., Tzimoulis, A., Zhao, Z.: Unified correspondence as a proof-theoretic tool. J. Log. Comput. (2016). doi:10.1093/logcom/exw022

38. Greco, G., Palmigiano, A.: Lattice logic properly displayed. Extended version: arXiv:1612.05930

39. Greco, G., Palmigiano, A.: Linear logic properly displayed (submitted). arXiv preprint arXiv:1611.04181

40. Huhn, H.P.: *n*-distributivity and some questions of the equational theory of lattices. In: Contributions to Universal Algebra, Colloq. Math. Soc. J. Bolyai North-Holland, vol. 17 (1997)

41. Lambek, J.: The mathematics of sentence structure. Am. Math. Mon. **65**(3), 154–170 (1958)

42. le Roux, C.: Correspondence theory in many-valued modal logics. Master's thesis, University of Johannesburg, South Africa (2016)

43. Ma, M., Zhao, Z.: Unified correspondence and proof theory for strict implication. J. Log. Comput. (2016). doi:10.1093/logcom/exw012. arXiv preprint arXiv:1604.08822

44. Palmigiano, A., Sourabh, S., Zhao, Z.: Jónsson-style canonicity for ALBA-inequalities. J. Log. Comput. (2015). doi:10.1093/logcom/exv041

45. Palmigiano, A., Sourabh, S., Zhao, Z.: Sahlqvist theory for impossible worlds. J. Log. Comput. (2016). doi:10.1093/logcom/exw014

46. Paoli, F.: Substructural Logics: A Primer, vol. 13. Springer, Heidelberg (2013)

47. Sambin, G., Battilotti, G., Faggian, C.: Basic logic: reflection, symmetry, visibility. J. Symb. Log. **65**(3), 979–1013 (2014)

48. von Plato, J., Negri, S.: Proof systems for lattice logic. Math. Struct. Comput. Sci. **14**(4), 507–526 (2014)

49. Troelstra, A., Schwichtenberg, H.: Basic Proof Theory. Cambridge University Press, Cambridge (2000)

50. Wansing, H.: Displaying Modal Logic. Kluwer, Dordrecht (1998)

Shift Registers Fool Finite Automata

Bjørn Kjos-Hanssen[✉]

University of Hawai'i at Mānoa, Honolulu, USA
bjoernkh@hawaii.edu

Abstract. Let x be an m-sequence, a maximal length sequence produced by a linear feedback shift register. We show that x has maximal subword complexity function in the sense of Allouche and Shallit. We show that this implies that the nondeterministic automatic complexity $A_N(x)$ is close to maximal: $n/2 - A_N(x) = O(\log^2 n)$, where n is the length of x. In contrast, Hyde has shown $A_N(y) \le n/2 + 1$ for all sequences y of length n.

1 Introduction

Linear feedback shift registers, invented by Golomb [2], may be "the most-used mathematical algorithm idea in history", used at least 10^{27} times in cell phones and other devices [7]. They are particularly known as a simple way of producing pseudorandom output sequences called m-sequences. However, thanks to the Berlekamp–Massey algorithm [5], one can easily find the shortest LFSR that can produce a given sequence x. The length of this LFSR, the *linear complexity* of x, should then be large for a truly pseudorandom sequence, but is small for m-sequences. In this article we show that using a different complexity measure, *automatic complexity*, the pseudorandomness of m-sequences can be measured and, indeed, verified.

While our computer results in Sect. 3 concern the linear case specifically, our theoretical results in Sect. 2 concern the following natural abstraction of the usual notion of feedback shift register [1].

Definition 1. *Let $[q] = \{0, \dots, q-1\}$. A q-ary k-stage combinatorial shift register (CSR) is a mapping*

$$\Lambda : [q]^k \to [q]^k$$

such that there exists $F : [q]^k \to [q]$ such that for all x_i,

$$\Lambda(x_0, \dots, x_{k-1}) = (x_1, x_2, \dots, x_{k-1}, F(x_0, x_1, \dots, x_{k-1})).$$

The function F is called the feedback function *of Λ.*

Definition 2. *An infinite sequence $x = x_0 x_1 \dots$ is eventually periodic if there exist integers M and $N > 0$ such that for all $n > M$, $x_n = x_{n-N}$. The least N for which there exists such an M is the* eventual period *of x.*

© Springer-Verlag GmbH Germany 2017
J. Kennedy and R.J.G.B. de Queiroz (Eds.): WoLLIC 2017, LNCS 10388, pp. 170–181, 2017.
DOI: 10.1007/978-3-662-55386-2_12

Definition 3. *For any k-stage CSR Λ and any word x of length $\geq k$, the* period *of Λ upon processing x is the eventual period of the sequence $\Lambda^t(x_0, \ldots, x_{k-1})$, $0 \leq t < \infty$.*

Lemma 4. *Let k and q be positive integers. Let Λ be a q-ary k-stage CSR. Let $x = x_0 x_1 \ldots$ be an infinite sequence produced by Λ. Then x is eventually periodic, and the period of Λ upon processing x exists and is finite.*

Proof. The infinite sequence $\Lambda^t(x_0, \ldots, x_{k-1})$ for $0 \leq t < \infty$ takes values in the finite set $[q]^k$. Thus, by the pigeonhole principle, there exist M and $N > 0$ with

$$\Lambda^M(x_0, \ldots, x_{k-1}) = \Lambda^{M-N}(x_0, \ldots, x_{k-1}).$$

Let $n > M$. Then

$$\begin{aligned}
(x_n, \ldots, x_{n+k-1}) &= \Lambda^n(x_0, \ldots, x_{k-1}) \\
&= \Lambda^{n-M} \Lambda^M(x_0, \ldots, x_{k-1}) \\
&= \Lambda^{n-M} \Lambda^{M-N}(x_0, \ldots, x_{k-1}) \\
&= \Lambda^{n-N}(x_0, \ldots, x_{k-1}) \\
&= (x_{n-N}, \ldots, x_{n-N+k-1}),
\end{aligned}$$

hence $x_n = x_{n-N}$.

We can now define LFSRs and m-sequences. As our computer results concern binary sequences, we take $q = 2$. However, a higher level of generality would also be possible.

Definition 5. *Suppose a k-stage CSR Λ produces the infinite output $x = x_0 x_1 \ldots$ and its feedback function is a linear transformation of $[q]$ when viewed as the finite field $\mathbb{F}_q$, where $q = 2$. Then Λ is a linear feedback shift register (LFSR). Suppose the period P of Λ upon processing x is $2^k - 1$. Then $x_0 \ldots x_{P-1}$ is called an m-sequence (or maximal length sequence, or PN (pseudo-noise) sequence).*

If m-sequences are pseudo-random in some sense then they should have high, or at least *not unusually low*, complexity according to some measure. In 2015, Jason Castiglione (personal communication) suggested that *automatic complexity* might be that measure.

Definition 6 ([3,6]). *Let $L(M)$ be the language recognized by the automaton M. Let x be a finite sequence.*

- *The (deterministic)* automatic complexity *of x is the least number $A(x)$ of states of a deterministic finite automaton M such that*

$$L(M) \cap \{0,1\}^n = \{x\}.$$

- *The* nondeterministic automatic complexity *$A_N(x)$ is the minimum number of states of a nondeterministic finite automaton (NFA) M accepting x such that there is only one accepting path in M of length $|x|$.*
- *The* non-total deterministic automatic complexity *$A^-(x)$ is defined like $A(x)$ but without requiring totality of the transition function.*

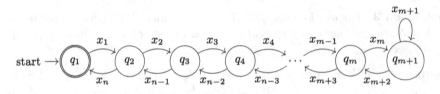

Fig. 1. A nondeterministic finite automaton that only accepts one sequence $x = x_1x_2x_3x_4 \cdots x_n$ of length $n = 2m + 1$.

As totality can always be achieved by adding at most one extra "dead" state, we have

$$A_N(x) \leq A^-(x) \leq A(x) \leq A^-(x) + 1.$$

Theorem 7 (Hyde [3]). *The nondeterministic automatic complexity $A_N(x)$ of a sequence x of length n satisfies*

$$A_N(x) \leq \lfloor n/2 \rfloor + 1.$$

Figure 1 gives a hint to the proof of Theorem 7 in the case where n is odd. Theorem 7 is sharp [3], and experimentally we find that about 50% of all binary sequences attain the bound. Thus, to "fool" finite automata this bound should be attained or almost attained.

2   Main Result for FSRs

Our strategy will be to prove that if a sequence has low complexity, then it contains repeated parts, forcing any shift register producing it to be in the same state (including memory contents) at two distinct points in the sequence.

We first introduce some automata theoretic notions that may not have standard names in the literature.

Definition 8.

- *A* state sequence *is a sequence of states visited upon processing of an input sequence by a finite automaton.*
- *An* abstract NFA *is an NFA without edge labels.*
- *The* abstract NFA M induced by *a state sequence $s = s_0 \dots s_n$ is defined as follows. The states of M are the states appearing in s. The transitions of M are $s_i \rightarrow s_{i+1}$ for each $0 \leq i < n$.*
- *A state sequence $s = s_0 \dots s_n$ is* path-unique *if the abstract NFA induced by s has only one path of length $|s|$ from s_0 to s_n, namely s.*

We use the interval notation $s_{[i,j]} = s_i s_{i+1} \dots s_{j-1} s_j$ and we concatenate as follows: $s_{[i,j]} {}^\frown s_{[j,k]} = s_{[i,j]} s_{[j,k]} = s_{[i,k]}$.

Lemma 9. *Let $s = s_0 \dots s_n$ be a path-unique state sequence. Suppose that $i \leq j \leq k$ are positive integers such that $s_i = s_j = s_k$, and $s_t \neq s_i$ for all $t \in [i, k] \setminus \{i, j, k\}$. Then $s_{[i,j]} = s_{[j,k]}$.*

Proof. By uniqueness of path, $s_{[i,k]} = s_{[i,j]}s_{[j,k]} = s_{[j,k]}s_{[i,j]}$, so one of $s_{[i,j]}$ and $s_{[j,k]}$ is a prefix of the other. But considering the position of the second occurrence of s_i in $s_{[i,k]}$, we can conclude $s_{[i,j]} = s_{[j,k]}$.

Definition 10. *Let $s = s_0 \ldots s_n$ be a path-unique state sequence and let $0 \le i \le n$. The period of s_i in s is defined to be $\min\{k - j : s_k = s_j = s_i, j < k\}$, if s_i occurs at least twice in s, and to be ∞, otherwise.*

An illustration of periods is given in Fig. 2.

Lemma 11. *Let $s = s_0 \ldots s_n$ be a path-unique state sequence. If $i \le j$ and $t > 0$ are integers such that $s_i = s_{i+t}$ and $s_j = s_{j+t}$, then $s_j \in \{s_i, \ldots, s_{i+t}\}$.*

Proof. Let M be the abstract NFA induced by s. We proceed by induction on the $k = k_j$ such that $j - t \in [i + (k - 1)t, i + kt]$, which exists since $t > 0$. If $k \le 0$ then $j \le i + t$ and we are done. So suppose $s_{j'} \in \{s_i, \ldots, s_{i+t}\}$ for each j' with $k_{j'} < k_j$. Both of the following state sequences of length $n + 1$ are accepting for M:

$$\hat{s} = s_{[0,i]} \frown \qquad s_{[i+t,j]} \frown s_{[j,j+t]} \frown s_{[j,n]},$$
$$s = s_{[0,i]} \frown s_{[i,i+t]} \frown s_{[i+t,j]} \frown \qquad s_{[j,n]}.$$

Since s is path-unique, $s = \hat{s}$, and so $s_j = \hat{s}_j = s_{i+(j-(i+t))} = s_{j-t}$. Since $k_{j-t} = k_j - 1 < k_j$, by induction $s_{j-t} \in \{s_i, \ldots, s_{i+t}\}$, giving $s_j \in \{s_i, \ldots, s_{i+t}\}$, as desired.

Lemma 12. *For each path-unique state sequence s, each number t is the period of at most t states in s.*

Proof. We may of course assume $t < \infty$. Fix i and suppose t is the period of s_i. Let us count how many states s_j there can be such that t is the period of s_j. Since $t < \infty$, s_i appears at least twice in s. Thus, either

- $i + t \le n$ and $s_i = s_{i+t}$, or
- $0 \le i - t$ and $s_i = s_{i-t}$.

By Lemma 11, either

- s_j is among the states in $s_{[i,i+t]}$ and $s_i = s_{i+t}$, or
- s_j is among the states in $s_{[i-t,i]}$ and $s_i = s_{i-t}$,

respectively. Either way, there are only at most t choices of such s_j.

Lemma 13. *Let Q be a positive integer. Let $f : \{1, \ldots, Q\} \to \mathbb{N}$ be a function such that $1 \le f(1)$ and $f(i) < f(i + 1)$ for each i. Then $i \le f(i)$ for each i.*

Proof.

$$f(i) - f(1) = \sum_{j=1}^{i-1} f(j + 1) - f(j) \ge \sum_{j=1}^{i-1} 1 = i - 1,$$

so $f(i) + 1 \ge i + f(1) \ge i + 1$.

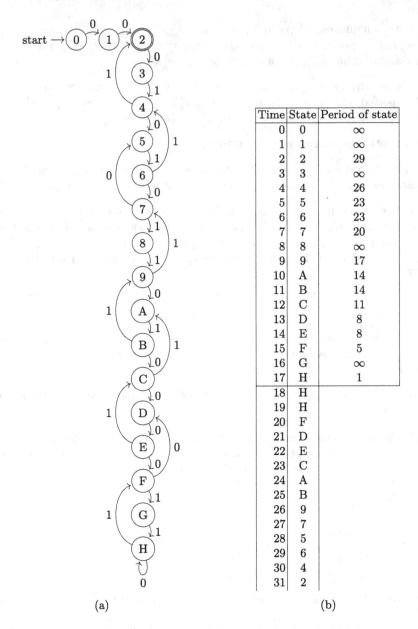

Time	State	Period of state
0	0	∞
1	1	∞
2	2	29
3	3	∞
4	4	26
5	5	23
6	6	23
7	7	20
8	8	∞
9	9	17
10	A	14
11	B	14
12	C	11
13	D	8
14	E	8
15	F	5
16	G	∞
17	H	1
18	H	
19	H	
20	F	
21	D	
22	E	
23	C	
24	A	
25	B	
26	9	
27	7	
28	5	
29	6	
30	4	
31	2	

(a)   (b)

Fig. 2.   An   optimal   deterministic   automaton,   witness   to
$A^-(0001010110100001100100111110111)$ = 18, and its times, states, and periods. There is only 1 state with period 1, and in general at most ℓ states with period ℓ.

Definition 14. *Let α be a word of length n, and let α_i be the ith letter of α for $1 \leq i \leq n$. We define the uth power of α for certain values of $u \in \mathbb{Q}_{\geq 0}$ (the set of nonnegative rational numbers) as follows.*

- If $u = 0$ then α^u is the empty word.
- If $u > 0$ is an integer then the power α^u is defined inductively by $\alpha\,\alpha^{u-1}$, where juxtaposition denotes concatenation.
- If $u = v + k/n$ where $0 < k < n$, and k is an integer, then α^u denotes $\alpha^v \alpha_1 \ldots \alpha_k$.

As an example of Definition 14, we have $ABBA^{1.5} = ABBAAB$.

Lemma 15. *Let* $f : \mathbb{N} \to \mathbb{N}$, $n \geq 0$, *and* $u \in \mathbb{Q}_{\geq 0}$. *Suppose that all* u^{th} *powers* α^u *within a sequence* x *of length* n *satisfy* $u \leq f(|\alpha|)$, *where* f *is non-increasing. Let* s *be a path-unique state sequence. Let* $q_1, \ldots, q_Q$ *be a list of states of* s *ordered by increasing period. Let* a_i *be the number of occurrences of* q_i *in* s. *Let* M_s *be the abstract NFA induced by* s.

Suppose moreover that x *and* s *are related as follows:* x *is the input read along the unique accepting path of length* $|x|$ *of some NFA* M *which is obtained from* M_s *by assigning one label to each edge.*

Let

$$(b_1, b_2, \ldots) = (f(1) + 1, f(2) + 1, f(2) + 1, \ldots,$$
$$\underbrace{f(i) + 1, \ldots, f(i) + 1}_{i \text{ times}}, \ldots).$$

Then $a_i \leq b_i$ *for each* i.

Proof. For each $1 \leq i \leq Q$, let ℓ_i be the period of q_i. (For instance, we could have $(\ell_1, \ell_2, \ldots) = (3, 4, 4, 4, 4, 5, 6)$.) If q_i occurs $u+1$ times then by Lemma 9 it occurs during the processing of a uth power α^u where $|\alpha| = \ell_i$. Thus q occurs at most $f(\ell_i) + 1$ times, i.e., $a_i \leq f(\ell_i) + 1$. By Lemma 12, the sequence $(\ell_1, \ell_2, \ell_3, \ldots)$ is a subsequence of the sequence $(1, 2, 2, 3, 3, 3, \ldots)$ hence by Lemma 13, dominates it pointwise. And so $a_i \leq f(\ell_i) + 1 \leq b_i$. $\qquad\blacksquare$

In particular, Lemma 15 tells us that if x is square-free then each state can occur at most twice, which was observed by Shallit and Wang [6].

Lemma 16. *Let* $s = s_0 \ldots s_n$ *be a state sequence. Let* $q_1, \ldots, q_Q$ *be the distinct states appearing in* s, *in any order. Let* $a_i \geq 1$ *be the number of times* q_i *occurs. Let* $T = n + 1 = |s| = \sum_{i=1}^Q a_i$. *Let* $Q_0 \leq Q$ *and let* $g : \mathbb{Z}_{\geq 0} \to \mathbb{Z}_{\geq 0}$. *If* $a_i \leq g(i)$ *for all* $1 \leq i \leq Q$, *and* $g(i) = 2$ *for all* $Q_0 < i < \infty$, *with* $T_0 := \sum_{i=1}^{Q_0} g(i) \leq T$, *then*

$$Q \geq Q_0 + \left\lceil \frac{T - T_0}{2} \right\rceil.$$

Proof. Let w be such that $T - T_0 \in \{2w+1, 2w+2\}$, i.e., $w = \lceil (T - T_0)/2 \rceil - 1$. Then we want to show $Q \geq Q_0 + w + 1$. If $Q < Q_0 + w + 1$ then $Q \leq Q_0 + w$ and then

$$T = \sum_{i=1}^Q a_i \leq \sum_{i=1}^{Q_0} g(i) + \sum_{i=Q_0+1}^{Q_0+w} 2 = T_0 + 2w,$$

so $2w + 1 \leq T - T_0 \leq 2w$, a contradiction.

Lemma 17. *Let k be a positive integer. Let Λ be a k-stage CSR. Let $x = x_0 x_1 \ldots$ be an infinite sequence produced by Λ. Let P be the period of Λ upon processing x. Suppose a sequence α of length $\ell < P$ is repeated u times consecutively within x, i.e., α^u is a contiguous subsequence of x.*
Then $u < k/\ell + 1$, i.e., $u \leq \lceil \frac{k}{\ell} \rceil$, i.e., $u \leq f(|\alpha|)$ where $f(a) = \lceil k/a \rceil$.

Proof. Suppose to the contrary that x contains a block

$$x_j \ldots x_{j+\ell u - 1} = z_1 \ldots z_{\ell u} = y_1 \ldots y_\ell y_1 \ldots y_\ell \ldots$$

with u many blocks of length ℓ, where $(u-1)\ell \geq k$, i.e., $\ell + k \leq \ell u$. Let $q \geq 0$ and $r \geq 0$ be such that $k = q\ell + r$. We have

$$\Lambda^j(x_0, \ldots, x_{k-1}) = (x_j \ldots x_{j+k-1}) = (z_1, \ldots, z_k)$$

$$= \overbrace{(y_1 \ldots y_\ell) \ldots (y_1 \ldots y_\ell)}^{q \text{ times}} y_1 \ldots y_r = (z_{\ell+1}, \ldots, z_{\ell+k})$$

$$= (x_{j+\ell} \ldots x_{j+\ell+k-1}) = \Lambda^{j+\ell}(x_0, \ldots, x_{k-1}).$$

So x is eventually periodic with period $N \leq a < P$, a contradiction.

Theorem 18. *Let x be an m-sequence and let $n = |x|$. Then $n/2 - A_N(x) = O(\log^2(n))$.*

Proof. Note that if x is produced by a k-stage CSR Λ, then the period P of Λ upon processing x is just $P = n$.

Let $Q = A_N(x)$. Thus Q is the number of states of an NFA M with only one accepting path s of length n, accepting x along that path. Let $q_1, \ldots, q_Q$ be the states of M ordered by increasing period within s. Let a_i be the number of occurrences of q_i.

By Lemma 17, if x contains α^u where $1 \leq |\alpha| \leq k < P$, then $u \leq f(|\alpha|)$ where $f(a) = \lceil k/a \rceil$, a non-increasing function. By Lemma 15, each $a_i \leq b_i$, where

$$(b_1, b_2, \ldots) = (f(1) + 1, f(2) + 1, f(2) + 1, \ldots,$$
$$\underbrace{f(i) + 1, \ldots, f(i) + 1}_{i \text{ times}}, \ldots).$$

Let $T = n+1 = |s| = \sum_{i=1}^{Q} a_i$. Let $g(i) = \max\{b_i, 2\}$. Let Q_0 be the least integer such that $b_i \leq 2$ for all $i > Q_0$. Then since $f(k) + 1 = \lceil \frac{k}{k} \rceil + 1 = 2$ and since $2k(k-1) \leq n+1$,

$$T_0 := \sum_{i=1}^{Q_0} g(i) = \sum_{i=1}^{Q_0} b_i \leq \sum_{i=1}^{k-1} i \left(\left\lceil \frac{k}{i} \right\rceil + 1 \right)$$

$$\leq \sum_{i=1}^{k-1} i \left(\frac{k}{i} + 2 \right) = k(k-1) + k(k-1)$$

$$= 2k(k-1) \leq n+1 = T,$$

and $g(i) = 2$ for all $i > Q_0$. Hence by Lemma 16, $Q \geq Q_0 + Q_1$, where $Q_1 = \lceil \frac{T-T_0}{2} \rceil$. Note that Q_1 is the minimum number of twos whose sum is at least $T - T_0$. (For instance, if $T - T_0 = 2w + 1$, say, then $Q_1 = w + 1 = \lceil \frac{T-T_0}{2} \rceil$.)

Thus

$$\overbrace{\left(\left\lceil \frac{k}{1} \right\rceil + 1 \right) + 2 \left(\left\lceil \frac{k}{2} \right\rceil + 1 \right) + \cdots + (k-1) \left(\left\lceil \frac{k}{k-1} \right\rceil + 1 \right)}^{Q_0 \text{ many terms}}$$

$$+ \overbrace{2 + 2 + \ldots}^{Q_1 \text{ many terms}} \geq n + 1.$$

Clearly $Q_0 = \sum_{i=1}^{k-1} i = k(k-1)/2$. Now $T_0 + (T - T_0) = n + 1$, $T_0 \leq 2k(k-1)$, and $2Q_1 \geq T - T_0$, so

$$2k(k-1) + 2Q_1 \geq n + 1, \qquad Q_1 \geq \frac{n+1}{2} - k(k-1),$$

and

$$Q \geq Q_0 + Q_1 \geq \frac{k(k-1)}{2} + \frac{n+1}{2} - k(k-1)$$

$$= \frac{n+1}{2} - \frac{k(k-1)}{2}.$$

Thus

$$A_N(x) \geq \frac{n+1}{2} - \frac{\log_2(n+1)(\log_2(n+1) - 1)}{2}.$$

3   Computer Results

3.1   Linear FSRs

Theorem 19. *Let $x \in \{0,1\}^n$ be an m-sequence, where $n = 2^k - 1$, $k \leq 5$. Then $A_N(x) = \lfloor n/2 \rfloor + 1$.*

Theorem 19 was verified in 36 h using a Python script.

Theorem 20. *There exists a sequence x with $A^-(x) - A_N(x) \geq 2$. In fact, there is an m-sequence x with $A^-(x) - A_N(x) = 2$.*

Proof. Let $x = 000101011010001100100111110111$. A computer run showed that $A^-(x) \geq 18$. The production of this sequence by an LFSR with 5 bits is shown in detail in Fig. 3. Figure 2 can be used to verify that $A^-(x) \leq 18$. According to Theorem 19, $A_N(x) = 16$.

We also found another m-sequence y for $k = 5$ with $A^-(y) = 17$. Thus not every m-sequence has maximal A^--complexity:

$$\begin{matrix} \oplus & \oplus & \oplus & & \oplus \\ \hline 0 & 1 & 0 & 0 & 0 \\ 1 & 0 & 1 & 0 & 0 \\ 0 & 1 & 0 & 1 & 0 \\ 1 & 0 & 1 & 0 & 1 \\ 1 & 1 & 0 & 1 & 0 \\ 0 & 1 & 1 & 0 & 1 \\ 1 & 0 & 1 & 1 & 0 \\ 0 & 1 & 0 & 1 & 1 \\ 0 & 0 & 1 & 0 & 1 \\ 0 & 0 & 0 & 1 & 0 \\ 0 & 0 & 0 & 0 & 1 \\ 1 & 0 & 0 & 0 & 0 \\ 1 & 1 & 0 & 0 & 0 \\ 0 & 1 & 1 & 0 & 0 \\ 0 & 0 & 1 & 1 & 0 \\ 1 & 0 & 0 & 1 & 1 \\ 0 & 1 & 0 & 0 & 1 \\ 0 & 0 & 1 & 0 & 0 \\ 1 & 0 & 0 & 1 & 0 \\ 1 & 1 & 0 & 0 & 1 \\ 1 & 1 & 1 & 0 & 0 \\ 1 & 1 & 1 & 1 & 0 \\ 1 & 1 & 1 & 1 & 1 \\ 0 & 1 & 1 & 1 & 1 \\ 1 & 0 & 1 & 1 & 1 \\ 1 & 1 & 0 & 1 & 1 \\ 1 & 1 & 1 & 0 & 1 \\ 0 & 1 & 1 & 1 & 0 \\ 0 & 0 & 1 & 1 & 1 \\ 0 & 0 & 0 & 1 & 1 \\ 1 & 0 & 0 & 0 & 1 \\ 0 & 1 & 0 & 0 & 0 \end{matrix}$$

$$\begin{bmatrix} 1 & 1 & 1 & 0 & 1 \\ 1 & 0 & 0 & 0 & 0 \\ 0 & 1 & 0 & 0 & 0 \\ 0 & 0 & 1 & 0 & 0 \\ 0 & 0 & 0 & 1 & 0 \end{bmatrix} \begin{bmatrix} 0 \\ 1 \\ 0 \\ 0 \\ 0 \end{bmatrix} = \begin{bmatrix} 1 \\ 0 \\ 1 \\ 0 \\ 0 \end{bmatrix}$$

(a) One LFSR step as a matrix multiplication.

(b) Producing an m-sequence.

Fig. 3. The operation of a linear feedback shift register producing the sequence from Theorem 20.

Theorem 21. *There is an m-sequence x and a sequence y with $|x| = |y|$ such that $A^-(x) < A^-(y)$.*

Conjecture 22. *There is an m-sequence x and a sequence y with $|x| = |y|$ such that $A_N(x) < A_N(y)$.*

Using our current algorithm and implementation, the calculation of $A_N(x)$ for m-sequences x of length $2^6 - 1$ is unfortunately out of reach.

3.2 Nonlinears FSRs

For $k = 3$ there are two possible feedback functions that give an injective function with a single cycle,

$$F(p, q, r) = p + pq + r + 1 \text{ and } F(p, q, r) = q + pq + r + 1 \mod 2.$$

One of them gives the output 00011101, which has $A_N(00011101) = 4$ and so is *not* maximally A_N-complex.

4 Complexity Function of a Language

Definition 23. *A word w is a* factor, *or* contiguous subsequence, *of a word v if $v = awb$ for some words a, b. For a finite word x, x^∞ is the infinite word satisfying $x x^\infty = x^\infty$. For a finite or infinite word u, $p_u(k)$ is the number of distinct factors of u of length k. The cyclic subword complexity of x is*

$$(p_{x^\infty}(1), \ldots, p_{x^\infty}(|x|)).$$

The plain subword complexity of x is

$$(p_x(1), \ldots, p_x(|x|)).$$

In general, neither of maximum subword complexity and maximum A_N-complexity implies the other.

Theorem 24. *Maximum subword complexity can be characterized as follows.*

1. *The cyclic subword complexity of a b-ary word is pointwise bounded above by*

$$(b^1, b^2, \ldots, b^t, n, n, \ldots, n)$$

   where t is maximal such that $b^t \leq n$.
2. *This upper bound is realized by m-sequences when $n = b^k - 1$, $k \geq 0$, $b = 2$.*

Proof. (1) is because both

$$(b^1, b^2, \ldots, b^n)$$

and

$$(n, n, \ldots, n)$$

are upper bounds, and the pointwise minimum of two upper bounds is an upper bound. To prove (2), we need to show that an m-sequence x has

$$(p_{x^\infty}(1), \ldots, p_{x^\infty}(n))$$

$$= (b^1, b^2, \ldots, b^{k-1}, b^k - 1, b^k - 1, \ldots, b^k - 1),$$

that is,

$$p_{x^\infty}(i) = \begin{cases} b^i & \text{if } i \leq k - 1, \\ b^k - 1 & \text{if } k \leq i \leq b^k - 1. \end{cases}$$

It suffices to show $p_{x^\infty}(k) = b^k - 1$, since

(i) this gives $p(i) = b^i$ for $i \leq k-1$ (only one string of length k is missing, so all strings of shorter length must be present; since any missing string of shorter length would give at least $b > 1$ missing strings of length k), and

(ii) $p(i)$ is monotonically increasing with i (if two words have distinct prefixes of a certain length, then the strings are distinct).

The statement $p(k) = b^k - 1$ when $b = 2$ follows easily from a note labelled (4) in [4], namely

"4. A sliding window of length k, passed along an m-sequence for $2^k - 1$ positions, will span every possible k-bit number, except all zeros, once and only once. That is, every state of a k-bit state register will be encountered, with the exception of all zeros."

This statement (4) surely is already implicit in Golomb's monograph. In any case, it is almost immediate from the fact that m bits are saved in the state and the sequence is maximum-length.

5   The No-Long-High-Powers Property

Definition 25. *Let $k \geq 0$ and $u \in \mathbb{Q}_{\geq 0}$. The no-long-high-power (NLHP or "no LHP") property of a sequence x of length $2^k - 1$ says that if a word α of length $s < 2^k - 1$ is such that α^u is a factor of x^∞, then $u < k/s + 1$.*

By Lemma 17, m-sequences have the NLHP property.

Theorem 26. *Let x be a word of length $n = 2^k - 1$. The following are equivalent:*

(i) x has the NLHP property.
(ii) x has maximal cyclic subword complexity.

Proof. Recall that when $n = 2^k - 1$ then the maximal subword complexity is realized by m-sequences and is

$$(2, 4, 8, \ldots, 2^{k-1}, n, n, \ldots, n). \tag{1}$$

Let us prove that (ii) implies (i). Suppose that w^u is contained cyclically (using only a single trip through the cycle) in x. We need to show that $|w^{u-1}| < k$ where $n = 2^k - 1$. It's just that if there is an LHP then there are two positions giving the same subword, thereby reducing one of the n's in (1) to $n - 1$.

Let us now prove that (i) implies (ii). We need to show that if, say, 01 is not a factor of x^∞ then there are so many occurrences of 00, 10, 11 as factors as to make an LHP. If there is no 01 then there are many strings of length k that are missing, and so some are repeated. Thus, if $p_{x^\infty}(s) < 2^s$ for $s < t$ then also $p_{x^\infty}(t) < 2^t$. Thus, $p_{x^\infty}(k-1) < 2^{k-1}$. And then we can argue that $p_{x^\infty}(k) < 2^k - 1$, as well. So by the Pigeonhole Principle some word of length k is repeated and hence there is an LHP.

Corollary 27. *Theorem 18 applies to any word of maximal subword complexity (when the length is $2^k - 1$).*

Corollary 27 is of interest because of the following result.

Theorem 28. *Words of maximal subword complexity do not in general have maximum A_N-complexity.*

Proof. It is easily checked that already at length 6, we have a string of maximal subword complexity but not maximal A_N-complexity: namely 001100.

Thus, while experimentally our computer results suggest that m-sequences always have maximal A_N-complexity, our main theoretical result Theorem 18 show that m-sequences have fairly high A_N-complexity also applies to some sequences that demonstrably do not have maximal A_N-complexity.

Acknowledgments. This work was partially supported by a grant from the Simons Foundation (#315188 to Bjørn Kjos-Hanssen). This material is based upon work supported by the National Science Foundation under Grant No. 1545707.

References

1. Gammel, B.M., Göttfert, R.: Linear filtering of nonlinear shift-register sequences. In: Ytrehus, Ø. (ed.) WCC 2005. LNCS, vol. 3969, pp. 354–370. Springer, Heidelberg (2006). doi:10.1007/11779360_28
2. Golomb, S.W.: Shift register sequences. With portions co-authored by Welch, L.R., Goldstein, R.M., Hales, A.W. Holden-Day Inc, San Francisco (1967)
3. Hyde, K., Kjos-Hanssen, B.: Nondeterministic automatic complexity of overlap-free and almost square-free words. Electron. J. Comb. **22**(3), 18 (2015). Paper 3.22
4. New Wave Instruments. Linear feedback shift registers: Implementation, m-sequence properties, feedback tables (2010). http://www.newwaveinstruments.com/resourc es/articles/m_sequence_linear_feedback_shift_register_lfsr.htm#M-Sequence%20Pro perties
5. Massey, J.L.: Shift-register synthesis and BCH decoding. IEEE Trans. Inf. Theory **IT-15**, 122–127 (1969)
6. Shallit, J., Wang, M.-W.: Automatic complexity of strings. J. Autom. Lang. Comb. **6**(4), 537–554 (2001). 2nd Workshop on Descriptional Complexity of Automata, Grammars and Related Structures (London, ON, 2000)
7. Wolfram, S.: Solomon Golomb (1932–2015). http://blog.stephenwolfram.com/2016/ 05/solomon-golomb-19322016/

The Lambek Calculus with Iteration: Two Variants

Stepan Kuznetsov[✉]

Steklov Mathematical Institute of RAS, Moscow, Russia
sk@mi.ras.ru

Abstract. Formulae of the Lambek calculus are constructed using three binary connectives, multiplication and two divisions. We extend it using a unary connective, positive Kleene iteration. For this new operation, following its natural interpretation, we present two lines of calculi. The first one is a fragment of infinitary action logic and includes an omega-rule for introducing iteration to the antecedent. We also consider a version with infinite (but finitely branching) derivations and prove equivalence of these two versions. In Kleene algebras, this line of calculi corresponds to the *-continuous case. For the second line, we restrict our infinite derivations to cyclic (regular) ones. We show that this system is equivalent to a variant of action logic that corresponds to general residuated Kleene algebras, not necessarily *-continuous. Finally, we show that, in contrast with the case without division operations (considered by Kozen), the first system is strictly stronger than the second one. To prove this, we use a complexity argument. Namely, we show, using methods of Buszkowski and Palka, that the first system is Π_1^0-hard, and therefore is not recursively enumerable and cannot be described by a calculus with finite derivations.

Keywords: Lambek calculus · Positive iteration · Infinitary action logic · Cyclic proofs

1 The Infinitary Lambek Calculus with Positive Iteration

The Lambek calculus **L** [12] deals with formulae that are built using three connectives, · (product), \, and / (left and right divisions). These connectives enjoy a natural interpretation as operations on formal languages (completeness shown by Pentus [17]). There are, however, also other interesting and well-respected operations on formal languages, and it is quite natural to try to extend **L** by adding these operations as new connectives.

One of the most common of such operations is *iteration,* or *Kleene star:* for a language M over an alphabet Σ its iteration is defined as follows:

$$M^* = \{u_1 \ldots u_n \mid n \geq 0, u_i \in M\}.$$

This work is supported by the Russian Science Foundation under grant 16-11-10252.

J. Kennedy and R.J.G.B. de Queiroz (Eds.): WoLLIC 2017, LNCS 10388, pp. 182–198, 2017.
DOI: 10.1007/978-3-662-55386-2_13

As one can notice, M^* always includes the empty word, ε. The original Lambek calculus, however, obeys so-called *Lambek's non-emptiness restriction,* that is, the empty sequence is never allowed in **L** (this restriction is motivated by linguistic applications; from the algebraic point of view, this means that we're considering residuated semigroups instead of residuated monoids). Therefore, throughout this paper we consider a modified version of Kleene star, called *positive iteration:*

$$M^+ = \{u_1 \ldots u_n \mid n \geq 1, u_i \in M\} = M^* - \{\varepsilon\}.$$

In this paper, we introduce several extensions of the Lambek calculus with this new connective, establish connections between them, and prove some complexity bounds.

Formulae of the Lambek calculus with positive iteration, usually called *types,* are built from a countable set of variables (primitive types) $\mathrm{Pr} = \{p_1, p_2, p_3, \ldots\}$ using three binary connectives, $\cdot$, $\backslash$, and $/$, and one unary connective, $^+$ (written in the postfix form, A^+). The set of all types is denoted by Tp. Types are denoted by capital Latin letters; capital Greek letters stand for finite linearly ordered sequences of types.

Derivable objects are *sequents* of the form $\Pi \to A$, where $A \in \mathrm{Tp}$ and Π is a *non-empty* finite sequence of types.

Now let's define the first calculus for positive iteration, $\mathbf{L}^+_\omega$. The axioms and the rules for $\cdot$, $\backslash$, and $/$ are the same as in the original Lambek calculus **L**:

$$\frac{}{A \to A} \text{ (ax)}$$

$$\frac{A, \Pi \to B}{\Pi \to A \backslash B} (\to \backslash), \text{ where } \Pi \text{ is non-empty} \qquad \frac{\Pi \to A \quad \Gamma, B, \Delta \to C}{\Gamma, \Pi, A \backslash B, \Delta \to C} (\backslash \to)$$

$$\frac{\Pi, A \to B}{\Pi \to B/A} (\to /), \text{ where } \Pi \text{ is non-empty} \qquad \frac{\Pi \to A \quad \Gamma, B, \Delta \to C}{\Gamma, B/A, \Pi, \Delta \to C} (/ \to)$$

$$\frac{\Gamma \to A \quad \Delta \to B}{\Gamma, \Delta \to A \cdot B} (\to \cdot) \qquad \frac{\Gamma, A, B, \Delta \to C}{\Gamma, A \cdot B, \Delta \to C} (\cdot \to)$$

For $^+$, this calculus includes a countable set of right rules:

$$\frac{\Pi_1 \to A \quad \ldots \quad \Pi_n \to A}{\Pi_1, \ldots, \Pi_n \to A^+} (\to {}^+)_n, \text{ for } n \geq 1$$

and one left rule

$$\frac{\Gamma, A, \Delta \to C \quad \Gamma, A, A, \Delta \to C \quad \Gamma, A, A, A, \Delta \to C \quad \ldots}{\Gamma, A^+, \Delta \to C} ({}^+ \to)_\omega$$

This rule is an ω-*rule,* or an *infinitary* rule. Application of such a rule makes the proof tree infinite. This is somewhat unpleasant from the computational point of view, but, as we show later on, it appears to be inevitable.

The rules $(\to {}^+)_n$ and $({}^+ \to)_\omega$ come from the rules for iteration in *infinitary action logic,* $\mathbf{ACT}_\omega$ [4]. Our system $\mathbf{L}^+_\omega$ differs from $\mathbf{ACT}_\omega$ in the following two points.

1. $\mathbf{L}_\omega^+$ enriches the "pure" (multiplicative) Lambek calculus $\mathbf{L}$, while $\mathbf{ACT}_\omega$ is based on the full Lambek calculus $\mathbf{FL}$, including also additive conjunction ($\wedge$) and disjunction ($\vee$). This means that complexity lower bounds for $\mathbf{L}_\omega^+$ are stronger results than lower bounds for $\mathbf{ACT}_\omega$.
2. In contrast to $\mathbf{ACT}_\omega$, in $\mathbf{L}_\omega^+$ we have Lambek's non-emptiness restriction, and therefore use positive iteration instead of Kleene star.

The cut rule of the form

$$\frac{\Pi \to A \quad \Gamma, A, \Delta \to C}{\Gamma, \Pi, \Delta \to C} \ (\text{cut})$$

is admissible in $\mathbf{L}_\omega^+$. This fact is proved by the same transfinitary cut-elimination procedure, as presented by Palka [14] for $\mathbf{ACT}_\omega$ (for a restricted fragment of $\mathbf{L}_\omega^+$ cut elimination was independently shown by Ryzhkova [21]).

The admissibility of (cut) yields the fact that the rules $(\to \backslash)$, $(\to /)$, $(\cdot \to)$, and, most interestingly, $(^+ \to)_\omega$ are invertible.

The Lambek calculus $\mathbf{L}$, defined by axioms (ax) and rules $(\to \backslash)$, $(\backslash \to)$, $(\to /)$, $(/ \to)$, $(\to \cdot)$, and $(\cdot \to)$, is a conservative fragment of $\mathbf{L}_\omega^+$. Cut elimination for $\mathbf{L}$ was known already by Lambek [12].

The calculus $\mathbf{L}_\omega^+$ defined in this section is sound with respect to the intended interpretation on formal languages, where $^+$ is interpreted as positive iteration:

$$M^+ = \{u_1 \ldots u_n \mid n \geq 1, u_i \in M\},$$

and the Lambek connectives are interpreted in the same way as for $\mathbf{L}$:

$$M \cdot N = \{uv \mid u \in M, v \in N\},$$
$$M \backslash N = \{u \in \Sigma^+ \mid (\forall v \in M)\, vu \in N\},$$
$$N/M = \{u \in \Sigma^+ \mid (\forall v \in M)\, uv \in N\}.$$

The arrow, $\to$, is interpreted as the subset relation.

Completeness with respect to this interpretation is an open problem.

2 Π_1^0-completeness of $\mathbf{L}_\omega^+$

In this section we prove that derivability in $\mathbf{L}_\omega^+$ is Π_1^0- (co-r.e.-) hard. Basically, we follow the same strategy as Buszkowski [4], namely, encoding the totality problem for context-free grammars. Our construction, however, is more involved: instead of embedding context-free grammars into the Lambek environment as Ajdukiewicz – Bar-Hillel basic categorical grammars, we use another translation by Safiullin [22] which yields a categorial grammar that assigns *exactly one type* to each letter of the alphabet. This trick allows us to avoid using additive operations, and prove the complexity lower bound for the extension of the original, purely multiplicative Lambek calculus $\mathbf{L}$. For the purely multiplicative fragment of $\mathbf{ACT}_\omega$, the lower complexity bound was left as an open problem by

Buszkowski in [4]. Here we solve not that problem exactly, but its version with Lambek's restriction.

Throughout this paper, all languages do not contain the empty word. Accordingly, all context-free grammars do not contain ε-rules. By TOTAL$^+$ we denote the set of all context-free grammars $\mathcal{G}$ such that the language generated by $\mathcal{G}$ is the set of all non-empty words, Σ^+. The problem TOTAL$^+$ is Π_1^0-hard. Indeed, as shown in [4], TOTAL, the totality problem for context-free grammars possibly using the empty word, reduces to TOTAL$^+$. In its turn, TOTAL itself is known to be undecidable, and standard proofs of this fact actually yield more: they reduce a well-known Σ_1^0- (r.e.-) complete problems, e.g., Post's correspondence problem [7, Theorem 9.22] or halting problem for Turing machines [5, Example 5.43], to the *complement* of TOTAL. This makes TOTAL and TOTAL$^+$ themselves Π_1^0-complete.

We can further restrict ourselves to context-free grammars over a two-letter alphabet, $\{b, c\}$. Denote the ε-free totality problem over $\{b, c\}$ by TOTAL$_2^+$. The original problem TOTAL$^+$ is reduced to TOTAL$_2^+$ in the following way. Let $\mathcal{G}$ be a context-free grammar that defines a language $\mathcal{L}(\mathcal{G})$ over $\Sigma = \{a_0, a_1, \ldots, a_n\}$. The homomorphism $h: a_i \mapsto b^i c$ is a one-to-one correspondence between Σ^+ and $\{u \in \{b, c\}^+ \mid u$ ends on c and doesn't contain b^{n+1} as a subword$\}$. Now we can computably transform $\mathcal{G}$ into a new context-free grammar $\mathcal{G}'$ for the language $h(\mathcal{L}(\mathcal{G})) \cup \{u \in \{b, c\}^+ \mid u$ ends on b or contains b^{n+1} as a subword$\}$ over $\{b, c\}$. Clearly, $\mathcal{G} \in$ TOTAL$^+$ $\iff$ $\mathcal{G}' \in$ TOTAL$_2^+$. This establishes the necessary reduction and Π_1^0-hardness of TOTAL$_2^+$.

Finally, we consider the *alternation problem* for context-free grammars over $\{b, c\}$, denoted by ALT$_2$. A context-free grammar $\mathcal{G}$ belongs to ALT$_2$ if the language it generates includes the language $(\{b\}^+\{c\}^+)^+ = \{b^{m_1}c^{k_1} \ldots b^{m_n}c^{k_n} \mid n \geq 1, m_i \geq 1, k_j \geq 1\}$ (as a subset). Clearly, ALT$_2$ is also Π_1^0-hard by reduction of TOTAL$_2^+$, since $M = \{b, c\}^+ \iff \{b\} \cdot M \cdot \{c\} \supseteq (\{b\}^+\{c\}^+)^+$.

Now we need an encoding of context-free grammars in the Lambek calculus. A *Lambek categorial grammar with unique type assignment* over the alphabet $\{a_1, \ldots, a_n\}$ consists of $(n + 1)$ types (without the $^+$ connective) $A_1, \ldots, A_n, H$ (H is called the *target* type), and a word $w = a_{i_1} \ldots a_{i_m}$ belongs to the language generated by this grammar iff the sequent $A_{i_1}, \ldots, A_{i_m} \to H$ is derivable in **L**. These grammars have the same expressive power as context-free grammars (without ε-rules).

Theorem 1 (A. Safiullin, 2007). *For every context-free language there exists, and can be effectively constructed from the original context-free grammar, a Lambek categorial grammar with unique type assignment* [22].

The inverse translation, from Lambek categorial grammars to context-free grammars, is also available due to Pentus [15] (in this paper we don't need it). In order to make this paper logically self-contained, we revisit Theorem 1 and give its full proof in the Appendix (in Safiullin's paper [22], the proof is only briefly sketched).

Now we're ready to prove the main result of this section.

Theorem 2. *The derivability problem for $\mathbf{L}_\omega^+$ is Π_1^0-complete.*

Proof. The fact that this problem belongs to class Π_1^0 (the upper bound) is established by the same argument as for $\mathbf{ACT}_\omega$ in [14].

To prove Π_1^1-hardness of the derivability problem in $\mathbf{L}_\omega^+$ (the lower bound), we encode ALT$_2$. For every context-free grammar $\mathcal{G}$ over $\{b, c\}$ we algorithmically construct an $\mathbf{L}_\omega^+$-sequent $E \to H$ such that

$$\mathcal{G} \in \text{ALT}_2 \iff E \to H \text{ is derivable in } \mathbf{L}_\omega^+.$$

First we apply Theorem 1 to $\mathcal{G}$ and obtain a Lambek categorial grammar with unique type assignment. In this case, it consists of three types, B, C, and H. Next, let $E = (B^+ \cdot C^+)^+$.

Now, since the $(\cdot \to)$ and $(^+ \to)$ rules are invertible, the sequent $E \to H$ is derivable in $\mathbf{L}_\omega^+$ iff for any positive natural numbers $n, m_1, \ldots, m_n, k_1, \ldots, k_n$ the sequent $B^{m_1}, C^{k_1}, \ldots, B^{m_n}, C^{k_n} \to H$ is derivable in $\mathbf{L}_\omega^+$, and, since it doesn't contain $^+$, by conservativity also in the Lambek calculus $\mathbf{L}$. By definition of Lambek grammar, this is equivalent to $b^{m_1}c^{k_1}\ldots b^{m_n}c^{k_n} \in \mathcal{L}(\mathcal{G})$. Therefore, $E \to H$ is derivable iff the language generated by $\mathcal{G}$ includes all words of the form $b^{m_1}c^{k_1}\ldots b^{m_n}c^{k_n}$, i.e., $\mathcal{G} \in \text{ALT}_2$.

3   The Calculus with Infinite Derivation Branches

In this section we define $\mathbf{L}_\infty^+$, another infinitary calculus that extends $\mathbf{L}$ with positive iteration, in the spirit of sequent systems with non-well-founded derivations for other logics [2,13,24]. Compared to $\mathbf{L}_\omega^+$, $\mathbf{L}_\infty^+$ has a finite number of rules and each rule has a finite number of premises. The tradeoff is that now derivation trees are allowed to have infinite depth.

The Lambek part (rules for $\backslash$, $/$, and $\cdot$) is taken from $\mathbf{L}$. The rules for positive iteration are as follows:

$$\frac{\Pi \to A}{\Pi \to A^+} \; (\to {}^+)_1 \qquad \frac{\Pi_1 \to A \quad \Pi_2 \to A^+}{\Pi_1, \Pi_2 \to A^+} \; (\to {}^+)_\mathrm{L}$$

$$\frac{\Gamma, A, \Delta \to C \quad \Gamma, A, A^+, \Delta \to C}{\Gamma, A^+, \Delta \to C} \; ({}^+ \to)_\mathrm{L}$$

As said before, we allow infinitely deep derivations. For the cut-free version, any trees with possibly infinite paths are allowed, but for the calculus with (cut) one has to be extremely cautious. Clearly, allowing arbitrary infinite proofs would yield dead circles without actually using rules for $^+$:

$$\frac{p \to p \quad \dfrac{\cdots}{\dfrac{p \to p \quad p \to q}{p \to q} \; \text{(cut)}} \; \text{(cut)}}{p \to q} \; \text{(cut)}$$

Such "derivations" should be ruled out. There are, however, trickier cases like the following:

$$\cfrac{\cfrac{\cfrac{p \to p \quad p^+ \to p^+}{p, p^+ \to p^+} (\to {}^+)_L \quad \cfrac{\vdots}{p^+ \to p} (^+ \to)_L}{p, p^+ \to p} (\text{cut})}{\cfrac{p \to p \qquad\qquad\qquad}{p^+ \to p} (^+ \to)_L}$$

Here in the only infinite path we can see an infinite number of $(^+ \to)$ applications. However, the resulting sequent, $p^+ \to p$, is not valid under the formal language interpretation (e.g., $\{a\}^+ \not\subseteq \{a\}$) and therefore should not be derivable.

For the calculus with (cut), we impose the following constraint on the infinite derivation tree: *in each infinite path there should be an infinite number of applications of* $(^+ \to)_L$ *with **the same** active occurrence of* A^+ (the occurrence is tracked by individuality from bottom to top), cf. [2, Definition 5.5].

In our example that "derives" $p^+ \to p$, the occurrence of p^+ that is active in the lower application of $(^+ \to)_L$ tracks to the *left* premise, and the p^+ that goes further to the infinite path is *another* occurrence generated by cut. For the cut-free system, this constraint holds automatically.

Also notice that the rules in $\mathbf{L}^+_\infty$ are asymmetric: we don't introduce the rules where A appears to the right of A^+. Yet, this calculus is equivalent to the symmetric system $\mathbf{L}^+_\omega$ (Proposition 1). A motivation for this asymmetry is explained in the end of Sect. 4.

We generalize both $\mathbf{L}^+_\infty$ and $\mathbf{L}^+_\omega$ by adding the additive disjunction, $\vee$, governed by the following rules:

$$\cfrac{\Gamma, A_1, \Delta \to C \quad \Gamma, A_2, \Delta \to C}{\Gamma, A_1 \vee A_2, \Delta \to C} (\vee \to) \qquad \cfrac{\Gamma \to A_i}{\Gamma \to A_1 \vee A_2} (\to \vee)$$

and denote the extensions by $\mathbf{L}^+_\infty(\vee)$ and $\mathbf{L}^+_\omega(\vee)$ respectively.

The cut-free calculi $\mathbf{L}^+_\omega(\vee)$ and $\mathbf{L}^+_\infty(\vee)$ (and, therefore, their conservative fragments $\mathbf{L}^+_\omega$ and $\mathbf{L}^+_\infty$) are equivalent.

Proposition 1. *A sequent is derivable in* $\mathbf{L}^+_\omega(\vee)$ *iff it is derivable in* $\mathbf{L}^+_\infty(\vee)$.

Proof (sketch of). The *"only if"* part is trivial: the ω-rule is derivable in $\mathbf{L}^+_\infty(\vee)$ and so are the $(\to {}^+)_n$ rules. All other rules are the same.

For the *"if"* part, we make use of the $*$-elimination result by Palka [14]. We consider the *n-th negative mapping* that replaces any negative occurrence of A^+ (polarity is defined as usual) by $A \vee A^2 \vee \ldots \vee A^n$ and show that if a sequent is derivable in $\mathbf{L}^+_\infty(\vee)$, than all its negative mappings are also derivable. In the negative mapping, however, there are no negative occurrences of $^+$, and therefore its cut-free derivation doesn't have infinite branches. Moreover, we replace each $(\to {}^+)_L$ rule application with the following subderivation:

$$\cfrac{\cfrac{\Pi_1 \to A \quad \Pi_2 \to A^+}{\Pi_1, \Pi_2 \to A \cdot A^+} \quad \cfrac{A, A^+ \to A^+}{A \cdot A^+ \to A^+}}{\Pi_1, \Pi_2 \to A^+} (\text{cut})$$

The sequent $A, A^+ \to A^+$ is derivable in $\mathbf{L}_\omega^+(\vee)$, using the ω-rule. Thus, the negative mapping of the original is derivable in $\mathbf{L}_\omega^+(\vee)$ using cut, and, by cut elimination, has a cut-free derivation. Then we go backwards and show, following the argument of Palka [14], that the original sequent is derivable in $\mathbf{L}_\omega^+(\vee)$.

4   The Cyclic Calculus

Now let's consider the following example:

$$
\cfrac{p, p \backslash p \to p \quad \cfrac{\cfrac{p \to p \quad p, (p \backslash p)^+ \to p}{p, p \backslash p, (p \backslash p)^+ \to p} \ (\backslash \to)}{p, (p \backslash p)^+ \to p}}{\cfrac{(p \backslash p)^+ \to p \backslash p}{} } \ (\to \backslash)
$$

We see that actually we don't have to develop the derivation tree further, since the sequent $p, (p \backslash p)^+ \to p$ on top already appears lower in the derivation, and now this tree can be built up to an infinite one in a regular way.

We define the notion of *cyclic proof* as done in [24, 25] (for GL, the Gödel – Löb logic) and call this system $\mathbf{L}_{\mathrm{circ}}^+$. In contrast to the situation with GL, however, here $\mathbf{L}_{\mathrm{circ}}^+$ is *strictly weaker* than $\mathbf{L}_\omega^+$ ($\mathbf{L}_\infty^+$) due to complexity reasons. Indeed, $\mathbf{L}_\omega^+$ is Π_1^0-hard, while in $\mathbf{L}_{\mathrm{circ}}^+$ derivations are finite and the derivability problem is recursively enumerable (belongs to Σ_1^0). This is true even in the signature without $\vee$.

For the extension of $\mathbf{L}_{\mathrm{circ}}^+$ with additive disjunction, we show that the cyclic system $\mathbf{L}_{\mathrm{circ}}^+(\vee)$ is equivalent to the corresponding variant of *action logic* considered by Pratt [19], Kozen [10], and Jipsen [9]. The difference is due to Lambek's non-emptiness restriction and the use of positive iteration instead of Kleene star.

Formally, cyclic derivations are defined as follows. The system $\mathbf{L}_{\mathrm{circ}}^+(\vee)$ has the same axioms and rules as $\mathbf{L}_\infty^+(\vee)$, but infinite derivations are not allowed. Instead, for each application of the $(^+ \to)_{\mathrm{L}}$ rule that yields $\Gamma, A^+, \Delta \to B$ we trace the active occurrence of A^+ upwards and are allowed to stop if we again get the same sequent, $\Gamma, A^+, \Delta \to B$ with the same occurrence of A^+. This sequent is *backlinked* to the original one, forming a cycle. The cut rule is also allowed. Note that in the bottom of each cycle we always have the $(^+ \to)_{\mathrm{L}}$ rule with the active occurrence of A^+ which is traced through the cycle, thus satisfying the constraint needed for infinite derivations with cut. Clearly, every cyclic derivation can be expanded into an infinite one. On the other hand, the cyclic system $\mathbf{L}_{\mathrm{circ}}^+$ is not equivalent to $\mathbf{L}_\infty^+$ due to complexity reasons.

This system $\mathbf{L}_{\mathrm{circ}}^+$ appears to have much in common with various *coinductive* proof systems [1, 8, 11, 18, 20]. These connections are worth further investigation.

The cyclic system $\mathbf{L}_{\mathrm{circ}}^+(\vee)$ happens to be equivalent to a non-sequential calculus $\mathbf{ACT}^+$ defined below, which is the positive iteration variant of the axioms for action algebras by Pratt [19]:

$$A \to A \quad (A \cdot B) \cdot C \to A \cdot (B \cdot C) \quad A \cdot (B \cdot C) \to (A \cdot B) \cdot C$$

$$\frac{A \to C/B}{A \cdot B \to C} \quad \frac{A \cdot B \to C}{A \to C/B} \quad \frac{B \to A \backslash C}{A \cdot B \to C} \quad \frac{A \cdot B \to C}{B \to A \backslash C}$$

$$\frac{A \to B \quad B \to C}{A \to C} \quad \frac{A \to B_i}{A \to B_1 \vee B_2} \quad \frac{A_1 \to B \quad A_2 \to B}{A_1 \vee A_2 \to B}$$

$$A \vee (A^+ \cdot A^+) \to A^+ \qquad \frac{A \vee (B \cdot B) \to B}{A^+ \to B}$$

The rules for $\backslash$, $/$, and $\cdot$ correspond to the non-sequential formulation of the Lambek calculus [12].

Lemma 1. *The following rule is admissible in* $\mathbf{ACT}^+$:

$$\frac{A \to C \quad C \cdot A \to C}{A^+ \to C}$$

This lemma is actually a modification of a well-known alternative formulation of the calculus for action logic (connecting it to Kleene algebra). The difference, again, is in using positive iteration instead of Kleene star.

Proof. The second premise yields $A \to C \backslash C$, and since $(C \backslash C) \cdot (C \backslash C) \to C \backslash C$ is derivable, we get $A \vee ((C \backslash C) \cdot (C \backslash C)) \to C \backslash C$, and therefore $A^+ \to C \backslash C$ and then $C \to C/A^+$. By transitivity with $A \to C$ this yields $A \to C/A^+$, and therefore $A \cdot A^+ \to C$. Combining this with $A \to C$, we get $A \vee (A \cdot A^+) \to C$, and it is sufficient to show $A^+ \to A \vee (A \cdot A^+)$. Denote $A \vee (A \cdot A^+)$ by B. We have $A \to B$ and also $B \cdot B \to B$. Indeed, using distributivity conditions: $(E \vee F) \cdot G \leftrightarrow (E \cdot G) \vee (E \cdot G)$ and $G \cdot (E \vee F) \leftrightarrow (G \cdot E) \vee (G \cdot F)$, that are derivable in $\mathbf{ACT}^+$, we replace $B \cdot B$ with $(A \cdot A) \vee (A \cdot A \cdot A^+) \vee (A \cdot A^+ \cdot A) \vee (A \cdot A^+ \cdot A \cdot A^+)$, and applying the axiom for $^+$ and monotonicity, we see that all four disjuncts here yield $A \cdot A^+$, and therefore B. Hence, by the rule for $^+$, we obtain $A^+ \to B$.

Lemma 2. *The following rule is admissible in* $\mathbf{ACT}^+$:

$$\frac{A \to C \quad A^2 \to C \quad \dots \quad A^k \to C \quad A^k \cdot C \to C}{A^+ \to C}$$

Lemma 2 is essential for emulating cyclic reasoning in the non-sequential calculus $\mathbf{ACT}^+$. The k parameter corresponds to the number of $(^+ \to)_L$ applications in the cycle.

Proof. First we prove that

$$A^+ \to A \vee A^2 \vee \dots \vee A^k \vee (A^k)^+ \vee (A^k)^+ \cdot A \vee \dots \vee (A^k)^+ \cdot A^{k-1}$$

is derivable in this calculus. We denote the right-hand side of this formula by B and show $A \to B$ and $B \cdot A \to B$ (this yields $A^+ \to B$ by Lemma 1). The first is trivial. For the second, using distributivity conditions, we replace $B \cdot A$ with

$$A^2 \vee A^3 \vee \dots \vee A^k \vee A^{k+1} \vee (A^k)^+ \cdot A \vee (A^k)^+ \cdot A^2 \vee \dots \vee (A^k)^+ \cdot A^k \vee (A^k)^+ \cdot A^{k+1}.$$

All types in this long disjunction, except A^{k+1} and $(A^k)^+ \cdot A^{k+1}$, belong to the disjunction B (and therefore yield B). For the two exceptions we have the following: $A^{k+1} \to (A^k)^+ \cdot A$ and $(A^k)^+ \cdot A^{k+1} \to (A^k)^+ \cdot A$.

Now we prove the lemma itself by deriving $B \to C$. To do this, we need to show $H \to C$ for any disjunct H in B. For $H = A, \ldots, H = A^k$ this is stated in the premises. Since that $(C/C) \cdot (C/C) \to C/C$ is derivable and $A^k \to C/C$ follows from the last premise, we get $A^k \vee ((C/C) \cdot (C/C)) \to (C/C)$, and therefore $(A^k)^+ \to C/C$. Thus, $(A^k)^+ \cdot C \to C$, then $C \to (A^k)^+ \backslash C$, and by transitivity with $A^i \to C$ we get $(A^k)^+ \cdot A^i \to C$ for any $i = 1, \ldots, k - 1$. It remains to show $(A^k)^+ \to C$. We have $(A^k)^+ \cdot A^k \to C$ and also $A^k \to C$ as a premise. One can easily prove $(A^k)^+ \to A^k \vee ((A^k)^+ \cdot A^k)$ and thus establish $(A^k)^+ \to C$.

Finally, by transitivity from $A^+ \to B$ and $B \to C$ we obtain $A^+ \to C$.

Theorem 3. *A sequent (of the form $E \to F$) is derivable in $\mathbf{L}_{\mathrm{circ}}^+(\vee)$ iff it is derivable in $\mathbf{ACT}^+$.*

Proof. The "if" part is easier. The rules operating Lambek connectives ($\cdot$, $/$, and $\backslash$) can be emulated in the sequential calculus due to Lambek [12]. The rules for $\vee$ in $\mathbf{ACT}^+$ directly correspond to the rules for $\vee$ in $\mathbf{L}_{\mathrm{circ}}^+(\vee)$.

The following cyclic derivation yields $A^+ \to B$ from $A \to B$ and $B, B \to B$, thus establishing the rule for $^+$ from $\mathbf{ACT}^+$:

$$
\cfrac{A \to B \quad \cfrac{A \to B \quad \cfrac{\cfrac{\vdots}{A^+ \to B} \quad B, B \to B}{B, A^+ \to B}\ (\mathrm{cut})}{A, A^+ \to B}\ (\mathrm{cut})}{A^+ \to B}\ (^+\!\to)_{\mathrm{L}}
$$

The track of A^+ goes through the cycle, and the $(^+\!\to)_{\mathrm{L}}$ rule is applied to it at every round.

Finally, for $A^+ \cdot A^+ \to A^+$, we first derive $A \cdot A^+ \to A^+$ (using $(\to^+)_{\mathrm{L}}$ and $(\cdot \to)$), and then, following Pratt [19], transform it into $A^+ \cdot A^+ \to A^+$:

$$
\cfrac{\cfrac{\cfrac{A, A^+ \to A^+}{A \to A^+ \backslash A^+} \quad A^+ \backslash A^+, A^+ \backslash A^+ \to A^+ \backslash A^+}{A \vee ((A^+ \backslash A^+) \cdot (A^+ \backslash A^+)) \to A^+ \backslash A^+}}{\cfrac{A^+ \to A^+ \backslash A^+}{A^+ \cdot A^+ \to A^+}}
$$

In this derivation we've used other rules of $\mathbf{ACT}^+$, which were previously shown to be valid in $\mathbf{L}_{\mathrm{circ}}^+(\vee)$. Together with $A \to A^+$ (derivable using $(\to{}^+)_1$), this yields the last axiom of $\mathbf{ACT}^+$, $A \vee (A^+ \cdot A^+) \to A^+$.

For *"only if" part*, we first replace all cycles in the $\mathbf{L}_{\mathrm{circ}}^+(\vee)$ derivation by applications of the rule from Lemma 2. We proceed by induction on the number

of cycles. For the induction step, let the derivation end with an application of
$(^+ \to)_L$, involved in a cycle. Let k be the number of applications of $(^+ \to)_L$ to
the active occurrence of A^+ that is tracked along this cycle. Let the goal sequent
be $\Gamma, A^+, \Delta \to B$; the same sequent appears on top of the cycle:

$$\Gamma, A^+, \Delta \to B$$
$$\vdots$$
$$\frac{\Gamma, A, \Delta \to B \quad \Gamma, A, A^+, \Delta \to B}{\Gamma, A^+, \Delta \to B} \ (^+ \to)_L$$

Let $C = \Gamma \backslash B / \Delta$ (if Γ or Δ contains more than one formula, we add $\cdot$'s between
them; if Γ or Δ is empty, we omit the corresponding division). The sequent
$\Gamma, C, \Delta \to B$ is derivable in the Lambek calculus. Then we go down the cycle
path, replacing the active A^+ with A^i, C. We start with $i = 0$ and increase i each
time we come across $(^+ \to)_L$ applied to the active A^+. After this substitution,
this application becomes trivial: instead of

$$\frac{\Gamma', A, \Delta' \to B' \quad \Gamma', A, A^+, \Delta' \to B'}{\Gamma', A^+, \Delta' \to B'} \ (^+ \to)_L$$

we get

$$\frac{\Gamma', A, A^i, C, \Delta' \to B'}{\Gamma', A^{i+1}, C, \Delta' \to B'}$$

and actually forget about the left premise of the rule. All other rules remain
valid. In the end, this gives us $\Gamma, A^k, C, \Delta \to B$, or $A^k \cdot C \to C$. Moreover,
the derivation of this sequent was obtained by substitution and cutting some
branches from the original derivation, and therefore contains less cycles. By
induction, we can suppose that $A^k \cdot C \to C$ was derived without cycles, using
the rule from Lemma 2.

Next, for an arbitrary j from 1 to k, we go upwards along the trace of the
active A^+ and find the j-th application of $(^+ \to)_L$:

$$\frac{\Gamma', A, \Delta' \to B' \quad \Gamma', A, A^+, \Delta' \to B'}{\Gamma', A^+, \Delta' \to B'}$$

Now we cut off the right (cyclic) derivation branch and replace A^+ in the goal
with A. Next, we trace it down back to the original sequent, replacing A^+ with
A^i. The index i starts from 1 and gets increased each time we pass through the
$(^+ \to)_L$ rule with the active A^+. Again, these applications trivialize, all other
rules remain valid. In the end, we get $\Gamma, A^j, \Delta \to B$ derivable with a less number
of cycles. This yields $A^j \to C$.

Finally, having $A \to C$, $A^2 \to C$, ..., $A^k \to C$, and $A^k \cdot C \to C$, we apply
Lemma 2 and obtain $A^+ \to C$. Using cut, we invert $(\cdot \to)$, $(\to /)$, and $(\to \backslash)$,
decompose C and arrive at the original goal sequent $\Gamma, A^+, \Delta \to B$.

This finishes the non-trivial part of the proof: now we have a normal, non-cyclic derivation, and it remains to show that other rules of $\mathbf{L}^+_{\mathrm{circ}}(\vee)$ used in it are admissible in $\mathbf{ACT}^+$. (Formally speaking, the languages of $\mathbf{L}^+_{\mathrm{circ}}(\vee)$ and $\mathbf{ACT}^+$ are different. In $\mathbf{ACT}^+$, instead of sequents of the form $A_1, \ldots, A_n \to B$, we consider $A_1 \cdot \ldots \cdot A_n \to B$.)

The rules for Lambek connectives ($\backslash$, $/$, and $\cdot$), and also the cut rule, are admissible in $\mathbf{ACT}^+$ due to Lambek [12]. The rules for $\vee$ correspond directly. Finally, the $(\to{}^+)_1$ and $(\to{}^+)_L$ are validated as follows (here we use previously validated Lambek rules):

$$
\cfrac{\Pi \to A \quad \cfrac{\cfrac{A \to A}{A \to A \vee (A^+ \cdot A^+)} \quad A \vee (A^+ \cdot A^+) \to A^+}{A \to A^+}}{\Pi \to A^+}
$$

$$
\cfrac{\cfrac{\Pi_1 \to A}{\Pi_1 \to A^+} \quad \Pi_2 \to A^+}{\Pi_1, \Pi_2 \to A^+ \cdot A^+} \quad \cfrac{A^+ \cdot A^+ \to A^+ \cdot A^+ \quad \cfrac{A^+ \cdot A^+ \to A \vee (A^+ \cdot A^+) \quad A \vee (A^+ \cdot A^+) \to A^+}{A^+ \cdot A^+ \to A^+}}{}
$$
$$
\Pi_1, \Pi_2 \to A^+
$$

Note that, despite the fact that the calculus for $\mathbf{ACT}^+$ is symmetric, the asymmetry in the rules of $\mathbf{L}^+_{\mathrm{circ}}(\vee)$ is essential for our reasoning, because if we allow both left and right rules for $^+$, the rule from Lemma 2, that is used to emulate cyclic derivation, would transform into

$$
\cfrac{A \to C \quad A^2 \to C \quad \ldots \quad A^k \to C \quad A^\ell \cdot C \cdot A^{k-\ell} \to C}{A^+ \to C}
$$

and for this rule we don't know whether it is admissible in $\mathbf{ACT}^+$.

5   Further Work and Open Questions

In this section we summarize the questions that are still (to the author's best knowledge) unsolved.

1. Though we don't claim cut elimination for $\mathbf{L}^+_\infty$ in this paper, it looks plausible that it could be proven using continuous cut elimination (cf. [13,23]). For $\mathbf{L}^+_{\mathrm{circ}}$, however, the problem looks harder, since if one unravels the cyclic derivation into an infinite one and eliminates cut, the resulting derivation could be not cyclic anymore.
2. In this paper we use complexity arguments to show that $\mathbf{L}^+_\omega$ is strictly more powerful than any its subsystem with finite derivations. This doesn't yield any examples of concrete sequents derivable in $\mathbf{L}^+_\omega$ and not derivable, say, in $\mathbf{L}^+_{\mathrm{circ}}$. Constructing such examples is yet an open problem.

3. We don't know whether the rule in the end of Sect. 4 is admissible if $\mathbf{ACT}^+$. If yes, we could allow both left and right rules for $^+$ is cyclic derivations, and this system would be still equivalent to $\mathbf{ACT}^+$.

4. Safiullin's construction (see Appendix) essentially uses Lambek's non-emptiness restriction. The question whether any context-free language can be generated by a categorial grammar with unique type assignment, based on the variant of the Lambek calculus allowing empty left-hand sides of sequents, is still open. From our perspective, a positive answer to this question (maybe, by modification of Safiullin's construction) would immediately yield Π_1^0-hardness of the Lambek calculus allowing empty left-hand sides of sequents, enriched with Kleene star (but without additive conjunction and disjunction), thus solving a problem posed by Buszkowski [4].

5. An open (and, in the view of the sophisticatedness of Pentus' completeness proof [17], very hard) question is the completeness of $\mathbf{L}_\omega^+$ w.r.t. language interpretation (see Sect. 1). A partial completeness result, for the fragment where $^+$ is allowed only in the denominators of $\backslash$ and $/$, was obtained by Ryzhkova [21], using Buszkowski's canonical model construction [3].

Acknowledgments. The author is grateful to Arnon Avron, Lev Beklemishev, Michael Kaminski, Max Kanovich, Glyn Morrill, Fedor Pakhomov, Mati Pentus, Nadezhda Ryzhkova, Andre Scedrov, Daniyar Shamkanov, and Stanislav Speranski for fruitful discussions. The author is also grateful to the anonymous referees for their comments.

Appendix: Safiullin's Construction Revisited

Theorem 1 by Safiullin is a crucial component of our Π_1^0-hardness proof for $\mathbf{L}_\omega^+$. Unfortunately, Safiullin's paper [22] is very brief and, moreover, includes this theorem (which is probably the most interesting result of that paper) as a side-effect of a more complicated construction. This makes it very hard to follow Safiullin's ideas and arrive at a complete proof. Therefore, in this Appendix we present Safiullin's proof clearly and in detail.

In this Appendix, the $^+$ connective is never used, and Tp stands for the set of types constructed from primitive ones using $\cdot$, $\backslash$, and $/$.

Define the *top* of a Lambek type in the following way: $\mathrm{top}(q) = q$ for $q \in \mathrm{Pr}$; $\mathrm{top}(A \backslash B) = \mathrm{top}(B/A) = \mathrm{top}(B)$. Note that the $A \cdot B$ case is missing. Thus, not every type has a top.

For types with tops, the $(\to \cdot)$ rule is invertible (proof by induction):

Lemma 3. *If all types of Π have tops and $\Pi \to A_1 \cdot \ldots \cdot A_n$ is derivable in $\mathbf{L}$, then $\Pi = \Pi_1, \ldots, \Pi_n$ and $\Pi_i \to A_i$ is derivable for every $i = 1, \ldots, n$.*

If a sequent of the form $\Pi \to q$, $q \in \mathrm{Pr}$, has a cut-free derivation in $\mathbf{L}$, trace the occurrence of q back to the axiom of the form $q \to q$, and then trace the left q back to its occurrence in Π. This occurrence of q will be called the *principal* occurrence (for different derivations, the principal occurrences could differ).

Lemma 4. *The principal occurrence has the following properties:*

1. *if all types in Π have tops, then the principal occurrence is one of them;*
2. *if in a derivation of $\Pi, q, \Phi \to q$ the occurrence of q between Π and Φ is principal, then Π and Φ are empty;*
3. *if in a derivation of $\Pi, q/A, \Phi \to q$ the occurrence of q in q/A is principal, then Π is empty;*
4. *if in a derivation of $\Pi, A \backslash q, \Phi \to q$ the occurrence of q in $A \backslash q$ is principal, then Φ is empty.*

Proof. For statement 1, proceed by induction on derivation. For statements 2–4, suppose the contrary and also proceed by induction on derivation.

Lemma 5. *If all types of Π have tops, and these tops are not q, then $\Pi \to q/q$ is not derivable in $\mathbf{L}$.*

Proof. Since $(\to /)$ is invertible, we get $\Pi, q \to q$, and by Lemma 4 Π should be empty. But $\to q/q$ is not derivable due to Lambek's restriction.

Lemma 6. *If in a derivation of $q/A, \Phi \to q$ the leftmost occurrence of q is principal or in a derivation of $\Phi, A \backslash q \to q$ the rightmost occurrence of q is principal, then $\Phi \to A$ is derivable.*

Proof. Induction on the derivation.

Lemma 7. *If $\Pi, q/q, \Phi \to q/q$ is derivable in $\mathbf{L}$, all types from Π and Φ have tops, and these tops are not q, then Π and Φ are empty.*

Proof. Again, by inverting $(\to /)$ we get $\Pi, q/q, \Phi, q \to q$. The rightmost q cannot be principal, because otherwise $\Pi, q/q, \Phi$ is empty (Lemma 4). The second possibility is the top of q/q. Then, again by Lemma 4, Π is empty, and by Lemma 6 $\Phi, q \to q$ is derivable. Since tops of Φ are not q, the rightmost occurrence of q is principal. By Lemma 4, Φ is empty.

By $\mathbb{F}$ we denote the free group generated by the set of primitive types Pr. For every $A \in$ Tp we define its interpretation in this free group, $[\![A]\!]$, as follows: $[\![q]\!] = q$ for $q \in$ Pr; $[\![A \cdot B]\!] = [\![A]\!][\![B]\!]$; $[\![A \backslash B]\!] = [\![A]\!]^{-1}[\![B]\!]$; $[\![B/A]\!] = [\![B]\!][\![A]\!]^{-1}$. If $[\![A]\!]$ is the unit of $\mathbb{F}$, A is called a *zero-balance type*.

The *primitive type* count, $\#_q(A)$, for $q \in$ Pr and $A \in$ Tp, is defined as follows: $\#_q(q) = 1$; $\#_q(q') = 0$, if $q' \in$ Pr and $q' \neq q$; $\#_q(A \cdot B) = \#_q(A) + \#_q(B)$; $\#_q(A \backslash B) = \#_q(B/A) = \#_q(B) - \#_q(A)$. Notice that if A is a zero-balance type, then $\#_q(A) = 0$ for every $q \in$ Pr.

If the sequent $A_1, \dots, A_n \to B$ is derivable in $\mathbf{L}$, then it is *balanced*, namely, $\#_q(B) = \#_q(A_1) + \dots + \#_q(A_n)$ for every $q \in$ Pr, and $[\![A_1]\!] \dots [\![A_n]\!] = [\![B]\!]$.

Theorem 4 (M. Pentus, 1994). *If $[\![A_1]\!] = [\![A_2]\!] = \dots = [\![A_n]\!]$, then there exists such $B \in$ Tp, that all sequents $A_1 \to B$, $A_2 \to B$, $\dots$, $A_n \to B$ are derivable in $\mathbf{L}$ [16].*

For a set of zero-balance types $\mathcal{U} = \{A_1, \ldots, A_n\}$, we construct an ersatz of their additive disjunction, $A_1 \vee \ldots \vee A_n$, in the following way. In the notations for types, we sometimes omit the multiplication sign, $\cdot$, if this doesn't lead to misunderstanding. Let u, t, and s be fresh primitive types, not occurring in A_i. By Theorem 4, there exist such types F and G that the following sequents are derivable for all $i = 1, \ldots, n$:

$$(t/t)A_i(t/t)\ldots(t/t)A_n(t/t) \to F, \qquad (t/t)A_1(t/t)\ldots(t/t)A_i(t/t) \to G.$$

Now let

$$E = (t/t)A_1(t/t)A_2(t/t)\ldots(t/t)A_n(t/t),$$

$$B = E\left(((u/F)\backslash u)\backslash(t/t)\right), \qquad C = ((t/t)/(u/(G\backslash u)))\, E.$$

We omit the multiplication sign, $\cdot$, if this doesn't lead to misunderstanding.
Finally, $\mathrm{is}(\mathcal{U}) = ((s/E)\cdot B)\backslash s/C$.

Lemma 8. *For each $A_i \in \mathcal{U}$, the sequent $A_i \to \mathrm{is}(\mathcal{U})$ is derivable in **L**.*

Proof. The derivation is straightforward.

Lemma 9. *If the sequent $\Pi \to \mathrm{is}(\mathcal{U})$ is derivable in **L**, all types in Π have tops, and these tops are not s or t, then for some $A_j \in \mathcal{U}$ the sequent $B_2, \Pi, C_1 \to A_j$, where B_2 is either empty or is a type such that $B = B_2$ or $B = B_1 \cdot B_2$ for some B_1, and C_1 is either empty or is a type such that $C = C_1$ or $C = C_1 \cdot C_2$ for some C_2 (up to associativity of $\cdot$).*

Using the invertibility of $(\cdot \to)$, we replace $\cdot$'s in B_2 and C_1 by commas, and thus consider them as sequences of types that have tops. Actually, we want them to be empty, and it will be so in our final construction.

Proof. Let $\Pi \to \mathrm{is}(\mathcal{U})$ be derivable. Then one can derive $(s/E), B, \Pi, C \to s$, and then by Lemma 6 we get $B, \Pi, C \to E$ (since the leftmost s is the only top s, and it is the principal occurrence). Recall that $E = (t/t)A_1 \ldots (t/t)A_n(t/t)$ and apply Lemma 3. It is sufficient so show that, after decompositon, the whole Π comes to one part of the left-hand side of the sequent. Suppose the contrary, then locate the principal occurrence of t (it should be in B). Then proceed by induction: finally we run out of t's in B and get a contradiction.

Proof (of Theorem 1). Given a context-free grammar $\mathcal{G}$ without ε-rules, we need to construct an equivalent Lambek grammar with unique type assignment. Let $\Sigma = \{a_1, \ldots, a_\mu\}$ be the alphabet, $\mathcal{N} = \{N_0, N_1, N_2, \ldots, N_\nu\}$ be the set of non-terminal symbols of $\mathcal{G}$, N_0 is the starting symbol.

First we algorithmically transform $\mathcal{G}$ into Greibach normal form [6] with rules of the following three forms: $N_i \Rightarrow a_j N_k N_\ell$, $N_i \Rightarrow a_j N_k$, or $N_i \Rightarrow a_j$.

Now we construct the Lambek grammar. Let Pr include distinct primitive types $p, p_1, \ldots, p_\nu, r, u, t$, and s. For each $i = 0, \ldots, \nu$ let $H_i = p/((p_i/p_i)\cdot p)$

(this type corresponds to the non-terminal N_i). Next, for each $j = 1, \ldots, \mu$, we form a set $\mathcal{U}_j$ in the following way:

$$\text{add } K_{i,k,\ell} = r/\big((H_k \cdot H_\ell \cdot (p_i/p_i)) \backslash r\big) \text{ for each rule } N_i \Rightarrow a_j N_k N_\ell,$$
$$\text{add } K_{i,k} = r/\big((H_k \cdot (p_i/p_i)) \backslash r\big) \text{ for each rule } N_i \Rightarrow a_j N_k,$$
$$\text{add } K_i = r/\big((p_i/p_i) \backslash r\big) \text{ for each rule } N_i \Rightarrow a_j.$$

Now let $D_j = \text{is}(\mathcal{U}_j)$ and $A_j = p/(D_j \cdot p)$ be the type corresponding to a_j. For the target type H we take H_0. Our **claim** is that $a_{i_1} \ldots a_{i_n} \in \mathcal{L}(\mathcal{G})$ iff the sequent $A_{i_1}, \ldots, A_{i_n} \to H_0$ is derivable in **L**.

For the easier *"only if" part*, we prove a more general statement: if $\gamma \in (\mathcal{N} \cup \Sigma)^+$ can be generated from N_m in $\mathcal{G}$, then the sequent $\Gamma \to H_m$ is derivable in **L**, where Γ is a sequence of types corresponding to letters of γ, A_j for $a_j \in \Sigma$ and H_i for $N_i \in \mathcal{N}$. To establish this, it is sufficient to prove that the following sequents are derivable (each step of the context-free generation maps to a (cut) with the corresponding sequent):

$$A_j, H_k, H_\ell \to H_i \text{ for each rule } N_i \Rightarrow a_j N_k N_\ell,$$
$$A_j, H_k \to H_i \text{ for each rule } N_i \Rightarrow a_j N_k,$$
$$A_j \to H_i \text{ for each rule } N_i \Rightarrow a_j.$$

Consider sequents of the first type (the second and the third types are handled similarly). Since $D_j = \text{is}(\mathcal{U}_j)$ and $K_{i,k,\ell} = r/\big((H_k \cdot H_\ell \cdot (p_i/p_i)) \backslash r\big) \in \mathcal{U}_j$, we have $K_{i,k,\ell} \to D_j$ by Lemma 8. Then the derivation is as follows:

$$\cfrac{\cfrac{\cfrac{H_k, H_\ell, p_i/p_i, (H_k \cdot H_\ell \cdot (p_i/p_i)) \backslash r \to r}{H_k, H_\ell, p_i/p_i \to K_{i,k,\ell}} \qquad K_{i,k,\ell} \to D_j}{\cfrac{H_k, H_\ell, p_i/p_i \to D_j}{\cfrac{H_k, H_\ell, p_i/p_i, p \to D_j \cdot p}{\cfrac{p/(D_j \cdot p), H_k, H_\ell, p_i/p_i, p \to p}{p/(D_j \cdot p), H_k, H_\ell \to p/((p_i/p_i) \cdot p)}}} \text{(cut)} \qquad p \to p \qquad\qquad p \to p}$$

For the *"if" part*, let $A_{i_1}, \ldots, A_{i_n} \to H_i$ be derivable and proceed by induction on the cut-free derivation (i is arbitrary here for induction; in the end $i = 0$). Since $H_i = p/((p_i/p_i) \cdot p)$ and $(\to /)$ and $(\cdot \to)$ are invertible, we get $A_{i_1}, \ldots, A_{i_n}$, $p_i/p_i, p \to p$. Locate the principal occurrence of p. By Lemma 4, it is the p in $A_{i_1} = p/(D_{i_1} \cdot p)$, and by Lemma 6 the sequent $A_{i_2}, \ldots, A_{i_n}, p_i/p_i, p \to D_{i_1} \cdot p$ is derivable. Let $j = i_1$. Since all our types have tops, apply Lemma 3.

Case 1 (good). The sequent decomposes into $A_{i_2}, \ldots, A_{i_n}, p_i/p_i \to D_j$ and $p \to p$. Consider the former sequent. Tops on the left side are p and p_i, we can apply Lemma 9 and get $B_2, A_{i_2}, \ldots, A_{i_n}, p_i/p_i, C_1 \to K$ for some $K \in \mathcal{U}_j$.

Let's prove that B_2 and C_1 in this case are empty. Suppose $K = K_{i',k,\ell}$ (the cases of $K_{i',k}$ and $K_{i'}$ are handled similarly). Then, by invertibility of $(\to /)$, we get $B_2, A_{i_2}, \ldots, A_{i_n}, p_i/p_i, C_1, (H_k \cdot H_\ell \cdot (p_{i'}/p_{i'})) \backslash r \to r$. Now locate the principal occurrence of r.

Subcase 1.1. The principal occurrence of r is the rightmost one. By Lemma 6, we get $B_2, A_{i_2}, \ldots, A_{i_n}, p_i/p_i, C_1 \to H_k \cdot H_\ell \cdot (p_{i'}/p_{i'})$. Apply Lemma 3. First, by Lemma 5, $i = i'$, otherwise there's no p_i in tops of the left-hand side. Next, the part of the left-hand side that yields (p_i/p_i), by Lemma 7, contains only (p_i/p_i). Therefore, C_1 is empty. Now, for some part Π we have $\Pi \to H_k$, and, decomposing H_k, we get $\Pi, p_i/p_i, p \to p$. By Lemma 4, the principal occurrence of p is not the rightmost one. Since B_2 doesn't have p in tops, Π should include also some of the $A_{i_2}, \ldots$, and the principal p is the top of one of them. But, since A_m is of the form $p/\ldots$, by Lemma 4 the part to the left of this A_m, and, therefore, B_2 should be empty.

Subcase 1.2. The principal occurrence of r is somewhere in B_2 or C_1, in a type $K \in \mathcal{U}_j$. By Lemma 4, it is then the leftmost occurrence of r, because K has the form $r/\ldots$. This rules out the possibility of it being in C_1 (we definitely have p_i/p_i to the left of it). If it is in B_2, again by Lemma 6, we get $B_2', A_{i_2}, \ldots, A_{i_n}, p_i/p_i, C_1 \to H_{k'} \cdot H_{\ell'} \cdot (p_i/p_i)$ ($H_{k'}$ and $H_{\ell'}$ are optional), and we're in the same situation, as Subcase 1.1. Thus, B_2' and C_1 should be empty. However, B_2' now should contain the last type of B, $(((u/F) \backslash u)/(t/t))$. Contradiction. Subcase 1.2 impossible.

Now we have $A_{i_2}, \ldots, A_{i_n}, p_i/p_i \to H_k \cdot H_\ell \cdot (p_i/p_i)$ (the only choice for the principal r now is the rightmost one, and we've applied Lemma 6). By Lemma 3, we get $A_{i_2}, \ldots, A_{i_z} \to H_k$, $A_{i_{z+1}}, \ldots, A_{i_n} \to H_\ell$, $p_i/p_i \to p_i/p_i$ (the last left side is p_i/p_i alone by Lemma 7).

Apply induction hypothesis. In the context-free grammar, we now have $N_k \Rightarrow^* a_{i_2} \ldots a_{i_z}$ and $N_\ell \Rightarrow^* a_{i_{z+1}} \ldots a_{i_n}$. Since $i' = i$, we also have the rule $N_i \Rightarrow a_j N_k N_\ell$ (N_k and N_ℓ are optional) in the grammar (since the corresponding K type was in $\mathcal{U}_j$). Thus, $N_i \Rightarrow a_j a_{i_2} \ldots a_{i_z} a_{i_{z+1}} \ldots a_{i_n}$. Recall that $j = i_1$.

Case 2 (bad). The sequent decomposes in another way, yielding $A_{i_2}, \ldots, A_{i_z} \to D_j$ and $\ldots, p_i/p_i, p \to p$. Again, by Lemma 9, we get $B_2, A_{i_2}, \ldots, A_{i_z}, C_1 \to K$ for some $K \in \mathcal{U}_j$, and further $B_2, A_{i_2}, \ldots, A_{i_z}, C_1, (H_{k'} \cdot H_{\ell'} \cdot (p_{i'}/p_{i'})) \backslash r \to r$. Now we locate the principal occurrence of r and proceed as in Case 1. The only difference, however, is that now there is no p_i/p_i in the left-hand side, and for that reason derivation fails by Lemma 5. Thus, Case 2 is impossible.

References

1. Brandt, M., Henglein, F.: Coinductive axiomatization of recursive type equality and subtyping. Fundam. Inf. **33**(4), 309–338 (1998)
2. Brotherston, J., Simpson, A.: Sequent calculi for induction and infinite descent. J. Log. Comput. **21**(6), 1177–1216 (2011)
3. Buszkowski, W.: Compatibility of a categorial grammar with an associated category system. Zeitschr. Math. Log. Grundl. Math. **28**, 229–238 (1982)
4. Buszkowski, W.: On action logic: equational theories of action algebras. J. Log. Lang. Comput. **17**(1), 199–217 (2007)
5. Du, D.-Z., Ko, K.-I.: Problem Solving in Automata, Languages, and Complexity. Wiley, New York (2001)

6. Greibach, S.A.: A new normal-form theorem for context-free phrase structure grammars. J. ACM **12**(1), 42–52 (1965)
7. Hopcroft, J.E., Motwani, R., Ullman, J.D.: Introduction to Automata Theory, Languages, and Computation, 2nd edn. Addison-Wesley, Boston (2001)
8. Jaffar, J., Santosa, A.E., Voicu, R.: A coinduction rule for entailment of recursively defined properties. In: Stuckey, P.J. (ed.) CP 2008. LNCS, vol. 5202, pp. 493–508. Springer, Heidelberg (2008). doi:10.1007/978-3-540-85958-1_33
9. Jipsen, P.: From semirings to residuated Kleene algebras. Stud. Log. **76**, 291–303 (2004)
10. Kozen, D.: On action algebras. In: van Eijck, J., Visser, A. (eds.) Logic and Information Flow, pp. 78–88. MIT Press, Cambridge (1994)
11. Kozen, D., Silva, A.: Practical coinduction. Math. Struct. Comp. Sci. 1–21 (2016). FirstView. https://doi.org/10.1017/S0960129515000493
12. Lambek, J.: The mathematics of sentence stucture. Am. Math. Mon. **65**(3), 154–170 (1958)
13. Mints, G.E.: Finite investigations on transfinite derivations. J. Math. Sci. **10**(4), 548–596 (1978)
14. Palka, E.: An infinitary sequent system for the equational theory of *-continuous action lattices. Fundam. Inform. **78**(2), 295–309 (2007)
15. Pentus, M.: Lambek grammars are context-free. In: Proceedings of LICS 1993, pp. 429–433 (1993)
16. Pentus, M.: The conjoinability relation in Lambek calculus and linear logic. J. Log. Lang. Inf. **3**(2), 121–140 (1994)
17. Pentus, M.: Models for the Lambek calculus. Ann. Pure Appl. Log. **75**(1–2), 179–213 (1995)
18. Pous, D., Sangiorgi, D.: Enhancements of coinductive proof methods. In: Advanced Topics in Bisimulation and Coinduction. Cambridge University Press (2011)
19. Pratt, V.: Action logic and pure induction. In: Eijck, J. (ed.) JELIA 1990. LNCS, vol. 478, pp. 97–120. Springer, Heidelberg (1991). doi:10.1007/BFb0018436
20. Roşu, G., Lucanu, D.: Circular coinduction: a proof theoretical foundation. In: Kurz, A., Lenisa, M., Tarlecki, A. (eds.) CALCO 2009. LNCS, vol. 5728, pp. 127–144. Springer, Heidelberg (2009). doi:10.1007/978-3-642-03741-2_10
21. Ryzhkova, N.S.: Properties of the categorial dependencies calculus. Diploma Paper, Moscow State University (2013, unpublished). (in Russian)
22. Safiullin, A.N.: Derivability of admissible rules with simple premises in the Lambek calculus. Moscow Univ. Math. Bull. **62**(4), 168–171 (2007)
23. Savateev, Y., Shamkanov, D.: Cut-elimination for the modal Grzegorczyk logic via non-well-founded proofs. In: Kennedy, J., de Queiroz, R.J.G.B. (eds.) WoLLIC 2017. LNCS, vol. 10388, pp. 321–335. Springer, Berlin (2017). doi:10.1007/978-3-662-55386-2_z
24. Shamkanov, D.S.: Circular proofs for the Gödel - Löb provability logic. Math. Notes **96**(4), 575–585 (2014)
25. Shamkanov, D.S.: A realization theorem for the Gödel - Löb provability logic. Sbornik: Math. **207**(9), 1344–1360 (2016)

Multi-type Display Calculus for Semi De Morgan Logic

Giuseppe Greco[1], Fei Liang[1,3(✉)], M. Andrew Moshier[2], and Alessandra Palmigiano[1,4]

[1] Delft University of Technology, Delft, The Netherlands
[2] Chapman University, Orange, CA, USA
[3] Institute of Logic and Cognition, Sun Yat-sen University, Guangzhou, China
liangf25@mail2.sysu.edu.cn
[4] University of Johannesburg, Johannesburg, South Africa

Abstract. We introduce a proper multi-type display calculus for semi De Morgan logic which is sound, complete, conservative, and enjoys cut-elimination and subformula property. Our proposal builds on an algebraic analysis of semi De Morgan algebras and applies the guidelines of the multi-type methodology in the design of display calculi.

1 Introduction

Semi De Morgan logic (**SDM**), introduced in an algebraic setting by Sankappanavar [14], is a very well known example of a paraconsistent logic [19], and is designed to capture the salient features of intuitionistic negation in a paraconsistent setting. Semi De Morgan algebras form a variety of normal distributive lattice expansions (cf. [11, Definition 9]), and are a common abstraction of De Morgan algebras and distributive pseudocomplemented lattices. Semi De Morgan logic has been studied from a duality-theoretic perspective (Hobby [15]), from the perspective of canonical extensions (Palma [17]), and from a proof-theoretic perspective (Ma and Liang [16]). Related to the proof-theoretic perspective, the G3-style sequent calculus introduced in [16] is shown to be cut-free. However, the proof of cut elimination is quite involved, due to the fact that, along with the standard introduction rules for conjunction and disjunction, this calculus includes also introduction rules under the scope of structural connectives. These difficulties can be explained by the fact that the axiomatization of **SDM** is not analytic inductive in the sense of [11, Definition 55]. The calculus introduced in the present paper is motivated by the need to overcome these difficulties. We introduce a proper multi-type display calculus (cf. [12, Definition A.1.]) for semi De Morgan logic which is sound, complete, conservative, and enjoys cut-elimination and subformula property. Our proposal builds on an algebraic analysis of semi De Morgan algebras, and applies the guidelines of the multi-type methodology, introduced in [5,7] and further developed in [1,6,8,12,13].

On the base of the calculus introduced in the present paper, an infinite class of axiomatic extensions of semi De Morgan logic[1] can be endowed with proper display

---

A. Palmigiano—This research is supported by the NWO Vidi grant 016.138.314, the NWO Aspasia grant 015.008.054, and a Delft Technology Fellowship awarded to the second author in 2013.

[1] Namely, all the axiomatic extensions of SDM defined by axioms the translation of which, given as in Sect. 4, is analytic inductive (cf. [11, Definition 55]).

© Springer-Verlag GmbH Germany 2017
J. Kennedy and R.J.G.B. de Queiroz (Eds.): WoLLIC 2017, LNCS 10388, pp. 199–215, 2017.
DOI: 10.1007/978-3-662-55386-2_14

calculi each of which is sound and complete w.r.t. its corresponding class of perfect algebras, conservative, and enjoys cut-elimination and subformula property. Notice that for each such axiomatic extension, the package of basic properties mentioned above follow immediately from the general theory of proper multi-type calculi.[2] Therefore, this calculus provides a proof-theoretic environment which is suitable to complement, from a proof-theoretic perspective, the investigations on the lattice of axiomatic extensions of SDM, which have been mainly developed using the tools of universal algebra (cf. [2,3]), as well as on the connections between the lattices of axiomatic extensions of SDM and of De Morgan logic.

Structure of the Paper. In Sect. 2, we report on the Hilbert-style presentation of SDM, and discuss why this axiomatization is not amenable to the standard treatment of display calculi. In Sect. 3, we define the algebraic environment which motivates our multi-type approach. Then we introduce the multi-type semantic environment and define translations between the single-type and the multi-type languages for SDM. In Sect. 4, we discuss how equivalent analytic (multi-type) reformulations can be given of non-analytic (single-type) axioms of SDM. In Sect. 5, we introduce the display calculus for SDM, and in Sect. 6, we discuss its soundness, completeness, conservativity, cut elimination and subformula property.

2   Preliminaries

2.1   Hilbert Style Presentation of Semi De Morgan Logic

Fix a denumerable set Atprop of propositional variables, let p denote an element in Atprop. The language $\mathcal{L}$ of SDM over Atprop is defined recursively as follows:

$$A ::= p \mid \top \mid \bot \mid \neg A \mid A \wedge A \mid A \vee A$$

Definition 1. *The Hilbert style presentation of De Morgan logic, denoted* H.DM, *consists of the following axioms*

$$\bot \vdash A, \quad A \vdash \top, \quad A \vdash A, \quad A \wedge B \vdash A, \quad A \wedge B \vdash B, \quad A \vdash A \vee B, \quad B \vdash A \vee B,$$

$$\neg A \wedge \neg B \vdash \neg(A \vee B), \quad \neg(A \wedge B) \vdash \neg A \vee \neg B, \quad A \wedge (B \vee C) \vdash (A \wedge B) \vee (A \wedge C)$$

and the following rules:

$$\frac{A \vdash B \quad B \vdash C}{A \vdash C} \quad \frac{A \vdash B \quad A \vdash C}{A \vdash B \wedge C} \quad \frac{A \vdash B \quad C \vdash B}{A \vee C \vdash B} \quad \frac{A \vdash \neg B}{B \vdash \neg A} \quad \frac{\neg A \vdash B}{\neg B \vdash A}$$

Definition 2. *The Hilbert style presentation of semi De Morgan logic, denoted* H.SDM, *consists of the following axioms:*

---

[2] The calculus introduced in [16] does not perform as well with axiomatic extensions, both in the sense that some which are captured by the multi-type environment introduced in the present paper cannot be accounted for, and in the sense that, even when accounted for by means of structural rules, cut elimination does not straightforwardly transfer to the resulting calculus.

$\perp \vdash A, \quad A \vdash \top, \quad \neg\top \vdash \perp, \quad \top \vdash \neg\perp, \quad A \vdash A, \quad A \wedge B \vdash A, \quad A \wedge B \vdash B, \quad A \vdash A \vee B,$

$B \vdash A \vee B, \quad \neg A \vdash \neg\neg\neg A, \quad \neg\neg\neg A \vdash \neg A, \quad \neg A \wedge \neg B \vdash \neg(A \vee B), \quad \neg(A \vee B) \vdash \neg A \wedge \neg B,$

$\neg\neg A \wedge \neg\neg B \vdash \neg\neg(A \wedge B), \quad \neg\neg(A \wedge B) \vdash \neg\neg A \wedge \neg\neg B, \quad A \wedge (B \vee C) \vdash (A \wedge B) \vee (A \wedge C)$

and the following rules:

$$\frac{A \vdash B \quad B \vdash C}{A \vdash C} \quad \frac{A \vdash B \quad A \vdash C}{A \vdash B \wedge C} \quad \frac{A \vdash B \quad C \vdash B}{A \vee C \vdash B} \quad \frac{A \vdash B}{\neg B \vdash \neg A}$$

In [11], a characterization is given of the logics that can be properly displayed. Namely, they are exactly those logics for which there exists a Hilbert-style presentation which only consists of axioms which are analytic inductive (cf. [11, Definition 55]). It is easy to verify that $\neg A \vdash \neg\neg\neg A$, $\neg\neg\neg A \vdash \neg A$, $\neg\neg A \wedge \neg\neg B \vdash \neg\neg(A \wedge B)$, and $\neg\neg(A \wedge B) \vdash \neg\neg A \wedge \neg\neg B$ in H.SDM are not analytic inductive. To our knowledge, no equivalent axiomatization has been introduced for semi De Morgan logic using only analytic inductive axioms. This provides the motivation for circumventing this difficulty by introducing a proper multi-type display calculus for semi De Morgan logic.

2.2   De Morgan and Semi De Morgan Algebras

In the present subsection, we recall the definitions of the algebras from which the logics of the previous subsection arise.

Definition 3. $\mathbb{D} = (D, \cap, \cup, ^*, 1, 0)$ *is a* De Morgan algebra *(DM-algebra) if:*

D1 $(L, \cap, \cup, 1, 0)$ *is a bounded distributive lattice;*
D2 $0^* = 1, 1^* = 0;$
D3 $(a \cup b)^* = a^* \cap b^*$ *for all* $a, b \in D;$
D4 $(a \cap b)^* = a^* \cup b^*$ *for all* $a, b \in D;$
D5 $a = a^{**}$ *for every* $a \in D.$

Definition 4. $\mathbb{A} = (A, \wedge, \vee, ', \top, \perp)$ *is a* semi De Morgan algebra *(SM-algebra) if:*

S1 $(A, \wedge, \vee, 1, 0)$ *is a bounded distributive lattice;*
S2 $\perp' = \top, \top' = \perp;$
S3 $(a \vee b)' = a' \wedge b'$ *for all* $a, b \in A;$
S4 $(a \wedge b)'' = a'' \wedge b''$ *for all* $a, b \in A;$
S5 $a' = a'''$ *for every* $a \in A.$

An *SM-algebra* is a *DM-algebra* if

D. $a \vee b = (a' \wedge b')' = (a \vee b)''$ *for all* $a, b \in A.$

Theorem 1 (Completeness). H.SDM *(resp. H.DM) is complete with respect to the class of SM-algebras (resp. DM-algebras).*

Definition 5. *A distributive lattice* $\mathbb{A}$ *is* perfect *(cf.* [9]*) if* $\mathbb{A}$ *is complete, completely distributive and completely join-generated by the set* $J^\infty(\mathbb{A})$ *of its completely join-irreducible elements (as well as completely meet-generated by the set* $M^\infty(\mathbb{A})$ *of its completely meet-irreducible elements).*

A DM-algebra $\mathbb{D}$ *is* perfect *if its lattice reduct is a perfect distributive lattice, and the negation operator satisfies the following distributive laws:*

$$(\bigvee X)^* = \bigwedge X^* \qquad (\bigwedge X)^* = \bigvee X^*$$

A *lattice homomorphism* $h : \mathbb{L} \to \mathbb{L}'$ is *complete if it satisfies the following properties for each $X \subseteq \mathbb{L}$:*

$$h(\bigvee X) = \bigvee h(X) \qquad h(\bigwedge X) = \bigwedge h(X).$$

3   Towards a Multi-type Presentation: Algebraic Analysis

In the present section, we introduce the algebraic environment which justifies semantically the multi-type approach to semi De Morgan logic we present in Sect. 5. In the next subsection, we define the kernel of a semi De Morgan algebra (or *SM-algebra*, cf. Definition 4) and show that it can be endowed with a structure of De Morgan algebra (or *DM-algebra*, cf. Definition 3). Then we define two maps between the kernel and the lattice reduct of any semi De Morgan algebra. These are the main components of the heterogeneous semi De Morgan algebras which we introduce in Subsect. 3.2, where we also show that semi De Morgan algebras can be equivalently presented in terms of heterogeneous semi De Morgan algebras. In Subsect. 3.3, we apply results pertaining to the theory of canonical extensions to heterogenous semi De Morgan algebras.

3.1   The Kernel of SM-Algebra

Since the negation $'$ of any SM-algebra $\mathbb{L}$ is order-reversing, the map $'' : L \to L$ is order-preserving. Let $K := \{a'' \in L \mid a \in L\}$. Define $h : L \twoheadrightarrow K$ and $e : K \hookrightarrow L$ by the assignments $a \mapsto a''$ and $\alpha \mapsto \alpha$, respectively.

Lemma 1. *Let $\mathbb{L}, K, h, e$ be defined as above. Then for all $\alpha \in K$ and $a \in L$,*

$$h(e(\alpha)) = \alpha \qquad and \qquad e(h(a)) = a''.$$

Definition 6. *For any SM-algebra $\mathbb{L} = (L, \wedge, \vee, \top, \bot, ')$, let the kernel of $\mathbb{L}$ be the algebra $\mathbb{K}_\mathbb{L} = (K, \cap, \cup, ^*, 1, 0)$ defined as follows:*

K1. $K := \mathsf{Range}(h)$, where $h : L \twoheadrightarrow K$ is defined by letting $h(a) = a''$ for any $a \in L$;
K2. $\alpha \cup \beta := h((e(\alpha) \vee e(\beta))'')$ for all $\alpha, \beta \in K$;
K3. $\alpha \cap \beta := h(e(\alpha) \wedge e(\beta))$ for all $\alpha, \beta \in K$;
K4. $1 := h(\top)$;
K5. $0 := h(\bot)$;
K6. $\alpha^* := h(e(\alpha)')$.

Proposition 1. *For any SM-algebra $\mathbb{L}$,*

1. *the kernel $\mathbb{K}_\mathbb{L}$, defined as above, is a DM-algebra.*
2. *h is a homomorphism from $\mathbb{L}$ onto $\mathbb{K}$, and for all $\alpha, \beta \in K$,*

$$e(\alpha) \wedge e(\beta) = e(\alpha \cap \beta) \qquad e(1) = \top \qquad e(0) = \bot.$$

In what follows, we will drop the subscript of the kernel whenever it does not cause confusion.

3.2 Heterogeneous SM-Algebras as Equivalent Presentations of SM-Algebras

Definition 7. *A heterogeneous SDM-algebra (HSM-algebra) is a tuple* $(\mathbb{L}, \mathbb{A}, e, h)$ *satisfying the following conditions:*

H1 $\mathbb{L}$ *is a bounded distributive lattice;*
H2 $\mathbb{A}$ *is a De Morgan lattice;*
H3 $e : \mathbb{A} \hookrightarrow \mathbb{L}$ *is an order embedding, and for all* $\alpha_1, \alpha_2 \in \mathbb{A}$,

$$e(\alpha_1) \wedge e(\alpha_2) = e(\alpha_1 \cap \alpha_2) \quad and \quad e(1) = \top \quad and \quad e(0) = \bot$$

H4 $h : \mathbb{L} \twoheadrightarrow \mathbb{A}$ *is a lattice homomorphism;*
H5 $h(e(\alpha)) = \alpha$ *for every* $\alpha \in \mathbb{A}$.[3]

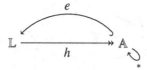

A HSM-algebra is perfect *if:*

1. both $\mathbb{L}$ *and* $\mathbb{A}$ *are perfect (see Definition 5);*
2. e is an order-embedding and is completely meet-preserving;
3. h is a complete homomorphism.

Proposition 2. *For any SM-algebra* $\mathbb{A}$, *the tuple* $(\mathbb{L}, \mathbb{K}, e, h)$ *is a heterogeneous SM-algebra, where* $\mathbb{L}$ *is the lattice reduct of* $\mathbb{A}$ *and* $\mathbb{K}$ *is the kernel of* $\mathbb{A}$ *(cf. Definition 6).*

Proposition 3. *If* $(\mathbb{L}, \mathbb{D}, e, h)$ *is a heterogeneous SM-algebra, then* $\mathbb{L}$ *can be endowed with the structure of SM-algebra defining* $' : \mathbb{L} \to \mathbb{L}$ *by* $a' := e(h(a)^*)$ *for every* $a \in \mathbb{L}$. *Moreover,* $\mathbb{D} \cong \mathbb{K}$.

Definition 8. *For any SM-algebra* $\mathbb{A}$, *we let* $\mathbb{A}^+ = (\mathbb{L}, \mathbb{K}, h, e)$, *where:*

· $\mathbb{L}$ *is the lattice reduct of* $\mathbb{A}$;
· $\mathbb{K}$ *is the kernel of* $\mathbb{A}$;
· $e : \mathbb{K} \hookrightarrow \mathbb{L}$ *is defined by* $e(\alpha) = \alpha$ *for all* $\alpha \in \mathbb{K}$;
· $h : \mathbb{L} \twoheadrightarrow \mathbb{K}$ *is defined by* $h(a) = a''$ *for all* $a \in \mathbb{L}$;

For any HSM-algebra $\mathbb{H}$, *we let* $\mathbb{H}_+ = (\mathbb{L}, ')$ *where:*

· $\mathbb{L}$ *is the distributive lattice of* $\mathbb{H}$;
· $' : \mathbb{L} \to \mathbb{L}$ *is defined by the assignment* $a \mapsto e(h(a)^*)$ *for all* $a \in \mathbb{L}$.

Proposition 4. *For any SM-algebra* $\mathbb{A}$ *and any HSM-algebra* $\mathbb{H}$:

$$\mathbb{A} \cong (\mathbb{A}^+)_+ \quad and \quad \mathbb{H} \cong (\mathbb{H}_+)^+.$$

---

[3] Condition H5 implies that h is surjective and e is injective.

3.3 Canonical Extensions of Heterogeneous SM-Algebras

Canonicity in the multi-type environment is used both to provide complete semantics for a large class of axiomatic extensions of the basic logic (SDM in the present case), and to prove the conservativity of its associated display calculus (cf. Sect. 6.3). In what follows, we let $\mathbb{L}^\delta$ and $\mathbb{D}^\delta$ denote the canonical extensions of the distributive lattice $\mathbb{L}$ and of the De Morgan algebra $\mathbb{D}$, respectively, and e^π and h^δ denote the π-extensions of e and h^4, respectively. We refer to [9] for the relevant definitions.

Lemma 2. *If* $(\mathbb{L}, \mathbb{D}, e, h)$ *is an HSM-algebra, then* $(\mathbb{L}^\delta, \mathbb{D}^\delta, e^\pi, h^\delta)$ *is a perfect HSM-algebra.*

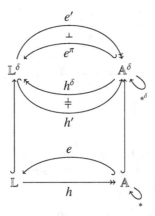

In perfect HSM-algebras, the adjoint(s) of each operation exist(s). This guarantees the soundness of every display rule in the calculus D.SDM. Moreover, in Sect. 6.1, we prove that perfect HSM-algebras are a sound semantics for D.SDM.

4 Multi-type Hilbert Style Presentation for Semi De Morgan Logic

The results of Sect. 3.2 show that HSM-algebras are equivalent presentations of semi De Morgan algebras, and motivate from a semantic perspective the syntactic shift we take in the present section, from single-type language to multi-type language. Indeed, heterogeneous algebras provide a natural interpretation for the following multi-type language $\mathcal{L}_{\mathrm{MT}}$ consisting of terms of types DL and DM.

$$\mathsf{DL} \ni A ::= p \mid e(\alpha) \mid \top \mid \bot \mid A \wedge A \mid A \vee A$$

$$\mathsf{DM} \ni \alpha ::= h_1(A) \mid h_2(A) \mid 1 \mid 0 \mid {\sim}\alpha \mid \neg\alpha \mid \alpha \cup \alpha \mid \alpha \cap \alpha$$

---

[4] The order-theoretic properties of h guarantee that the σ-extension and the π-extension of h coincide. This is why we use h^δ to denote the resulting extension.

The interpretation of $\mathcal{L}_{MT}$-terms into HSM-algebras is defined as the easy generalization of the interpretation of propositional languages in universal algebra; namely, DL-terms (resp. DM-terms) are interpreted in the distributive lattice (resp. De Morgan algebra) component of HSM-algebras. For the sake of modularity, we choose to split the lattice homomorphism h into its finitely join-preserving personality[5] h_1 and its finitely meet-preserving personality h_2. Similarly, $\neg$ is the join-reversing personality of $*$, and $\sim$ is its meet-reversing personality.

The toggle between SM-algebras and heterogeneous algebras (cf. Sect. 3.2) is reflected syntactically by the translations $(\cdot)^t, (\cdot)_t : \mathcal{L} \to \mathcal{L}_{MT}$ defined as follows:

$$
\begin{aligned}
p^t &= p & p_t &= p \\
\top^t &= \top & \top_t &= \top \\
\perp^t &= \perp & \perp_t &= \perp \\
(A \wedge B)^t &= A^t \wedge B^t & (A \wedge B)_t &= A_t \wedge B_t \\
(A \vee B)^t &= A^t \vee B^t & (A \vee B)_t &= A_t \vee B_t \\
(\neg A)^t &= e\sim h_2 A_t & (\neg A)_t &= e\neg h_1 A^t
\end{aligned}
$$

The translations above are compatible with the toggle between SM-algebras and their associated heterogeneous algebras. Indeed, recall that $\mathbb{L}^+$ denotes the heterogeneous algebra associated with the SM-algebra $\mathbb{L}$ (cf. Definition 8). The following proposition is proved by a routine induction on $\mathcal{L}$-formulas.

Proposition 5. *For all $\mathcal{L}$-formulas A and B and every SM-algebra $\mathbb{L}$,*

$$
\mathbb{L} \models A \leq B \quad \textit{iff} \quad \mathbb{L}^+ \models A^t \leq B_t.
$$

We are now in a position to translate the axioms of SDM into $\mathcal{L}_{MT}$.

$$
\neg\neg(p \wedge q) = \neg\neg p \wedge \neg\neg q \rightsquigarrow
\begin{cases}
e\sim h_1 e\neg h_2(p \wedge q) \leq e\neg h_1 e\sim h_2(p) \wedge e\neg h_1 e\neg h_2(q) & (i) \\
e\neg h_1 e\neg h_2(p) \wedge e\neg h_1 e\neg h_2(q) \leq e\neg h_1 e\sim h_2(p \wedge q) & (ii)
\end{cases}
$$

$$
\neg p = \neg\neg\neg p \rightsquigarrow
\begin{cases}
e\sim h_2(p) \leq e\neg h_1 e\sim h_2 e\neg h_1(p) & (iii) \\
e\sim h_2 e\neg h_1 e\sim h_2(p) \leq e\neg h_1(p) & (iv)
\end{cases}
$$

$$
\neg(p \vee q) = \neg p \wedge \neg q \rightsquigarrow
\begin{cases}
e\sim h_2(p \vee q) \leq e\neg h_1(p) \wedge e\neg h_1(q) & (v) \\
e\sim h_2(p) \wedge e\sim h_2(q) \leq e\neg h_1(p \vee q) & (vi)
\end{cases}
$$

$$
\neg\perp = \top \rightsquigarrow
\begin{cases}
e\sim h_2\perp \leq \top & (vii) \\
\top \leq e\neg h_1\perp & (viii)
\end{cases}
$$

$$
\neg\top = \perp \rightsquigarrow
\begin{cases}
e\sim h_2\top \leq \perp & (viiii) \\
\perp \leq e\neg h_1\top & (x)
\end{cases}
$$

---

[5] In what follows, when talking about the *personalities* of an operation, we mean that we consider two copies of the same map, and attribute to each copy only part of the order theoretic properties of the original map.

Notice that the defining identities of HSM-algebras (cf. Definition 7) can be expressed as *analytic inductive* $\mathcal{L}_{\mathrm{MT}}$-inequalities (cf. Definition 11). Hence, these inequalities can be used to generate the analytic rules of the calculus introduced in Sect. 5, with a methodology analogous to the one introduced in [11]. As we will discuss in Sect. B, the inequalities (i)–(x) are derivable in the calculus obtained in this way.

5 Proper Display Calculus for Semi De Morgan Logic

In the present section, we introduce a proper multi-type display calculus for semi De Morgan logic. The language manipulated by this calculus has types DL and DM, and is built up from structural and operational (aka logical) connectives. In the tables of Sect. 5.1, each structural connective corresponding to a logical connective which belongs only to the family $\mathcal{F}$ (resp. $\mathcal{G}$) defined in Sect. A is denoted by decorating that logical connective with ˆ (resp. ˇ). Each logical connective which belongs to both $\mathcal{F}$ and $\mathcal{G}$ is split into two connectives, one capturing its $\mathcal{F}$-personality and the other its $\mathcal{G}$-personality (cf. Footnote 5). The two personalities are assigned the same structural connective, decorated with ˜. The interpretation of each structural connective decorated with ˇ (resp. ˆ) coincides with that of the corresponding logical connective when occurring in succedent (resp. precedent) position; structural connectives decorated with ˜ are interpreted as the $\mathcal{F}$-personality (resp. $\mathcal{G}$-personality) of their associated logical connective when occurring in precedent (resp. succedent) position.[6]

5.1 Language

– Structural and operational terms:

$$
\mathsf{DL}\begin{cases} A ::= p \mid \top \mid \bot \mid \Box\alpha \mid A \wedge A \mid \overset{\text{'}}{A} \vee A \\[4pt] X ::= \hat{\top} \mid \check{\bot} \mid \tilde{\bullet}\Gamma \mid X \,\hat{\wedge}\, X \mid X \,\check{\vee}\, X \mid X \succ\!\!- X \mid X \rightarrow X \end{cases}
$$

$$
\mathsf{DM}\begin{cases} \alpha ::= \Box A \mid \Diamond A \mid 1 \mid 0 \mid \sim\!\alpha \mid \neg\alpha \mid \alpha \cap \alpha \mid \alpha \cup \alpha \\[4pt] \Gamma ::= \hat{1} \mid \check{0} \mid \tilde{*}\Gamma \mid \tilde{\delta}X \mid \Gamma \,\hat{\wedge}\, \Gamma \mid \Gamma \,\check{\cup}\, \Gamma \mid \Gamma \succ\!\!- \Gamma \mid \Gamma -\!\!\prec \Gamma \end{cases}
$$

---

[6] For any sequent $x \vdash y$, we define the signed generation trees $+x$ and $-y$ by labelling the root of the generation tree of x (resp. y) with the sign $+$ (resp. $-$), and then propagating the sign to all nodes according to the polarity of the coordinate of the connective assigned to each node. Positive (resp. negative) coordinates propagate the same (resp. opposite) sign to the corresponding child node. Then, a substructure z in $x \vdash y$ is in *precedent* (resp. *succedent*) *position* if the sign of its root node as a subtree of $+x$ or $-y$ is $+$ (resp. $-$).

- Interpretation of structural connectives as their logical counterparts[7]

DL						DM						DL → DM	DM → DL	DM → DL		DL → DM
$\hat\top$	$\hat\wedge$	$\succ$	$\mathrm{Ⅰ}$	$\check\vee$	$\dot\to$	$\hat{\mathrm{i}}$	$\hat\cap$	$\tilde\succ$	$\hat{\mathrm{o}}$	$\check\cup$	$\tilde\supset$	$\tilde{*}$	$\tilde\delta$	$\bullet$	$\breve{\mathrm{o}}$	$\hat\blacklozenge$
$\top$	$\wedge$	$(\succ)$	$\bot$	$\vee$	$(\to)$	1	$\cap$	$(\supset)$	0	$\cup$	$(-\supset)$	$\sim$	$\neg$	$\Diamond$	$\Box$	$(\Diamond)$	$(\Box)$	$\Box$	$(\blacklozenge)$

- Algebraic interpretation of logical connectives as operations in perfect HSM-algebras (see Lemma 2; we let h_1' and h_2' denote the $\mathcal{F}$-personality and $\mathcal{G}$-personality of h', respectively.)

DL						DM								DL → DM	DM → DL	DM → DL		DL → DM	
$\top$	$\wedge$	$(\succ)$	$\bot$	$\vee$	$(\to)$	1	$\cap$	$(\supset)$	0	$\cup$	$(-\supset)$	$\sim$	$\neg$	$\Diamond$	$\Box$	$(\Diamond)$	$(\Box)$	$\Box$	$\blacklozenge$
$\top$	$\wedge$	$\succ$	$\bot$	$\vee$	$\to$	1	$\cap$	$\supset$	0	$\cup$	$-\supset$	$\sim$	$\neg$	h_1	h_2	h_1'	h_2'	e	e'

5.2 Rules

- Single-type display structural rules

$$\text{DL} \qquad\qquad\qquad \text{DM}$$

$$\text{res}\ \frac{X \hat\wedge Y \vdash Z}{Y \vdash X \dot\to Z} \qquad \frac{X \vdash Y \check\vee Z}{Y \succ X \vdash Z}\ \text{res} \qquad\qquad \text{adj}\ \frac{\tilde{*}\Gamma \vdash \Delta}{\tilde{*}\Delta \vdash \Gamma} \qquad \frac{\Gamma \vdash \tilde{*}\Delta}{\Delta \vdash \tilde{*}\Gamma}\ \text{adj}$$

$$\text{res}\ \frac{\Gamma \hat\cap \Delta \vdash \Theta}{\Delta \vdash \Gamma \,\tilde\supset\, \Theta} \qquad \frac{\Gamma \vdash \Delta \check\cup \Theta}{\Delta \,\tilde\succ\, \Gamma \vdash \Theta}\ \text{res}$$

- Single-type structural rules

$$\text{DL} \qquad\qquad\qquad\qquad \text{DM}$$

$$\text{Id}\ \frac{}{p \vdash p} \qquad \frac{X \vdash A \qquad A \vdash Y}{X \vdash Y}\ \text{Cut} \qquad\qquad \frac{\Gamma \vdash \alpha \qquad \alpha \vdash \Delta}{X \vdash \Delta}\ \text{Cut}$$

$$\hat\top\ \frac{X \vdash Y}{X \hat\wedge \hat\top \vdash Y} \qquad \frac{X \vdash Y}{X \vdash Y \check\vee \mathrm{Ⅰ}}\ \mathrm{Ⅰ} \qquad\qquad \hat{\mathrm{i}}\ \frac{\Gamma \vdash \Delta}{\Gamma \hat\cap \hat{\mathrm{i}} \vdash \Delta} \qquad \frac{\Gamma \vdash \Delta}{\Gamma \vdash \Delta \check\cup \breve{\mathrm{o}}}\ \breve{\mathrm{o}}$$

$$\text{E}\ \frac{X \hat\wedge Y \vdash Z}{Y \hat\wedge X \vdash Z} \qquad \frac{X \vdash Y \check\vee Z}{X \vdash Z \check\vee Y}\ \text{E} \qquad\qquad \text{E}\ \frac{\Gamma \hat\cap \Delta \vdash \Theta}{\Delta \hat\cap \Gamma \vdash \Theta} \qquad \frac{\Gamma \vdash \Delta \check\cup \Theta}{\Gamma \vdash \Theta \check\cup \Delta}\ \text{E}$$

$$\text{A}\ \frac{(X \hat\wedge Y) \hat\wedge Z \vdash W}{X \hat\wedge (Y \hat\wedge Z) \vdash Z} \qquad \frac{X \vdash (Y \check\vee Z) \check\vee W}{X \vdash Y \check\vee (Z \check\vee W)}\ \text{A} \qquad\qquad \text{A}\ \frac{(\Gamma \hat\cap \Delta) \hat\cap \Theta \vdash W}{\Gamma \hat\cap (\Delta \hat\cap \Theta) \vdash \Theta} \qquad \frac{\Gamma \vdash (\Delta \check\cup \Theta) \check\cup W}{\Gamma \vdash \Delta \check\cup (\Theta \check\cup W)}\ \text{A}$$

$$\text{W}\ \frac{X \vdash Y}{X \hat\wedge Z \vdash Y} \qquad \frac{X \vdash Y}{X \vdash Y \check\vee Z}\ \text{W} \qquad\qquad \text{W}\ \frac{\Gamma \vdash \Delta}{\Gamma \hat\cap \Theta \vdash \Delta} \qquad \frac{\Gamma \vdash \Delta}{\Gamma \vdash \Delta \check\cup \Theta}\ \text{W}$$

$$\text{C}\ \frac{X \hat\wedge X \vdash Y}{X \vdash Y} \qquad \frac{X \vdash Y \check\vee Y}{X \vdash Y}\ \text{C} \qquad\qquad \text{C}\ \frac{\Gamma \hat\cap \Gamma \vdash \Delta}{\Gamma \vdash \Delta} \qquad \frac{\Gamma \vdash \Delta \check\cup \Delta}{\Gamma \vdash \Delta}\ \text{C}$$

$$\frac{\Gamma \vdash \Delta}{\tilde{*}\Delta \vdash \tilde{*}\Gamma}\ \text{cont}$$

---

[7] In the synoptic table, the operational symbols which occur only at the structural level will appear between round brackets.

- Single-type operational rules

DL DM

$$\top\,\frac{\hat{\top}\vdash X}{\top\vdash X}\qquad \frac{}{\hat{\top}\vdash\top}\,\top \qquad\qquad 1\,\frac{\hat{1}\vdash\Gamma}{1\vdash\Gamma}\qquad \frac{}{\hat{1}\vdash 1}\,1$$

$$\bot\,\frac{}{\bot\vdash\check{\bot}}\qquad \frac{X\vdash\check{\bot}}{X\vdash\bot}\,\bot \qquad\qquad 0\,\frac{}{0\vdash\check{0}}\qquad \frac{X\vdash\check{0}}{\Gamma\vdash 0}\,0$$

$$\wedge\,\frac{A\,\hat{\wedge}\,B\vdash X}{A\wedge B\vdash X}\qquad \frac{X\vdash A\qquad Y\vdash B}{X\,\hat{\wedge}\,Y\vdash A\wedge B}\,\wedge \qquad\qquad \cap\,\frac{\alpha\,\hat{\cap}\,\beta\vdash\Gamma}{\alpha\cap\beta\vdash\Gamma}\qquad \frac{\Gamma\vdash\alpha\qquad\Delta\vdash\beta}{\Gamma\,\hat{\cap}\,\Delta\vdash\alpha\cap\beta}\,\cap$$

$$\vee\,\frac{A\vdash X\qquad B\vdash Y}{A\vee B\vdash X\,\check{\vee}\,Y}\qquad \frac{X\vdash A\,\check{\vee}\,B}{X\vdash A\vee B}\,\vee \qquad\qquad \cup\,\frac{\alpha\vdash\Gamma\qquad\beta\vdash\Delta}{\alpha\cup\beta\vdash\Gamma\,\check{\cup}\,\Delta}\qquad \frac{\Gamma\vdash\alpha\,\check{\cup}\,\beta}{\Gamma\vdash\alpha\cup\beta}\,\cup$$

$$\sim\,\frac{\circledast\alpha\vdash\Gamma}{\sim\alpha\vdash\Gamma}\qquad \frac{\alpha\vdash\Gamma}{\circledast\Gamma\vdash\sim\alpha}\,\sim$$

$$\neg\,\frac{\Gamma\vdash\alpha}{\neg\alpha\vdash\circledast\Gamma}\qquad \frac{\Gamma\vdash\circledast\alpha}{\Gamma\vdash\neg\alpha}\,\neg$$

- Multi-type display structural rules

$$\mathrm{adj}\,\frac{X\vdash\check{\square}\,\Gamma}{\blacklozenge X\vdash\Gamma} \qquad\qquad \frac{\check{\lozenge} X\vdash\Gamma}{X\vdash\bullet\Gamma}\,\mathrm{adj}$$

- Multi-type structural rules

$$\check{\lozenge}\,\frac{X\vdash Y}{\check{\lozenge} X\vdash\check{\lozenge} Y} \qquad\qquad \bullet\,\frac{\bullet\Gamma\vdash\bullet\Delta}{\Gamma\vdash\Delta}$$

$$\check{\lozenge}\check{\square}\,\frac{\Gamma\vdash\check{\lozenge}\,\check{\square}\,\Delta}{\Gamma\vdash\Delta} \qquad\qquad \blacklozenge\hat{1}\,\frac{\hat{1}\vdash\Gamma}{\blacklozenge\hat{\top}\vdash\Gamma}$$

$$\check{\square}\check{0}\,\frac{X\vdash\check{\square}\,\check{0}}{X\vdash\check{\bot}}$$

- Multi-type operational rules

$$\lozenge\,\frac{\check{\lozenge} A\vdash\Gamma}{\lozenge A\vdash\Gamma} \qquad\qquad \frac{\Gamma\vdash A}{\check{\lozenge} X\vdash\lozenge A}\,\lozenge$$

$$\square\,\frac{A\vdash\Gamma}{\square A\vdash\check{\lozenge}\Gamma} \qquad\qquad \frac{\Gamma\vdash\check{\lozenge} A}{\Gamma\vdash\square A}\,\square$$

$$\square\,\frac{A\vdash X}{\square A\vdash\check{\square} Y} \qquad\qquad \frac{X\vdash\check{\square} A}{X\vdash\square A}\,\square$$

6 Properties

6.1 Soundness

In the present subsection, we outline the verification of the soundness of the rules of D.SDM w.r.t. the semantics of *perfect* HSM-algebras (cf. Definition 7). The first step consists in interpreting structural symbols as logical symbols according to their (precedent or succedent) position, as indicated at the beginning of Sect. 5. This makes it possible to interpret sequents as inequalities, and rules as quasi-inequalities. For example, the rules on the left-hand side below are interpreted as the quasi-inequalities on the right-hand side:

$$\frac{X \vdash Y}{\bar{\circ}X \vdash \bar{\circ}Y} \quad \rightsquigarrow \quad \forall a \forall b[a \leq b \Rightarrow \Diamond a \leq \Box b]$$

$$\frac{\Gamma \vdash \bar{\circ}\,\check{\circ}\,\Delta}{\Gamma \vdash \Delta} \quad \rightsquigarrow \quad \forall \alpha \forall \beta[\alpha \leq \Box\Box\beta \Leftrightarrow \alpha \leq \beta]$$

The verification of the soundness of the rules of D.SDM then consists in verifying the validity of their corresponding quasi-inequalities in perfect HDM-algebras. The verification of the soundness of pure-type rules and of the introduction rules following this procedure is routine, and is omitted. The validity of the quasi-inequalities corresponding to multi-type structural rules follows straightforwardly from the observation that the quasi-inequality corresponding to each rule is obtained by running the algorithm ALBA (cf. Sect. 3.4 [11]) on one of the defining inequalities of HSM-algebras.[8] For instance, the soundness of the first rule above is due to h being order-preserving; the soundness of the invertible rule is due to condition H5.

6.2 Completeness

Let us translate sequents $A \vdash B$ in the language of H.SDM into sequents $A^\tau \vdash B_\tau$ in the language of D.SDM by means of the following translations:

$$
\begin{array}{ll}
p^\tau ::= p & p_\tau ::= p \\
\top^\tau ::= \top & \top_\tau ::= \top \\
\bot^\tau ::= \bot & \bot_\tau ::= \bot \\
(A \wedge B)^\tau ::= A^\tau \wedge B^\tau & (A \wedge B)_\tau ::= A_\tau \wedge B_\tau \\
(A \vee B)^\tau ::= A^\tau \vee B^\tau & (A \vee B)_\tau ::= A_\tau \vee A_\tau \\
(\neg A)^\tau ::= \Box{\sim}\,\Box\, A_\tau & (\neg A)_\tau ::= \Box\neg\Diamond A^\tau
\end{array}
$$

The translations of the axioms and rules of H.SDM are derivable in D.SDM.

Proposition 6. *For every $A \in$ H.SDM, the sequent $A^\tau \vdash A_\tau$ is derivable in D.SDM.*

Proof. see Sect. B.

Proposition 7. *For every $A, B \in$ H.SDM, if $A \vdash B$ is derivable in H.SDM, then $A^\tau \vdash B_\tau$ is derivable in D.SDM.*

Proof. Sect. B.

---

[8] Indeed, as discussed in [11], the soundness of the rewriting rules of ALBA only depends on the order-theoretic properties of the interpretation of the logical connectives and their adjoints and residuals. The fact that some of these maps are not internal operations but have different domains and codomains does not make any substantial difference.

6.3 Conservativity

To argue that the calculus introduced in Sect. 5 is conservative w.r.t. H.SDM (cf. Sect. 2), we follow the standard proof strategy discussed in [10, 11]. Let $\vdash_{\text{H.SDM}}$ denote the syntactic consequence relation arising from H.SDM, and $\models_{\text{HSM}}$ denote the semantic consequence relation arising from (perfect) HSM-algebras. We need to show that, for all formulas A and B of the original language of H.SDM, if $A^\tau \vdash B_\tau$ is a D.SDM-derivable sequent, then $A \vdash_{\text{H.SDM}} B$. This claim can be proved using the following facts: (a) the rules of D.SDM are sound w.r.t. perfect HSM-algebras (cf. Sect. 6.1); (b) H.SDM is complete w.r.t. (perfect) SM-algebras (cf. Theorem 1); and (c) (perfect) SM-algebras are equivalently presented as (perfect) HSM-algebras (cf. Sect. 3.2), so that the semantic consequence relations arising from each type of structures preserve and reflect the translation (cf. Proposition 5). Then, let A, B be formulas of the original H.SDM-language. If $A^\tau \vdash B_\tau$ is a D.SDM-derivable sequent, then, by (a), $A^\tau \models_{\text{HSM}} B_\tau$. By (c), this implies that $A \models_{\text{SM}} B$, where $\models_{\text{SM}}$ denotes the semantic consequence relation arising from SM-algebras. By (b), this implies that $A \vdash_{\text{H.SDM}} B$, as required.

6.4 Cut Elimination and Subformula Property

In the present section, we briefly sketch the proof of cut elimination and subformula property for D.SDM. As discussed earlier on, proper display calculi have been designed so that the cut elimination and subformula property can be inferred from a meta-theorem, following the strategy introduced by Belnap for display calculi. The meta-theorem to which we will appeal for D.SDM was proved in [6].

All conditions in [6, Theorem 4.1] except C_8' are readily satisfied by inspecting the rules. Condition C_8' requires to check that reduction steps are available for every application of the cut rule in which both cut-formulas are principal, which either remove the original cut altogether or replace it by one or more cuts on formulas of strictly lower complexity. In what follows, we only show C_8' for the unary connectives.

Pure DM-type connectives:

$$
\cfrac{\cfrac{\vdots\,\pi_1}{\cfrac{\Gamma \vdash \tilde{*}\alpha}{\Gamma \vdash \neg\alpha}}}{\Gamma \vdash \tilde{*}\Delta}
\qquad
\cfrac{\cfrac{\vdots\,\pi_2}{\Delta \vdash \alpha}\quad\cfrac{\vdots\,\pi_1}{\cfrac{\Delta \vdash \alpha}{\neg\alpha \vdash \tilde{*}\Delta}}}{}
\qquad
\rightsquigarrow
\qquad
\cfrac{\cfrac{\vdots\,\pi_1}{\cfrac{\Gamma \vdash \tilde{*}\alpha}{\alpha \vdash \tilde{*}\Gamma}}\quad\cfrac{\Delta \vdash \tilde{*}\Gamma}{}}{\Gamma \vdash \tilde{*}\Delta}
$$

The cases for $\sim\alpha$ is standard and similar to the one above.

Multi-type connectives:

$$
\cfrac{\cfrac{\vdots\,\pi_1}{X \vdash \breve{\square}\alpha}\quad\cfrac{\vdots\,\pi_2}{\square\alpha \vdash \breve{\square}\Delta}}{X \vdash \breve{\square}\Delta}
\qquad
\rightsquigarrow
\qquad
\cfrac{\cfrac{\cfrac{\vdots\,\pi_1}{X \vdash \breve{\square}\alpha}}{\blacklozenge X \vdash \alpha}\quad\cfrac{\vdots\,\pi_2}{\alpha \vdash \Delta}}{\cfrac{\blacklozenge X \vdash \Delta}{X \vdash \breve{\square}\Delta}}
$$

$$\dfrac{\dfrac{\vdots \pi_1}{\Gamma \vdash \delta A} \qquad \dfrac{\vdots \pi_2}{A \vdash X}}{\dfrac{\Gamma \vdash \Box A \qquad \Box A \vdash \delta X}{\Gamma \vdash \delta X}} \qquad \rightsquigarrow \qquad \dfrac{\dfrac{\dfrac{\vdots \pi_1}{\Gamma \vdash \delta A}}{\bullet \Gamma \vdash A} \qquad \dfrac{\vdots \pi_2}{A \vdash X}}{\dfrac{\bullet \Gamma \vdash X}{\Gamma \vdash \delta X}}$$

The cases for $\diamond A$ is standard and similar to the one above.

A   Analytic Inductive Inequalities

In the present section, we specialize the definition of *analytic inductive inequalities* (cf. [11]) to the multi-type language $\mathcal{L}_{MT}$, in the types DL and DM, defined in Sect. 4 and reported below for the reader's convenience.

$$DL \ni A ::= p \mid e(\alpha) \mid \top \mid \bot \mid A \wedge A \mid A \vee A$$
$$DM \ni \alpha ::= h_1(A) \mid h_2(A) \mid 1 \mid 0 \mid \sim\alpha \mid \neg\alpha \mid \alpha \cup \alpha \mid \alpha \cap \alpha$$

We will make use of the following auxiliary definition: an *order-type* over $n \in \mathbb{N}$ is an n-tuple $\epsilon \in \{1, \partial\}^n$. For every order type ϵ, we denote its *opposite* order type by ϵ^∂, that is, $\epsilon^\partial(i) = 1$ iff $\epsilon(i) = \partial$ for every $1 \leq i \leq n$. The connectives of the language above are grouped together into the families $\mathcal{F} := \mathcal{F}_{DL} \cup \mathcal{F}_{DM} \cup \mathcal{F}_{MT}$ and $\mathcal{G} := \mathcal{G}_{DL} \cup \mathcal{G}_{DM} \cup \mathcal{G}_{MT}$ defined as follows:

$$\begin{aligned} \mathcal{F}_{DL} &:= \varnothing & \mathcal{G}_{DL} &= \varnothing \\ \mathcal{F}_{DM} &:= \{\sim\} & \mathcal{G}_{DM} &:= \{\neg\} \\ \mathcal{F}_{MT} &:= \{h_1\} & \mathcal{G}_{MT} &:= \{e, h_2\} \end{aligned}$$

For any $f \in \mathcal{F}$ (resp. $g \in \mathcal{G}$), we let $n_f \in \mathbb{N}$ (resp. $n_g \in \mathbb{N}$) denote the arity of f (resp. g), and the order-type ϵ_f (resp. ϵ_g) on n_f (resp. n_g) indicate whether the ith coordinate of f (resp. g) is positive ($\epsilon_f(i) = 1$, $\epsilon_g(i) = 1$) or negative ($\epsilon_f(i) = \partial$, $\epsilon_g(i) = \partial$). The order-theoretic motivation for this partition is that the algebraic interpretations of $\mathcal{F}$-connectives (resp. $\mathcal{G}$-connectives), preserve finite joins (resp. meets) in each positive coordinate and reverse finite meets (resp. joins) in each negative coordinate.

For any term $s(p_1, \ldots p_n)$, any order type ϵ over n, and any $1 \leq i \leq n$, an ϵ-*critical node* in a signed generation tree of s is a leaf node $+p_i$ with $\epsilon(i) = 1$ or $-p_i$ with $\epsilon(i) = \partial$. An ϵ-*critical branch* in the tree is a branch ending in an ϵ-critical node. For any term $s(p_1, \ldots p_n)$ and any order type ϵ over n, we say that $+s$ (resp. $-s$) *agrees with* ϵ, and write $\epsilon(+s)$ (resp. $\epsilon(-s)$), if every leaf in the signed generation tree of $+s$ (resp. $-s$) is ϵ-critical. We will also write $+s' \prec *s$ (resp. $-s' \prec *s$) to indicate that the subterm s' inherits the positive (resp. negative) sign from the signed generation tree $*s$. Finally, we will write $\epsilon(s') \prec *s$ (resp. $\epsilon^\partial(s') \prec *s$) to indicate that the signed subtree s', with the sign inherited from $*s$, agrees with ϵ (resp. with ϵ^∂).

Definition 9 (Signed Generation Tree). *The* positive *(resp.* negative*) generation tree of any $\mathcal{L}_{MT}$-term s is defined by labelling the root node of the generation tree of s with the sign + (resp. −), and then propagating the labelling on each remaining node as* *follows: For any node labelled with $\ell \in \mathcal{F} \cup \mathcal{G}$ of arity n_ℓ, and for any $1 \leq i \leq n_\ell$, assign the same (resp. the opposite) sign to its ith child node if $\epsilon_\ell(i) = 1$ (resp. if $\epsilon_\ell(i) = \partial$). Nodes in signed generation trees are* positive *(resp.* negative*) if are signed + (resp. −).*

Definition 10 (Good branch). *Nodes in signed generation trees will be called* Δ-*adjoints, syntactically left residual (SLR), syntactically right residual (SRR), and syntactically right adjoint (SRA), according to the specification given in Table 1. A branch in a signed generation tree $*s$, with $* \in \{+, -\}$, is called a* good branch *if it is the concatenation of two paths P_1 and P_2, one of which may possibly be of length 0, such that P_1 is a path from the leaf consisting (apart from variable nodes) only of PIA-nodes[9], and P_2 consists (apart from variable nodes) only of Skeleton-nodes.*

Table 1. Skeleton and PIA nodes.

Skeleton	PIA
Δ-adjoints	SRA
$+ \quad \vee \quad \cup \quad \wedge \quad \cap$	$+ \quad \wedge \quad \cap \quad h_2 \quad \neg \quad e$
$- \quad \wedge \cap \vee \cup$	$- \quad \vee \quad \cup \quad h_1 \quad \sim$
SLR	SRR
$+ \quad \wedge \quad \cap \quad h_1 \quad \sim$	$+ \quad \vee \quad \cup$
$- \quad \vee \quad \cup \quad h_2 \quad \neg \quad e$	$- \quad \wedge \quad \cap$

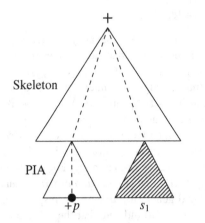
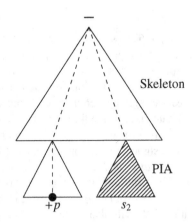

Definition 11 (Analytic inductive inequalities). *For any order type ϵ and any irreflexive and transitive relation $<_\Omega$ on $p_1, \ldots p_n$, the signed generation tree $*s$ ($* \in \{-, +\}$) of an $\mathcal{L}_{MT}$ term $s(p_1, \ldots p_n)$ is* analytic (Ω, ϵ)-inductive *if*

---

[9] For an expanded discussion on this definition, see [18, Remark 3.24] and [4, Remark 3.3].

1. *every branch of* $*s$ *is good (cf. Definition 10);*
2. *for all* $1 \leq i \leq n$, *every SRR-node occurring in any* ϵ-*critical branch with leaf* p_i *is of the form* $\circledast(s,\beta)$ *or* $\circledast(\beta,s)$, *where the critical branch goes through* β *and*
   (a) $\epsilon^{\partial}(s) \prec *s$ *(cf. discussion before Definition 10), and*
   (b) $p_k <_{\Omega} p_i$ *for every* p_k *occurring in* s *and for every* $1 \leq k \leq n$.

We will refer to $<_{\Omega}$ *as the* dependency order *on the variables. An inequality* $s \leq t$ *is* analytic (Ω, ϵ)-inductive *if the signed generation trees* $+s$ *and* $-t$ *are analytic* (Ω, ϵ)-*inductive. An inequality* $s \leq t$ *is* analytic inductive *if is analytic* (Ω, ϵ)-*inductive for some* Ω *and* ϵ.

In each setting in which they are defined, analytic inductive inequalities are a subclass of inductive inequalities (cf. [11]). In their turn, inductive inequalities are *canonical* (that is, preserved under canonical extensions, as defined in each setting).

B   Completeness

Proposition 8. *For every* $A \in$ H.SDM, *the sequent* $A^{\tau} \vdash A_{\tau}$ *is derivable in D.SDM.*

Proof. By induction on $A \in \mathcal{L}$. The proof of base cases: $A := \top$, $A := \bot$ and $A := p$, is straightforward, and is omitted.

Inductive cases: $A := B \wedge C$ and $A := B \vee C$ can be proved by the induction hypothesis and (w), (c), ($\vee$) and ($\wedge$) rules. As to $A := \neg B$,

$$
\begin{array}{c}
\text{ind.hyp.} \\ \hline
\tilde{\delta}\dfrac{B^{\tau} \vdash B_{\tau}}{\tilde{\delta}B^{\tau} \vdash \tilde{\delta}B_{\tau}} \\ \hline
\diamondsuit B^{\tau} \vdash \tilde{\delta}B_{\tau} \\ \hline
\diamondsuit B^{\tau} \vdash \Box B_{\tau} \\ \hline
\tilde{*}\Box B_{\tau} \vdash \tilde{*}\diamondsuit B^{\tau} \\ \hline
\tilde{*}\Box B_{\tau} \vdash \neg\diamondsuit B^{\tau} \\ \hline
\sim \Box B_{\tau} \vdash \neg\diamondsuit B^{\tau} \\ \hline
\Box\sim\Box B_{\tau} \vdash \breve{\Box}\neg\diamondsuit B^{\tau} \\ \hline
\Box\sim\Box B_{\tau} \vdash \Box\neg\diamondsuit B^{\tau}
\end{array} \quad \text{cont}
$$

Proposition 9. *For every* $A, B \in$ H.SDM, *if* $A \vdash B$ *is derivable in H.SDM, then* $A^{\tau} \vdash B_{\tau}$ *is derivable in D.SDM.*

Proof. As page limited, we just show an example: $\neg\neg A \wedge \neg\neg B \vdash \neg\neg(A \wedge B)$ $\rightsquigarrow$ $\Box\sim\Box\Box\neg\diamondsuit A \wedge \Box\sim\Box\Box\neg\diamondsuit B \vdash \Box\neg\diamondsuit\Box\sim\Box(A \wedge B)$.

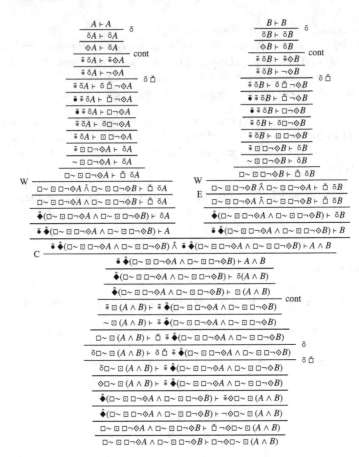

References

1. Bílková, M., Greco, G., Palmigiano, A., Tzimoulis, A., Wijnberg, N.: The logic of resources and capabilities (submitted). arXiv preprint arXiv:1608.02222
2. Celani, S.A.: Distributive lattices with a negation operator. Math. Logic Q. **45**(2), 207–218 (1999)
3. Celani, S.A.: Representation for some algebras with a negation operator. Contrib. Discrete Math. **2**(2), 205–213 (2007)
4. Conradie, W., Palmigiano, A.: Algorithmic correspondence and canonicity for non-distributive logics. J. Logic Comput. arXiv preprint arXiv:1603.08515 (forthcoming)
5. Frittella, S., Greco, G., Kurz, A., Palmigiano, A.: Multi-type display calculus for propositional dynamic logic. J. Logic Comput. **26**(6), 2067–2104 (2016)
6. Frittella, S., Greco, G., Kurz, A., Palmigiano, A., Sikimić, V.: Multi-type sequent calculi. In: Indrzejczak, A., Kaczmarek, J., Zawidski, M. (eds.) Proceedings Trends in Logic XIII, vol. 13, pp. 81–93 (2014)
7. Frittella, S., Greco, G., Kurz, A., Palmigiano, A., Sikimić, V.: A multi-type display calculus for dynamic epistemic logic. J. Logic Comput. **26**(6), 2017–2065 (2016)
8. Frittella, S., Greco, G., Palmigiano, A., Yang, F.: A multi-type calculus for inquisitive logic. In: Väänänen, J., Hirvonen, Å., de Queiroz, R. (eds.) WoLLIC 2016. LNCS, vol. 9803, pp. 215–233. Springer, Heidelberg (2016). doi:10.1007/978-3-662-52921-8_14

9. Gehrke, M., Harding, J.: Bounded lattice expansions. J. Algebra **238**(1), 345–371 (2001)
10. Greco, G., Kurz, A., Palmigiano, A.: Dynamic epistemic logic displayed. In: Grossi, D., Roy, O., Huang, H. (eds.) LORI 2013. LNCS, vol. 8196, pp. 135–148. Springer, Heidelberg (2013). doi:10.1007/978-3-642-40948-6_11
11. Greco, G., Ma, M., Palmigiano, A., Tzimoulis, A., Zhao, Z.: Unified correspondence as a proof-theoretic tool. J. Logic Comput. (2016). doi:10.1093/logcom/exw022
12. Greco, G., Palmigiano, A.: Linear logic properly displayed (submitted). arXiv preprint: arXiv:1611.04184 (2016)
13. Greco, G., Palmigiano, A.: Lattice logic properly displayed. In: Proceedings WoLLIC 2017 (forthcoming)
14. Sankappanavar, H.P.: Semi-De Morgan algebras. J. Symb. Logic **52**(03), 712–724 (1987)
15. Hobby, D.: Semi-De Morgan algebras. Stud. Logica. **56**(1–2), 151–183 (1996)
16. Ma, M., Liang, F.: Sequent calculi for semi-De Morgan and De Morgan algebras (submitted). arXiv preprint: arXiv:1611.05231 (2016)
17. Palma, C.: Semi De Morgan algebras. Dissertation, The University of Lisbon (2005)
18. Palmigiano, A., Sourabh, S., Zhao, Z.: Sahlqvist theory for impossible worlds. J. Logic Comput. **27**(3), 775–816 (2017)
19. Priest, G.: Paraconsistent logic. In: Gabbay, D.M., Guenthner, F. (eds.) Handbook of Philosophical Logic, pp. 287–393. Springer, Heidelberg (2002)

Dependent Event Types

Zhaohui Luo[1]([✉]) and Sergei Soloviev[2]

[1] Royal Holloway, University of London, Egham, Surrey, UK
zhaohui.luo@hotmail.co.uk
[2] IRIT, Toulouse, France
Sergei.Soloviev@irit.fr

Abstract. This paper studies how dependent types can be employed for a refined treatment of event types, offering a nice improvement to Davidson's event semantics. We consider dependent event types indexed by thematic roles and illustrate how, in the presence of refined event types, subtyping plays an essential role in semantic interpretations. We consider two extensions with dependent event types: first, the extension of Church's simple type theory as employed in Montague semantics that is familiar with many linguistic semanticists and, secondly, the extension of a modern type theory as employed in MTT-semantics. The former uses subsumptive subtyping, while the latter uses coercive subtyping, to capture the subtyping relationships between dependent event types. Both of these extensions have nice meta-theoretic properties such as normalisation and logical consistency; in particular, we shall show that the former can be faithfully embedded into the latter and hence has expected meta-theoretic properties. As an example of applications, it is shown that dependent event types give a natural solution to the incompatibility problem (sometimes called the event quantification problem) in combining event semantics with the traditional compositional semantics, both in the Montagovian setting with the simple type theory and in the setting of MTT-semantics.

1 Introduction

The event semantics, whose study was initiated by Davidson [6] and further studied in its neo-Davidsonian turn (see [17] among others), has several notable advantages including Davidson's original motive to provide a satisfactory semantics for adverbial modifications. Dependent types, as those found in Modern Type Theories such as Martin-Löf's type theory [15] and UTT [11], provide a useful tool in formalising event types and a nice treatment of the event semantics.

In this paper, we shall study event types that may depend on thematic roles such as agents and patients of the events. For example, we can consider the type

Z. Luo—Partially supported by EU COST Action CA15123 and CAS/SAFEA International Partnership Program.

S. Soloviev—Also an associated researcher at ITMO Univ, St. Petersburg, Russia. Partially supported by EU COST Action CA15123 and Russian Federation Grant 074-U01.

© Springer-Verlag GmbH Germany 2017
J. Kennedy and R.J.G.B. de Queiroz (Eds.): WoLLIC 2017, LNCS 10388, pp. 216–228, 2017.
DOI: 10.1007/978-3-662-55386-2_15

$Evt_{AP}(a, p)$ of events whose agent and patient are a and p, respectively. We shall investigate subtyping relations between event types which include dependent types such as $Evt_{AP}(a, p)$ and the non-dependent type $Event$ of all events (the latter is found in the traditional setting). For example, it may be natural to have $Evt_{AP}(a, p) \le Evt_A(a)$, that is, the type of events with agent a and patient p is a subtype of that with agent a. With such subtyping relations in place, the semantics of verb phrases can now take the usual non-dependent types, as in the traditional setting, although dependent event types are considered.

It is shown that such dependent event types give a natural solution to the incompatibility problem in combining event semantics with the traditional Montague semantics [2,20] (sometimes called the event quantification problem [7]). When introducing events into formal semantics, one faces a problem, which is long-standing and has seemed intractable: it comes from the issue of scopes for two kinds of quantifiers – the existential quantifier over an event variable and the other quantifiers such as one that arises from a quantificational noun phrase (see Sect. 5 for examples). It is in general expected that the correct semantics is obtained when the event quantifier takes the lower scope, but the problem is that, even when the event quantifier takes a wider scope, which would give an incorrect semantics, the resulting semantic formula is still well-formed formally. This has led to many proposals such as that considered by Champollion [2] or the related Scope Domain Principle proposed by Landman [9], but all of them are rather ad hoc. Dependent event types will solve this problem: they give a solution where the correct semantics are accepted while the incorrect ones are excluded by typing because they would be ill-typed and hence formally illegal.

Dependent event types (DETs) were first considered in an example in [1] to study linguistic coercions in formal semantics, where types of events are indexed by their agents: $Evt(h)$ is the type of events conducted by h: $Human$. In this paper, we shall study event types dependent on thematic roles in event semantics both in the traditional Montague semantics [16] and in formal semantics in modern type theories (MTT-semantics, for short) [4,13,18]. For the former, we extend Church's simple type theory [5], as employed in Montague semantics that is familiar with many linguistic semanticists, by means of dependent event types, resulting in the system C_e, where the subtyping relationships between DETs are captured by subsumptive subtyping. For the latter, we extend an MTT (in particular, the type theory UTT [11]) with DETs whose subtyping relationships are reflected by coercive subtyping [12,14], resulting in the type system UTT[E]. Both of these extensions have nice meta-theoretic properties such as normalisation and logical consistency; in particular, we shall show that C_e can be faithfully embedded into UTT[E] and hence has desirable properties.

The rest of the paper is organised as follows. In Sect. 2, we shall describe the basics of DETs, introducing notations and examples. Subtyping between event types is described in Sect. 3, where we show, for example, how VPs can take the traditional non-dependent type, while we consider DETs. The formal systems C_e and $UTT[E]$ and the embedding of C_e in UTT[E] are studied in Sect. 4. Section 5 considers the solution of the event quantification problem by means of DETs:

Sect. 5.1 shows examples in the Montagovian setting and Sect. 5.2 considers it in MTT-semantics. The concluding section briefly discusses the future work.

2   Dependent Event Types

In the Davidsonian event semantics in the traditional Montagovian setting [6,17], there is only one type *Event* of all events. For example, the sentence (1) is interpreted as (2):

(1) John kissed Mary passionately.
(2) $\exists e\colon Event.\ kiss(e)\ \&\ agent(e, j)\ \&\ patient(e, m)\ \&\ passionate(e)$

where in (2), *Event* is the type of all events, *kiss*, *passionate*: $Event \rightarrow \mathbf{t}$ are predicates over events, and *agent*, *patient*: $Event \rightarrow \mathbf{e} \rightarrow \mathbf{t}$ are relations between events and entities.[1] Please note that, in the above neo-Davidson's semantics (2), adverbial modifications and thematic role relations are all propositional conjuncts in parallel with the verb description, an advantageous point as compared with an interpretation without events.

We propose to consider refined types of events. Rather than a single type *Event* of events, we introduce types of events that are dependent on some parameters. For instance, an event type can be dependent on agents and patients. Let *Agent* and *Patient* be the types of agents and patients, respectively. Then, for $a\colon Agent$ and $p\colon Patient$, the dependent type

$$Evt_{AP}(a, p)$$

is the type of events whose agents are a and whose patients are p. With such dependent event types, the above sentence (1) can now be interpreted as:[2]

(3) $\exists e\colon Evt_{AP}(j, m).\ kiss(e)\ \&\ passionate(e)$

Note that, besides other things we are going to explain below, we do not need to consider the relations *agent* and *patient* as found in (2) because they can now be 'recovered' from typing. For example, for $a\colon Agent$ and $p\colon Patient$, we

---

[1] In logical formulas or lambda-expressions, people often omit the type labels of events and entities: for example, (2) would just be written as $\exists e.\ kiss(e)\ \&\ agent(e, j)\ \&\ patient(e, m)\ \&\ passionate(e)$, since traditionally there are only one type of events and one type of entities; we shall put in the type labels explicitly. Another note on notations is: $\mathbf{e}$ and $\mathbf{t}$ in boldface stand for the type of entities and the type of truth values, respectively, as in MG, while e and t not in boldface stand for different things (for example, e would usually be used as a variable of an event type).

[2] Please note here that, for $kiss(e)$ and $passionate(e)$ to be well-typed, the type of event e must be the same as the domain of *kiss* and *passionate* – see the next section about subtyping, which allows them to be well-typed.

can define functions $\text{AGENT}_{AP}[a,p]$ and $\text{PATIENT}_{AP}[a,p]$ such that, for any event $e \colon Evt_{AP}(a,p)$, $\text{AGENT}_{AP}[a,p](e) = a$ and $\text{PATIENT}_{AP}[a,p](e) = p$.[3]

The parameters of dependent event types are usually names of thematic roles such as agents and patients. Formally, the dependent event types are parameterised by objects of types $A_1, \ldots, A_n$. Event types with n parameters are called n-ary event types. In this paper, we shall only consider n-ary event types with $n = 0, 1, 2$:

- When $n = 0$, the event type, usually written as *Event*, has no parameters. *Event* corresponds to the type of all events in the traditional setting.
- When $n = 1$, we only consider $Evt_A(a)$ and $Evt_P(p)$, where $a \colon Agent$ and $p \colon Patient$; i.e., these are event types dependent on agents a and those dependent on patients p. For example, if John is an agent with interpretation j, $Evt_A(j)$ is the type of events whose agents are John.
- When $n = 2$, we only consider $Evt_{AP}(a,p)$ for $a \colon Agent$ and $p \colon Patient$, i.e., the event type dependent on agent a and patients p. For example, if agent John and patient Mary, $Evt_{AP}(j,m)$ is the type of events whose agents and patients are John and Mary, respectively (cf., the example (3) above).

Introducing dependent event types has several advantages. In this paper, we shall detail one of them, that is, it gives a natural solution to the event quantification problem – see Sect. 5. Before doing that, we shall first consider in Sect. 3 the subtyping relationship between event types which, among other things, simplifies the semantic interpretations of VPs in the semantics with dependent event types, and then in Sect. 4 the formal systems that underlie the proposed semantic treatments and their meta-theoretic properties.

3   Subtyping Between Event Types

Event types have natural subtyping relationships between them. For example, an event whose agent is a and patient is p is an event with agent a. In other words, for $a \colon Agent$ and $p \colon Patient$, the type $Evt_{AP}(a,p)$ is a subtype of $Evt_A(a)$. If we only consider the event types *Event*, $Evt_A(a)$, $Evt_P(p)$ and $Evt_{AP}(a,p)$ (cf., the last section), they have the following subtypnig relationships:

$$Evt_{AP}(a,p) \leq Evt_A(a) \leq Event$$

$$Evt_{AP}(a,p) \leq Evt_P(p) \leq Event$$

which can be depicted as Fig. 1.

Formally, the subtyping relationship obeys the following rule (called subsumption rule):

$(*)$
$$\frac{a \colon A \quad A \leq B}{a \colon B}$$

---

[3] Formally, we have $\text{AGENT}_{AP}[a,p] = \lambda e : Evt_{AP}(a,p).a$, of type $Evt_{AP}(a,p) \to Agent$. Usually we simply write, for example, $\text{AGENT}_{AP}(e)$ for $\text{AGENT}_{AP}[a,p](e)$.

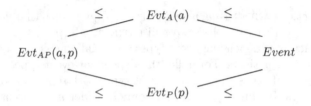

Fig. 1. Subtyping between event types with a: *Agent* and p: *Patient*.

It is also reflexive and transitive. The underlying type theory for formal semantics can be extended by dependent event types together with the subtyping relations. The underlying type theory can either be the simple type theory [5] in the Montagovian semantics or a Modern Type Theory such as UTT [11] in MTT-semantics as considered in, for example, [13]. If the former, extending it with dependent event types results in the formal system C_e with subsumptive subtyping, and if the latter, the resulting theory is UTT[E] with coercive subtyping whose basic coercion relationships in E characterise the subtyping relationships between event types (see Sect. 4 for more details).[4]

The incorporation of subtyping between event types is not only natural but plays an essential role in semantic interpretations. This can best be explained by considering how verb phrases are interpreted. In the neo-Davidson's event semantics (with only *Event* as the type of events), a verb phrase is interpreted as a predicate over events, as the following example shows.

(4) *talk*: *Event* → **t**.
(5) John talked loudly.
(6) $\exists e$: *Event*. $talk(e)$ & $loud(e)$ & $agent(e, j)$

With dependent event types such as $Evt_A(j)$, how can we interpret *talk* and (5)? In analogy, the desired semantics of (5) would be (7), where the agent of the event e can be obtained as $\text{AGENT}_A(e) = j$:

(7) $\exists e$: $Evt_A(j)$. $talk(e)$ & $loud(e)$

However, if *talk* is of type *Event* → **t**, $talk(e)$ would be ill-typed since e is of type $Evt_A(j)$, not of type *Event*. Is (7) well-typed? The answer is, if we do not have subtyping, it is not. But, if we have subtyping as described above, it is! To elaborate, because e: $Evt_A(j) \leq Event$, $talk(e)$ is well-typed by the subsumption rule (∗). Similarly, we have $loud$: *Event* → **t** and, therefore, $loud(e)$ is well-typed for e: $Evt_A(j) \leq Event$ as well.

To summarise, the subtyping relations have greatly simplified the event semantics in the presence of refined dependent event types.

---

[4] It may be worth mentioning that, in the setting of MTT-semantics, coercive subtyping [12,14] is used and, for uniformity, we may adopt coercive subtyping rather than subsumptive subtyping, although in general subsumptive subtyping is simpler.

Remark 1. The subtyping relations also facilitate a natural relationship between the functions such as AGENT$_{AP}$ and AGENT$_A$ (see Sect. 2 and Footnote 3). For example, because of the subtyping relations as depicted in Fig. 1, for $e\colon Evt_{AP}(a,p) \leq Evt_A(a)$, we have, by definition: AGENT$_{AP}[a,p](e)$ = AGENT$_A[a](e) = a$.

4   The Underlying Systems C_e and UTT[E]

In this section, we describe the formal systems C_e and UTT[E]: C_e extends the simple type theory [5] and UTT[E] extends the modern type theory UTT [11], both with dependent event types and their subtyping relationships as informally described in Sects. 2 and 3.[5] C_e is the underlying type theory when we consider formal semantics in the traditional Montagovian setting (as familiar by most of the linguistic semanticists) and UTT[E] when we consider formal semantics in a modern type theory (see, for example, [4,13]). We also outline the construction of an embedding of C_e into UTT[E] that shows that, like UTT[E], C_e has nice meta-theoretic properties such as normalisation and logical consistency.[6]

4.1   The Types System C_e

We shall first explain what a context is and what a judgement is in the system C_e, and then describe the rules of C_e.

Contexts. A context is a sequence of entries either of the form $x\colon A$ or of the form P *true*. Informally, the former assumes that the variable x be of type A and the latter that the proposition P be true. Only valid contexts are legal and context validity is governed by the following rules:

$$\frac{}{\langle\rangle \ valid} \qquad \frac{\Gamma \vdash A \ type \quad x \notin FV(\Gamma)}{\Gamma, \ x : A \ valid} \qquad \frac{\Gamma \vdash P\colon \mathbf{t}}{\Gamma, \ P \ true \ valid}$$

where $\langle\rangle$ is the empty sequence and $FV(\Gamma)$ is the set of free variables in Γ defined as: (1) $FV(\langle\rangle) = \emptyset$; (2) $FV(\Gamma, x : A) = FV(\Gamma) \cup \{x\}$; (3) $FV(\Gamma, P \ true) = FV(\Gamma)$.

Judgements. Judgements are sentences in C_e, whose correctness are governed by the inference rules below. In C_e, there are five forms of judgements:

- Γ *valid*, which means that Γ is a valid context (the rules of deriving context validity are given above).

---

[5] A notational remark: in C_e, C stands for 'Church' and e for 'event'. The notation UTT[E] comes from the work of coercive subtyping (see, for example, [14]) where T[C] denotes type theory T extended by coercive subtyping whose basic subtyping are given as the set C of subtyping judgements.

[6] This section is rather formal and, for a reader less interested in formal matters, its details might be safely skipped if one wishes so.

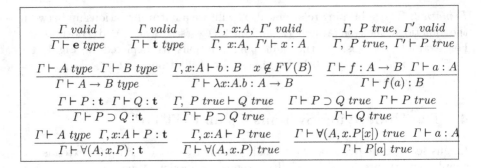

$$\frac{\Gamma \ valid}{\Gamma \vdash \mathbf{e} \ type} \qquad \frac{\Gamma \ valid}{\Gamma \vdash \mathbf{t} \ type} \qquad \frac{\Gamma, \ x{:}A, \ \Gamma' \ valid}{\Gamma, \ x{:}A, \ \Gamma' \vdash x : A} \qquad \frac{\Gamma, \ P \ true, \ \Gamma' \ valid}{\Gamma, \ P \ true, \ \Gamma' \vdash P \ true}$$

$$\frac{\Gamma \vdash A \ type \quad \Gamma \vdash B \ type}{\Gamma \vdash A \to B \ type} \qquad \frac{\Gamma, x{:}A \vdash b : B \quad x \notin FV(B)}{\Gamma \vdash \lambda x{:}A.b : A \to B} \qquad \frac{\Gamma \vdash f : A \to B \quad \Gamma \vdash a : A}{\Gamma \vdash f(a) : B}$$

$$\frac{\Gamma \vdash P : \mathbf{t} \quad \Gamma \vdash Q : \mathbf{t}}{\Gamma \vdash P \supset Q : \mathbf{t}} \qquad \frac{\Gamma, \ P \ true \vdash Q \ true}{\Gamma \vdash P \supset Q \ true} \qquad \frac{\Gamma \vdash P \supset Q \ true \quad \Gamma \vdash P \ true}{\Gamma \vdash Q \ true}$$

$$\frac{\Gamma \vdash A \ type \quad \Gamma, x{:}A \vdash P : \mathbf{t}}{\Gamma \vdash \forall(A, x.P) : \mathbf{t}} \qquad \frac{\Gamma, x{:}A \vdash P \ true}{\Gamma \vdash \forall(A, x.P) \ true} \qquad \frac{\Gamma \vdash \forall(A, x.P[x]) \ true \quad \Gamma \vdash a : A}{\Gamma \vdash P[a] \ true}$$

Fig. 2. Rules for Church's STT.

- $\Gamma \vdash A \ type$, which means that A is a type under context Γ.
- $\Gamma \vdash a : A$, which means that a is an object of type A under context Γ.
- $\Gamma \vdash P \ true$, which means that P is a true proposition under context Γ.
- $\Gamma \vdash A \leq B$, which means that A is a subtype of B under context Γ.

Inference Rules. The inference rules for C_e consist of:

1. Rules for context validity (the three rules above);
2. Figure 2: the rules for Church's simple type theory including those for (1) the basic types $\mathbf{e}$ and $\mathbf{t}$ of entities and truth values, (2) function types with β-conversion $((\lambda x : A.b[x])(a) \simeq b[a])$, and (3) logical formulas[7]; and
3. Figure 3: the rules for dependent event types including those for (1) dependent event types and (2) their subtyping relations, and (3) general subtyping rules including subsumption.

Some explanations of the rules are in order:

- In the λ-rule in Fig. 2, we have added a side condition $x \notin FV(B)$, i.e., x does not occur free in B. This is necessary because we have dependent event types like $Evt_A(a)$: for example, we need to forbid to derive $\Gamma \vdash (\lambda x : Agent.\lambda e : Evt_A(x).e) : Agent \to Evt_A(x) \to Evt_A(x)$ from $\Gamma, x : Agent \vdash (\lambda e : Evt_A(x).e) : Evt_A(x) \to Evt_A(x)$, where in the former judgement, x in $Agent \to Evt_A(x) \to Evt_A(x)$ would be a free variable that has not been declared in Γ. Note that, in Church's formulation [5], the side condition is not needed because, there, there are no dependent types (and x does not occur free in B for sure).
- In the rules in Fig. 3, since all of the judgements have the same contexts, we have omitted the contexts. For example, the first rule in its third row should have been, if written in full:

$$\frac{\Gamma \vdash a : Agent \quad \Gamma \vdash p : Patient}{\Gamma \vdash Evt_{AP}(a, p) \leq Evt_A(a)}$$

---

[7] We only consider the intuitionistic $\supset$ and $\forall$ here, omitting other operators including, in particular, those about, e.g. negation/classical logic in [5]. Also, we shall not assume extensionality.

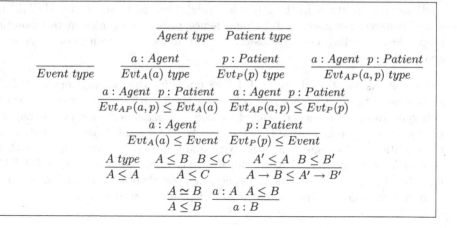

Fig. 3. Rules for dependent event types.

4.2   The Type System UTT[E]

The type theory UTT (Chap. 9 of [11]) is a dependent type theory with inductive types, type universes and higher-order logic. UTT is a typical Modern Type Theory (MTT) as employed in MTT-semantics [4,13] (actually, it is the MTT the first author and colleagues have employed in developing MTT-semantics). Its meta-theory was studied in the Ph.D. thesis by Goguen [8]. Coercive subtyping [12,14] has been developed by the authors and colleagues for modern type theories such as Martin-Löf's type theory and UTT.

Besides the type constructors in UTT as described in [11], UTT[E] has the following constant types and constant type families for dependent event types:

- $Entity: Type$
- $Agent,\ Patient: Type.$
- $Event: Type,$
  $Evt_A: (Agent)Type,$
  $Evt_P: (Patient)Type,$ and
  $Evt_{AP}: (Agent)(Patient)Type.$

The coercive subtyping relations in UTT[E] are given by subtyping judgements in E: they specify the subtyping relationships between dependent event types by means of the following parameterised constant coercions c_i $(i = 1, \ldots, 4)$ in E, where $a: Agent$ and $p: Patient$:

$$Evt_{AP}(a,p) \leq_{c_1[a,p]} Evt_A(a), \quad Evt_{AP}(a,p) \leq_{c_2[a,p]} Evt_P(p),$$

$$Evt_A(a) \leq_{c_3[a]} Event, \quad Evt_P(p) \leq_{c_4[p]} Event,$$

The coercions also satisfy the coherence condition $c_3[a] \circ c_1[a,p] = c_4[p] \circ c_2[a,p]$.

Based on the study in [14,21], it is straightforward to show that UTT[E] is a well-behaved extension of UTT and hence preserves its nice meta-theoretic properties, including Church-Rosser, subject reduction, strong normalisation, and logical consistency.

Remark 2. As mentioned above, UTT[E] underlies the development of MTT-semantics by the first author and colleagues [4,13]. In the recent trend of using rich type theories in formal semantics (see, for example, some of the papers in [3]), the development of MTT-semantics provides a full-blown alternative to the traditional Montague semantics with many advantages and has its further potentials to be developed in the future. It is worth remarking that UTT[E] underlies the event semantics in dependent type theories (or MTT-semantics with events) which contain, in particular, dependent event types.

4.3   Embedding of C_e into UTT[E]

In this subsection, we show that C_e can be faithfully embedded into UTT[E] and hence has nice meta-theoretic properties. The embedding of C_e into UTT[E] is defined as follows and it is faithful as the theorem below shows.

Definition 1 (embedding).   *The embedding $[\![\_]\!]$ from C_e to UTT[E] is inductively defined as follows:*[8]

1. *Constant types and dependent event types:*
   - $[\![e]\!]_\Gamma = Entity.$
   - $[\![t]\!]_\Gamma = Prop.$
   For the other constant types and dependent event types, they are mapped to the 'same' types in UTT[E], since we have overloaded their names. For example,
   - $[\![Agent]\!] = Agent$
   - $[\![Evt_A(a)]\!] = Evt_A([\![a]\!])$
2. *Non-constant terms:*
   - $[\![x]\!]_\Gamma = x$
   - $[\![A \to B]\!]_\Gamma = [\![A]\!]_\Gamma \to [\![B]\!]_\Gamma$
   - $[\![\lambda x : A.b]\!]_\Gamma = \lambda([\![A]\!]_\Gamma, T, [x : [\![A]\!]_\Gamma] [\![b]\!]_{\Gamma,x:A})$, *if* $[\![\Gamma, x : A]\!] \vdash [\![b]\!]_{\Gamma,x:A} : T$
   - $[\![f(a)]\!]_\Gamma = app(S, T, [\![f]\!]_\Gamma, [\![a]\!]_\Gamma)$, *if* $[\![\Gamma]\!] \vdash [\![f]\!]_\Gamma : S \to T$ *and* $[\![\Gamma]\!] \vdash [\![a]\!]_\Gamma : S_0$, *where* $[\![\Gamma]\!] \vdash S_0 \leq S$
   - $[\![P \supset Q]\!]_\Gamma = [\![P]\!]_\Gamma \supset [\![Q]\!]_\Gamma$
   - $[\![\forall(A, x.P)]\!]_\Gamma = \forall([\![A]\!]_\Gamma, [x : [\![A]\!]_\Gamma].[\![P]\!]_{\Gamma,x:A})$
3. *Contexts:*
   - $[\![\langle\rangle]\!] = \langle\rangle$ *(the empty context in UTT[E])*
   - $[\![\Gamma, x : A]\!] = [\![\Gamma]\!], x : [\![A]\!]_\Gamma$

---

[8] Formally, this is a partial function – it is only defined when certain conditions hold. The embedding theorem shows that the embedding is total for well-typed terms. Also, a notional note: we shall use S and T to stand for types in UTT[E] where function types are special cases of Π-types: for any types S and T, $S \to T = \Pi(S, [\_ : S]T)$.

– $[\![\Gamma,\ P\ true]\!] = [\![\Gamma]\!]$, $x\colon \mathbf{Prf}([\![P]\!]_\Gamma)$, where x does not occur free in $[\![\Gamma]\!]$.

The following theorem shows that the embedding is well-defined and faithful (in the sense of the theorem) and hence C_e has nice meta-theoretic properties (the corollary). Its proof is based on the embedding of Church's simple type theory into the calculus of constructions [10]. We omit the discussion of technical details, for otherwise we would have to detail the syntax and rules of UTT and coercive subtyping [11,14], except remarking that a key reason that the proof goes through is because the coercions to model subtyping for dependent event types are constants and coherent (see Sect. 4.2) and hence model subsumptive subtyping in C_e faithfully.

Theorem 2 (faithfulness). *The embedding in Definition 1 is defined for every well-typed term in C_e and, furthermore, we have:*

1. *If Γ valid in C_e, then $[\![\Gamma]\!]$ valid in UTT[E].*
2. *If $\Gamma \vdash A$ type in C_e, then $[\![\Gamma]\!] \vdash [\![A]\!]: Type$ in UTT[E].*
3. *If $\Gamma \vdash a\colon A$ in C_e, then in UTT[E], $[\![\Gamma]\!] \vdash [\![a]\!]: T$ for some T such that $[\![\Gamma]\!] \vdash T \leq_d [\![A]\!]$ for some d.*
4. *If $\Gamma \vdash P$ true in C_e, then $[\![\Gamma]\!] \vdash p\colon \mathbf{Prf}([\![P]\!])$ for some p in UTT[E].*
5. *If $\Gamma \vdash A \leq B$ in C_e, then $[\![\Gamma]\!] \vdash [\![A]\!] \leq_c [\![B]\!]$ for some unique c in UTT[E].*

Corollary 3. *C_e inherits nice meta-theoretic properties from UTT[E], including strong normalisation and logical consistency.*

Remark 3. Instead of the embedding method we have described here, one may consider a more direct approach to metatheory of C_e by directly showing that it has nice properties such as Church-Rosser and strong normalisation (as suggested by an anonymous reviewer). However, we think the above is simpler, which is of course a subjective view, and also demonstrates a generic approach to such meta-theoretic studies.

5 Event Quantification Problem

It is known that, when considering (neo-)Davidsonian event semantics where existential quantifiers for event variables are introduced, there is a problem in dealing with the scopes of the quantifiers when other quantificational phrases are involved. It has been argued that there is some incompatibility between event semantics and the traditional compositional semantics [2,20]. De Groote and Winter [7] have called this as the *event quantification problem* (EQP for short).

Consider the following sentence (8) which, under the traditional event semantics with $bark\colon Event \to \mathbf{t}$, could have two possible interpretations (9) and (10), where (10) is incorrect.

(8) No dog barks.
(9) $\neg\exists x\colon \mathbf{e}.\ dog(x)\ \&\ \exists e\colon Event.\ bark(e)\ \&\ agent(e,x)$
(10) (#) $\exists e\colon Event.\ \neg\exists x\colon \mathbf{e}.\ dog(x)\ \&\ bark(e)\ \&\ agent(e,x)$

Formally, the incorrect interpretation is acceptable just as the correct one: (10) is a legal formula. In order to avoid such incorrect interpretations as (10), people have made several proposals (see, for example, [2,20]) which involve, for instance, consideration of quantification not over events but over sets of events [2], or some informal (and somewhat *ad hoc*) principles whose adherence would disallow the incorrect interpretations (see, for example, the related Scope Domain Principle proposed by Landman [9]).

We shall study this with dependent event types as informally studied in Sects. 2 and 3, both in the Montagovian setting (i.e., in C_e as described in Sect. 4.1) and in the MTT-semantics (i.e., in UTT[E] as described in Sect. 4.2). It is shown that, with dependent event types, the incorrect semantics are blocked as illegal since they are ill-typed.

5.1 EQP in Montague Semantics with Dependent Event Types

In the Montagovian setting with dependent event types (formally, C_e in Sect. 4.1), this problem is solved naturally and *formally* – the incorrect semantic interpretations are excluded because they are ill-typed (in the empty context, where semantic interpretations of whole sentences like (8) are considered).

For example, (8) will be interpreted as (11), while the 'incorrect' interpretation (12) is not available (the formula (12) is ill-typed because x in $Evt_A(x)$, outside the scope of second/bound x (although intuitively it refers to it), is a free variable without being declared.)

(11) $\neg \exists x : \mathbf{e}.\ (dog(x)\ \&\ \exists e : Evt_A(x).\ bark(e))$
(12) $(\#)\ \exists e : Evt_A(x).\ \neg\exists x : \mathbf{e}.\ dog(x)\ \&\ bark(e)$

This offers a natural solution to the event quantification problem. Compared with existing solutions with informal ad hoc principles such as those mentioned above, our solution comes naturally as a 'side effect' of introducing dependent event types: it is formally disciplined and natural.

5.2 EQP in MTT-semantics with Dependent Event Types

In this paper, we have focussed on extending the traditional Montague semantics with dependent event types (formally, C_e), since the simple type theory is what the most semanticists are familiar with. One can also extend the MTT-semantics [4,13] with dependent event types (formally, UTT[E], if we use UTT for MTT-semantics) and hence consider such refined event semantics in the setting of MTT-semantics. Here, we give an example to show how this is done.

Still consider the sentence (8): No dog barks. In the MTT-semantics, where CNs are interpreted as types (rather than predicates), the verb bark is given a dependent type as its semantics:

(13) $bark : \Pi x : Dog.\ Evt_A(x) \to Prop$
   It is also the case that the correct semantics (14) for (8) is legal (well-typed), while the incorrect one (15) is not:

(14) $\neg\exists x\colon Dog.\ \exists e\colon Evt_A(x).\ bark(x,e)$

(15) (#) $\exists e\colon Evt_A(x).\ \neg\exists x\colon Dog.\ bark(x,e)$

Note that (15) is ill-typed for two reasons now: the first x is a variable not assumed anywhere and the term $bark(x,e)$ is ill-typed as well.

Employing dependent event types in the Montagovian semantics (i.e., in C_e as described in Sect. 4.1), would still leave a small possibility of some formally legal but incorrect semantics. For instance, one might consider the following semantics for (8):

(16) (#) $\exists e\colon Event.\ \neg\exists x\colon \mathbf{e}.\ dog(x)\ \&\ bark(e)$

Note that, although (16) is incorrect, it is still well-typed because e is just an event, not an event with x as agent.[9] This, however, would not happen in the MTT-semantic setting where the type of the verb bark is the dependent type (13) and the following semantic sentence is ill-typed:

(17) (#) $\exists e\colon Event.\ \neg\exists x\colon Dog.\ bark(x,e),$

because $bark(x,e)$ is not well-typed (it requires e to be of type $Evt_A(x)$, not just of type $Event$).

6   Conclusion

In this paper, we have introduced dependent event types for formal semantics. Subtyping is shown to play an essential role in this setting. We have also considered how dependent event types naturally solve the event quantification problem in combining event semantics with the traditional compositional semantics.

The notion of event types as studied in this paper is *intensional*, rather than extensional. For instance, when considering inverse verb pairs such as **buy** and **sell**, one may think that the events in (18) and (19) are the same [19].

(18) John bought the book from Mary.

(19) Mary sold the book to John.

If one considers this from the angle of extensionality/intensionality, the buying event and the selling event in the above situation are extensionally the same, but intensionally different. More generally, this is related to how to understand the sameness of events in the setting with dependent event types. Work need be done to study event structures and relevant inference patterns.

Another interesting research topic is to study whether all thematic roles should be considered as parameters of dependent event types. Unlike agents and patients, some thematic roles considered in the literature may not be suitable to play the role of indexing dependent event types. In such cases, we would still propose that they should be formalised by means of logical predicates/relations. In the other direction, event types may depend on other entities other than thematic roles and further studies are called for to understand this better.

---

[9] Of course, one can argue that this is not intended since the agent is known, but formally, nothing prevents one from doing it.

Acknowledgement. Thanks go to Stergios Chatzikyriakidis, David Corfield, Koji Mineshima and Christian Retoré for helpful comments on this work.

References

1. Asher, N., Luo, Z.: Formalisation of coercions in lexical semantics. In: Sinn und Bedeutung, vol. 17, Paris (2012)
2. Champollion, L.: The interaction of compositional semantics and event semantics. Linguist. Philos. **38**, 31–66 (2015)
3. Chatzikyriakidis, S., Luo, Z. (eds.): Modern Perspectives in Type-Theoretical Semantics. Springer, Heidelberg (2017)
4. Chatzikyriakidis, S., Luo, Z.: Formal Semantics in Modern Type Theories. ISTE/Wiley (2018, to appear)
5. Church, A.: A formulation of the simple theory of types. J. Symb. Log. **5**(1), 56–68 (1940)
6. Davidson, D.: The logical form of action sentences. In: Rothstein, S. (ed.) The Logic of Decision and Action. University of Pittsburgh Press, Pittsburgh (1967)
7. de Groote, P., Winter, Y.: A type-logical account of quantification in event semantics. In: Logic and Engineering of Natural Language Semantics, vol. 11 (2014)
8. Goguen, H.: A typed operational semantics for type theory. Ph.D. thesis, University of Edinburgh (1994)
9. Landman, F.: Plurality. In: Lappin, S. (ed.) The Handbook of Contemporary Semantic Theory (1996)
10. Luo, Z.: A problem of adequacy: conservativity of calculus of constructions over higher-order logic. Technical report, LFCS report series ECS-LFCS-90-121, Department of Computer Science, University of Edinburgh (1990)
11. Luo, Z.: Computation and Reasoning: A Type Theory for Computer Science. Oxford University Press, Oxford (1994)
12. Luo, Z.: Coercive subtyping in type theory. In: Dalen, D., Bezem, M. (eds.) CSL 1996. LNCS, vol. 1258, pp. 275–296. Springer, Heidelberg (1997). doi:10.1007/3-540-63172-0_45
13. Luo, Z.: Formal semantics in modern type theories with coercive subtyping. Linguist. Philos. **35**(6), 491–513 (2012)
14. Luo, Z., Soloviev, S., Xue, T.: Coercive subtyping: theory and implementation. Inf. Comput. **223**, 18–42 (2012)
15. Martin-Löf, P.: Intuitionistic Type Theory. Bibliopolis, Berkeley (1984)
16. Montague, R.: Formal Philosophy. Yale University Press, New Haven (1974). Collected papers Ed. by R. Thomason
17. Parsons, T.: Events in the Semantics of English. MIT Press, Cambridge (1990)
18. Ranta, A.: Type-Theoretical Grammar. Oxford University Press, Oxford (1994)
19. Williams, A.: Arguments in Syntax and Semantics. Cambridge University Press, Cambridge (2015)
20. Winter, Y., Zwarts, J.: Event semantics and abstract categorial grammar. In: Kanazawa, M., Kornai, A., Kracht, M., Seki, H. (eds.) MOL 2011. LNCS (LNAI), vol. 6878, pp. 174–191. Springer, Heidelberg (2011). doi:10.1007/978-3-642-23211-4_11
21. Xue, T.: Theory and implementation of coercive subtyping. Ph.D. thesis, Royal Holloway, University of London (2013)

A Geometry of Interaction Machine for Gödel's System T

Ian Mackie[✉]

Department of Informatics, University of Sussex, Brighton, UK

Abstract. Gödel's System T is the simply typed lambda calculus extended with numbers and an iterator. The higher-order nature of the language gives it enormous expressive power—the language can represent all the primitive recursive functions and beyond, for instance Ackermann's function. In this paper we use System T as a minimalistic functional language. We give an interpretation using a data-flow model that incorporates ideas from the geometry of interaction and game semantics. The contribution is a reversible model of higher-order computation which can also serve as a novel compilation technique.

1 Introduction

We present a data-flow model of functional computation, where a single token (the run-time system) travels around a fixed network (the program). Computation begins with an empty token at the root of the network, and it ends when the token returns back to the root with the result. The token is deterministic (at any choice point in the network the token has enough information to proceed in a unique way) and reversible (the token can turn back on itself, to re-trace its steps exactly to undo the computation done).

Gödel's System T (see e.g., [6]) is an applied typed λ-calculus. It is a functional programming language supporting higher-order functions, pairs and projections, numbers and an iterator. It can express all the primitive recursive functions, and up to the so-called ϵ_0 functions. Ackermann's function is included in this set, so its expressive power is sufficiently large to make it interesting. We present a simpler version of System T, called Linear System T [2] with the same expressive power but with a syntax more suited to our needs.

The data-flow interpretation uses ideas developed for linear logic [4]. Specifically, the geometry of interaction (GOI), which models the dynamics of the logic using paths in networks. The GOI machine [9] was a concrete realisation of this idea, originally given for PCF, where an ad-hoc solution was given for base types. To make the data-flow idea work for System T, we need to provide a general reversible interpretation of base types (numbers), which are the data constructors, and also iterators. The need to provide a solution to these points distinguishes the work from others that are related to this, for example [1] gives a general theory of reversible computation through reversible combinators but does not deal with base types in this way. Our approach is also quite different

© Springer-Verlag GmbH Germany 2017
J. Kennedy and R.J.G.B. de Queiroz (Eds.): WoLLIC 2017, LNCS 10388, pp. 229–241, 2017.
DOI: 10.1007/978-3-662-55386-2_16

from other reversible functional (which are often first-order), or higher-order, languages, for instance the reversible SECD machine [8]. To summarise, the main contributions are:

- A (reversible) GOI-style model of computation for Gödel's System T.
- An implementation of the model that is a direct compilation into current hardware (essentially directly to assembly language).
- An interesting side effect of this work is an implementation technique that uses an exceptionally small (in terms of space) run-time system in some cases, thus can open up applications to embedded systems, for instance.

Overview. In the next section we give some background material on System T and the geometry of interaction. Section 3 gives the definitions of the token and network structure needed to model System T and a compilation into these networks. In Sect. 4 we briefly look at some properties, and in Sect. 5 we discuss some implementation aspects. Finally, we conclude in Sect. 6.

2   Background

We use a specific version of System T, which is equivalent to the standard presentation (see e.g., [6]), but has a linear syntax. We assume familiarity with the λ-calculus, and refer the reader to [3,7] for standard notations and concepts. In [2] it was shown that System T can be presented using the linear λ-calculus without losing any computational power. This linear System T was called System L. Essentially, that work illustrates that it is possible to duplicate and erase in System T either using the λ-calculus or the iterator. We simplify the presentation with a variant of System L, that includes numbers as primitives, and gives a simple reversible model. The set of terms is given by the following grammar:

$$t, u ::= x \mid \lambda x.t \mid tu \mid \langle t, u \rangle \mid \mathsf{let}\ \langle x, y \rangle = t\ \mathsf{in}\ u \mid n \mid \mathsf{S}t \mid \mathsf{iter}\ tuv$$

where n ranges over natural numbers, and x ranges over a finite set of variables. The **let** construct is a way of splitting the pair so that we have access to both components. Numbers are included, so that we can write n for $\mathsf{S}^n 0$. The typing rules (Fig. 1) show the valid terms; note that the type system captures the linearity constraints, for example x must occur in t in the abstraction rule. We can now write simple functions, for example: $\mathsf{add} = \lambda mn.\mathsf{iter}\ m(\lambda x.\mathsf{S}x)n$, $\mathsf{mult} = \lambda mn.\mathsf{iter}\ m(\mathsf{add}\ n)0$, $\mathsf{two} = \lambda fx.\mathsf{iter}\ 2fx$ and finally Ackermann's function: $\mathsf{ack} = \lambda mn.(\mathsf{iter}\ m(\lambda gu.\mathsf{iter}\ (\mathsf{S}u)g1)(\lambda x.\mathsf{S}x))n$.

For reference, we define the reduction rules, thus giving an operational semantics to the language. This is also useful for a correctness result of the token interpretation that we give later.

$$\frac{}{x : A \vdash x : A} \text{ (Axiom)}$$

$$\frac{\Gamma, x : A \vdash t : B}{\Gamma \vdash \lambda x.t : A \to B} (\to \text{Intro}) \qquad \frac{\Gamma \vdash t : A \multimap B \qquad \Delta \vdash u : A}{\Gamma, \Delta \vdash tu : B} (\to \text{Elim})$$

$$\frac{\Gamma \vdash t : A \qquad \Delta \vdash u : B}{\Gamma, \Delta \vdash \langle t, u \rangle : A \otimes B} (\otimes\text{Intro}) \qquad \frac{\Gamma \vdash t : A \otimes B \qquad x : A, y : B, \Delta \vdash u : C}{\Gamma, \Delta \vdash \mathbf{let}\ \langle x, y \rangle = t\ \mathbf{in}\ u : C} (\otimes\text{Elim})$$

$$\frac{}{\vdash n : \mathbb{N}} (\text{Num}) \qquad \frac{\Gamma \vdash t : \mathbb{N}}{\Gamma \vdash \mathsf{S}\, t : \mathbb{N}} (\text{Succ}) \qquad \frac{\Gamma \vdash t : \mathbb{N} \quad \Theta \vdash u : A \multimap A \quad \Delta \vdash v : A}{\Gamma, \Theta, \Delta \vdash \mathsf{iter}\ t\ u\ v : A} (\text{Rec})$$

Fig. 1. Type system

Definition 1 (Reduction). *Reduction can take place in any context:*

Name	Reduction		Condition
Succ	$\mathsf{S}n$	$\to n+1$	
Beta	$(\lambda x.t)v$	$\to t[v/x]$	$\mathsf{fv}(v) = \varnothing$
Let	$\mathbf{let}\ \langle x, y \rangle = \langle t, u \rangle\ \mathbf{in}\ v \to (v[t/x])[u/y]$		$\mathsf{fv}(t) = \mathsf{fv}(u) = \varnothing$
Rec1	$\mathsf{iter}\,(n+1)vu$	$\to v(\mathsf{iter}\,nvu)$	$\mathsf{fv}(v) = \varnothing$
Rec2	$\mathsf{iter}\,0vu$	$\to u$	$\mathsf{fv}(v) = \varnothing$

For simplicity, we assume substitution $t[v/x]$ is a meta-operation corresponding to the explicit substitution defined in [2]. We only consider evaluation of programs, that is closed terms of base type. In this way, all programs give a number as a result. The following main properties that we need for this paper (see [2] for the proofs) are:

Proposition 1.

- *Adequacy: If $\vdash t : \mathbb{N}$, then $t \to^* n$, for some n.*
- *Subject Reduction. If $\Gamma \vdash t : A$ and $t \to u$ then $\Gamma \vdash u : A$.*
- *Strong Normalisation: if $\Gamma \vdash t : A$, then t is strongly normalisable.*

Data-flow and the Geometry of Interaction. The starting point for the dataflow model comes from the geometry of interaction, which was first set up as a semantics for linear logic [5]. In [9] this is used as an implementation technique— essentially mapping the model to assembly language. This idea was extended to cover the language PCF. Here we use some similar ideas, but our focus is on building a simple data-flow model based on the linear calculus.

We build networks out of nodes and edges. Each node is labelled and has a fixed arity that specifies how many edges can be connected to it. The point of connection between the edge and the node is called a port. The ports of a node are ordered, and we use the label • to give the position of the first port to make clear the orientation. Edges connect two ports together (potentially on the same node) with the constraint that only one edge can be connected to each port. Edges may also connect to one port only in which case we say that it is a free edge. The networks that we build for our programs always have one unique free edge that we call the root of the network.

A token travelling on this network moves from port-to-port along the edges, and is transformed and directed by the nodes: the nodes are routing devices. Consider the following example network that uses two occurrences of a node m of arity three, connected as shown. In this example, t_0 is the starting token, and it is moving to the right. The node m will transform the token to t_1 and direct it towards the second node. The second m node will transform the token to t_2 and direct it along the edge as shown:

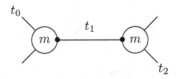

Depending on the port of arrival, the token will be inspected by the node, then modified and re-directed. In this example, t_0 is directed along the edge as shown, and it will be modified to remember where it came from: the right-hand edge of the m node. Token t_1 can now change direction and because it has information about where it came from, it can return to t_0 and forget the information it had. Equally, t_1 could continue moving to the right, and this time the second m node will do exactly the same transformation and use the information to direct it to t_2. A simple stack of left (l) and right (r) labels can be stored in the token to achieve this: when arriving from the left or right, push l or r on the stack. When there is a choice, go in the direction indicated by the top of the stack, and pop the stack. In the example above, if t_0 is the empty stack, t_1 will hold r and the token t_2 is again the empty stack. This travel is deterministic, and reversible.

In the next section we define the structure of the token, and define the different nodes that we need to encode System T. Each node is defined by giving the transformation and re-directions of the token.

3   Encoding

Here we define the nodes for the network together with the token interpretation. We also give a compilation of terms into networks, and computation is then a flow over the network with a token that stores the current state.

Definition 2 (Token). *A token (m, i, s, d) is defined by the following components:*

- *m is a stack that contains the elements l and r. This stack is used to navigate the lambda, application and pairs.*
- *i is a stack which keeps information about the iterator. It can contain elements that are either numbers or pairs of numbers.*
- *s is an array of stacks which keeps temporary information about the different iterators. The size of the array is known at compile time, as it depends on the number of iterators.*

– *d is the data stack that contains numbers and an additional element* $*$.

We write $\square$ *for the empty stack/array, and use the notation* $n : s$ *for an element* n *pushed onto the stack* s.

Each node has a flow associated with it, and operations that change the token. We will make this explicit when we define the nodes. The token contains information that is used to redirect it though the nodes of the graph in a deterministic way. Information stored in the token is the only information that is used for this. The data stack stores numbers, but also $*$ which can be thought of as a place-holder for a number that will be found later. When the data stack has a $*$ at the top of the stack, then this is a question: the token is looking for an answer to this question. When the token has a number at the top, then the token has an answer for the last un-answered question. When the initial question is answered, computation is complete. Some nodes will use the question/answer information to direct the node.

For a program, starting with the token $(\square, \square, \square, * : \square)$, every run will end with $(\square, \square, \square, n : \square)$, where n is the result. Any part of the computation can be reversed, and in particular, the whole computation can be reversed: a computation starting with $(\square, \square, \square, n : \square)$, will end with $(\square, \square, \square, * : \square)$.

Definition 3 (Network). *A network is a (not necessarily connected) graph built from a set of nodes and edges. A node has a name, a fixed number of ports, and in some cases also a value. The collection of nodes is fixed and defined below.*

We next define the nodes and the transformations for our networks. For each node α of arity n we label the ports $l_1, \ldots, l_n$:

We then describe the transformation by the convention that arriving from l_i and leaving from $\hat{l}_j$ (i.e., the port that is connected to the other end of l_j). For each operation f, we have the reverse operation f^*. We write these two functions as $ft = t'f^*$, which is just an abbreviation of $ft = t'$ and $f^*t' = t$.

Value Nodes. A number n is represented as the following network:

The data-flow for this node is from l to $\hat{l}$. There are two cases: if the token is a question it collects the value n, otherwise, the token is an answer, and it drops off the value n. We define operations n and n^* which are defined as $n(m, i, s, * : d) = (m, i, s, n : d)n^*$ to do these transformations.

Unary Functions. The successor node is represented as the net:

There are two data-flows for this node, and two cases for each depending on whether the token is a question or an answer. We explain two flows, and the other two are the reverses. If the token is a question and arrives at l_2, then we ask another question to find the argument of the function, thus we push $*$ onto the data stack and the token travels along the edge to $\hat{l}_1$. We define an operation s_1 to do this transformation. If the token is an answer and arrives at l_1, then we need to continue to $\hat{l}_2$, and apply the successor function. The operation s_2 does this transformation. These two operations, and inverses, are defined as: $s_1(m, i, s, * : d) = (m, i, s, * : * : d)s_1^*$ and $s_2(m, i, s, n : * : d) = (m, i, n+1 : d)s_2^*$.

Multiplexing. We use a multiplexing node for abstraction, application, the pair and the let constructs.

There are four different data-flows for m, and all the operations only alter the m component of the token. Arriving at l_1, we need to decide which way to go. If the top of the stack m stack is l, we apply the l^* transformation and travel along the edge to $\hat{l}_2$. Otherwise, the top of the stack is r, so we apply the r^* transformation and travel along the edge to $\hat{l}_3$. Arriving at l_2 or l_3, the token travels along the edge to $\hat{l}_1$, and we need to remember which side we came from. The operations l and r do this transformation, respectively. These functions are defined as: $l(m, i, s, d) = (l : m, i, s, d)l^*$ and $r(m, i, s, d) = (r : m, i, s, d)r^*$.

Iterator. We next need the nodes that will allow for the encoding of the iterator. The first one will find the number of times we need to iterate.

There are four different data-flows for this node, and the operations alter the d and i components of the token. Arriving at l_1, there are two cases. If the token is an answer, then the token travels along the edge to $\hat{l}_3$ and we apply an operation called i_2. Otherwise, the token is a question, and it travels along the edge to $\hat{l}_2$ and we apply the i_1^* operation. Arriving at l_2, the token travels along the edge to $\hat{l}_1$, and we apply the i_1 operation. Finally, arriving at l_3, the token travels along the edge to $\hat{l}_1$, and we apply the i_2^* operation. These functions are defined as $i_1(m, i, s, d) = (m, i, s, * : d)i_1^*$ and $i_2(m, i, s, n : d) = (m, n : i, s, d)i_2^*$.

Counter. To model the iteration process in a reversible way, we need something similar to the multiplexing node, but having the ability to count.

There are four different data-flows for this node, and the operations only alter the i component of the token. Arriving at l_1, if the top of the i stack is 0, we apply the z^* transformation and travel along the edge to $\hat{l}_2$. Otherwise, the top of the stack is non-zero, so we apply the s^* transformation and travel along the edge to $\hat{l}_3$. Arriving at l_2 or l_3 the token travels along the edge to $\hat{l}_1$, and we apply the operation z or s, respectively. Counters will always come in pairs: we call them c and c', and the prime operations are essentially the same, but operate on the other component of the pair. These functions are defined as:

$$
\begin{aligned}
z(m, c : i, s, d) &= (m, (0, c) : i, s, d)z^* \\
z'(m, c : i, s, d) &= (m, (c, 0) : i, s, d)z'^* \\
s(m, (c_1, c_2) : i, s, d) &= (m, (c_1 + 1, c_2) : i, s, d)s^* \\
s'(m, (c_1, c_2) : i, s, d) &= (m, (c_1, c_2 + 1) : i, s, d)s'^*
\end{aligned}
$$

The final node that we need captures the scope of an iterator.

There are two data-flows possible for this kind of node. Arriving at l_1, the token travels towards $\hat{l}_2$, and we apply the operation t^*. Arriving at l_2, the token travels along the edge to $\hat{l}_1$, and we apply the operation t. The collection of functions are defined as $t_j(m, (c_1, c_2) : i, s, d) = (m, i, push(j, s, (c_1, c_2)), d)t_j^*$, where the operation *push* updates the array of stacks at position j ($1 \le j \le n$):

$$
push(j, [s_1, \ldots, s_n], (c_1, c_2)) = [s_1, \ldots, (c_1, c_2) : s_j, \ldots, s_n]
$$

This completes the definition of the nodes that we need for our networks.

Compilation. The compilation $\mathcal{T}(\cdot)$ of terms into networks is given using the nodes introduced above. A term t with $\mathsf{fv}(t) = \{x_1, \ldots, x_n\}$ will be translated as a network $\mathcal{T}(t)$ with the root edge at the top, and n free edges corresponding to the free variables of the term.

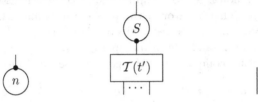

The labelling of free edges is just for convenience, and is not part of the system. We proceed by induction over the structure of the term being translated.

Numbers, Functions and Variables. When t is a number, say n, then $\mathcal{T}(n)$ is given by the first network below. When t is the unary function $\mathsf{S}\ t'$, then $\mathcal{T}(t)$ is given by the middle diagram, where we use node S. Finally, when t is a variable, say x, then $\mathcal{T}(t)$ is translated into a wire, as shown right-most below.

Abstraction, Application, Pairs and Let. If t is an abstraction, say $\lambda x.t'$, then $\mathcal{T}(t)$ is translated into the network on the left below. We connect the (necessarily unique) occurrence of the variable x to the binding λ (m node). We have assumed, without loss of generality that x is the leftmost free variable of the term t'. If t is uv, then $\mathcal{T}(t)$ is given by the second diagram below. If t is $\langle u, v \rangle$, then $\mathcal{T}(t)$ is given by the third network below. Because of the linearity constraints, there are no common free variables in u and v. Finally, if t is **let** $\langle x, y \rangle = u$ **in** v, then $\mathcal{T}(t)$ is given by the network on the right—we have assumed that x and y are the right-most two free variables in v. The other free variables of v are represented as a line struck through.

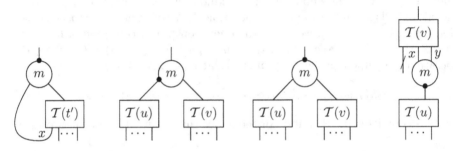

Recursor. If t is iter uvw, then $\mathcal{T}(t)$ is given by the following network.

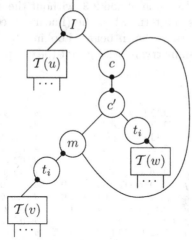

The compilation of the iterator uses several different nodes. The two counter nodes work together to maintain the iterations. The I node is responsible for navigating the token to the number, then passing that value to the counters. The t_i nodes are used to record the state of the iterators if the token leaves the scope of the iterator, where a new index i is associated to each iterator.

There are a number of ways that a network can be simplified at compile time, but we leave the details for another occasion.

Examples. We show several examples to illustrate how this style of computation works. The first one is the program $(\lambda x.Sx)3$, thus the successor function applied to 3. The network generated through the compilation of this term is given below, where we have labelled the edges for reference. The initial token (which is the same for any program of base type) is $(\Box, \Box, \Box, * : \Box)$, and the program counter starts in the network at the root (labelled with a). The token then travels along the sequence b, c, d, etc. We show the edge together with the token at that place.

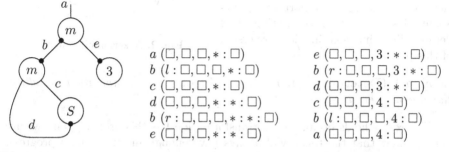

$a\ (\Box, \Box, \Box, * : \Box)$ $\qquad$ $e\ (\Box, \Box, \Box, 3 : * : \Box)$

$b\ (l : \Box, \Box, \Box, * : \Box)$ $\qquad$ $b\ (r : \Box, \Box, \Box, 3 : * : \Box)$

$c\ (\Box, \Box, \Box, * : \Box)$ $\qquad$ $d\ (\Box, \Box, \Box, 3 : * : \Box)$

$d\ (\Box, \Box, \Box, * : * : \Box)$ $\qquad$ $c\ (\Box, \Box, \Box, 4 : \Box)$

$b\ (r : \Box, \Box, \Box, * : * : \Box)$ $\qquad$ $b\ (l : \Box, \Box, \Box, 4 : \Box)$

$e\ (\Box, \Box, \Box, * : * : \Box)$ $\qquad$ $a\ (\Box, \Box, \Box, 4 : \Box)$

The token arrives back to the root with the answer 4. At any point during the computation we can use the information in the token to turn back and re-trace the computation—deterministically—to the start.

The next example is iter $2(\lambda x.S\ x)3$, which computes the addition of 2 and 3. (This results in a simplified version of add 2 3, without the need for abstraction and application; it is the same term when the (linear) β-reductions have been done). The resulting network is shown below. Looking at the structure of the execution trace reveals a symmetry, reflecting that the computation is reversible.

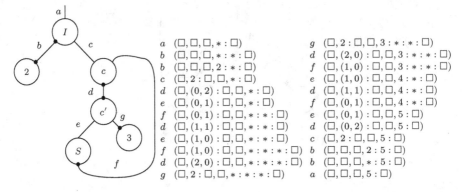

$$
\begin{array}{ll}
a & (\square,\square,\square,* : \square) \\
b & (\square,\square,\square,* : * : \square) \\
b & (\square,\square,\square,2 : * : \square) \\
c & (\square,2 : \square,\square,* : \square) \\
d & (\square,(0,2) : \square,\square,* : \square) \\
e & (\square,(0,1) : \square,\square,* : \square) \\
f & (\square,(0,1) : \square,\square,* : * : \square) \\
d & (\square,(1,1) : \square,\square,* : * : \square) \\
e & (\square,(1,0) : \square,\square,* : * : \square) \\
f & (\square,(1,0) : \square,\square,* : * : * : \square) \\
d & (\square,(2,0) : \square,\square,* : * : * : \square) \\
g & (\square,2 : \square,\square,* : * : * : \square)
\end{array}
\qquad
\begin{array}{ll}
g & (\square,2 : \square,\square,3 : * : * : \square) \\
d & (\square,(2,0) : \square,\square,3 : * : * : \square) \\
f & (\square,(1,0) : \square,\square,3 : * : * : \square) \\
e & (\square,(1,0) : \square,\square,4 : * : \square) \\
d & (\square,(1,1) : \square,\square,4 : * : \square) \\
f & (\square,(0,1) : \square,\square,4 : * : \square) \\
e & (\square,(0,1) : \square,\square,5 : \square) \\
d & (\square,(0,2) : \square,\square,5 : \square) \\
c & (\square,2 : \square,\square,5 : \square) \\
b & (\square,\square,\square,2 : 5 : \square) \\
b & (\square,\square,\square,* : 5 : \square) \\
a & (\square,\square,\square,5 : \square)
\end{array}
$$

The final example, shown in Fig. 2, gives the compilation of Ackermann's function applied to two arguments: *ack* 2 3. We have applied several optimisations to this example which remove some of the nodes that are not needed, and we have also performed a linear β-reduction. We will not attempt a trace of this example, as the number of operations is quite large (see next section). The example is just to show what a more elaborate program looks like in this setting.

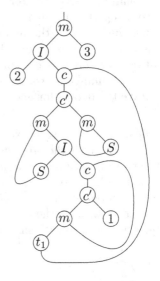

Fig. 2. Ackermann's function

4   Properties

Here we show some important results about this method of computing. Full details will be included in a longer version of this paper.

Lemma 1 (Determinism). *For a closed term $t : \mathbb{N}$, computation is bidirectional and deterministic.*

This result holds because each transformation has this property, and we are able to show that the token will always have enough information to progress. Computation is therefore reversible: at any point we can turn back and undo the computation. For instance, the second example in the last section can be started with the token $(\square,\square,\square,4 : \square)$, and the computation will be performed in a deterministic way to give the final result $(\Box, \Box, \Box, * : \Box)$ as required. To show that this notion of computation is correct, then we need a way of relating the token flow with an operational semantics.

Theorem 1 (Correctness). *Let $t : \mathbb{N}$ be a closed term. $t \rightarrow^* n$ iff there is a run in the network $\mathcal{T}(t)$: $(\Box, \Box, \Box, * : \Box) \rightarrow^* (\Box, \Box, \Box, n : \Box)$.*

Using this result we can give the main result about reduction:

Theorem 2. *For each reduction rule $t \rightarrow u$, the initial and final states of the token are the same.*

5   Discussion: Implementation

We give a few implementation details for our data-flow model on traditional hardware by giving a compilation for System T directly to assembly language. The data stack d can be separated out as a number stack and a question stack. The question stack can then be implemented using a register, using 0 for question and 1 for answer. To push and pop values on this stack, we can then use simple register shifts. Knowing if the current operation is a question or an answer is then just a bit test. We can also reduce the size of the question stack in some cases. For instance, when we ask for the result of a successor applied to an argument, then we can use the same question for the argument as the result. The other stacks can be mapped onto memory in standard ways. Depending on the number of iterators (known at compile time), we can map some of the structure to registers or blocks of memory.

The compilation procedure is to build a network, then compile this network to instructions. This is done by compiling each node as a block of code. We show the example below for the case of the m nodes, which are the simplest case (we refer the reader to the diagram of the m nodes with the labels to help understand the labels, and we use a register $R0$ to hold the m stack). Edges are then just linking information, instructing how to get to the next block of code.

$l2$: lsl $R0$ $l3$: lsl $R0$ $l1$: lsr $R0$
    br $\widehat{l_1}$ inc $R0$ be $\widehat{l_2}$
                     br $\widehat{l_1}$ br $\widehat{l_3}$

Thus for each node we need to generate instructions that will implement the appropriate collection of functions associated to it. In some cases, this will mean interrogating the question stack, and then doing an operation depending on that value. All these functions however map quite directly to assembly level.

There are many improvements and optimisations that can be made to the compilation. For example, if the number of times a function is being iterated is known, then we do not need to push the value on the stack if we leave through a t node when compiling the base value, as we know that this value will always be the same—a single stack location is all that is needed. This can be identified statically. There are other little improvements that can be made also, but it is not yet clear if any of these have a significant impact on performance.

We have implemented all the ideas in this paper, and it is worth mentioning that the execution times are long, but the run-time memory usage can be surprisingly small. We give a couple of illustrative examples in the table below.

Program	Result	i	d	jumps
two two two two ($\lambda x.Sx$) 0	65536	16	2	92M
ack 3 2	29	3	15	2.4M
ack 2 24	51	2	26	537M
ack 3 3	61	3	31	4.7M

We show the example program (using the terms given in Sect. 2) together with the result, the maximum height of the iterator stacks, and the maximum height of the data stack. Since two integers are needed for each element of the iterator stack, and each data element just one, this means that $2i + d$ is the total number of memory space used. The table also shows the (approximate) number of jumps (units are millions) in the network. Computation can take a long time, but little memory is used.

6   Conclusion

We have given a geometry of interaction style data-flow implementation of a simple language built from the linear λ-calculus extended with a recursor operator. Depending on how we constrain the use of the recursor, this language is rich enough to capture all primitive recursive functions or more generally Gödel's System T. Current work in this area includes developing the ideas to richer languages (in particular to include other data-structures), and developing new compilation techniques. Various program transformations can be applied to change the kind of runs in interesting ways, and computations can be significantly shortened by jumping in the network. The work has been implemented, and benchmark tests show that significant computations can be performed using very little run-time memory, thus there is potential for application in embedded systems in addition to giving a reversible implementation of a seemingly non-reversible language.

References

1. Abramsky, S.: A structural approach to reversible computation. Theoret. Comput. Sci. **347**, 441–464 (2005)
2. Alves, S., Fernández, M., Florido, M., Mackie, I.: Gödel's system T revisited. Theoret. Comput. Sci. **411**(11–13), 1484–1500 (2010)
3. Barendregt, H.P.: The Lambda Calculus: Its Syntax and Semantics, Studies in Logic and the Foundations of Mathematics, second, revised edn., vol. 103. North-Holland Publishing Company (1984)
4. Girard, J.Y.: Linear logic. Theoret. Comput. Sci. **50**(1), 1–102 (1987)

5. Girard, J.Y.: Geometry of interaction 1: interpretation of system F. In: Ferro, R., Bonotto, C., Valentini, S., Zanardo, A. (eds.) Logic Colloquium 88, Studies in Logic and the Foundations of Mathematics, vol. 127, pp. 221–260. North Holland Publishing Company, Amsterdam (1989)
6. Girard, J.Y., Taylor, P., Lafont, Y.: Proofs and Types. Cambridge University Press, New York, USA (1989)
7. Hankin, C.: An Introduction to Lambda Calculi for Computer Scientists, Texts in Computing, vol. 2. King's College Publications, London (2004)
8. Kluge, W.: A reversible SE(M)CD machine. In: Koopman, P., Clack, C. (eds.) IFL 1999. LNCS, vol. 1868, pp. 95–113. Springer, Heidelberg (2000). doi:10.1007/10722298_6
9. Mackie, I.: The geometry of interaction machine. In: Proceedings of the 22nd ACM Symposium on Principles of Programming Languages (POPL 1995), pp. 198–208. ACM Press, January 1995

Disjoint Fibring of Non-deterministic Matrices

Sérgio Marcelino[✉] and Carlos Caleiro

SQIG - Instituto de Telecomunicações, Dep. Matemática - Instituto Superior Técnico,
Universidade de Lisboa, Lisboa, Portugal
{smarcel,ccal}@math.tecnico.ulisboa.pt

Abstract. In this paper we give a first definitive step towards endowing the general mechanism for combining logics known as *fibring* with a meaningful and useful semantics given by *non-deterministic logical matrices (Nmatrices)*. We present and study the properties of two semantical operations: a unary operation of ω-*power* of a given Nmatrix, and a binary operation of *strict product* of Nmatrices with disjoint similarity types (signatures). We show that, together, these operations can be used to characterize the disjoint fibring of propositional logics, when each of these logics is presented by a single Nmatrix. As an outcome, we also provide a decidability and complexity result about the resulting fibred logic. We illustrate the constructions with a few meaningful examples.

1 Introduction

Fibring is a general and powerful mechanism for combining logics. Given its fundamental character, abstract formulation and compositional nature, fibring is a key ingredient of the general theory of universal logic [4]. The ubiquity of its underlying problems also justifies fibring as a valuable tool for the construction and analysis of complex logics, a subject of ever growing importance in application fields like software engineering and artificial intelligence (see [14]).

As entailed by Dov Gabbay's original formulation in [15,16], given two logics $\mathcal{L}_1$ and $\mathcal{L}_2$, fibring should combine $\mathcal{L}_1$ and $\mathcal{L}_2$ into the smallest logical system for the combined language that extends both $\mathcal{L}_1$ and $\mathcal{L}_2$ [8]. The semantics of fibred logics has been the subject of much attention, but thus far we do not know of a generally satisfactory semantic counterpart of fibring that naturally relates models of the component logics with models of the combined logic. There have been several approaches to this problem, even leading to some interesting results, like sufficient conditions for conservativity, completeness preservation, or the finite model property [9]. However, these attempts are not satisfactory and, in particular, have reduced practical use. We can arguably divide these

C. Caleiro—Work done under the scope of Project UID/EEA/50008/2013 of Instituto de Telecomunicações, financed by the applicable framework (FCT/MEC through national funds and co-funded by FEDER-PT2020). The first author also acknowledges the FCT postdoctoral grant SFRH/BPD/76513/2011. This research is part of the MoSH initiative of SQIG at Instituto de Telecomunicações.

J. Kennedy and R.J.G.B. de Queiroz (Eds.): WoLLIC 2017, LNCS 10388, pp. 242–255, 2017.
DOI: 10.1007/978-3-662-55386-2_17

approaches into three categories: some are very specific and cover only particular cases, such as [12,20] for the fusion of modal logics; others are quite general, but at the expense of losing any sensitivity regarding the particular semantics given to the logics being combined [7,26,30] (this is due to the fact that, for completeness sake, the semantics considered end up neglecting the given semantics of the logics being combined, namely via *fullness* assumptions). Still other approaches explicitly combine the models of the particular semantics of the component logics, but involve semantical structures that are highly uncommon [27], or constructions that cannot be iterated [25]. For these reasons, general fibred semantics is still an open problem: how to combine, in the general case, two classes of models $\mathcal{M}_1$ (adequate for logic $\mathcal{L}_1$) and $\mathcal{M}_2$ (adequate for logic $\mathcal{L}_2$) into a class $\mathcal{M}_1 \star \mathcal{M}_2$, whose elements are built from the models in $\mathcal{M}_1$ and $\mathcal{M}_2$, providing an adequate semantics for $\mathcal{L}_1 \bullet \mathcal{L}_2$?

We have known for some time that this question cannot be given a simple answer when taking logical matrices as models, as is most common. For instance, we know that fibring two logics, each given by a single finite matrix, can result in a logic that cannot even be given by a single matrix [6,22]. Herein, for the moment focusing only in the particular case of disjoint fibring, i.e. when the logic being combined do not share any connectives, we will for the first time propose a meaningful and useful semantics semantics for fibring by using *non-deterministic logical matrices (Nmatrices)*. Nmatrices generalize matrices, the long standing reference for abstract logic semantics dating back to the beginning of 20th century (see [29]), by allowing a (non-empty) set of possible values whenever applying a connective to known values. The algebraic structures underlying Nmatrices are multi-algebras, a.k.a. hyper-algebras [10,18,23], and have been brought to the attention of the logic community in [1–3,11].

We present and study the properties of two operations on Nmatrices: a unary operation of ω-*power* of a given Nmatrix, and a binary operation of *strict product* of (non-trivial) Nmatrices with disjoint similarity types (signatures). Together, these operations can be used to characterize the disjoint fibring of propositional logics, when each of the logics is presented by a single Nmatrix. We start by identifying a fundamental property of a given Nmatrix, which we call *saturation*, that holds when one is able to refute simultaneously all formulas lying outside any given theory of the logic. Saturation guarantees that the Nmatrix provides exact semantic witnesses (valuations) to all theories of the logic. Then, we show that the strict product operation characterizes precisely the (disjoint) fibring of the logics presented by two given Nmatrices, as long as both are saturated. Finally, we show that the ω-power operation always yields a saturated Nmatrix which defines the same logic as the original Nmatrix. Note that the ω-power of a (deterministic) logical matrix is still deterministic. As an outcome, we also recover from [21] a decidability and complexity result about the fibred logic. Along the paper, we illustrate these constructions with a few meaningful examples.

The paper is structured as follows. Section 2 is a brief introduction to some necessary basic concepts. In Sect. 3 we recall the definition of Nmatrix semantics, and take advantage of its non-deterministic nature to define a natural strict product operation that is the core ingredient to capturing the essentials of fibred semantics, but which cannot be (finitely) mimicked by an operation on (deterministic) matrices. Then, in Sect. 4, we introduce the notion of a *saturated* Nmatrix, and show that the strict product of saturated Nmatrices precisely characterizes (disjoint) fibring (Theorem 1). We also show how to generally saturate Nmatrices by means of a ω-power operation (Theorem 2), and as a consequence we obtain the first recipe that captures the semantical side of (disjoint) fibring by suitably combining the models of the given logics (Corollary 1). Further, by identifying some (finite) sub-models of the combined semantics, we also recover some known decidability and complexity results about the fibred logic (Proposition 2). We wrap up, in Sect. 5, with an overview of the results obtained and a discussion of several future extensions.

2   Preliminaries

Logics. A *logic* is a tuple $\mathcal{L} = \langle \Sigma, \vdash \rangle$, where $\Sigma = \{\Sigma^{(n)} : n \in \mathbb{N}\}$ is a propositional *signature* ($\Sigma^{(n)}$ contains the n-ary connectives of Σ) and $\vdash \subseteq 2^{L_\Sigma(P)} \times L_\Sigma(P)$ is a structural (Tarskian) consequence relation over the language $L_\Sigma(P)$ (the absolutely free Σ-algebra over a given set of propositional variables P). Given $\Sigma' \supseteq \Sigma$, we consider $\mathcal{L}^{\Sigma'} = \langle \Sigma', \vdash^{\Sigma'} \rangle$ where $\Gamma \vdash^{\Sigma'} \varphi$ if and only if there exist $\Delta \cup \{\psi\} \subseteq L_\Sigma(P)$ and $\sigma : P \to L_{\Sigma'}(P)$ such that $\Delta \vdash \psi$, $\Delta^\sigma = \Gamma$ and $\psi^\sigma = \varphi$. A set $T \subseteq L_\Sigma(P)$ is called a $\mathcal{L}$-*theory* whenever T is closed for $\vdash$, that is $T^\vdash = \{\varphi : T \vdash \varphi\} = T$.

Example 1. $\mathcal{L}_{\mathsf{cnj}} = \langle \Sigma_{\mathsf{cnj}}, \vdash_{\mathsf{cnj}} \rangle$ *where* Σ_{cnj} *contains a single binary connective* $\wedge$ *and* $\vdash_{\mathsf{cnj}}$ *is the underlying consequence relation of the conjunction fragment of classical logic.* $\triangle$

Hilbert Systems. A *Hilbert calculus* is a pair $\mathcal{H} = \langle \Sigma, R \rangle$ where Σ is a signature, and $R \subseteq 2^{L_\Sigma(P)} \times L_\Sigma(P)$ is a set of *inference rules*. Given $\langle \Delta, \psi \rangle \in R$, we refer to Δ as the set of *premises* and to ψ as the *conclusion* of the rule. When the set of premises is empty, ψ is dubbed an *axiom*. An inference rule $\langle \Delta, \psi \rangle \in R$ is often denoted by $\frac{\Delta}{\psi}$, or simply by $\frac{\psi_1 \ldots \psi_n}{\psi}$ if $\Delta = \{\psi_1, \ldots, \psi_n\}$ is finite, or by $\frac{}{\psi}$ if $\Delta = \emptyset$.

Given $\Sigma \subseteq \Sigma'$ and $P \subseteq P'$, a Hilbert calculus $\mathcal{H} = \langle \Sigma, R \rangle$ induces a consequence relation $\vdash_{\mathcal{H}}$ on $L_{\Sigma'}(P')$ such that, for each $\Gamma \subseteq L_{\Sigma'}(P')$, $\Gamma^{\vdash_{\mathcal{H}}}$ is the least set that contains Γ and is closed for all applications of instances of the inference rules in R, that is, if $\frac{\Delta}{\psi} \in R$ and $\sigma : P \to L_{\Sigma'}(P')$ is such that $\Delta^\sigma \subseteq \Gamma^{\vdash_{\mathcal{H}}}$ then $\psi^\sigma \in \Gamma^{\vdash_{\mathcal{H}}}$. Of course, this definition induces a logic $\mathcal{L}^{\Sigma'} = \langle \Sigma', \vdash_{\mathcal{H}} \rangle$ over P' for each $\Sigma \subseteq \Sigma'$.

Example 2. $\mathcal{L}_{\mathsf{cnj}} = \mathcal{L}_{\mathcal{H}_{\mathsf{cnj}}}$ *where* $\mathcal{H}_{\mathsf{cnj}} = \langle \Sigma_{\mathsf{cnj}}, \{\frac{p \wedge q}{p}, \frac{p \wedge q}{q}, \frac{p \quad q}{p \wedge q}\} \rangle$. $\triangle$

Given $\mathscr{H} = \langle \Sigma, R \rangle$ and $\Sigma' \supseteq \Sigma$, note that $\mathcal{L}_{\mathscr{H}}^{\Sigma'} = \mathcal{L}_{\mathscr{H}'}$ where $\mathscr{H}' = \langle \Sigma', R \rangle$.

Fibring. Let $\mathcal{L}_1 = \langle \Sigma_1, \vdash_1 \rangle$ and $\mathcal{L}_2 = \langle \Sigma_2, \vdash_2 \rangle$ be two logics. The *fibring* of $\mathcal{L}_1$ and $\mathcal{L}_2$ is the smallest logic $\mathcal{L}_1 \bullet \mathcal{L}_2$ over the joint signature $\Sigma_{12} = \Sigma_1 \cup \Sigma_2$, with $\Sigma_{12}^{(n)} = \Sigma_1^{(n)} \cup \Sigma_2^{(n)}$ for all $n \in \mathbb{N}$, that extends both $\mathcal{L}_1$ and $\mathcal{L}_2$. Given Hilbert calculi $\mathscr{H}_1 = \langle \Sigma_1, R_1 \rangle$ and $\mathscr{H}_2 = \langle \Sigma_2, R_2 \rangle$ then $\mathcal{L}_1 \bullet \mathcal{L}_2 = \mathcal{L}_{\mathscr{H}_1 \bullet \mathscr{H}_2}$ where $\mathscr{H}_1 \bullet \mathscr{H}_2 = \langle \Sigma_{12}, R_1 \cup R_2 \rangle$. Clearly, besides joining the given signatures, which allows building so-called *mixed formulas*, the fibring of the two calculi consists in simply putting together their rules, thus allowing also *mixed reasoning*. When there are no shared connectives, i.e. $\Sigma_1 \cap \Sigma_2 = \emptyset$, or better $\Sigma_1^{(n)} \cap \Sigma_2^{(n)} = \emptyset$ for all $n \in \mathbb{N}$, the fibring is usually said to be *disjoint*. In the remainder of the paper we will focus our attention on disjoint fibring.

Example 3. Let $\mathcal{L}_{djn} = \langle \Sigma_{djn}, \vdash_{djn} \rangle$ where Σ_{djn} contains a single binary connective $\vee$, be the disjunction-only fragment of intuitionistic logic. It is known that $\mathcal{L}_{djn} = \mathcal{L}_{\mathscr{H}_{djn}}$ where $\mathscr{H}_{djn} = \langle \Sigma_{djn}, \{\frac{p}{p \vee q}, \frac{p \vee p}{p}, \frac{p \vee q}{q \vee p}, \frac{p \vee (q \vee r)}{(p \vee q) \vee r}\} \rangle$, and that this coincides with the disjunction fragment of classical logic, see [22,24]. The fibred logic $\mathcal{L}_{cnj} \bullet \mathcal{L}_{djn} = \langle \Sigma_{cnj} \cup \Sigma_{djn}, \vdash \rangle$ is strictly weaker than the conjunction-disjunction fragment of classical logic. Namely, $p \vee (p \wedge p) \nvdash p$, as was also noted in [19]. $\triangle$

3   Non-deterministic Matrices and Strict Products

The quest for an adequate semantical operation corresponding to the logical operation of fibring has been long, as we discussed above. In this section we will show how the additional freedom given by semantical non-determinism crucially captures the essential behavior of the fibring mechanism. We start by recalling non-deterministic matrices, as introduced in [1].

Non-deterministic Matrices (Nmatrices). A *Nmatrix* over a signature Σ is a tuple $\mathbb{M} = \langle A, \cdot_{\mathbb{M}}, D \rangle$ where[1] A is a set (of *truth-values*), $D \subseteq A$ is the set of *designated* values and, for each $c \in \Sigma^{(n)}$, $\cdot_{\mathbb{M}}$ gives the interpretation of c in $\mathbb{M}$, $c_{\mathbb{M}} : A^n \to 2^A \setminus \{\emptyset\}$. We shall refer to the set of *undesignated* values by $U = A \setminus D$. Henceforth, we will assume that we are working only with non-trivial Nmatrices, in the sense that $D \neq \emptyset$ and $U \neq \emptyset$. Clearly, this will only leave out some trivial uninteresting logics. A *valuation* over $\mathbb{M}$ is a function $v : L_\Sigma(P) \to A$ such that for all for each $c \in \Sigma^{(n)}$ and $\varphi_1, \dots, \varphi_n \in L_\Sigma(P)$, $v(c(\varphi_1, \dots, \varphi_n)) \in c_{\mathbb{M}}(v(\varphi_1), \dots, v(\varphi_n))$. We say that $\Gamma \models_{\mathbb{M}} \varphi$ if, for every valuation over $\mathbb{M}$, $v(\Gamma) \subseteq D$ then $v(\varphi) \in D$. It is well known that $\mathcal{L}_{\mathbb{M}} = \langle \Sigma, \models_{\mathbb{M}} \rangle$ is always a logic. Of course, the traditional notion of logical matrix is recovered by considering Nmatrices for which the image of each $c_{\mathbb{M}}$ is always a singleton (deterministic), as in the Nmatrix in the next example. It is also important to highlight that Nmatrices, as matrices, provide analytic semantics, in the sense that any partial valuation defined over a set of formulas closed under subformulas can be extended to a valuation. We say that $\mathbb{M}' = \langle A', \cdot_{\mathbb{M}'}, D' \rangle$ over signature

---

[1] $\langle A, \cdot_{\mathbb{M}} \rangle$ is a multi-algebra, see [10,18].

Σ', is a *sub-Nmatrix* of $\mathbb{M} = \langle A, \cdot_{\mathbb{M}}, D \rangle$ whenever $\Sigma' \subseteq \Sigma$, $A' \subseteq A$, $D' \subseteq D$ and for every n-ary $c \in \Sigma'$ and $\boldsymbol{a} \in (A')^n$ we have $c_{\mathbb{M}'}(\boldsymbol{a}) \subseteq c_{\mathbb{M}}(\boldsymbol{a})$. Of course, in this situation, we have that $\models_{\mathbb{M}'} \subseteq \models_{\mathbb{M}}$. Given $X \subseteq A$, we say that $\mathbb{M}'$ is the sub-Nmatrix of $\mathbb{M}$ *generated by* X if it is the smallest sub-Nmatrix of $\mathbb{M}$ whose carrier contains X.

(Finite) Nmatrices allow a natural tabular representation similar to the one for matrices, with the difference that in the table appear sets of elements instead of just elements.

Example 4. Let $\mathcal{L}_{@} = \langle \Sigma_{@}, \vdash \rangle$ be the smallest logic over the signature $\Sigma_{@}$ containing a single connective $@$ of any given arity, that is, $\Gamma \vdash \varphi$ iff $\varphi \in \Gamma$, for every $\Gamma \cup \{\varphi\} \subseteq L_{\Sigma_{@}}(P)$. It is easy to see that $\mathcal{L}_{@} = \mathcal{L}_{\mathcal{H}_{@}} = \mathcal{L}_{\mathbb{M}_{@}}$ where $\mathcal{H}_{@} = \langle \Sigma_{@}, \emptyset \rangle$ and $\mathbb{M}_{@} = \langle \{0,1\}, \tilde{\cdot}, \{1\} \rangle$, with $@$ interpreted freely. In case $@$ is binary, the table is as follows

$\tilde{@}$	0	1
0	$\{0,1\}$	$\{0,1\}$
1	$\{0,1\}$	$\{0,1\}$

This logic cannot be presented by a single finite matrix, nor even by any finite set of finite matrices [6]. △

If an Nmatrix is deterministic we simply present it as a matrix.

Example 5. $\mathcal{L}_{\text{cnj}} = \mathcal{L}_{\mathbb{M}_{\text{cnj}}}$ where $\mathbb{M}_{\text{cnj}} = \langle \{0,1\}, \tilde{\cdot}, \{1\} \rangle$, with

$\tilde{\wedge}$	0	1
0	0	0
1	0	1

△

We can now give a hint on how the non-deterministic semantic environment provided by Nmatrices allows us to capture very basic operations on logics in a simple and intuitive way. Take, for example, the very particular case of fibring consisting in adding an absolutely free new connective to a given logic $\mathcal{L}$. We know that the resulting logic cannot in general be presented by a single logical matrix, even if $\mathcal{L}$ is. That is the case, for instance, when adding a free nullary connective [6]. When we consider a non-nullary connective other problems emerge and, as shown in [6], we might obtain a non-finitely valued logic. So, when we try to add a new (free) connective $@$, of any arity, the resulting logic might only be characterizable by an infinite logical matrix, or in alternative by an infinite collection of finite matrices, even when $\mathcal{L}$ is given simply by a finite matrix. In the non-deterministic setting, however, it is obvious that one can simply extend any given (N)matrix $\mathbb{M} = \langle A, \cdot_{\mathbb{M}}, D \rangle$ defining $\mathcal{L}$ by letting the new connective range non-deterministically all over A, like in Example 4.

Proposition 1. *Let* $\mathbb{M} = \langle A, \cdot_{\mathbb{M}}, D \rangle$ *be an Nmatrix over signature* Σ *and consider an* n-ary *connective* $@ \notin \Sigma$. *Letting* $\Sigma' = \Sigma \cup \Sigma_{@}$, *we have that* $\mathcal{L}_{\mathbb{M}}^{\Sigma'} = \mathcal{L}_{\mathbb{M}'}$ *where* $\mathbb{M}' = \langle A, \cdot_{\mathbb{M}'}, D \rangle$ *and* $\cdot_{\mathbb{M}'}$ *extends* $\cdot_{\mathbb{M}}$ *by* $@_{\mathbb{M}'}(\boldsymbol{a}) = A$ *for all* $\boldsymbol{a} \in A^n$.

Proof. Given $\Gamma, \{\psi\} \subseteq L_{\Sigma'}(P)$, by definition, we have that $\Gamma \vdash \psi$ if and only if there exist $\Delta \cup \{\psi\} \subseteq L_{\Sigma}(P)$ and $\sigma : P \rightarrow L_{\Sigma'}(P)$ such that $\Delta \vdash \psi$, $\Delta^\sigma = \Gamma$ and $\psi^\sigma = \varphi$.

$\mathcal{L}_{\mathbb{M}}^{\Sigma'}$-soundness of $\mathbb{M}'$ follows easily from the fact that $\Gamma \vdash^{\Sigma'} \varphi$ implies there is $\Delta \cup \{\psi\} \subseteq L_{\Sigma}(P)$ and $\sigma : P \rightarrow L_{\Sigma'}(P)$ such that $\Delta \vdash \psi$, $\Delta^\sigma = \Gamma$ and $\psi^\sigma = \varphi$. Hence, given a valuation v over $\mathbb{M}'$ such that $v(\Gamma) \subseteq D$, there is valuation v' over $\mathbb{M}$, such that $v(\Gamma) = v(\Delta^\sigma) = v'(\Delta)$ where $v = v' \circ \sigma$. Therefore, $v'(\psi) = v(\psi^\sigma) = v(\varphi) \in D$ by $\mathcal{L}_{\mathbb{M}}^{\Sigma'}$-soundness of $\mathbb{M}$.

For completeness, consider that $\Gamma \models_{\mathbb{M}'} \psi$. Then, we know that $v(\Gamma) \subseteq D$ implies $v(\psi) \in D$ for every valuation v. In particular, this means that the implication holds regardless the value the formulas with @ as main connective. Let $\Delta \cup \{\psi\} \subseteq L_{\Sigma}(P)$ contain the most particular formulas such that there is $\sigma : P \rightarrow L_{\Sigma'}(P)$ satisfying $\Delta \vdash \psi$, $\Delta^\sigma = \Gamma$ and $\psi^\sigma = \varphi$ (in the sense that every other such inference must be an instance of this one, which was rigorously characterized in [21] using Σ-skeletons). As @ ranges all over A, this means that $\Delta \models_{\mathbb{M}} \varphi$, hence $\Delta \vdash \varphi$, and thus $\Gamma \vdash^{\Sigma'} \varphi$.   □

As we shall see, this phenomenon is a particular case of a general advantage offered by the non-deterministic character of Nmatrices. Let us close this section by presenting an operation between Nmatrices, which we call *strict product*, that generalizes the previous construction, and is also the basic ingredient of the fibred semantics we present in the following section.

Strict Product of Nmatrices. Let Σ_1 and Σ_2 be signatures with $\Sigma_1 \cap \Sigma_2 = \emptyset$. Given non-trivial Nmatrices $\mathbb{M}_1 = \langle A_1, \cdot_1, D_1 \rangle$ over Σ_1 and $\mathbb{M}_2 = \langle A_2, \cdot_2, D_2 \rangle$ over Σ_2, their *strict product* is the Nmatrix over $\Sigma_1 \cup \Sigma_2$

$$\mathbb{M}_1 \star \mathbb{M}_2 = \langle A_{12}, \cdot_{12}, D_{12} \rangle$$

where $A_{12} = (D_1 \times D_2) \cup (U_1 \times U_2)$, $D_{12} = D_1 \times D_2$ and

$$c_{12}((a_1, b_1), \ldots, (a_k, b_k)) = \begin{cases} \{(a, b_a) \in A_{12} : a \in c_1(a_1, \ldots, a_k)\} & \text{if } c \in \Sigma_1 \\ \{(a_b, b) \in A_{12} : b \in c_2(b_1, \ldots, b_k)\} & \text{if } c \in \Sigma_2 \end{cases} .$$

Note that the strict product operation only considers pairings between agreeing elements, either both designated or both undesignated. This dependency is highlighted when we write b_a and a_b. Further, if some of the given Nmatrices was trivial the operation could get ill-defined, as the sets of possible values for c_{12} above could be empty. It is also clear that if both $\mathbb{M}_1$ and $\mathbb{M}_2$ are finite, $\mathbb{M}_1 \star \mathbb{M}_2$ is also finite.

It is not hard to see that the strict product subsumes the construction in Proposition 1. Just observe that $\mathcal{L}_{\mathbb{M}}^{\Sigma'} = \mathcal{L}_{\mathbb{M}} \bullet \mathcal{L}_{\mathbb{M}@} = \mathcal{L}_{\mathbb{M} \star \mathbb{M}@}$.

Remark 1. Any valuation v over $\mathbb{M}_1 \star \mathbb{M}_2$ corresponds to two valuations, $\pi_1 \circ v$ over $\mathbb{M}_1^{\Sigma_{12}}$ and $\pi_2 \circ v$ over $\mathbb{M}_2^{\Sigma_{12}}$, where each π_i is the usual projection of elements of A_{12} in A_i. These two valuations are *compatible*, in the sense that, $\pi_1 \circ v(\varphi) \in D_1$ if and only if $\pi_2 \circ v(\varphi) \in D_2$, for every formula φ.

It is important to note that not every product of valuations gives rise to a valuation in this manner. Given valuations v_1 and v_2, over $\mathbb{M}_1^{\Sigma_{12}}$ and $\mathbb{M}_2^{\Sigma_{12}}$, $v_1 \times v_2$ is a valuation over $\mathbb{M}_1 \star \mathbb{M}_2$ if and only if v_1 and v_2 are compatible.

This reasoning can be extended to partial valuations over $\mathbb{M}_1^{\Sigma_{12}}$ and $\mathbb{M}_2^{\Sigma_{12}}$. Given v_1 and v_2 partial valuations defined over a set of formulas $\Delta \subseteq L_{\Sigma_{12}}(P)$ closed under subformulas, if v_1 and v_2 are compatible for the formulas in Δ then there exists a valuation v over $\mathbb{M}_1 \star \mathbb{M}_2$ such that $v(\psi) = (v_1(\psi), v_2(\psi))$ for all $\psi \in \Delta$. This is a consequence of the fact that the semantics over Nmatrices are *analytic*, in the sense that every partial valuation defined over a set of formulas closed under subformulas is extendable to a full valuation [1,2].

We close this section with two further examples.

Example 6. Consider the box-only fragment of modal logic **K**, that is, $\mathcal{L}_{box} = \langle \Sigma_{box}, \vdash_{box} \rangle = \mathcal{L}_{\mathcal{H}_{box}}$, where Σ_{box} contains a single unary connective $\Box$, with $\mathcal{H}_{box} = \langle \Sigma_{box}, \{\frac{p}{\Box p}\} \rangle$. It is easy to see that $\mathcal{L}_{box} = \mathcal{L}_{\mathbb{M}_{box}}$ with $\mathbb{M}_{box} = \langle \{0,1\}, \tilde{\cdot}, \{1\} \rangle$

$$
\begin{array}{c|cc}
\tilde{\Box} & 0 & 1 \\
\hline
& \{0,1\} & \{1\}
\end{array}
$$

Let us see that $\mathcal{L}_{cnj} \bullet \mathcal{L}_{box} = \langle \Sigma_{cnj} \cup \Sigma_{box}, \vdash \rangle = \mathcal{L}_{\mathbb{M}_{cnj} \star \mathbb{M}_{box}}$.
Applying the strict product operation we obtain

$$
\mathbb{M}_{cnj} \star \mathbb{M}_{box} = \langle \{(0,0), (1,1)\}, \tilde{\cdot}, \{(1,1)\} \rangle
$$

where
$$
\begin{array}{c|cc}
\tilde{\wedge} & (0,0) & (1,1) \\
\hline
(0,0) & (0,0) & (0,0) \\
(1,1) & (0,0) & (1,1)
\end{array}
\quad \text{and} \quad
\begin{array}{c|cc}
\tilde{\Box} & (0,0) & (1,1) \\
\hline
& \{(0,0),(1,1)\} & \{(1,1)\}
\end{array}
$$

It is not hard to see that $\Gamma^\vdash$ is the closure under applications of $\wedge$ and $\Box$ of the set $\Gamma^{\vdash_{cnj}}$, and that the equality above holds.   △

Example 7. Let us return to the fibring of classical conjunction with classical disjunction already considered in Example 3. Clearly $\mathcal{L}_{djn}$ is given by the disjunction reduct of the two-valued Boolean matrix, $\mathbb{M}_{djn} = \langle \{0,1\}, \tilde{\cdot}, \{1\} \rangle$, with

$$
\begin{array}{c|cc}
\tilde{\vee} & 0 & 1 \\
\hline
0 & 0 & 1 \\
1 & 1 & 1
\end{array}
$$

Let us see that $\mathcal{L}_{cnj} \bullet \mathcal{L}_{djn} \neq \mathcal{L}_{\mathbb{M}_{cnj} \star \mathbb{M}_{djn}}$. Applying the strict product operation to the respective matrices we obtain

$$
\mathbb{M}_{cnj} \star \mathbb{M}_{djn} = \langle \{(0,0), (1,1)\}, \tilde{\cdot}, \{(1,1)\} \rangle
$$

with $\tilde{\wedge}$ as in Example 6 and with

$$
\begin{array}{c|cc}
\tilde{\vee} & (0,0) & (1,1) \\
\hline
(0,0) & (0,0) & (1,1) \\
(1,1) & (1,1) & (1,1)
\end{array}
$$

It is clear that $\mathbb{M}_{cnj} \star \mathbb{M}_{djn}$ is (a renaming of) the $\wedge\vee$-reduct of the 2-valued Boolean matrix. Hence, its logic is just the conjunction-disjunction of classical logic and therefore the inequality above holds. $\triangle$

In the next section we will solve this discrepancy, and characterize exactly why strict products do not adequately capture disjoint fibring in all cases.

4   Saturation and the Semantics of Disjoint Fibring

In this section we will show how to use strict products to overcome the difficulties observed earlier and capture disjoint fibring in all cases. For the purpose, we need to be able to understand the differences underlying the two examples above. The key ingredient is the following notion of *saturation*.

Saturated Nmatrices. An Nmatrix $\mathbb{M} = \langle A, \cdot_{\mathbb{M}}, D \rangle$ over Σ is *saturated* if for every non-trivial $\mathcal{L}_{\mathbb{M}}$-theory T there is a valuation v over $\mathbb{M}$ such that $v(\psi) \in D$ if and only if $\psi \in T$.

In Example 7, where the strict product does not coincide with fibring, it is easy to see that the two-valued classical matrix for disjunction $\mathbb{M}_{djn}$ is not saturated. In particular, $p \notin \{p \vee q\}^{\vdash_{djn}}$ and $q \notin \{p \vee q\}^{\vdash_{djn}}$ but it is clear that there is no valuation v such that $v(p \vee q) = 1$ and $v(p) = v(q) = 0$. On the other hand, the (N)matrices of Example 6 are both saturated, which actually explains, as shown by the following theorem, why the strict product captures the fibring of the logics given by the two (N)matrices.

Theorem 1. *Let Σ_1 and Σ_2 be signatures with $\Sigma_1 \cap \Sigma_2 = \emptyset$. Given non-trivial saturated Nmatrices $\mathbb{M}_1 = \langle A_1, \cdot_1, D_1 \rangle$ over Σ_1 and $\mathbb{M}_2 = \langle A_2, \cdot_2, D_2 \rangle$ over Σ_2, we have that $\mathbb{M}_1 \star \mathbb{M}_2$ is saturated and $\mathcal{L}_{\mathbb{M}_1} \bullet \mathcal{L}_{\mathbb{M}_2} = \mathcal{L}_{\mathbb{M}_1 \star \mathbb{M}_2}$.*

Proof. $\mathcal{L}_1 \bullet \mathcal{L}_2$-soundness of $\mathbb{M}_1 \star \mathbb{M}_2$ follows easily from the fact that given v valuation over $\mathbb{M}_1 \star \mathbb{M}_2$, we have that for $i = 1, 2$, $\pi_i \circ v$ are compatible valuations over $\mathbb{M}_i^{\Sigma_{12}}$, mentioned in Remark 1. As, $\Gamma \vdash_i \varphi$ and $v(\Gamma) \subseteq D_{12}$ imply that $\pi_i(\Gamma) \subseteq D_i$, and so, $\pi_i(\varphi) \in D_i$, and therefore $v(\varphi) \in D_{12}$.

We now prove this combination mechanism preserves saturation which also implies that $\mathcal{L}_1 \bullet \mathcal{L}_2 \subseteq \mathcal{L}_{\mathbb{M}_1 \star \mathbb{M}_2}$. Let T be a $\mathcal{L}_1 \bullet \mathcal{L}_2$-theory, we will show there is a valuation v over $\mathbb{M}_1 \star \mathbb{M}_2$, such that $v(\psi) \in D$ if and only if $\psi \in T$. We have that for $i \in \{1, 2\}$ is a $\mathcal{L}_i^{\Sigma_{12}}$-theory. It is not hard to see that if $\mathbb{M}_i$ is saturated, then, so it is $\mathbb{M}_i^{\Sigma_{12}}$. Hence, we can pick, for $i \in \{1, 2\}$, valuations over $\mathbb{M}_i^{\Sigma_{12}}$, such that $v_i(\psi) \subseteq D_i$ if and only if $\psi \in T$. Hence, these v_1 and v_2 are compatible, and by Remark 1, we obtain that $v : L_{\Sigma_{12}}(P) \to A$ defined as $v(\varphi) = (v_1(\varphi), v_2(\varphi))$ is a valuation over $\mathbb{M}_1 \star \mathbb{M}_2$ satisfying exactly the formulas in T. $\square$

Next, we will see what can be done when given Nmatrices that are not saturated to start with.

The ω-Power Operation on (N)matrices. Given a (N)matrix $\mathbb{M} = \langle A, \cdot_{\mathbb{M}}, D \rangle$ over signature Σ, the ω-power of $\mathbb{M}$ is the (N)matrix $\mathbb{M}^\omega = \langle A^\omega, \cdot_\omega, D^\omega \rangle$, where $A^\omega = \{\alpha : \langle \alpha(n) \in A \rangle_{n<\omega}\}$, $D^\omega = \{\alpha : \langle \alpha(n) \in D \rangle_{n<\omega}\}$ and, for each $c \in \Sigma$,

$$c_\omega(\alpha_1, \ldots, \alpha_k) = \{\alpha \in A^\omega : \alpha(n) \in c_{\mathbb{M}}(\alpha_1(n), \ldots, \alpha_k(n)) \text{ for all } n < \omega\}.$$

It is important to stress that the ω-power of a deterministic matrix is still deterministic, which can be simply observed from the definition. The deterministic case has already been considered, namely in [13, Prop. 4.73], where it is shown that powers of a deterministic matrix yield the same logic. We generalize that result, and show that this simple operation actually does what we need. It transforms any Nmatrix in an equivalent saturated Nmatrix.

Theorem 2. *For every non-empty Nmatrix $\mathbb{M}$, $\mathbb{M}^\omega$ is saturated and $\mathcal{L}_{\mathbb{M}} = \mathcal{L}_{\mathbb{M}^\omega}$.*

Proof. First we show that $\mathbb{M}^\omega$ is $\mathcal{L}_{\mathbb{M}}$-sound. Let $\Gamma \vdash_{\mathbb{M}} \varphi$ and v be a valuation over $\mathbb{M}^\omega$ such that $v(\Gamma) \subseteq D^\omega$. Then, $n < \omega$ we have that $v(\Gamma)(n) \subseteq D$. Hence, by $\mathcal{L}_{\mathbb{M}}$-soundness of $\mathbb{M}$, for all $n < \omega$, $v(\varphi)(n) \in D$, hence $v(\varphi) \in D^\omega$.

To see that $\mathcal{L}_{\mathbb{M}} = \mathcal{L}_{\mathbb{M}^\omega}$ and $\mathbb{M}^\omega$ is saturated, it is enough to show that, for any consistent $\mathcal{L}_{\mathbb{M}}$-theory T, there is v such that $v(\psi) \in D$ if and only if $\psi \in T$. Let e be an enumeration of the formulas not in T (a surjective function $e : \omega \to L_\Sigma(P) \setminus T$). We can pick for each $\psi \notin T$, a valuation over $\mathbb{M}$, v_ψ, such that $v_\psi(T) \subseteq D$ and $v(\psi) \notin D$. Let v be the valuation over $\mathbb{M}^\omega$ defined as $v(\varphi) = \{\alpha : \alpha(n) = v_{e(n)}(\varphi)\}$. Of course, $v(T)(n) \in D$ for every $n < \omega$, and so $v(T) \subseteq D^\omega$. It is clear that for $\psi \notin T$ and for n such that $e(n) = \psi$, we have that $v(\psi)(n) \notin D$, hence, since e is an enumeration of Δ we have that $v(\psi) \notin D^\omega$. $\square$

Unfortunately, the ω-power operation is not sensitive to whether the input Nmatrix is itself saturated or not. In case the input Nmatrix is saturated, its ω-power will nevertheless have a larger cardinality, and thus the operation is not idempotent. We will further discuss this question in the concluding section.

At this point, it is worth illustrating the ω-power construction.

Example 8. We revisit the fibring of the conjunction and disjunction fragments of classical logic from Example 7. As the matrix $\mathbb{M}_{djn} = \langle \{0,1\}, \tilde{\cdot}, \{1\} \rangle$ is not saturated, we shall consider its ω-power, defined as the matrix

$$\mathbb{M}_{djn}^\omega = \langle \{0,1\}^\omega, \tilde{\cdot}, \{1\}^\omega \rangle \text{ where } \alpha \hat{\tilde{\vee}} \beta = \langle \alpha(n) \tilde{\vee} \beta(n) \rangle_{n<\omega}.$$

Letting $\mathbf{1} = \{1\}^\omega$, $\hat{\mathbf{0}} = \{0,1\}^\omega \setminus \mathbf{1}$, allows us to present

$$\mathbb{M}_{cnj} \star \mathbb{M}_{djn}^\omega = \langle \{0\} \times \hat{\mathbf{0}} \cup \{1\} \times \mathbf{1}, \overline{\tilde{\cdot}}, \{1\} \times \mathbf{1} \rangle$$

succinctly as follows

$$(a, \alpha) \overline{\wedge} (b, \beta) = \{(a \tilde{\wedge} b, \eta) : \eta \in \mathbf{1} \text{ if } a \tilde{\wedge} b = 1 \text{ and } \eta \in \hat{\mathbf{0}} \text{ otherwise}\}$$

and

$$(a, \alpha) \overline{\vee} (b, \beta) = (c, \alpha \hat{\tilde{\vee}} \beta) \text{ where } c = \begin{cases} 1 & \text{if } \alpha \hat{\tilde{\vee}} \beta = 1^\omega \\ 0 & \text{otherwise} \end{cases}.$$

Note that $\hat{\vee}$ is deterministic, as $\check{\vee}$ is too. Moreover, as $\mathbf{1}$ is a singleton, the non-determinism of $\mathbb{M}_{\mathsf{cnj}} \star \mathbb{M}_{\mathsf{djn}}^{\omega}$ is concentrated only on conjunctions involving the undesignated values in $\{0\} \times \hat{\mathbf{0}}$. By invoking Theorems 1 and 2, we can conclude that $\mathcal{L}_{\mathsf{cnj}} \bullet \mathcal{L}_{\mathsf{djn}} = \mathcal{L}_{\mathbb{M}_{\mathsf{cnj}} \star \mathbb{M}_{\mathsf{djn}}^{\omega}}$. $\quad\triangle$

A wrap up of the previous results offers us a saturated semantics for disjoint fibring of logics of single Nmatrices in all cases.

Corollary 1. *Let Σ_1 and Σ_2 be signatures with $\Sigma_1 \cap \Sigma_2 = \emptyset$. Given non-trivial Nmatrices $\mathbb{M}_1 = \langle A_1, \cdot_1, D_1 \rangle$ over Σ_1 and $\mathbb{M}_2 = \langle A_2, \cdot_2, D_2 \rangle$ over Σ_2, we have that $\mathcal{L}_{\mathbb{M}_1} \bullet \mathcal{L}_{\mathbb{M}_2} = \mathcal{L}_{\mathbb{M}_1^{\omega} \star \mathbb{M}_2^{\omega}}$. Furthermore, $\mathbb{M}_1^{\omega} \star \mathbb{M}_2^{\omega}$ is saturated.*

We conclude this section with another illustrating example: the combination of the negation and disjunction fragments of intuitionistic logic.

Example 9. Let us consider the negation-only fragment of intuitionistic logic $\mathcal{L}_{\mathsf{ineg}}$. It is not hard see that $\mathcal{L}_{\mathsf{ineg}} = \mathcal{L}_{\mathcal{H}_{\mathsf{ineg}}} = \mathcal{L}_{\mathbb{M}_{\mathsf{ineg}}}$ where Σ_{ineg} has a single unary connective $\neg$, $\mathcal{H}_{\mathsf{ineg}} = \langle \Sigma_{\mathsf{ineg}}, \{\frac{p \; \neg p}{q}, \frac{p}{\neg\neg p}, \frac{\neg\neg\neg p}{\neg p}\} \rangle$, and $\mathbb{M}_{\mathsf{ineg}} = \{\{0, \frac{1}{2}, 1\}, \tilde{\neg}, \{1\}\}$ is the reduct of the 3-valued Gödel algebra $\mathbb{G}_3$ (see [17]), defined by

$$\begin{array}{c|ccc} \tilde{\neg} & 0 & \frac{1}{2} & 1 \\ \hline & 1 & 0 & 0 \end{array}$$

Let also $\mathcal{L}_{\mathsf{djn}} = \langle \Sigma_{\mathsf{djn}}, \vdash_{\mathsf{djn}} \rangle$ be the disjunction fragment of intuitionistic logic considered in Example 3.

We want to provide a semantics to $\mathcal{L}_{\mathsf{ineg}} \bullet \mathcal{L}_{\mathsf{djn}} = \langle \Sigma, \vdash \rangle$. We already know that $\mathbb{M}_{\mathsf{djn}}$ is not saturated, and we have presented its ω-power in Example 8. It is not difficult to see that $\mathbb{M}_{\mathsf{ineg}}$ also fails to be saturated. Clearly $\not\vdash \neg p \vee \neg\neg p$ since $\mathcal{L}_{\mathsf{ineg}} \bullet \mathcal{L}_{\mathsf{djn}}$ is, by definition, not stronger than the negation-disjunction fragment of intuitionistic logic. However, it is not possible to define a single valuation over the matrix $\mathbb{M}_{\mathsf{ineg}}$ that gives undesignated elements to both $\neg p$ and $\neg\neg p$ (which would be crucial in order to have a strict product semantics that would not entail $\neg p \vee \neg\neg p$).

The ω-power of the matrix is $\mathbb{M}_{\mathsf{ineg}}^{\omega} = \langle \check{\mathbf{0}} \cup \mathbf{1}, \tilde{\neg}, \{1\}^{\omega} \rangle$, where $\mathbf{1} = \{1\}^{\omega}$ and $\check{\mathbf{0}} = \{0, \frac{1}{2}, 1\}^{\omega} \setminus \mathbf{1}$, with

$$\tilde{\neg} \alpha = \langle \tilde{\neg} \alpha(n) \rangle_{n < \omega}$$

According to Corollary 1, the fibred logic $\mathcal{L}_{\mathsf{ineg}} \bullet \mathcal{L}_{\mathsf{djn}}$ is characterized by the Nmatrix $\mathbb{M}_{\mathsf{ineg}}^{\omega} \star \mathbb{M}_{\mathsf{djn}}^{\omega} = \langle \check{\mathbf{0}} \times \hat{\mathbf{0}} \cup \mathbf{1} \times \mathbf{1}, \overline{\neg\cdot\cdot}, \mathbf{1} \times \mathbf{1} \rangle$, where $\hat{\mathbf{0}} = \{0, 1\}^{\omega} \setminus \mathbf{1}$, and which can be succintly presented as follows

$$\overline{\neg}((\alpha, \beta)) = \{(\tilde{\neg}\alpha, \eta) : \eta \in \mathbf{1} \text{ if } \tilde{\neg}\alpha = 1^{\omega} \text{ and } \eta \in \hat{\mathbf{0}} \text{ otherwise}\}$$

and

$$(\alpha_1, \beta_1) \overline{\vee} (\alpha_2, \beta_2)) = \{(\eta, \alpha\hat{\vee}\beta) : \eta \in \mathbf{1} \text{ if } \alpha\hat{\vee}\beta = 1^{\omega} \text{ and } \eta \in \check{\mathbf{0}} \text{ otherwise}\}.$$

Non-determinism is more spread out here than in the previous example. $\quad\triangle$

4.1   Matryoshkas, Decidability and Complexity

Given an Nmatrix $\mathbb{M} = \langle A, \tilde{\cdot}, D \rangle$, it is clear that its ω-power $\mathbb{M}^\omega$ can be seen as the limit of a succession of matryoshka-style finite powers $\mathbb{M}^n$ for $n < \omega$. Each $\mathbb{M}^n$ can be understood as the sub-Nmatrix of $\mathbb{M}^\omega$ generated by the elements $\alpha \in A^\omega$ such that $\alpha(m) = \alpha(n)$ for all $m > n$. Clearly, $\mathbb{M}^n$ is finite whenever $\mathbb{M}$ is. Moreover, this is indeed a matryoshka-style sequence in the sense that $\mathbb{M}^n$ is a sub-Nmatrix $\mathbb{M}^k$ for $n < k$.

The next result takes advantage of these finite power matrices to provide finite counterexamples in the fibred logic (a form of finite model property).

Lemma 1. *Let Σ_1 and Σ_2 be signatures with $\Sigma_1 \cap \Sigma_2 = \emptyset$. Given non-trivial Nmatrices $\mathbb{M}_1 = \langle A_1, \cdot_1, D_1 \rangle$ over Σ_1 and $\mathbb{M}_2 = \langle A_2, \cdot_2, D_2 \rangle$ over Σ_2, for all finite $\Gamma \cup \{\varphi\} \subseteq L_{\Sigma_{12}}(P)$ we have that $\Gamma \vdash_{12} \varphi$ if and only if $\Gamma \models_{\mathbb{M}_1^n \star \mathbb{M}_2^n} \varphi$ with $n = \mathsf{size}(\Gamma \cup \{\varphi\})$.*

Proof. Soundness is obvious since $\mathbb{M}_1^n \star \mathbb{M}_2^n$ is a generated sub-Nmatrix of $\mathbb{M}_1^\omega \star \mathbb{M}_2^\omega$. Conversely, if $\Gamma \nvdash_{12} \varphi$ then we know that $\Gamma \nvdash_i \varphi$ for both $i = 1, 2$. According to Remark 1, we can guarantee the existence of a valuation over $\mathbb{M}_1^n \star \mathbb{M}_2^n$ by providing compatible partial valuations v_1 and v_2 over $(\mathbb{M}_1^n)^{\Sigma_{12}}$ and $(\mathbb{M}_2^n)^{\Sigma_{12}}$ defined over $\mathsf{sub}(\Gamma \cup \{\varphi\})$. The reasoning in Theorem 1 can be adapted to obtain v_1 and v_2 in these conditions such that $v_i(\psi) \in D_i^n$ if and only $\Gamma \vdash_{12} \psi$ for every $\psi \in \Gamma \cup \{\varphi\}$. $\square$

It is well known that deciding finite inferences in the logic defined by a finite Nmatrix is decidable, and in *coNP* as a byproduct of the analyticity of the Nmatrix semantics.

Proposition 2. *Let Σ_1 and Σ_2 be signatures with $\Sigma_1 \cap \Sigma_2 = \emptyset$, and let $\mathbb{M}_1 = \langle A_1, \cdot_1, D_1 \rangle$ over Σ_1 and $\mathbb{M}_2 = \langle A_2, \cdot_2, D_2 \rangle$ over Σ_2 be non-trivial (not necessarily saturated) finite Nmatrices. If $\mathcal{L}_{\mathbb{M}_1} \bullet \mathcal{L}_{\mathbb{M}_2} = \langle \Sigma_{12}, \vdash_{12} \rangle$ then the problem of deciding whether $\Gamma \vdash_{12} \psi$ for finite $\Gamma \cup \{\varphi\} \subseteq L_{\Sigma_{12}}(P)$ is in coNP.*

Proof. Immediate from Lemma 1 and the observation above.

5   Concluding Remarks

We have provided the first meaningful and useful semantics for disjoint fibring using non-deterministic logical matrices. The crucial operation of strict product cannot, in general, be finitely mimicked using logical matrices. Our characterization also takes advantage of the ω-power construction, which enables enriching a (N)matrix with enough valuations to characterize exactly every theory of the logic.

There is a wealth of directions in which this work can be further pursued.

First and foremost, we envisage a semantics for fibring in general, beyond the disjoint case. This will be a landmark of a long line of research. Namely, we aim at extending the strict product operation to so-called *PNmatrices* [3] (where the

P stands for *partiality*). PNmatrices are a smooth generalization of Nmatrices where the set of possible values yielded by applying a connective to other values is allowed to be empty.

A rigorous categorial study of the strict product operation is also in hand, in the spirit of our original perspective of characterizing fibring as a universal construction. For the moment, just note that the operation was baptized here already having in mind that the useful notion of homomorphism will have to be *strict* (i.e., preserving both designated and undesignated values). A word is due here with respect to the juxtaposition semantics for fibring proposed in [25]. Despite the various problems identified before, the idea underlying juxtaposition is very naturally captured by our strict product operation.

The ω-power operation allows us to close for intersections the set of valuations defined by a given Nmatrix. In a technical sense, this saturation mechanism can be understood as a practical version of the fullness requirements from [30], and others. It is worth taking a closer look at finite powers too, as they relate to finitary forms of saturation. Finite saturation is in general not preserved by fibring, but can be useful in practice as illustrated by our complexity result.

Our ω-power construction is by no means the unique way to achieve saturation, and it is certainly worth exploring alternatives, namely taking advantage of some possible trade-off between non-determinism (which it does not use) and a more manageable cardinality. Overall, we also want to generalize these results to logics defined by classes of Nmatrices, instead of a single Nmatrix. This will require, in particular, a more sophisticated saturation process.

Saturation is interesting in itself, and deeply connected with the multiple-conclusion approach to logic [28]. Indeed, the need for saturation and its preservation by the strict product operation highlights the fact that fibring indeed corresponds to combining the smallest multiple-conclusion logics associated to the logics given. This analysis also paves the way for a study of the analogous fibring operation for multiple-conclusion logics.

Finally, we expect that a thorough understanding of fibring, even beyond the propositional case, shall be able to contribute to a modular semantics for Hilbert calculi that may bring a better understanding of their proof-theory. Learning from the many interesting results about the semantics of sequent calculi [1–3,5] will certainly play an important role. Another fundamental corpus of knowledge that needs to be well studied and brought to this effort is the long track of work on multialgebras, e.g. on the quotient representation theorem by Grätzer [18].

References

1. Avron, A., Lev, I.: Non-deterministic multiple-valued structures. J. Logic Comput. **15**(3), 241–261 (2005)
2. Avron, A., Zamansky, A.: Non-deterministic semantics for logical systems. In: Gabbay, D.M., Guenthner, F. (eds.) Handbook of Philosophical Logic. Handbook of Philosophical Logic, vol. 16, pp. 227–304. Springer, Netherlands (2011). doi:10. 1007/978-94-007-0479-4_4

3. Baaz, M., Lahav, O., Zamansky, A.: Finite-valued semantics for canonical labelled calculi. J. Autom. Reasoning **51**(4), 401–430 (2013)
4. Béziau, J.-Y.: The challenge of combining logics. Logic J. IGPL **19**(4), 543 (2011)
5. Bongini, M., Ciabattoni, A., Montagna, F.: Proof search and co-np completeness for many-valued logics. Fuzzy Sets Syst. **292**, 130–149 (2016)
6. Marcelino, S., Caleiro, C., Rivieccio, U.: Characterizing finite-valuedness. Technical report, SQIG - Instituto de Telecomunicações and IST - U Lisboa, Portugal (2017). Submitted for publication, 2017. http://sqig.math.ist.utl.pt/pub/CaleiroC/17-CMR-finval.pdf
7. Caleiro, C., Ramos, J.: From fibring to cryptofibring: a solution to the collapsing problem. Logica Univers. **1**(1), 71–92 (2007)
8. Caleiro, C., Sernadas, A.: Fibring logics. In: Béziau, J.-Y. (ed.) Universal Logic: An Anthology (From Paul Hertz to Dov Gabbay), pp. 389–396. Birkhäuser (2012)
9. Coniglio, M., Sernadas, A., Sernadas, C.: Preservation by fibring of the finite model property. J. Logic Comput. **21**(2), 375–402 (2011)
10. Corsini, P., Leoreanu, V.: Applications of Hyperstructure Theory. Advances in Mathematics. Springer, Heidelberg (2009)
11. Crawford, J.M., Etherington, D.W.: A non-deterministic semantics for tractable inference. In: Mostow, J., Rich, C. (eds.) Proceedings of the Fifteenth National Conference on Artificial Intelligence and Tenth Innovative Applications of Artificial Intelligence Conference, AAAI 98, IAAI 98, 26–30 July 1998, Madison, Wisconsin, USA, pp. 286–291. AAAI Press/The MIT Press (1998)
12. Fine, K., Schurz, G.: Transfer theorems for stratified multimodal logic. In: Copeland, J. (ed.) Logic and Reality: Proceedings of the Arthur Prior Memorial Conference, pp. 169–123. Cambridge University Press (1996)
13. Font, J.M.: Abstract Algebraic Logic. An Introductory Textbook. College Publications, London (2016)
14. FroCoS. The International Symposium on Frontiers of Combining Systems. http://frocos.cs.uiowa.edu
15. Gabbay, D.: Fibred semantics and the weaving of logics part 1: modal and intuitionistic logics. J. Symbolic Logic **61**(4), 1057–1120 (1996)
16. Gabbay, D.: Fibring Logics. Oxford Logic Guides, vol. 38. Clarendon Press, Wotton-under-Edge (1999)
17. Gottwald, S.: A Treatise on Many-Valued Logics. Studies in Logic and Computation. Research Studies Press, Baldock (2001)
18. Grätzer, G.: A representation theorem for multi-algebras. Arch. Math. **13**, 452–456 (1962)
19. Humberstone, L.: Béziau on And and Or, pp. 283–307. Springer International Publishing, Cham (2015)
20. Kracht, M., Wolter, F.: Properties of independently axiomatizable bimodal logics. J. Symbolic Logic **56**(4), 1469–1485 (1991)
21. Marcelino, S., Caleiro, C.: Decidability and complexity of fibred logics without shared connectives. Logic J. IGPL **24**(5), 673–707 (2016)
22. Marcelino, S., Caleiro, C.: On the characterization of fibred logics, with applications to conservativity and finite-valuedness. Journal of Logic and Computation (2016). https://doi.org/10.1093/logcom/exw023
23. Marty, F.: Sur une generalization de la notion de group. In Proceedings of the 8th Congres des Mathematiciens Scandinave, pp. 45–49 (1934)
24. Rautenberg, W.: 2-element matrices. Stud. Logica. **40**(4), 315–353 (1981)
25. Schechter, J.: JUXTAPOSITION: a new way to combine logics. Rev. Symbolic Logic **4**, 560–606 (2011)

26. Sernadas, A., Sernadas, C., Caleiro, C.: Fibring of logics as a categorial construction. J. Logic Comput. **9**(2), 149–179 (1999)
27. Sernadas, A., Sernadas, C., Rasga, J., Coniglio, M.: On graph-theoretic fibring of logics. J. Log. Comput. **19**(6), 1321–1357 (2009)
28. Shoesmith, D., Smiley, T.: Multiple-Conclusion Logic. Cambridge University Press, Cambridge (1978)
29. Wójcicki, R.: Theory of Logical Calculi. Kluwer, Dordrecht (1988)
30. Zanardo, A., Sernadas, A., Sernadas, C.: Fibring: completeness preservation. J. Symbolic Logic **66**, 414–439 (2001)

Generalized Relations in Linguistics and Cognition

Bob Coecke, Fabrizio Genovese, Martha Lewis, and Dan Marsden$^{(\boxtimes)}$

Department of Computer Science, University of Oxford, Oxford, UK
{bob.coecke,fabrizio.genovese,martha.lewis,
daniel.marsden}@cs.ox.ac.uk

Abstract. Categorical compositional models of natural language exploit grammatical structure to calculate the meaning of sentences from the meanings of individual words. This approach outperforms conventional techniques for some standard NLP tasks. More recently, similar compositional techniques have been applied to conceptual space models of cognition.

Compact closed categories, particularly the category of finite dimensional vector spaces, have been the most common setting for categorical compositional models. When addressing a new problem domain, such as conceptual space models of meaning, a key problem is finding a compact closed category that captures the features of interest.

We propose categories of generalized relations as source of new, practical models for cognition and NLP. We demonstrate using detailed examples that phenomena such as fuzziness, metrics, convexity, semantic ambiguity and meaning that varies with context can all be described by relational models. Crucially, by exploiting a technical framework described in previous work of the authors, we also show how we can combine multiple features into a single model, providing a flexible family of new categories for categorical compositional modelling.

1 Introduction

Distributional models of language describe the meaning of a word using co-occurrence statistics derived from corpus data. A central question with these models is how to combine meanings of individual words, in order to understand phrases and sentences. Categorical compositional models of natural language [15] address this problem, providing a principled approach to combining the meanings of words to form the meanings of sentences, by exploiting their grammatical structure. They also outperform conventional techniques for some standard NLP tasks [23,29].

Distributional models of language can be thought of as "process theories" [16] A process theory consists of a graphical language for reasoning about composite systems of abstract processes, and a categorical semantics modelling the application domain. A particularly important class of categorical models are the compact closed categories, which come equipped with an elegant graphical calculus. Process theoretic models built upon compact closed categories have been

© Springer-Verlag GmbH Germany 2017
J. Kennedy and R.J.G.B. de Queiroz (Eds.): WoLLIC 2017, LNCS 10388, pp. 256–270, 2017.
DOI: 10.1007/978-3-662-55386-2_18

successfully exploited in many application areas, including quantum computation [1], signal flow graphs [11], control theory [2], Markov processes [4], electrical circuits [3] and even linear algebra [43].

Recently [9], the categorical compositional approach to meaning has been applied to the conceptual space models of human cognition introduced in [21,22]. When addressing a new application domain, it is necessary to identify a compact closed category with mathematical structure compatible with the application phenomena of interest.

Amongst the compact closed categories the hypergraph categories [20] are a particularly well behaved class of practical interest. In [33] we presented a flexible parameterized mathematical framework for constructing hypergraph categories. We view this framework as a practical tool for building new models in a principled manner, by varying the parameter choices according to the needs of the application domain. These models are based upon generalizing the well understood notion of a binary relation, providing a concrete and intuitive setting for model development.

In the present work we demonstrate, via extensive examples, that categories of generalized relations present an attractive setting for constructing new models of language and cognition. We emphasize the intuitive interpretation of the models under construction, and their connections to concrete ideas in computation, NLP and further afield. These examples are structured as follows:

- In Sect. 3 we introduce relations with generalized truth values, and exploit them to model features such as distances, forces, connectivity and fuzziness. Relations with generalized truth values are well known in the mathematical community, but seem to have received little attention from the perspective of compositional semantics, with the recent exception of [19].
- In Sect. 4 we generalize relations in another direction, considering relations that respect algebraic structure. These relations can capture features such as convexity, which is important in conceptual spaces models [21,22]. In this case, we recover a model first used in [9], originally constructed in an ad-hoc manner using techniques from monad theory and the theory of regular categories. Importantly, we then show that we can combine generalized truth values with relations respecting algebraic structure, providing conceptual space models with access to distance measures.
- In Sect. 5 we view spans as generalized "proof aware" relations in which the apex of the span contains witnesses to relatedness between the domain and codomain. Spans can be extended to support generalized truth values, and to respect algebraic structure. Exploiting a combination of these features, we construct a new model of semantic ambiguity in conceptual space models of natural language, in which different proof witnesses allow us to vary how strongly different words are related, depending on how they are interpreted.
- The previous examples were essentially built upon the category of sets. Our techniques can be applied with different choices of ambient topos. In Sect. 6, as a practical example of this feature, we use presheaf toposes to build models

in which meanings can vary with context, such as the progress of time or states of the world.

All of our models are preorder enriched, providing a natural candidate for modelling semantic entailment or hyponymy [5,6]. Preorder enrichment also means we can consider internal monads within our various categories of relations. We emphasize the importance of these internal monads throughout our discussions. They provide access to important structured objects such as preorders, generalized metric spaces and ultrametric spaces, and similar well behaved relationships when we combine various modelling features.

2   Compositional Models of Meaning

The grammatical structure of natural language can be modelled using Lambek's pregroup grammars [31].

Definition 1. *A* **pregroup** *is a tuple* $(X, \cdot, 1, (-)^l, (-)^r, \leq)$ *where* $(X, \cdot, 1, \leq)$ *is a partially ordered monoid, or pomonoid, and* $(-)^r, (-)^l$ *are unary functions of type* $X \to X$ *such that for all* $x \in X$ *the following conditions hold,*

$$1 \leq x \cdot x^l \qquad x^l \cdot x \leq 1 \qquad 1 \leq x^r \cdot x \qquad x \cdot x^r \leq 1$$

We say that x **reduces** *to* x' *if* $x \leq x'$.

A grammar is typically described using the free pregroup over some set of basic types. For example, we may consider the free pregroup of the set $\{n, s\}$ where n and s are basic types for nouns and sentences respectively. More complex terms are then built up using the algebraic operations, for example the type of a transitive verb is $n^r s n^l$. We can calculate the type of a phrase by composing the types of the individual terms using the monoid multiplication. For example, the phrase "mice eat cheese" has type $n(n^r s n^l)n$. A composite term is then a well typed sentence if its type reduces to the sentence type. For example:

$$n(n^r s n^l)n = (nn^r)s(n^l n) \leq s(n^l n) \leq s$$

and so "mice eat cheese" is a well typed sentence. In this way, pregroups give us access to the *compositional* features of language.

On the other hand, distributional models [40] of the *meaning* of words in natural language are built using vector space models automatically derived from co-occurrence statistics in a large corpus of text. The key observation of the categorical compositional approach to natural language is that both pregroups and the category of finite dimensional real vector spaces carry the same categorical structure, that of an autonomous category.

Definition 2. *A monoidal category* $\mathcal{V}$ *has* **left/right duals** *if every object has an internal left/right adjoint when* $\mathcal{V}$ *is regarded as a one object bicategory. An* **autonomous category** *is a monoidal category in which every object has both left and right duals. A* **compact closed category** *is a symmetric monoidal category in which every object has right duals.*

A straightforward application oriented introduction to monoidal categories and compact closed categories can be found in [17].

This observation can be exploited to derive the meanings of sentences from the meanings of words. We fix a strong monoidal functor from a pregroup describing grammatical structure to the category of finite dimensional vector spaces. This functor maps type reductions to linear maps, allowing us to automatically derive the meaning of a sentence from its constituent parts. Clearly, this approach can be seen as an instance of functorial semantics. By varying the domain and preserved structure we can consider different categorial grammars [14]. By varying the codomain we can consider different models, as has been important in recent work broadening the scope to mathematical models of cognition [9,10]. When varying the category of meanings, it is desirable to remain within the domain of compact closed categories, in order to exploit connections with previous linguistic developments, and to retain access to their powerful graphical calculus.

The question then becomes: *How can we find or construct compact closed categories with desirable mathematical properties?* This is the question we explore in this paper. In fact, our constructions produce a subclass of compact closed categories, referred to as hypergraph categories [20,30], and so this is where we shall focus our attention.

Definition 3. *A **hypergraph category** is a symmetric monoidal category in which every object is equipped with a choice of special commutative Frobenius algebra, coherently with the monoidal structure.*

Details of the notion of a Frobenius algebra, and linguistic applications including modelling relative pronouns can be found in [38,39]. If I is the monoidal unit, we will occasionally refer to morphisms of types $I \to X$ and $X \to I$ as the **states** and **effects** of X. Morphisms of type $I \to I$ are referred to as **numbers**.

Example 1. The category **Rel** of sets and binary relations between them can be given the structure of a hypergraph category. The monoidal structure is given by forming Cartesian products of sets. A state of a set X is a subset of X and the numbers are the Boolean truth values. The Frobenius algebra is given by the copying relation $x \sim (x,x) : X \to X \times X$, the deletion relation $x \sim * : X \to I$, and their converses.

All the compact closed categories discussed in this paper will be hypergraph categories, generalizing Example 1 along different axes of variation.

3   Generalized Truth Values

A binary relation $R : A \to B$ between sets can be identified with a characteristic function of type $A \times B \to \{\top, \bot\}$ mapping the related pairs of elements to $\top$. It is fruitful to consider generalizing the codomain of such characteristic functions to a set Q, thought of as a collection of truth values. We can then consider functions of the form $A \times B \to Q$ as generalized relations, with truth values in Q. In order for the corresponding binary relations to have satisfactory notions of identities and composition, the set Q must carry the structure of a quantale.

Definition 4 (Quantale). *A **quantale** is a join complete partial order Q with a monoid structure $(\otimes, k)$ satisfying the following distributivity axioms, for all $a, b \in Q$ and $A, B \subseteq Q$:*

$$a \otimes \left[\bigvee B\right] = \bigvee \{a \otimes b \mid b \in B\} \qquad \left[\bigvee A\right] \otimes b = \bigvee \{a \otimes b \mid a \in A\}$$

*A quantale is said to be **commutative** if its monoid structure is commutative.*

All the quantales encountered in this paper will be commutative. We introduce some examples of importance in later developments.

Example 2. The **Boolean quantale** is given by the two element complete Boolean algebra $\mathbf{B} = \{\top, \bot\}$, with the join and multiplication given by the join and meet in the Boolean algebra.

Example 3. The **Lawvere quantale L** is given by the chain $[0, \infty]$ of extended positive reals with the *reverse* ordering, hence minima in $[0, \infty]$ provide the joins of the quantale, and the monoid structure is given by addition.

Example 4. The quantale **F** has again the extended positive reals with reverse order as its partial order, but now with max as the monoid multiplication.

Example 5. The **interval quantale I** is given by the ordered interval $[0, 1]$ with minima as the monoid structure.

For a quantale Q, the Q-relations form a category $\mathbf{Rel}(Q)$ with composition and identities[1]

$$(S \circ R)(a, c) = \bigvee_b R(a, b) \otimes S(b, c) \qquad 1_A(a, b) = \bigvee \{k \mid a = b\}$$

If Q is a commutative quantale, $\mathbf{Rel}(Q)$ carries a symmetric monoidal structure, with the tensor product of objects given by the cartesian product of sets, and the action on relations given for $R : A \to C$ and $S : B \to D$ by

$$(R \otimes S)(a, b, c, d) = R(a, c) \otimes S(b, d)$$

The singleton set is the monoidal unit. A key observation from the perspective of this paper is:

Theorem 1. $\mathbf{Rel}(Q)$ *is compact closed with respect to this monoidal structure.*

Now that we have described how Q-relations compose, we can consider computational interpretations for our example choices of quantale.

---

[1] The slightly unusual formulation of identities·is to avoid definition by cases. This means they can be interpreted in the internal language of an arbitrary topos.

Example 6. The relations over the Lawvere quantale **L** can be thought of as describing costs. The value $R(a, b)$ describes the cost of converting a into b. A cost of 0 means they are maximally related and can be freely inter-converted. A cost of ∞ indicates completely unrelated values, that cannot be converted between each other for finite cost. The value $(S \circ R)(a, c)$ describes the cheapest way of converting a into some b, and then converting that b into c, and adds the associated costs. If we perform two conversions in parallel $(R \otimes R')(a, a', b, b')$ describes the sum of the two individual conversion costs.

In this setting, we can think of a state $I \to A$ as giving a table of costs for acquiring the resources in A, and similarly an effect $A \to I$ is a table of costs for disposing of resources in A.

Example 7. The quantale **F** has the same underlying set as the Lawvere quantale, but its different algebraic structure leads to a very different interpretation. We think of $R(a, b)$ as the peak force required to move a to b. The value given by the composite $(S \circ R)(a, c)$ then describes optimum peak force we will require to move a to c. For example if we can convert a to b with one unit of force, and then move b to c for two units of force, then the peak force required is two units. An alternative procedure converting a to b' for zero units of cost, and then converting b' to c for 2.5 units of cost has a peak cost of 2.5 units, so we would prefer the first procedure to minimize our peak effort. Similarly, the truth value $(R \otimes R')(a, a', b, b')$ gives the peak force required to complete both conversions, assuming these costs are independently incurred.

As with Example 6, we can think of states and effects as tables of acquisition and elimination forces.

Example 8. We can interpret ordinary relations over the Boolean quantale as modelling connectivity. $R(a, b)$ tells us that a is connected to b, composition tells us that we can chain connections together, and the tensor product tells us that we can connect pairs of elements together using a pair of connections between their components. Generalizing to the interval quantale, we now think of $R(a, b)$ as a "connection strength" between a and b. The composite $(S \circ R)(a, c)$ gives the best connection quality that we can achieve in two steps via B. Similarly, the parallel composite $(R \otimes R')(a, a', b, b')$ gives a conservative judgment of the connection quality we can achieve simultaneously between both a and b and a' and b' as the lower of the two individual connection strengths. States describe the "transmission strength" with which signals enter the system from the environment, and effects describe the "reception quality" when consuming output signals.

Alternatively, we could view relations over **I** as fuzzy relations, with states and effects sets with fuzzy membership, and fuzzy predicates. Graded membership is widely used in cognitive science, for example in [8, 18, 24, 25, 37]. Concepts such as 'tall' have no crisp boundary and are better modelled using grades of membership. Although human concept use does not obey fuzzy logic [35], fuzzy relations may prove useful.

Rel(Q) is partial order-enriched if we order relations pointwise with respect to the underlying quantale order. It therefore makes sense to consider internal monads in **Rel**(Q) as interesting "structured objects". An internal monad on an object in a partially ordered category is an endomorphism R satisfying:

$$(R \circ R) \subseteq R, \qquad 1_A \subseteq R \tag{1}$$

Example 9. If we specialize condition (1) to **Rel**(**L**), it is equivalent to:

$$R(a, b) + R(b, c) \geq R(a, c), \qquad 0 = R(a, a)$$

We therefore consider these internal monads as describing *generalized metric spaces*. This observation is important in the field of monoidal topology [26].

As before, we can also interpret our internal monad as giving a well behaved collection of conversion costs between resources. Converting a resource to itself is free, and converting a resource via an intermediate state is at least as expensive as taking the direct route. Similarly, if we consider **Rel**(**F**) the conditions of (1) become:

$$\max(R(a, b), R(b, c)) \geq R(a, c), \qquad 0 = R(a, a)$$

and we can therefore see such internal monads as *generalized ultrametric spaces*. Again, the interpretation in terms of maximum force requirements extends to a sensible interpretation of these axioms.

Example 10. Internal monads in the category of ordinary relations are preorders on their underlying set. The generalization to the interval quantale then gives a fuzzy generalization of the notion of preorder. We can also apply our intuition in terms of connection strengths. Reflexivity tells us that every element can be perfectly connected to itself. Transitivity tell us that the optimal connection strength available is always at least as good as connecting via an intermediate node.

4   Incorporating Convexity

Up to this point, the domain and codomain of our relations have been sets. If we fix an algebraic structure (Σ, E) with set of operations Σ and equations between terms E, we can define a notion of binary relation between these algebras.

Definition 5. *An **algebraic** Q-**relation** of type $A \to B$ is an ordinary Q-relation R between the underlying sets, such that for each operation $\sigma \in \Sigma$ of arity n the following inequation holds in the quantale order:*

$$R(a_1, b_1) \otimes \ldots \otimes R(a_n, b_n) \leq R(\sigma(a_1, \ldots, a_n), \sigma(b_1, \ldots, b_n))$$

As shown in [33], algebraic Q-relations form a hypergraph category:

Theorem 2. *For commutative quantale Q and algebraic signature (Σ, E) there is a hypergraph category* **Rel**$_{(\Sigma, E)}$(Q) *with objects (Σ, E)-algebras and morphisms algebraic Q-relations.*

In the conceptual spaces literature, convexity is conceptually important. In [9] this convexity was captured using relations between convex algebras. We refer to [9] and the extended paper [10] for explicit modelling of toy computations of composed concepts in this category.

These convex algebras can be described as the Eilenberg-Moore algebras of the finite distribution monad. They can in fact be presented by a family Σ_c of binary operations

$$+^p, \quad p \in (0,1)$$

satisfying suitable axioms. We can read $x +^p y$ as "choose x with probability p and y with probability $(1-p)$". By considering algebraic **B**-relations over this signature, we can construct a category isomorphic to the category **ConvexRel** of convex relations from [9]. By changing our quantale of truth values, we can go further than this.

Proposition 1. *In the category of convex* **L**-*relations, the internal monads are generalized metric spaces satisfying the additional axioms for* $p \in (0,1)$:

$$R(a_1, b_1) + R(a_2, b_2) \geq R(a_1 +^p b_1, a_2 +^p b_2)$$

So internal monads in the category of convex relations over the Lawvere quantale are generalized metric spaces that interact well with formation of convex mixtures. The usual distance on $\mathbb{R}^n$ is an example of such a metric.

As shown in [33], every quantale homomorphism $h : Q_1 \to Q_2$ induces a strict monoidal functor of type $\mathbf{Rel}_{(\Sigma,E)}(Q_1) \to \mathbf{Rel}_{(\Sigma,E)}(Q_2)$. If the quantale morphism is injective, this functor is faithful. In particular, the mapping $\bot \mapsto \infty; \top \mapsto 0$ is an injective quantale homomorphism from the Boolean to the Lawvere quantale. This means we can find the ordinary Boolean binary relations as a monoidal subcategory of the category $\mathbf{Rel(L)}$. This presents some flexible modelling possibilities. If U and V are two subsets of a set X, they induce two states $U, V : I \to X$ in $\mathbf{Rel(B)}$. If we consider the number $V° \circ U$, where $R°$ denotes relational converse, it evaluates to true if and only if $U \cap V \neq \emptyset$.

Proposition 2. *If* $U, V \subseteq X$ *and* d *is an internal monad in* $\mathbf{Rel(L)}$, *the composite* $V° \circ d \circ U$ *is the infimum of the distances between elements in* U *and* V.

This gives us the greatest lower bound on the distances between elements in U and V, providing a finer grain measure of similarity than can conventionally be achieved in relational models. We note that as distances are in general asymmetric, the number $U° \circ d \circ V$ may give a different measure of similarity. Similarly, we can find the ordinary Boolean convex relations within the category of **L**-valued convex relations, presenting analogous opportunities for performing calculations with discrete convex relations, and then measuring their separation on a continuum of values.

Such asymmetric distance measures are of practical use in cognitive science applications. A fundamental concept in psychology is that of similarity, which can be used as the basis of concept formation. Similarity between objects or concepts can be explained by locating objects in some sort of conceptual or feature space, and modelling similarity as a function of distance, for example in [42]. However, judgements of similarity are not necessarily symmetric [45]. In one study examining the similarity between pairs of countries, participants are asked to choose between statements 'Country A is similar to country B' or 'Country B is similar to country A'. In all cases, a majority of participants preferred the statement where the latter country was considered more prominent.

5   Proof Relevance

A **span** S of sets, between sets A and B, is a set X and a pair of functions $X \xrightarrow{p_1} A$ and $X \xrightarrow{p_2} B$. Paralleling the notation for relations, we will write

$$S_x(a,b) := x \in X \ \wedge \ p_1(x) = a \ \wedge \ p_2(x) = b$$

We can think of such a span as a *proof relevant relation* in which $S_x(a,b)$ tells us that x witnesses that a and b are related. In a computational linguistics or cognition application where relations may have been derived automatically from data in some way, we can exploit these proof witnesses to track evidence for our beliefs that certain relationships hold.

Sets and spans between them form a hypergraph category **Span** with composition given by pullback, and tensor product induced by a choice of products[2]. In fact, as we did for relations, we can extend these spans with algebraic structure and a choice of truth values in a partially ordered monoid. We no longer require full quantale structure on our truth values, as multiple proof witnesses mean we don't need to choose a single representative truth value when composing relations.

Definition 6. *For an algebraic signature* (Σ, E) *and pomonoid* Q *an* **algebraic** Q-**span** *of type* $A \to B$ *between* (Σ, E)-*algebras is a span* $A \xleftarrow{p_1} X \xrightarrow{p_2} B$ *between the underlying objects, with a* **characteristic morphism** $\chi : X \to Q$. *We require that the algebraic structure is respected in that for all* $\sigma \in \Sigma$, *with arity* n:

$$\bigwedge_{1 \le i \le n} (p_1(x_i) = a_i \wedge p_2(x_i) = b_i) \Rightarrow \bigotimes_{1 \le i \le n} \chi(x_i) \le \chi(\sigma(x_1, ..., x_n))$$

Intuitively, these are intensional relations in which proof witnesses are weighted by a truth value, and the relations respect the algebraic structure. As shown in [33], algebraic Q-spans also form a hypergraph category:

Theorem 3. *For commutative pomonoid* Q *and algebraic signature* (Σ, E) *there is a hypergraph category* **Span**$_{(\Sigma, E)}(Q)$ *with objects* (Σ, E)-*algebras and morphisms algebraic* Q-*spans.*

---

[2] In fact, in order for composition to be associative, it is necessary to work with equivalence classes of spans. It is sufficient to consider representatives, and we do so to avoid distracting technicalities.

For algebraic Q span S we define

$$S_x^q(a,b) := x \in X \ \wedge \ p_1(x) = a \ \wedge \ p_2(x) = b \ \wedge \ \chi(x) = q$$

We then read $S_x^q(a,b)$ as telling us that x witnesses that a and b are related with strength q. In fact, we can order algebraic Q-spans in a manner similar to that for relations, but accounting for proof witnesses.

Definition 7. *For pomonoid Q, we define a preorder on algebraic Q-spans by setting $(X_1, f_1, g_1, \chi_1) \subseteq (X_2, f_2, g_2, \chi_2)$ if there is a* **Set***-monomorphism $\varphi :$ $X_1 \to X_2$ such that $f_1 = f_2 \circ \varphi$, $g_1 = g_2 \circ \varphi$ and $\forall x . \chi_1(x) \leq \chi_2(\varphi(x))$.*

The ordering accounts pointwise for strengths of relatedness in a natural way. The requirement that the function φ in Definition 7 is a monomorphism ensures that even if our truth values are trivial, we take account of the "number" of proof witnesses available.

As internal monads provided interesting objects in the setting of relations, we should consider them in the span setting as well.

Proposition 3. *An internal monad on A in* **Span(L)** *is an* **L**-*span $S : A \to A$ such that if $S_x^p(a_1, a_2)$ and $S_y^q(a_2, a_3)$ we can choose an element $\varphi(x, y)$ of the apex such that $S_{\varphi(x,y)}^r(a_1, a_3)$ and $p + q$ is greater than r in the usual ordering on the real numbers. Furthermore, we can do this in a way such that the assignment φ is injective.*

So internal **L**-span monads further generalize metric spaces to incorporate multiple possible distances, which we can think of as describing different paths between points. We now outline a new practical application of spans in models of language.

Example 11 (Semantic Ambiguity via Spans). In natural language, we often encounter ambiguous situations. For example the word "bank" can refer to either a "river bank" or a "financial bank". A compositional account of semantic ambiguity was presented in [36], using mathematical models of incomplete information from quantum theory. The techniques applied implicitly assume meanings are built upon a vector space model, to which we apply Selinger's CPM construction [41] to yield a new category of ambiguous meanings. The CPM construction can also be applied to categories of relations, but in this case it does not provide a satisfactory model of ambiguity [34].

An alternative approach to ambiguity in relational models is to use spans. We consider how the ambiguous word "bank" is related to the word "water"

- In the "river bank" context, we would expect a strong relationship
- In the "financial bank" context, we would expect a weaker relationship

By using spans rather than relations, we can introduce two different proof witnesses for the different contexts under consideration. By choosing our quantale of truth values to be the Lawvere quantale **L**, we can attach a different choice of distance to each of these choices. As we compose spans to describe the meanings of phrases and sentences, the proof witnesses will keep track of the different possible relationships in play.

6   Variable Contexts

Our definitions of algebraic Q-relations and algebraic Q-spans are constructive. This means that Theorems 2 and 3 continue to hold for any elementary topos, as proved in [33]. Standard sources on topos theory are [12,27,28,32]. We will write $\mathbf{Rel}^{\mathcal{E}}_{(\Sigma, E)}(Q)$ and $\mathbf{Span}^{\mathcal{E}}_{(\Sigma, E)}(Q)$ for the categories of spans and relations, to make the choice of topos $\mathcal{E}$ explicit. This generalization has practical implications if we move to different choices of background topos.

Definition 8. *Let $\mathcal{C}$ be a small category. A* **presheaf** *on $\mathcal{C}$ is a functor of type $\mathcal{C}^{op} \to \mathbf{Set}$. Presheaves and natural transformations between them form a topos, denoted $\mathbf{Set}^{\mathcal{C}^{op}}$. For presheaf X over a preorder, we will write X_i for the set in the image under X of element i of the preorder, and $X_{i,j}$ for the image of $j \leq i$ under X.*

Presheaves can be interpreted as *sets varying with context*. This is exactly the perspective we shall adopt in our examples. To exploit our generalized span construction, we need to describe internal pomonoids in presheaf categories.

Lemma 1. *A commutative partially ordered monoid in a presheaf category $\mathbf{Set}^{\mathcal{C}^{op}}$ is a presheaf Q such that for each $\mathcal{C}$-object x and $\mathcal{C}$-morphism f, $Q(x)$ is a commutative pomonoid and $Q(f)$ is a pomonoid morphism in $\mathbf{Set}$. See [28, D1.2.14].*

Example 12 (Temporal dependence). In Example 11 we modelled ambiguity using multiple proof witnesses to describe different interpretations of words. We now investigate the description of time dependent ambiguous relationships, by exploiting spans over presheaves. To do so, we consider presheaves over the partial order $\mathbb{N} = 0 \leftarrow 1 \leftarrow 2...$ having objects natural numbers. We view these presheaves as sets varying in time. We assume our notion of truth is fixed, and so we will consider $\mathbf{Span}^{\mathbf{Set}^{\mathbb{N}^{op}}}(\mathbf{L})$, where $\mathbf{L}$ is the constant presheaf on the pomonoid underlying the Lawvere quantale. An $\mathbf{L}$-span between presheaves X and Y then consists of natural transformations $p_1 : X \Rightarrow A$ and $p_2 : X \Rightarrow B$, and a characteristic natural transformation $\chi : X \Rightarrow \mathbf{L}$. We see naturality as a consistency condition between the relationships described by proof witnesses, as they move forward in time. As our pomonoid is constant, $\chi_i(x) = \chi_j(X_{i,j}(x))$, so the truth value associated with a proof witness is preserved through time. Intuitively, in this model, a steadily increasing collection of relationships hold over time.

Example 13 (Perspective Dependence). In Example 12, the truth object was fixed in all contexts. We now examine a brief example in which our notion of truth is context dependent. Consider two agents. Agent 0 has a binary view of the world, relationships either hold or they don't. Agent 1 has a richer view incorporating different strengths of relation in the unit interval. Consider presheaves on the category $\mathcal{C}$ with a single non-trivial arrow $0 \leftarrow 1$. We define an internal pomonoid Q with $Q(0) = \mathbf{B}$, $Q(1) = \mathbf{I}$ and $Q_{0,1}$ the canonical pomonoid morphism between the Boolean and interval quantales. Now if we consider a Q-span between constant presheaves A and B with apex an arbitrary presheaf X, we can think of it as follows. Each element of X_0 relates two elements $a \in A$ and $b \in B$ with strength 0 or 1. The structure of X then forces that X_1 contains a witness relating those two elements with the same strength. As X_1 encodes the views of the more powerful agent, it may describe additional relationships, now with strengths weighted in the interval $[0, 1]$.

If we wish to consider algebraic Q-relations over an arbitrary topos things are more delicate since internal quantales cannot be defined pointwise. Nevertheless there are standard sources of internal commutative quantales, for example:

- If $\mathcal{C}$ is a groupoid and Q is a commutative quantale in **Set**, then Q can be lifted to an internal commutative quantale in $\mathbf{Set}^{\mathcal{C}^{op}}$.
- The subobject classifier Ω of a topos is an internal locale, and therefore an internal commutative quantale.

We conclude by establishing the relationship between our framework of generalized relations and the standard notion of the category of relations over a regular category. This will involve the internal locale given by the subobject classifier.

Definition 9. *A category $\mathcal{C}$ is* **regular** *if it is finitely complete, every kernel pair has a coequalizer and regular epimorphisms are stable under pullback.*

There is standard construction of a category of relations $\mathbf{Rel}(\mathcal{C})$ of a regular category $\mathcal{C}$, see for example [13]. For the category **Set** for example, this construction recovers exactly the usual category of binary relations. As we have been constructing categories of relations in this paper, it would be interesting to know how this relates to the relations of a regular category. Every topos is regular, and in fact for any algebraic theory (Σ, E), the category of internal (Σ, E)-algebras in a regular category [7], meaning we can consider the impact of algebraic structure. In fact, the resulting category of relations is equivalent to the one produced by our construction with the subobject classifier as the object of truth values.

Theorem 4. *Let $\mathcal{E}$ be a topos, Ω its subobject classifier and (Σ, E) an algebraic signature. The category $\mathbf{Rel}^{\mathcal{E}}_{(\Sigma, E)}(\Omega)$ resulting from the algebraic Q-relations construction is equivalent to the category of internal relations over the regular category of internal (Σ, E)-algebras in $\mathcal{E}$.*

In this way, we see that relations over suitable regular categories are a special case of our construction.

7   Conclusion

We have demonstrated that categories of generalized relations present a flexible modelling tool for categorical compositional models of natural language and cognition. We presented various potential models worthy of further investigation, capturing features such as fuzziness, distances, convexity, ambiguity and context sensitivity, and showed how these features can be used in combination within a generic framework. One natural direction for further work would be empirical investigation of the compatibility of these theoretical models with concrete applications. Another one would be to investigate whether the techniques in [44] can be used to build models with either non-commutative or typed quantales, known as quantaloids.

Acknowledgments. This work was funded by AFSOR grant "Algorithmic and Logical Aspects when Composing Meanings" and FQXi grant "Categorical Compositional Physics".

References

1. Abramsky, S., Coecke, B.: A categorical semantics of quantum protocols. In: Proceedings of the 19th Annual IEEE Symposium on Logic in Computer Science, pp. 415–425. IEEE (2004)
2. Baez, J.C., Erbele, J.: Categories in control. Theory Appl. Categ. **30**(24), 836–881 (2015)
3. Baez, J.C., Fong, B.: A compositional framework for passive linear networks. arXiv preprint arXiv:1504.05625 (2015)
4. Baez, J.C., Fong, B., Pollard, B.S.: A compositional framework for Markov processes. J. Math. Phys. **57**(3), 033301 (2016)
5. Bankova, D., Coecke, B., Lewis, M., Marsden, D.: Graded entailment for compositional distributional semantics. arXiv preprint arXiv:1601.04908 (2015)
6. Bankova, D.: Comparing meaning in language and cognition - p-hypononymy, concept combination, asymmetric similarity. Master's thesis, University of Oxford (2015)
7. Barr, M.: Exact categories. Exact Categories and Categories of Sheaves. LNM, vol. 236, pp. 1–120. Springer, Heidelberg (1971). doi:10.1007/BFb0058580
8. Barsalou, L.W.: Ideals, central tendency, and frequency of instantiation as determinants of graded structure in categories. J. Exp. Psychol. Learn. Mem. Cogn. **11**(4), 629 (1985)
9. Bolt, J., Coecke, B., Genovese, F., Lewis, M., Marsden, D., Piedeleu, R.: Interacting conceptual spaces. In: Kartsaklis, D., Lewis, M., Rimell, L. (eds.) Proceedings of the 2016 Workshop on Semantic Spaces at the Intersection of NLP, Physics and Cognitive Science, SLPCS@QPL 2016, Glasgow, Scotland, 11 June 2016. EPTCS, vol. 221, pp. 11–19 (2016). http://dx.doi.org/10.4204/EPTCS.221.2
10. Bolt, J., Coecke, B., Genovese, F., Lewis, M., Marsden, D., Piedeleu, R.: Interacting conceptual spaces I: Grammatical composition of concepts. arXiv preprint arXiv:1703.08314 (2017)
11. Bonchi, F., Sobocinski, P., Zanasi, F.: Full abstraction for signal flow graphs. ACM SIGPLAN Not. **50**(1), 515–526 (2015)
12. Borceux, F.: Handbook of Categorical Algebra: Volume 3, Categories of Sheaves. Cambridge University Press, Cambridge (1994)
13. Borceux, F.: Handbook of Categorical Algebra: Volume 2, Categories and Structures, vol. 2. Cambridge University Press, Cambridge (1994)
14. Coecke, B., Grefenstette, E., Sadrzadeh, M.: Lambek vs. Lambek: functorial vector space semantics and string diagrams for lambek calculus. Ann. Pure Appl. Logic **164**(11), 1079–1100 (2013)

15. Coecke, B., Sadrzadeh, M., Clark, S.: Mathematical foundations for distributed compositional model of meaning. Lambek festschrift. Linguist. Anal. **36**, 345–384 (2010)
16. Coecke, B., Kissinger, A.: Picturing Quantum Processes. A First Course in Quantum Theory and Diagrammatic Reasoning. Cambridge University Press (2017, forthcoming)
17. Coecke, B., Paquette, E.O.: Categories for the practising physicist. In: Coecke, B. (ed.) New Structures for Physics, pp. 173–286. Springer, Heidelberg (2010)
18. Dale, R., Kehoe, C., Spivey, M.J.: Graded motor responses in the time course of categorizing atypical exemplars. Mem. Cogn. **35**(1), 15–28 (2007)
19. Dostal, M., Sadrzadeh, M.: Many valued generalised quantifiers for natural language in the DisCoCat model. Technical report, Queen Mary University of London (2016)
20. Fong, B.: The algebra of open and interconnected systems. Ph.D. thesis, University of Oxford (2016)
21. Gärdenfors, P.: Conceptual Spaces: The Geometry of Thought. MIT Press, Cambridge (2004)
22. Gärdenfors, P.: The Geometry of Meaning: Semantics Based on Conceptual Spaces. MIT Press, Cambridge (2014)
23. Grefenstette, E., Sadrzadeh, M.: Experimental support for a categorical compositional distributional model of meaning. In: The 2014 Conference on Empirical Methods on Natural Language Processing, pp. 1394–1404 (2011). arXiv:1106.4058
24. Hampton, J.A.: Disjunction of natural concepts. Mem. Cogn. **16**(6), 579–591 (1988)
25. Hampton, J.A.: Overextension of conjunctive concepts: evidence for a unitary model of concept typicality and class inclusion. J. Exp. Psychol. Learn. Mem. Cogn. **14**(1), 12 (1988)
26. Hofmann, D., Seal, G.J., Tholen, W.: Monoidal Topology: A Categorical Approach to Order, Metric, and Topology, vol. 153. Cambridge University Press, Cambridge (2014)
27. Johnstone, P.T.: Sketches of an Elephant: A Topos Theory Compendium, vol. 1. Oxford University Press, Oxford (2002)
28. Johnstone, P.T.: Sketches of an Elephant: A Topos Theory Compendium, vol. 2. Oxford University Press, Oxford (2002)
29. Kartsaklis, D., Sadrzadeh, M.: Prior disambiguation of word tensors for constructing sentence vectors. In: The 2013 Conference on Empirical Methods on Natural Language Processing, pp. 1590–1601. ACL (2013)
30. Kissinger, A.: Finite matrices are complete for (dagger-)hypergraph categories. arXiv preprint arXiv:1406.5942 (2014)
31. Lambek, J.: Type grammar revisited. In: Lecomte, A., Lamarche, F., Perrier, G. (eds.) LACL 1997. LNCS, vol. 1582, pp. 1–27. Springer, Heidelberg (1999). doi:10.1007/3-540-48975-4_1
32. MacLane, S., Moerdijk, I.: Sheaves in Geometry and Logic: A First Introduction to Topos Theory. Springer Science & Business Media, Heidelberg (2012)
33. Marsden, D., Genovese, F.: Custom hypergraph categories via generalized relations. In: CALCO 2017 (2017, to appear)
34. Marsden, D.: A graph theoretic perspective on CPM(Rel). In: Heunen, C., Selinger, P., Vicary, J. (eds.) Proceedings 12th International Workshop on Quantum Physics and Logic, QPL 2015, Oxford, UK, 15–17 July 2015. EPTCS, vol. 195, pp. 273–284 (2015). http://dx.doi.org/10.4204/EPTCS.195.20
35. Osherson, D.N., Smith, E.E.: Gradedness and conceptual combination. Cognition **12**(3), 299–318 (1982)

36. Piedeleu, R., Kartsaklis, D., Coecke, B., Sadrzadeh, M.: Open system categorical quantum semantics in natural language processing. In: Moss, L.S., Sobocinski, P. (eds.) 6th Conference on Algebra and Coalgebra in Computer Science, CALCO 2015. LIPIcs, vol. 35, pp. 270–289. Schloss Dagstuhl - Leibniz-Zentrum fuer Informatik (2015)

37. Rosch, E., Mervis, C.B.: Family resemblances: studies in the internal structure of categories. Cogn. Psychol. **7**(4), 573–605 (1975)

38. Sadrzadeh, M., Clark, S., Coecke, B.: The Frobenius anatomy of word meanings I: subject and object relative pronouns. J. Logic Comput. **23**(6), ext044 (2013)

39. Sadrzadeh, M., Clark, S., Coecke, B.: The Frobenius anatomy of word meanings II: possessive relative pronouns. J. Logic Comput. **26**(2), exu027 (2014)

40. Schütze, H.: Automatic word sense discrimination. Comput. Linguist. **24**(1), 97–123 (1998)

41. Selinger, P.: Dagger compact closed categories and completely positive maps. Electron. Not. Theor. Comput. Sci. **170**, 139–163 (2007)

42. Shepard, R.N., et al.: Toward a universal law of generalization for psychological science. Science **237**(4820), 1317–1323 (1987)

43. Sobocinski, P.: Graphical linear algebra. Mathematical blog. https://graphicallinearalgebra.net/

44. Stubbe, I.: Categorical structures enriched in a quantaloid: categories and semicategories. Ph.D. thesis, Université Catholique de Louvain (2003)

45. Tversky, A.: Features of similarity. Psychol. Rev. **84**(4), 327 (1977)

Concrete Mathematics. Finitistic Approach to Foundations

Marcin Mostowski[1]([⊠]) and Marek Czarnecki[2]([⊠])

[1] Institute of Philosophy, Department of Logic, Jagiellonian University,
Gołębia 24, Room 24, 31-007 Cracow, Poland
marcin.mostowski@uj.edu.pl

[2] Institute of Philosophy, Department of Logic, Warsaw University,
Krakowskie Przedmieście 3, Room 111, 00-927 Warsaw, Poland
m.czarnecki2@uw.edu.pl

Abstract. We discuss the idea of concrete mathematics inspired by Hilbert's idea of finitistic mathematics as the part of mathematics not engaged into actual infinity. We explicate it as the part of mathematics based on Δ_2^0 arithmetical concepts. The explication is justified by equivalence of Δ_2^0 definability with algorithmic learnability (an epistemic argument) and with FM–representability (representability in finite models, an ontological argument).

We show that the essential part of classical mathematics can be interpreted in the concrete framework. We claim that current mathematics is a social game of proving theorems on some axiomatic set theoretic background. On the other hand, concrete mathematics is the reality on which our mathematical experience is based. This is what makes the game intersubjective. Nevertheless, this game is one of the most efficient methods of building our mathematical knowledge.

Keywords: Concrete mathematics · FM–representability · Mathematical truth · Potential infinity · Actual infinity · Foundations of mathematics

1 Introduction

We discuss the idea of finitistic foundations of mathematics which is partially motivated and inspired by Hilbert's lecture *On the infinite* [13]. According to Hilbert we have finitistic mathematics – the part of mathematics which does not depend on actual infinity[1] and which is well founded in our mathematical experience[2]. The remaining part essentially employing actual infinity was called

This work was funded by the Polish National Science Centre grant number 2013/11/B/HS1/04168.

[1] The distinction between potential and actual infinity is due to Aristotle, see [1].

[2] Another source of inspiration would be Leopold Kronecker's view on foundations of mathematics, see [17] and a few famous remarks elsewhere. Unfortunately Kronecker never gave any systematic presentation of his views on foundations. Nevertheless, they are coherent, and probably they influenced Hilbert's idea.

J. Kennedy and R.J.G.B. de Queiroz (Eds.): WoLLIC 2017, LNCS 10388, pp. 271–280, 2017.
DOI: 10.1007/978-3-662-55386-2_19

by him ideal mathematics. The name ideal refers to the method used in algebra in his times, called the method of ideal elements. The method consists in adding entities to algebraic structures filling some gaps e.g. in a ring without sufficient number of prime elements we can add them by some natural algebraic construction, for the sake of obtaining a version of fundamental theorem of arithmetic for this extended ring. Actual infinity can be considered as a useful fiction which smooths the mathematical universe making it more regular and as it appeared later also essentially easier.

We share Hilbert's view that the finitistic part of mathematics is well defined and independent of any presentation e.g. in an axiomatic way. On the other hand truth for the ideal part of mathematics depends on various assumptions, some of them are contradictory with some others[3].

The cited Hilbert's lecture is one of the early formulations of so called the Hilbert's program – the research program of grounding the ideal part of mathematics by its reduction to the finitistic part. This is not the topic of our considerations. However the Hilbert's program was the most influential research project in foundations of mathematics in the 20th century. Over time the vague notions used by Hilbert have been explicated. In [30] Tait suggested to identify the finitistic part of mathematics with the part of mathematics expressible in the *primitive recursive arithmetic* (PRA). This explication became popular thanks to Friedman and Simpson (see e.g. [28]) and their research program called *reverse mathematics*. Although it is convenient to consider a theory as weak as PRA in the context of reverse mathematics, we believe that Hilbert's finitistic part of mathematics, i.e. the part not involving actual infinity, exceeds PRA. **However, in this paper we do not relate in any way to Tait's work. We do not consider proof theory at all. We rather present a suitable ontology for mathematics without actual infinity.** Therefore we use the term *concrete mathematics* which was introduced in [12] by Knuth et al. for describing the part of mathematics based on computations, algorithms and constructions.

We claim that the concrete mathematics can be determined in terms of computations. On the other hand ideal mathematics is based on axioms and developed by proofs. Finding and verifying proofs is a social game. Axioms are necessary for organizing proofs in a well founded way. Nevertheless axioms can be chosen in many different ways. Currently, the most accepted axiomatic background is ZFC, Zermelo-Fraenkel set theory with the axiom of choice. Traditionally various schools of philosophy of mathematics differentiate by choosing the assumptions for the part of ideal mathematics which are allowed. However all the possible metaphysics of mathematics have a common part which is determined by our computational experience. We claim that it consists essentially with Δ_2^0 notions. In the following section we justify this claim.

---

[3] Contemporarily we know that the ideal part of mathematics is essentially undetermined. However, in 1926 Hilbert was not aware of this fact.

2   The Boundary of Concrete Mathematics

In this section we present two different approaches to setting the limit for our computational experience. Such a limit can be understood as the boundary of concrete mathematics. Notions that surpass this limit cannot be objects of direct mathematical experience and therefore some external assumptions have to be made about them. Such assumptions may vary between different metaphysics of mathematics.

Both considered approaches are related to the Shoenfield limit lemma. Before we can state it, recall the notion of computable approximation of a set.

Definition 1 (Computable approximation). *Let $A \subseteq \mathbb{N}^r$. We say that a function $f \colon \mathbb{N}^{r+1} \longrightarrow \{0,1\}$ computably approximates A if f is total, computable and for all $a_1, \ldots, a_r \in \mathbb{N}$ it holds that:*

- *$\lim_{t \to \infty} f(t, a_1, \ldots, a_r) = 1$ if and only if $A(a_1, \ldots, a_r)$,*
- *$\lim_{t \to \infty} f(t, a_1, \ldots, a_r) = 0$ if and only if $\neg A(a_1, \ldots, a_r)$.*

We say that A is computably approximable if there exists $f \colon \mathbb{N}^{r+1} \longrightarrow \{0,1\}$ which computably approximates A.

The limit lemma establishes equivalence between notions which are computable with recursively enumerable oracles i.e. those whose degrees of unsolvability are $\leqslant \mathbf{0}'$ and notions which are computably approximable.

Theorem 1 (Limit lemma, Shoenfield, [27]). *Let $A \subseteq \mathbb{N}^r$. The following are equivalent:*

- *$\deg(A) \leqslant \mathbf{0}'$,*
- *A is computably approximable.*

The first discussed approach is epistemological. It originates from considerations on how learning processes may look like and what does it mean to learn e.g. a language. This was studied independently by Gold [11] and Putnam [25]. They point that learning processes are computable and that a learner can make mistakes (proceeds by trial and error) until she or he finally fixes on the correct answer. Therefore, the mathematization of algorithmic learnability is basically the same as in the definition of computable approximability. The function f is the computable learning procedure for the notion A. If a learner is learning according to f, then for every instance $(a_1, \ldots, a_r)$ she or he can change mind whether $A(a_1, \ldots, a_r)$ holds or not, but only finitely many times. There is some point in time t_0 and a truth value e such that for all $t \geqslant t_0$ it holds that $f(t, a_1, \ldots, a_r) = f(t_0, a_1, \ldots, a_r)$ – this is the point when the learner has learned that $A(a_1, \ldots, a_r)$ if $e = 1$, and $\neg A(a_1, \ldots, a_r)$ if $e = 0$.

The following theorem shows the limit for complexity of notions which can be algorithmically learned.

Theorem 2 (Algorithmic learnability theorem, Gold [11], Putnam [25]). *Let $A \subseteq \mathbb{N}^r$. The following are equivalent:*

– *A is algorithmically learnable,*
– $A \in \Delta_2^0$.

Therefore we can only learn Δ_2^0 notions. Thus Δ_2^0 notions seem to be the limit of possible computational experience.

Another argument supporting this claim is due to Epstein. In [5], in his survey on degrees of unsolvability, Epstein justifies why we should particularly study notions with degrees of unsolvability $\leqslant 0'$. These notions, by the limit lemma, can be approximated by uniform computable constructions which are *truly constructions, not constructions "relative to".* Epstein claims that they *have the same flavor as arguments in other areas of finite mathematics, such as number theory or graph theory.* He also claims that *the center of the subject* is *the tension between the finite and the infinite.* We think that this *tension* comes from the possibility of representing these notions without actual infinity – by means of potential infinity only. Gauss [6] describes infinity as a *façon de parler* of talking about limits and this is the only infinity available: infinity without real infinity.

We proceed to another approach of searching for the limit of our computational experience. This is an ontological approach by Mostowski, see [20–22] for more detailed presentation. He studies meaningfulness of notions in a world without actual infinity. Imagine that there are only finitely many objects in the world, but their number is not bounded. Every time a new finite batch of objects can be added. This is a potentially infinite world. It is finite in every moment in time, but it also has no bounds. We may identify objects of such a world with natural numbers – understood as a potentially infinite set.

The model of such a world is called an FM–domain. Here we present an FM–domain of the standard model of arithmetic (in a relational vocabulary) $\underline{\mathbb{N}} = (\mathbb{N}, R_+, R_\times)$.

Definition 2 (Standard FM–domain). *For $k \in \mathbb{N}$ let $\underline{\mathbb{N}}_k$ be the finite initial segment of $\underline{\mathbb{N}}$ of size k with the greatest element distinguished i.e.*

$$\underline{\mathbb{N}}_k = (\{0, \ldots, k-1\}, R_+ {\restriction}_{\{0,\ldots,k-1\}}, R_\times {\restriction}_{\{0,\ldots,k-1\}}, k-1).$$

The standard FM–domain is $FM(\underline{\mathbb{N}}) = \{\underline{\mathbb{N}}_k : k = 1, 2, \ldots\}$.

FM–domains serve for models of potentially infinite worlds. We would say that a sentence is true in a potentially infinite world if it is true in all its approximations which are *big enough.* Therefore, we allow it to be false in some finite number of models – we concentrate on its asymptotic truth. Hence, we consider sl–semantics (sl for *sufficiently large*) for FM–domains.

Definition 3 (sl–semantics). *Let $a_1, \ldots, a_r \in \mathbb{N}$ and let $\varphi(x_1, \ldots, x_r)$ be a formula. We say that elements $a_1, \ldots, a_r$ sl–satisfy φ and we write*

$$FM(\underline{\mathbb{N}}) \models_{sl} \varphi[a_1, \ldots, a_r],$$

if it holds that

$$\exists k \in \mathbb{N} \, \forall n \geqslant k \, \underline{\mathbb{N}}_n \models \varphi[a_1, \ldots, a_r].$$

The motivation for considering sl–semantics is that potentially infinite world grows boundlessly. We do not care if some phenomena appear only on some finitely many early stages. We are interested in the properties that fix at some point. We want to characterize the notions that are meaningful in potentially infinite world i.e. those which are representable with respect to sl–semantics for FM–domain in positive and negative cases. Such notions are said to be FM–representable.

Definition 4 (FM–representability). *Let $A \subseteq \mathbb{N}^r$. We say that A is FM–represented by a formula $\varphi(x_1, \ldots, x_r)$ if for every $a_1, \ldots, a_r \in \mathbb{N}$:*

- *$A(a_1, \ldots, a_r)$ if and only if $\mathrm{FM}(\underline{\mathbb{N}}) \models_{sl} \varphi[a_1, \ldots, a_r]$,*
- *$\neg A(a_1, \ldots, a_r)$ if and only if $\mathrm{FM}(\underline{\mathbb{N}}) \models_{sl} \neg\varphi[a_1, \ldots, a_r]$.*

We say that A is FM–representable if there exists a formula φ which FM–represents A.

Theorem 3 (FM–representability theorem, Mostowski, [19]).
Let $A \subseteq \mathbb{N}^r$. The following are equivalent:

- *A is FM–representable,*
- *$A \in \Delta_2^0$.*

By the FM–representability theorem notions which are meaningful in a world without actual infinity are exactly those which are Δ_2^0. Starting with another essentially different approach we – again – get Δ_2^0 notions as a distinguished class.

What all of the approaches described here have in common is considering limits. Algorithmic learning is basically learning in the limit and sl–semantics are semantics in the limit. But these are the limits of potentially infinite processes and therefore they belong to concrete mathematics.

3   What Can Be Done in Concrete Mathematics?

Currently mathematics works as a social game of proving and checking activities. Nevertheless no real mathematician can accept her or his activity as a game with arbitrarily defined rules. This is because she or he wishes to prove *true* theorems. Therefore the rules of the game should conform with some mathematical experience. We claim that this common mathematical experience is exactly concrete mathematics.

In this section we consider some examples of what can be done in the concrete framework. Firstly let us mention the paper by Jan Mycielski "Analysis without actual infinity" [24]. He considers there an interpretation of basic concepts of mathematical analysis by finite approximations. It conforms quite well with what we really do in applications. It is surprising because currently we learn these notions in an axiomatic framework and because the concrete interpretation may seem impossible since mathematical analysis deals with real numbers – an essentially actually infinite set.

Another example of a concrete approach is in Gödel's paper [7], where the completeness theorem for first order logic is shown. The main part of his argument consists of the construction of a sequence of finite models[4] $\mathbf{A}_0, \mathbf{A}_1, \mathbf{A}_2, \ldots$ which approximates the truth of an irrefutable Π_2^0-sentence $\varphi = \forall \overline{x} \exists \overline{y} \psi(\overline{x}, \overline{y})$, where ψ is quantifier-free. These models form a chain and therefore can be summed to obtain an infinite model $\bigcup_{k \in \mathbb{N}} \mathbf{A}_k$ which satisfies φ. For every $k \in \mathbb{N}$ and for every tuple $\overline{a}$ from the domain of $\mathbf{A}_k$ we can easily find $n \geqslant k$ and a tuple $\overline{b}$ from the domain of $\mathbf{A}_n$ such that $\mathbf{A}_n \models \psi[\overline{a}, \overline{b}]$. Therefore, the construction requires no actual infinity.

Later, in [16], Kleene shows that the completeness theorem is indeed a part of concrete mathematics. We can restate Kleene's version of the completeness theorem in the following form.

Theorem 4 (Kleene, [16]). *Every consistent countable axiomatic theory[5] has a concrete model.*

Kleene's theorem gives a background for the computable model theory. Later, in [14], Jockusch and Soare prove the low basis theorem which is considered the computability theoretical version of König's lemma. We say that the set A is low if the halting problem for computations with A as an oracle is recursive with recursively enumerable oracle i.e. $\deg(A)' \leqslant \mathbf{0}'$.

Theorem 5 (Jockusch and Soare, [14]). *Every computable infinite binary tree has a low infinite branch.*

For every consistent countable axiomatic theory the tree of its consistent complete extensions is computable. Whereas from consistent complete extension S of a theory T we easily obtain a computable in S model $\mathbf{M}$ of T[6]. We get an immediate corollary.

Corollary 1 (Jockusch and Soare, [14]). *Every computable consistent theory has a low model.*

Corollary 1 enables[7] to perform various model-theoretic constructions in concrete framework. In [4] Czarnecki discusses a few model-theoretic constructions from the classical monograph by Chang and Keisler [3] e.g. preservation theorems, Craig interpolation lemma and the Robinson joint consistency theorem. The constructions of chains and towers of models can be easily put into concrete framework. The sum of concrete elementary chain of concrete models is also a concrete model. Therefore, preservation theorems hold for computable theories. On the other hand the sum of an arbitrary concrete chain of concrete models need not to be a concrete model[8]. Similarly the final step of the model-theoretic

---

[4] The notion of a model was introduced later.

[5] By axiomatic theory we mean a theory with a finite presentation i.e. recursively axiomatizable theory.

[6] Here we not only get that the structure of $\mathbf{M}$ i.e. the universe and the relations is computable in S but the satisfaction relation in $\mathbf{M}$ is also computable in S.

[7] We need a slightly stronger version: every low consistent theory has a low model.

[8] Czarnecki requires concrete models to have both concrete structure and satisfaction relation.

construction in the proof of the Craig interpolation lemma fails in concrete framework: the model obtained from two concrete models with the same universe by
taking all their relations altogether may be a non-concrete model.

4   Representation Theorems

In this section we consider concrete versions of so called representation theorems.
We concentrate on two most classical cases, the representation theorems for
groups and boolean algebras.

Intuitively concrete groups are groups of permutations, and concrete boolean
algebras are boolean algebras of sets. Between these two theorems mathematical
community started to change its view on foundations. The work of Cayley [2] on
groups was published in 1854, the Stone theorem on boolean algebras [29] was
published in 1936.

In the meantime a new way of thinking started to be more and more popular,
particularly in algebra. This new way of thinking was presented in the book
Moderne Algebra by van der Waerden [31]. It was probably the first presentation
of a crucial mathematical theory inside of set theory, which was comprehensible
and acceptable by larger mathematical community.[9] Russell and Whitehead's
Principia Mathematica [26] was a crucial work in this direction, but it was too far
from mathematical practice of their time for influencing mathematical standards
immediately.

In the axiomatic framework a statement that all groups have a property W
is expressed in the form

$$\text{ZFC} \vdash \forall G(G \text{ satisfies axioms of groups } \Rightarrow \varphi_W(G)),$$

where $\varphi_W(G)$ is ZFC translation of $W(G)$.

In the concrete framework the same statement says that all concrete groups
have the property W. Originally groups were the first mathematical notion with
many interpretations. Nevertheless they were interpreted as groups of permutations. The Cayley theorem says that each abstract group, that is structure
$\mathbf{G} = (G, \circ, ^{-1}, e)$ satisfying the axioms of the group theory, is isomorphic with
some group of permutations. The construction is very simple. Let S_G be the
group of all permutations of G. Then we define $F: G \longrightarrow S_G$ as $(F(g))(x) = g \circ x$,
for all $g, x \in G$. So F is an isomorphism of $\mathbf{G}$ with some subgroup of S_G.

In the concrete framework a permutation group can be represented by two
sets X, Y of elements and names of permutations of X respectively, the ternary

---

[9] The book *General Topology* [15] published in 1955 by Kelley gives a presentation
of topological concepts in set theoretical framework. It gives explicitly axioms of
set theory assumed. Later on Chang and Keisler in their *Model Theory* [3] give an
equivalent set of axioms as the declared background of the theory. In both cases
it was so called Kelley–Morse set theory, shortly KM, which is essentially stronger
than ZFC. In many works published in these times and later it was clear that the
basic framework is ZFC or some stronger theory, e.g. KM, which was in this case
explicitly mentioned.

relation $\Phi(x, y) = z$ describing actions of elements from Y on elements of X. Then we obtain the following.

Theorem 6. *Each concrete abstract group* **G** *is isomorphic with some concrete group of permutations.*

Another famous *representation theorem* was given by Stone in [29] for boolean algebras. It says that each boolean algebra is isomorphic with some boolean algebra of sets (field of sets). Its proof is a little bit more subtle than the one for groups. We take all ultrafilters over a given boolean algebra as elements and we identify an element of the algebra with the set of ultrafilters containing it.

The proof of the Stone theorem can be easily reconstructed in a proper axiomatic framework, e.g. in ZFC. However it is not obvious how it can be reconstructed in the concrete framework. We say that sets X, Y and a relation $R \subseteq X \times Y$ represent a class of sets C if $C = \{b_R : b \in Y\}$, where $b_R = \{x \in X : R(x, b)\}$. Particularly X, Y, R represent a boolean algebra of sets **C** if it is of the form $\mathbf{C} = (C, \cap, \cup, -_X, \emptyset, X)$, where $-_X$ is the operation of complement to X. Intuitively X represents the universe (the set of all elements), Y represents a set of names for subsets of X, and finally R represents $\in$–relation between elements of X and Y.

Theorem 7. *For each concrete boolean algebra* $\mathbf{B} = (B, \wedge, \vee, -, 0, 1)$ *there are recursive in* **B***, sets* X*,* Y *and a relation* $R \subseteq X \times Y$ *representing a boolean algebra of sets isomorphic to* **B***.*

Proof. Let a concrete boolean algebra $\mathbf{B} = (B, \wedge, \vee, -, 0, 1)$ be given. We take $A = B - \{0\}$ and the natural enumeration of A:

$$a_0, a_1, a_2, \ldots.$$

The relation $R(a, 0)$ is false, for each $a \in A$. For other $b \in B$ we define $F(a_n, b) = a_n$ if $(F(a_0, b) \wedge F(a_1, b) \wedge \ldots \wedge F(a_{n-1}, b) \wedge b \wedge a_n) \neq 0$, otherwise we take $F(a_n, b) = -a_n$.

Now we define $R(a_n, b)$ if $F(a_n, b) = a_n$. By routine checking we see that the boolean algebra **B** is isomorphic to the boolean algebra of sets **C** represented by A, B and R.

From the construction it follows that F is recursive in **B**, therefore also A, B and R are recursive in **B**.

From Sect. 2 and Theorem 4 we obtain the following.

Theorem 8. *The first order theory of groups is complete with respect to concrete groups of permutations, in the sense that it is the set statements true exactly in all concrete groups of permutations.*

Similarly the first order theory of boolean algebras is complete with respect to concrete boolean algebras of sets.

5   Final Remarks

In this paper we show that an essential part of classical mathematics is mean-ingful in the concrete framework. We claim that concrete mathematics is our main mathematical experience determining mathematical truth in a way inde-pendent of our axiomatic assumptions. Nevertheless, we do not claim that our mathematics have to be reduced to any better defined structures. Axiomatic mathematics based on set theoretical assumptions give important and successful tools of getting mathematical truths. Let us observe that investigating border-lines of concrete mathematics cannot be done inside it.

Our mathematics in the current form is based on the idea of proof. It is one of human epistemological inventions which surprisingly preserves value of its results. Old mathematical proofs from ancient Greece are still valid. In the same time other scientific results in majority disappeared. Nevertheless, we know that this marvelous epistemic tool has its own restrictions. After Gödel's discovery [8] we know that proving technique is not sufficient. Therefore we know that we need other techniques of obtaining mathematical truth. Algorithmic learning seems to be a good candidate.

The great advantage of our axiomatic approach to mathematics is the exis-tence of a complete procedure for first order logic. In [23] it is shown that in a concrete framework we have a similar complete procedure for first order logic, but based on learning of finiteness.

References

1. Aristotle: Physics (Circa 350 BC), http://classics.mit.edu/Aristotle/physics.html. English translation by Hardie, R.P. and Gaye, R.K
2. Cayley, A.: On the theory of groups, as depending on the symbolic equation $\theta^n = 1$. In: Philosophical Magazine, vol. 7, pp. 40–47. Taylor & Francis, London (1854)
3. Chang, C.C., Keisler, H.J.: Model Theory, Studies in Logic and the Foundations of Mathematics Series, vol. 73. North-Holland, Amsterdam (1973)
4. Czarnecki, M.: Foundations of mathematics without actual infinity. Ph.D. thesis, University of Warsaw (2014)
5. Epstein, R.L.: Degrees of Unsolvability Structure and Theory. Lecture Notes in Mathematics, 1st edn. Springer, Heidelberg (1979)
6. Gauss, C.F., Schumacher, H.C., Peters, C.A.F.: Briefwechsel zwischen C.F. Gauss und H.C. Schumacher, vol. 2 in Briefwechsel zwischen C.F. Gauss und H.C. Schu-macher (1860)
7. Gödel, K.: Die Vollständigkeit der Axiome des logischen Funktionenkalküls. Monat-shefte für Mathematik und Physik **37**(1), 349–360 (1930). English translation in [9]
8. Gödel, K.: Über Formal Unentscheidbare Sätze der Principia Mathematica und Verwandter Systeme, I. Monatshefte für Math.u.Physik **38**, 173–198 (1931). Eng-lish translation in [10]
9. Gödel, K.: The completeness of the axioms of the functional calculus of logic. In: van Heijenoort, J. (ed.) From Frege to Gödel: A Source Book in Mathematical Logic, 1879–1931, pp. 582–591. Harvard University Press (2002)

10. Gödel, K.: On formally undecidable propositions of Principia mathematica and related systems i. In: van Heijenoort, J. (ed.) From Frege to Gödel: A Source Book in Mathematical Logic, 1879–1931, pp. 596–617. Harvard University Press (2002)
11. Gold, E.M.: Limiting recursion. J. Symbolic Logic 30, 28–48 (1965)
12. Graham, R.L., Knuth, D.E., Patashnik, O.: Concrete Mathematics: A Foundation for Computer Science, 2nd edn. Addison-Wesley Longman Publishing Co., Inc., Boston (1994)
13. Hilbert, D.: On the infinite. In: van Heijenoort, J. (ed.) From Frege to Gödel: A Source Book in Mathematical Logic, 1879–1931, pp. 367–392. Harvard University Press (2002)
14. Jockusch, C.G.J., Soare, R.I.: Classes and degrees of theories. Trans. Am. Math. Soc. 173, 33–56 (1972)
15. Kelley, J.L.: General Topology. Graduate texts in mathematics. Van Nostrand, New York City (1955)
16. Kleene, S.C.: Introduction to Metamathematics. North-Holland, Amsterdam (1952)
17. Kronecker, L.: Über den Zahlbegriff. Journal für die reine und angewandte Mathematik 101, 337–355 (1887). English translation in [18]
18. Kronecker, L.: On the concept of number. In: Ewald, W.B. (ed.) From Kant to Hilbert: A Source Book in the Foundations of Mathematics, vol. 2, pp. 947–955. OUP Oxford (2005)
19. Mostowski, M.: On representing concepts in finite models. Math. Logic Q. 47, 513–523 (2001)
20. Mostowski, M.: On representing semantics in finite models. In: Rojszczak, A., Cachro, J., Kurczewski, G. (eds.) Philosophical Dimensions of Logic and Science, pp. 15–28. Kluwer Academic Publishers (2003)
21. Mostowski, M.: Potential infinity and the church thesis. Fundamenta Informaticae 81(1–3), 241–248 (2007)
22. Mostowski, M.: Limiting recursion, FM-representability, and hypercomputations. In: Logic and Theory of Algorithms, Fourth Conference on Computability in Europe, CiE 2008, Local Proceedings, 15–20 June 2008
23. Mostowski, M.: Truth in the limit. Rep. Math. Logic 51, 75–89 (2016)
24. Mycielski, J.: Analysis without actual infinity. J. Symbolic Logic 46, 625–633 (1981)
25. Putnam, H.: Trial and error predicates and the solution to a problem of Mostowski. J. Symbolic Logic 30, 49–57 (1965)
26. Russell, B.A.W., Whitehead, A.N.: Principia mathematica, vol. I.-III. Cambridge University Press, Cambridge (1910–1913)
27. Shoenfield, J.R.: On degrees of unsolvability. Ann. Math. 69, 644–653 (1959)
28. Simpson, S.G.: Partial realizations of Hilbert's program. J. Symb. Log. 53(2), 349–363 (1988)
29. Stone, M.H.: The theory of representation for boolean algebras. Trans. Am. Math. Soc. 40(1), 37–111 (1936)
30. Tait, W.W.: Finitism. J. Philos. 78(9), 524–546 (1981)
31. van der Waerden, B.L.: Moderne Algebra: Unter Benutzung von Vorlesungen von E. Artin und E. Noether, vol. I.-II. Springer, Berlin (1930–1931)

Solovay's Completeness Without Fixed Points

Fedor Pakhomov[✉]

Steklov Mathematical Institute of Russian Academy of Sciences, Moscow, Russia
pakhfn@mi.ras.ru
http://www.mi.ras.ru/~pakhfn/

Abstract. In this paper we present a new proof of Solovay's theorem on arithmetical completeness of Gödel-Löb provability logic GL. Originally, completeness of GL with respect to interpretation of □ as provability in PA was proved by Solovay in 1976. The key part of Solovay's proof was his construction of an arithmetical evaluation for a given modal formula that made the formula unprovable in PA if it were unprovable in GL. The arithmetical sentences for the evaluations were constructed using certain arithmetical fixed points. The method developed by Solovay have been used for establishing similar semantics for many other logics. In our proof we develop new more explicit construction of required evaluations that doesn't use any fixed points in their definitions. To our knowledge, it is the first alternative proof of the theorem that is essentially different from Solovay's proof in this key part.

1 Introduction

The study of provability as a modality could be traced back to at least as early as Gödel work [Gö33]. Löb [Lö55] have proved a generalization of Gödel's Second Incompleteness Theorem that is now known as Löb's Theorem. In order to formulate his theorem Löb have stated conditions on provability predicates that are now known as Hilbert-Bernays-Löb derivability conditions. Despite Löb haven't mentioned the interpretation of a modality as a provability predicate there, his conditions essentially corresponded to the standard axiomatization of modal logic K4. Also note that arithmetical soundness of Gödel-Löb provability logic GL immediately follows from Löb's Theorem.

The axioms of modal system GL have first appeared in [Smi63]. Segerberg have shown that GL is Kripke-complete and moreover that it is complete with respect to the class of all finite transitive irreflexive trees [Seg71]. The arithmetical completeness of the system GL were established by Solovay [Sol76]. Solovay have proved that a modal formula φ is a theorem of GL iff for every arithmetical evaluation $f(x)$ the arithmetical sentence $f(\varphi)$ is provable in PA.

Latter modifications of Solovay's method were used in order to prove a lot of other similar results, we will mention just few of them. Japaridze have proved

F. Pakhomov—This work is supported by the Russian Science Foundation under grant 14-50-00005.

J. Kennedy and R.J.G.B. de Queiroz (Eds.): WoLLIC 2017, LNCS 10388, pp. 281–294, 2017.
DOI: 10.1007/978-3-662-55386-2_20

arithmetical completeness of polymodal provability logic GLP [Jap86]. Shavrukov [Sha88] and Berarducci [Ber90] have determined the interpretability logic of PA.

The key part of Solovay's proof was to show that in certain sense every finite GL-model is "embeddable" in arithmetic. Using the construction of "embeddings", it is easy to construct evaluations $f_\varphi(x)$ such that PA $\nvdash f_\varphi(\varphi)$, for all GL-unprovable modal formulas φ. In order to construct the "embeddings", Solovay have used Diagonal Lemma to define certain primitive-recursive function (Solovay function), for every finite GL Kripke model. Then, using the functions, Solovay have defined the sentences that constituted the "embeddings".

de Jongh, Jumelet, and Montagna have shown that GL is complete with respect to Σ_1-provability predicates for theories $T \supseteq I\Delta_0 + Exp$ [dJJM91]. Their proof have avoided the use of Solovay functions, however, their construction still "emulated" Solovay's approach using individual sentences constructed by Diagonal Lemma.

In a discussion on FOM (Foundation of Mathematics mailing list) Shipman have asked a question about important theorems that have "essentially" only one proof [Shi09]. The example of Solovay's theorem were provided by Sambin. To the author knowledge, up to the date there were no proofs of Solovay's theorem that have avoided the central idea of Solovay's proof—the Solovay's method of constructing required sentences in terms of certain fixed points.

We note that completeness of some extensions of GL with respect to interpretations of $\Box$ that are similar to formalized provability were proved by the completely different methods. Solovay in his paper [Sol76] have briefly mentioned a method of determining modal logics of several natural interpretations of $\Box$ in set theory, namely for the interpretations of $\Box$ as "to be true in all transitive models" and as "to be true in all models $\mathbf{V}_\kappa$, where κ is an inaccessible cardinal" (there are more detailed proofs in Boolos book [Boo95, Chap. 13]). A modification of the method also have been used to show completeness of wide variety of extensions of GL with respect to artificially defined (not Σ_1) provability-like predicates [Pak16].

In the paper we present a new approach to the proof of arithmetical completeness theorem for GL. We introduce a different method of "embedding" of finite GL Kripke models. As the result, the completeness of GL is achieved with the use of evaluations given by more explicitly constructed and more "natural" sentences (in particular, we do not rely on Diagonal Lemma in the construction). In order to avoid potential misunderstanding, we note that despite the sentences from evaluations are given explicitly, our proof rely on Gödel's Second Incompleteness Theorem and the results by Pudlák [Pud86] that were proved with the use of Diagonal Lemma.

Now we will give an example of unprovable GL-formula φ and an evaluation $f(x)$ provided by our proof such that PA $\nvdash f(\varphi)$. We consider the formula

$$\varphi \leftrightharpoons \Diamond v \to (\Diamond u \to \Diamond(v \wedge u)).$$

We use the following definitions for numerical functions in order to define the evaluation $f(x)$:

$$\exp(x) = 2^x, \quad \log(x) = \max(\{y \mid \exp(y) \le x\} \cup 0),$$

$$\exp^\star(x) = \underbrace{\exp(\exp(\dots \exp(0)\dots))}_{x \text{ times}}, \quad \log^\star(x) = \max(\{y \mid \exp^\star(y) \le x\} \cup 0)$$

(note that the functions $\exp^\star(x)$ and $\log^\star(x)$ are called *super exponentiation* and *super logarithmic* functions, respectively). The evaluation $f(x)$ is given as following:

$$f(v) \leftrightharpoons \exists x (\mathsf{Prf}(x, \ulcorner 0 = 1 \urcorner) \wedge \forall y < x (\neg \mathsf{Prf}(y, \ulcorner 0 = 1 \urcorner)) \wedge \log^\star(x) \equiv 0 \pmod 2),$$

$$f(u) \leftrightharpoons \exists x (\mathsf{Prf}(x, \ulcorner 0 = 1 \urcorner) \wedge \forall y < x (\neg \mathsf{Prf}(y, \ulcorner 0 = 1 \urcorner)) \wedge \log^\star(x) \equiv 1 \pmod 2).$$

We note that somewhat similar approach based on the parity of $\log^\star$ were used by Solovay in his letter to Nelson [Sol86]. Solovay proved that there are sentences F and G such that $I\Delta_0 + \Omega_1 + F$ and $I\Delta_0 + \Omega_1 + G$ are cut-interpretable in $I\Delta_0 + \Omega_1$, but $I\Delta_0 + \Omega_1 + F \wedge G$ isn't cut-interpretable in $I\Delta_0 + \Omega_1$. Also, Kotlarski in [Kot96] have used an explicit parity-based construction of a pair of sentences in order to give an alternative proof for Rosser's Theorem.

2 Preliminaries

Let us first define Gödel-Löb provability logic GL. The language of GL extends the language of propositional calculus with propositional constants ⊤ (truth) and ⊥ (false) by the unary modal connective □. GL have the following Hilbert-style deductive system:

1. axiom schemes of classical propositional calculus PC;
2. $\Box(\varphi \to \psi) \to (\Box\varphi \to \Box\psi)$;
3. $\Box(\Box\varphi \to \varphi) \to \Box\varphi$;
4. $\dfrac{\varphi \quad \varphi \to \psi}{\psi}$;
5. $\dfrac{\varphi}{\Box\varphi}$.

The expression $\Diamond\varphi$ is an abbreviation for $\neg\Box\neg\varphi$.

A set with a binary relation $(W, \prec)$ is called *irreflexive transitive tree* if

1. $\prec$ is a transitive irreflexive relation;
2. there is an element $r \in W$ that is called the *root* of $(W, \prec)$ such that the upward cone $\{a \mid r \prec a\}$ coincides with W;
3. for any element $w \in W$ the restriction of $\prec$ on the downward cone $\{a \mid a \prec w\}$ is a strict well-ordering order.

Segerberg [Seg71] have shown that the logic GL is complete with respect to the class of all finite irreflexive transitive trees.

Our proof relies on the results by Verbrugge and Visser [VV94] and indirectly on the results by Pudlák [Pud86]. This results are sensitive to details of formalization of some metamathematical notions. Thus unlike some other papers, where this kind of details could be safely be left unspecified, we will need to be more careful here.

We identify syntactical expressions with binary strings. We encode binary strings by positive integers numbers. A positive integer n of the form $1a_{k-1}\ldots a_0$ in binary notation encodes the binary string $a_{k-1}\ldots a_0$. We note that the binary logarithm $\log(n)$ of a number n coincides with the length of the binary string that the number n encodes. For a formula F the number n that encodes F is known as the *Gödel number* of F.

A *proof* of an arithmetical formula φ in an arithmetical theory T is a list of arithmetical formulas such that it ends with φ and every formula in the list is either an axiom of T, or is an axiom of predicate calculus, or is obtained by inference rules from previous formulas.

We will be interested in formalization of provability in the theory PA and its extensions by finitely many axioms. We take the standard axiomatization of PA (by axioms of Robinson arithmetic Q and the induction schema). We consider the natural axiomatization in arithmetic of the property of a number to be the Gödel number of some axiom of PA. For all extensions T of PA by finitely many axioms this gives us Δ_0-predicates $\mathsf{Prf}_\mathsf{T}(x,y)$ that are natural formalizations of "x is a proof of the formula with Gödel number y in the theory T" that is based on the definition of the notion of proof given above. And we obtain Σ_1-provability predicates

$$\mathsf{Prv}_\mathsf{T}(y) \leftrightharpoons \exists x \mathsf{Prf}_\mathsf{T}(x,y).$$

We will use effective binary numerals. The *n-th numeral* is defined as follows:

1. $\underline{0}$ is the term 0;
2. $\underline{1}$ is the term 1;
3. $\underline{2n}$ is the term $(1+1)\cdot\underline{n}$;
4. $\underline{2n+1}$ is the term $(1+1)\cdot\underline{n}+1$.

Clearly, the length of $\underline{n}$ is $\mathcal{O}(\log(n))$.

For an arithmetical formula F we denote by $\ulcorner \mathsf{F} \urcorner$ the n-th numeral, where n is the Gödel number of the formula F.

We denote by $\mathsf{Prv}(x)$ and $\mathsf{Prf}(x,y)$ the predicates $\mathsf{Prv}_\mathsf{PA}(x)$ and $\mathsf{Prf}_\mathsf{PA}(x,y)$.

An *arithmetical evaluation* is a function $f(x)$ from GL formulas to the sentences of the language of first-order arithmetic such that

1. $f(\varphi \wedge \psi) \leftrightharpoons f(\varphi) \wedge f(\psi)$;
2. $f(\varphi \vee \psi) \leftrightharpoons f(\varphi) \vee f(\psi)$;
3. $f(\neg\varphi) \leftrightharpoons \neg f(\varphi)$;
4. $f(\varphi \rightarrow \psi) \leftrightharpoons f(\varphi \rightarrow \psi)$;
5. $f(\top) \leftrightharpoons 0 = 0$;
6. $f(\bot) \leftrightharpoons 0 = 1$;
7. $f(\Box\varphi) \leftrightharpoons \mathsf{Prv}(\ulcorner f(\varphi) \urcorner)$.

Note that an arithmetical evaluation is uniquely determined by its values on propositional variables $u, v, \ldots$.

We will use $\top$, $\bot$, $\square$, and $\lozenge$ within arithmetical formulas: the expression $\top$ is an abbreviation for $0 = 0$, the expression $\bot$ is an abbreviation for $0 = 1$, the expression $\square F$ is an abbreviation for $\mathsf{Prv}(\ulcorner F \urcorner)$, and the expression $\lozenge F$ is an abbreviation for $\neg\mathsf{Prv}(\ulcorner \neg F \urcorner)$. The expressions of the form $\square^n F$ and $\lozenge^n F$ are abbreviations for $\underbrace{\square\square\ldots\square}_{n \text{ times}} F$ and $\underbrace{\lozenge\lozenge\ldots\lozenge}_{n \text{ times}} F$, respectively.

3   Proof of Solovay's Theorem

In the section we will just give a proof of "completeness part" of Solovay's theorem. Soundness of the logic GL essentially is due to Löb [Lö55] and we refer a reader to Boolos book [Boo95, Chap. 3] for a detailed proof.

Theorem 1. *If a modal formula φ is not provable in GL then there exists an arithmetical evaluation $f(x)$ such that $\mathsf{PA} \nvdash f(\varphi)$.*

Let us fix some modal formula φ that is not provable in GL. By Segerberg's result [Seg71], we can find a finite transitive irreflexive tree $\mathfrak{F} = (W, \prec)$ such that r is the root of $\mathfrak{F}$ and there is a model $\mathbf{M}$ on $\mathfrak{F}$ with $\mathbf{M}, r \nVdash \varphi$. For all the worlds a of $\mathfrak{F}$ we denote by $h(a)$ their "height":

$$h(a) = \sup(\{0\} \cup \{h(b) + 1 \mid a \prec b\}).$$

Let us assign arithmetical sentences C_a to all the worlds a of $\mathfrak{F}$. We put C_r to be $0 = 0$. We consider a non-leaf world a and assign sentences C_b to all its immediate successors b. Suppose $b_0, \ldots, b_n$ are all the immediate successors of a. We fix some enumeration $b_0, \ldots, b_n$ such that $h(b_n) = h(a) - 1$. For $i < n$ we put C_{b_i} to be the sentence

$$\exists x(\mathsf{Prf}_{\mathsf{PA}+\lozenge^{h(a)-1}\top}(x, \ulcorner 0 = 1 \urcorner) \wedge \forall y < x(\neg\mathsf{Prf}_{\mathsf{PA}+\lozenge^{h(a)-1}\top}(y, \ulcorner 0 = 1 \urcorner))$$
$$\wedge \log^{\star}(x) \equiv i \pmod{n+1}$$
$$\wedge \exists y < \exp(\exp(x))(\mathsf{Prf}_{\mathsf{PA}+\lozenge^{h(b_i)}\top}(y, \ulcorner 0 = 1 \urcorner))).$$

The sentence C_{b_n} is

$$\square^{h(a)}\bot \wedge \bigwedge_{i<n} \neg C_{b_i}.$$

Note that $\mathsf{PA} \vdash \neg(C_{b_i} \wedge C_{b_j})$, for $i \neq j$ and

$$\mathsf{PA} \vdash \square^{h(a)}\bot \leftrightarrow \bigvee_{i \leq n} C_{b_i}.$$

We note that all C_{b_i} are PA-equivalent to Σ_1-sentences: it is obvious for $i \neq n$ and C_{b_n} is equivalent to Σ_1-sentence since it states that there is a $\mathsf{PA}+\lozenge^{h(a)-1}\top$-proof of $0 = 1$ and in addition it states that the least $\mathsf{PA} + \lozenge^{h(a)-1}\top$-proof of $0 = 1$ satisfy certain $\Delta_0(\exp)$-property.

We assign sentences F_a to all the worlds a of $\mathfrak{F}$. The sentence F_a is

$$\bigwedge_{b \preceq a} C_b \wedge \Diamond^{h(a)} \top.$$

It is easy to see that the disjunction of all F_a's is provable in PA and any conjunction $F_a \wedge F_b$, for $a \neq b$, is disprovable in PA.

Lemma 1. *For any set of worlds A we have*

$$\mathsf{PA} + \Box^{h(r)+1} \bot \vdash \Diamond \left(\bigvee_{a \in A} F_a \right) \leftrightarrow \bigvee_{b, \exists a \in A (b \prec a)} F_b.$$

Let us first prove Theorem 1 using Lemma 1 and only then prove the lemma.

Proof. For a variable v we assign the evaluation $f(v)$:

$$\bigvee_{\mathbf{M}, a \Vdash v} F_a.$$

By induction on the length of modal formulas ψ we prove that

$$\mathsf{PA} + \Box^{h(r)+1} \bot \vdash f(\psi) \leftrightarrow \bigvee_{\mathbf{M}, a \Vdash \psi} F_a.$$

The only non-trivial case for the induction step is when the topmost connective of ψ is modality. Assume ψ is of the form $\Box \chi$. From inductive assumption we know that

$$\mathsf{PA} \vdash \Box^{h(r)+1} \bot \to \left(f(\chi) \leftrightarrow \bigvee_{\mathbf{M}, a \Vdash \chi} F_a \right).$$

We use Lemma 1:

$$\mathsf{PA} + \Box^{h(r)+1} \bot \vdash f(\Box \chi) \leftrightarrow \Box(f(\chi))$$

$$\leftrightarrow \Box(\Box^{h(r)+1} \bot \wedge f(\chi))$$

$$\leftrightarrow \Box \left(\Box^{h(r)+1} \bot \wedge \bigvee_{\mathbf{M}, a \Vdash \chi} F_a \right)$$

$$\leftrightarrow \Box \left(\bigvee_{\mathbf{M}, a \Vdash \chi} F_a \right)$$

$$\leftrightarrow \Box \left(\neg \bigvee_{\mathbf{M}, a \Vdash \neg \chi} F_a \right)$$

$$\leftrightarrow \neg \Diamond \left(\bigvee_{\mathbf{M}, a \Vdash \neg \chi} F_a \right)$$

$$\leftrightarrow \neg \bigvee_{\mathbf{M}, a \Vdash \Diamond \neg \chi} F_a.$$

$$\leftrightarrow \bigvee_{\mathbf{M}, a \Vdash \Box \chi} F_a.$$

Therefore,

$$PA + \Box^{h(r)+1}\bot \vdash f(\varphi) \leftrightarrow \bigvee_{\mathbf{M},a \Vdash \varphi} F_a.$$

Since $\mathbf{M}, r \not\Vdash \varphi$, we have $PA + \Box^{h(r)+1}\bot + F_r \vdash \neg f(\varphi)$. The sentence F_r is just equivalent to $\Diamond^{h(r)}\top$. Hence, by Gödel's Second Incompleteness Theorem for $PA + \Diamond^{h(r)}\top$, the theory $PA + \Box^{h(r)+1}\bot + F_r$ is consistent. Therefore, $\neg f(\varphi)$ is consistent with PA and thus $PA \not\vdash f(\varphi)$.

In order to prove Lemma 1, clearly, it will be enough to prove the following two lemmas:

Lemma 2. *For any world a from $\mathfrak{F}$, we have*

$$PA + \Box^{h(r)+1}\bot \vdash \Diamond F_a \to \bigvee_{b \prec a} F_b.$$

Proof. Let us reason in $PA + \Box^{h(r)+1}\bot$. Assume $\Diamond F_a$. We need to prove $\bigvee_{b \prec a} F_b$. Let us denote by $r = c_0 \prec c_1 \prec \ldots \prec c_n = a$ the maximal chain from r to a. Let us find the greatest k such that C_{c_k} holds.

Note that for any $1 \leq i \leq n$ the sentence $\Box^{h(c_{i-1})}\bot$ implies C_{c_i}. Indeed, $\Box^{h(c_{i-1})}\bot$ implies that C_c for some immediate successor c of c_{i-1}. But since C_c is Σ_1 and we assumed $\Diamond F_a$, we would have $\Diamond(F_a \wedge C_c)$, which is possible only for $c = c_i$.

By a simple check of cases $k = 0$ and $k \neq 0$ we obtain $\Box^{h(c_k)+1}\bot$. Therefore, for all $i < k$, we have $\Box^{h(c_i)}\bot$ and hence, for all $i \leq k$, the sentence C_{c_i} holds. From $\Box(F_a \to \Diamond^{h(a)}\top)$ and $\Diamond F_a$ we derive $\Diamond^{h(a)+1}\top$. Thus, $\neg C_a$ and hence $k < n$. Since $\Box^{h(c_k)}\bot$ implies $C_{c_{k+1}}$, we have $\Diamond^{h(c_k)}\top$. Therefore the sentence F_{c_k} holds and finally we derive $\bigvee_{b \prec a} F_b$.

Lemma 3. *For any worlds $a \prec b$, we have $PA + \Box^{h(r)+1}\bot \vdash F_a \to \Diamond F_b$.*

We will use model-theoretic methods in our proof of Lemma 3. More precisely, we will need to use within PA some facts that we will establish using model-theoretic methods. There is an approach to formalization in arithmetic of results obtained by model-theoretic methods that is based on the use of the systems of the second-order arithmetic. In particular there is a well-known system ACA_0 that is a conservative extension of PA. We will use the formalization of model-theoretic notions in systems of second-order arithmetic that could be found in Simpson book [Sim09, Sects. II.8 and IV.3].

The key model-theoretic result that we use is the Injecting Inconsistencies Theorem. We will use the version of the theorem that is a corollary of the version of the theorem that were proved by Visser and Verbrugge [VV94, Theorem 5.1]. Earlier similar results are due to Hájek, Solovay, Krajíček, and Pudlák [Há84, Sol89, KP89].

Definition 1. *Suppose $\mathfrak{M}$ is a model of* PA. *We denote by $\mathfrak{M} \upharpoonright a$ the structure with the domain $\{e \in \mathfrak{M} \mid \mathfrak{M} \models e \leq a\}$ the constant 0 and partial functions S, $+$, and $\cdot$ induced by $\mathfrak{M}$ on the domain. For two structures $\mathfrak{A}$ and $\mathfrak{B}$ with the constant 0 and (maybe) partial functions S, $+$, and $\cdot$ we write*

1. $\mathfrak{A} \subseteq \mathfrak{B}$ *if the domain of $\mathfrak{A}$ is a subset of the domain of $\mathfrak{B}$ and for any arithmetical term $t(x_1, \ldots, x_n)$ and elements $q_1, \ldots, q_n \in \mathfrak{A}$:*
   (a) *if p is the value of $t(q_1, \ldots, q_n)$ in $\mathfrak{B}$ and $p \in \mathfrak{A}$ then the value of $t(q_1, \ldots, q_n)$ is defined in $\mathfrak{A}$ and is equal to p,*
   (b) *if p is the value of $t(q_1, \ldots, q_n)$ in $\mathfrak{A}$ then the value of $t(q_1, \ldots, q_n)$ is defined in $\mathfrak{B}$ and is equal to p;*
2. $\mathfrak{A} = \mathfrak{B}$ *if $\mathfrak{A} \subseteq \mathfrak{B}$ and $\mathfrak{B} \subseteq \mathfrak{A}$.*

We note that the definition actually could also be applied to models of $\mathsf{I}\Delta_0$.

We will show in Appendix B that the following theorem is formalizable in ACA_0:

Theorem 2. *Let T be an extension of* PA *by finitely many axioms. Let $\mathsf{Con}_T(x)$ denote the formula $\forall y(\log(y) \leq x \to \neg\mathsf{Prf}_T(y, \ulcorner 0 = 1 \urcorner))$. Let $\mathfrak{M}$ be a nonstandard countable model of* T. *And let q, p be nonstandard elements of $\mathfrak{M}$ such that $\mathfrak{M} \models q \leq p$ and $\mathfrak{M} \models \mathsf{Con}_T(p^k)$, for all standard k. Then there exists a countable model $\mathfrak{N}$ of* T *such that $p \in \mathfrak{N}$ and*

1. $\mathfrak{M} \upharpoonright p = \mathfrak{N} \upharpoonright p$;
2. $\mathfrak{M} \upharpoonright \exp(p^k) \subseteq \mathfrak{N}$, *for all standard k;*
3. $\mathfrak{N} \models \neg\mathsf{Con}_T(p^q)$;
4. $\mathfrak{N} \models \mathsf{Con}_T(p^k)$, *for all standard k.*

Let us now prove Lemma 3 using the formalization of Theorem 2.

Proof. It would be enough to prove the lemma for the case when b is an immediate successor of a. Indeed, after that we will be able to derive $\Diamond^n \mathsf{F}_b$ for any b, $a \prec b$, where n is the length of the maximal chain from a to b; next we could conclude that we have the required $\Diamond \mathsf{F}_b$.

Now let us consider the case when b is an immediate successor of a and is b_k in our fixed order $b_0, \ldots, b_n$ of the immediate successors of a.

For the rest of the proof we reason in $\mathsf{ACA}_0 + \mathsf{F}_a + \Box^{h(r)+1}\bot$ in order to show that we have $\Diamond \mathsf{F}_b$; since ACA_0 is a conservative extension of PA, this will conclude the proof.

Since we have $\Diamond^{h(a)}\top$, we could construct a model $\mathfrak{M}$ of $\mathsf{PA} + \Diamond^{h(a)-1}\top$. Suppose $v \in \mathfrak{M}$ is the least $\mathsf{PA} + \Diamond^{h(a)-1}\top$-proof of $0 = 1$ in $\mathfrak{M}$, if there exists one and an arbitrary nonstandard number, otherwise. Note that since we have $\Diamond^{h(a)}\top$, the element v couldn't be standard. Next we find some nonstandard $u \in \mathfrak{M}$ such that

1. $\mathfrak{M} \models \exp(\exp(u)) < v$,
2. $\mathfrak{M} \models \log^\star(u+1) \equiv k - 1 \pmod{n+1}$,
3. $\mathfrak{M} \models \log^\star(u) \equiv k - 2 \pmod{n+1}$.

We can find u with this properties since we know that the functions $\exp(x)$ and $\exp^\star(x)$ are total on standard natural numbers and hence we know that the functions $\log(x)$ and $\log^\star(x)$ map nonstandard elements to nonstandard elements in $\mathfrak{M}$.

Now we apply Theorem 2 to the model $\mathfrak{M}$ with $p = u$ and $q = \log(u) + 1$. We obtain a model $\mathfrak{M}'$ of $\mathsf{PA} + \Diamond^{h(a)-1}\top$ such that $\mathfrak{M} \restriction u = \mathfrak{M}' \restriction u$ and there is the least $\mathsf{PA} + \Diamond^{h(a)-1}\top$-proof $d \in \mathfrak{M}'$ of $0 = 1$ such that

$$\mathfrak{M}' \models u + 1 < u^2 < \log(d) \leq u^{\log(u)+1} \leq \exp((\log(u) + 1)^2) < \exp(u).$$

Thus,
$$\mathfrak{M}' \models \log^\star(d) \equiv k \pmod{n+1}.$$

If $h(b) = h(a) - 1$, then we have constructed a model of $\mathsf{PA} + \mathsf{C}_b + \Diamond^{h(b)}\top$.

Assume $h(b) < h(a) - 1$. Clearly, there are no $\mathsf{PA} + \Diamond^{h(b)}\top$-proofs of $0 = 1$ in $\mathfrak{M}'$. We apply Theorem 2 to $\mathfrak{M}'$ with $p = d^{\log(d)+1}$ and $q = \log(d) + 1$. We obtain a model $\mathfrak{M}''$ of $\mathsf{PA} + \Diamond^{h(b)}\top$ such that

$$\mathfrak{M}' \restriction d^{\log(d)+1} = \mathfrak{M}'' \restriction d^{\log(d)+1},$$

there is a $\mathsf{PA} + \Diamond^{h(b)}\top$-proof of $0 = 1$ in $\mathfrak{M}''$ and for the least $\mathsf{PA} + \Diamond^{h(b)}\top$-proof $e \in \mathfrak{M}''$ of $0 = 1$ we have

$$\mathfrak{M}'' \models \log(e) \leq d^{(\log(d)+1)^2} \leq \exp((\log(d) + 1)^3) < \exp(d).$$

Since $\mathfrak{M}' \restriction d^{\log(d)+1} = \mathfrak{M}'' \restriction d^{\log(d)+1}$ and $\mathsf{Prf}(x, y)$ is a Δ_0 predicate, we see that d is the least $\mathsf{PA} + \Diamond^{h(a)-1}\top$-proof of $0 = 1$ in $\mathfrak{M}''$. Hence $\mathfrak{M}''$ is a model of $\mathsf{PA} + \mathsf{C}_b + \Diamond^{h(b)}\top$.

Thus, under no additional assumptions, we have a model of $\mathsf{PA} + \mathsf{C}_b + \Diamond^{h(b)}\top$. Since all C_c, for $c \preceq a$, are Σ_1-sentences, actually we have a model of $\mathsf{PA} + \mathsf{F}_b$. Therefore, $\Diamond \mathsf{F}_b$.

4   Conclusions

In the present paper we have gave a new method of constructing arithmetical evaluations of modal formulas from a given Kripke model and proved arithmetical completeness of GL with respect to provability in PA using the method. We consider the evaluations that have been constructed in the paper to be more "natural" than the evaluations provided by Solovay's proof.

We proved the theorem specifically for the standard provability predicate for PA. It is unclear to author, for what exact class of provability predicates our methods are applicable. The most essential limitation for our technique seems to be the fact that it relies on the formalized version of Theorem 2. It seems very likely that for theories that are stronger than PA one could apply our method with only minor adjustments. In particular, it seems that for a general result one would need to modify Prf-predicates while preserving Prv-predicate (up to provable equivalence) in order to ensure that [VV94, Theorem 5.1] is applicable.

For theories that are weaker than PA, there are more significant problems with adopting our technique. Namely, our technique essentially relies on formalized version of the Injecting Inconsistencies Theorem. And the proofs of stronger versions of this theorem [KP89, VV94] essentially rely on the Omitting Types Theorem. We have provided a proof of the Omitting Types Theorem in ACA$_0$ in Appendix A, but it is not clear whether it could be done in weaker systems. The author is not familiar with results that calibrate reverse mathematics strength of the required version of the Omitting Types Theorem. We note that reverse mathematics analysis of other version of Omitting Types Theorem have been done by Hirschfeldt et al. [HSS09], in particular from their results it follows that their version of the Omitting Types Theorem is not provable in WKL$_0$ but follows from RT$_2^2$ (and thus couldn't be equivalent to ACA$_0$ over RCA$_0$). But nevertheless, we conjecture that the same kind of evaluations as we have gave in Sect. 3 will provide completeness of GL for all finitely axiomatizable extensions of IΔ_0 + Exp.

Also, since the technique that were introduced in the paper is significantly different from Solovay's technique, it seems plausible that it may give some advantage for some open problems, for which Solovay's method have been the "default approach" before (see [BV06] for open problems in provability logic).

Acknowledgments. I want to thank David Fernández-Duque and Albert Visser for their questions that were an important part of the reason why I have started the research on the subject. And I want to thank Paula Henk, Vladimir Yu. Shavrukov, and Albert Visser for their useful comments on an early draft of the paper.

A   Formalization of the Omitting Types Theorem

In order to formalize Theorem 2 in ACA$_0$ we will first show that the Omitting Types Theorem is formalizable in ACA$_0$. We will adopt the proof from [CK90]. We remind a reader that we use the approach to formalization of model theory from Simpson book [Sim09].

Definition 2 (ACA$_0$). *Let* T *be a first-order theory and* $\Sigma = \Sigma(x_1, \ldots, x_n)$ *be a set of formulas of the language of* T *that have no free variables other than* $x_1, \ldots, x_n$. *We say that* T *locally omits* Σ *if for every formula* $\varphi(x_1, \ldots, x_n)$ *at least one of the following fails:*

1. *the theory* T $+ \varphi$ *is consistent;*
2. *for all* $\psi \in \Sigma$ *we have* T $\vdash \forall x_1, \ldots, x_n(\varphi \to \psi)$.

We say that a model $\mathfrak{M}$ *of* T *omits* Σ *if for any* $a_1, \ldots, a_n \in \mathfrak{M}$ *there is a formula* $\psi(x_1, \ldots, x_n) \in \Sigma$ *such that* $\mathfrak{M} \not\models \psi(a_1, \ldots, a_n)$.

Theorem 3 (ACA$_0$). *Suppose* T *is a consistent theory that locally omits the set of formulas* $\Sigma(x_1, \ldots, x_n)$. *Then there is a model* $\mathfrak{M}$ *of* T *that omits the set* Σ.

Proof. We will follow the proof of [CK90, Theorem 2.2.9] but make sure that our arguments could be carried out in ACA_0.

We will prove the theorem for $n = 1$, i.e. $\Sigma = \Sigma(x)$. The case $n > 1$ could be proved essentially the same way, but the notations would be more complicated.

We extend the language of T by fresh constants $c_0, c_1, \ldots$. We arrange all sentences of the extended language in a sequence $\varphi_0, \varphi_1, \ldots$ (since we work in ACA_0 the formulas are encoded by Gödel numbers and we could arrange them by their Gödel numbers). We will construct a sequence of finite sets of sentences

$$\emptyset = U_0 \subset U_1 \subset \ldots \subset U_m \subset \ldots$$

such that for every m we have the following:

1. U_m is consistent with T;
2. either $\varphi_m \in U_{m+1}$ or $\neg\varphi_m \in U_{m+1}$;
3. if φ_m is of the form $\exists x \psi(x)$ and $\varphi_m \in U_{m+1}$ then $\psi(c_p) \in U_{m+1}$, where c_p is the first c_i that doesn't occur in U_m or φ_m;
4. there is a formula $\chi(x) \in \Sigma$ such that $\neg\chi(c_m) \in U_{m+1}$.

We will give the definition that will determine unique sequence $U_0, U_1, \ldots$. We want to make sure that for our definition of the sequence $U_0, U_1, \ldots$, the property of a number x to be the code of the sequence $\langle U_0, U_1, \ldots, U_y \rangle$ is expressible by a formula without second-order quantifiers. If we will ensure this, then we will be able to construct a set that encodes the sequence $U_0, U_1, \ldots, U_m, \ldots$ using the arithmetic comprehension.

Let us define U_{m+1} in terms of U_m. If φ_m is consistent with $T \cup U_m$ then we put σ_m to be φ_m. Otherwise we put σ_m to be $\neg\varphi_m$. If σ_m is φ_m and is of the form $\exists x \psi(x)$ then we put ξ_m to be $\psi(c_p)$, where c_p is the first c_i that doesn't occur in U_m or φ_m. Otherwise, we put ξ_m to be equal to σ_m. We choose the formula $\chi(x)$ with the smallest Gödel number such that $\chi(x) \in \Sigma$ and $T \nvdash \bigwedge U_m \to \chi(c_m)$. We put $U_{m+1} = U_m \cup \{\xi_m, \sigma_m, \chi(c_m)\}$.

It is easy to see that for this definition, indeed, we could express by a formula without second-order quantifiers the property of a number x to be the code of the sequence $\langle U_0, U_1, \ldots, U_y \rangle$. By a trivial induction on y we could prove that for every y the said sequence exists and unique. Thus, we have obtained the sequence $U_0, U_1, \ldots, U_m, \ldots$ encoded by a set.

Now, using the definition of the sequence, we could easily prove that the sequence satisfy the conditions 1, 2, 3, and 4.

We consider the union $T \cup \bigcup_{i \in \mathbb{N}} U_i = T'$. By condition 1. the theory T' is consistent. By condition 2. the theory T' is complete. By condition 3. the theory T' gives the truth definition with Tarski conditions for a model with the domain $\{c_0, c_1, \ldots\}$; this gives us a model $\mathfrak{M}$ of T' with the domain $\{c_0, c_1, \ldots\}$. By condition 4. The model $\mathfrak{M}$ omits the set Σ.

B   Formalization of the Injecting Inconsistencies Theorem

Now we are going to check that Theorem 2 is provable in $\mathsf{ACA_0}$. Below we assume that a reader is familiar with the paper [VV94] and we will use some notions from the paper without giving the definitions here.

Theorem 4. *Let* $\mathsf{R} \subset \mathsf{I\Delta_0} + \Omega_1$ *be a finitely axiomatizable theory. Then* $\mathsf{ACA_0}$
proves the following:
Let $\mathsf{T} \supseteq \mathsf{I\Delta_0} + \Omega_1$ *be a* Σ_1^b*-axiomatized theory for which the small reflection principle is provable in* R. *Let* $\mathsf{Con_T}(x)$ *denote the formula* $\forall y (\log(y) \leq x \to \neg\mathsf{Prf_T}(y, \ulcorner 0 = 1 \urcorner))$. *Let* $\mathfrak{M}$ *be a non-standard model of* T *and let* c, a *be non-standard elements of* $\mathfrak{M}$ *such that* $\mathfrak{M} \models c \leq a$, $\exp(a^c) \in \mathfrak{M}$, *and* $\mathfrak{M} \models \mathsf{Con_T}(a^k)$, *for all standard* k. *Then there exists a model* $\mathfrak{K}$ *of* T *such that* $a \in \mathfrak{K}$ *and*

1. $\mathfrak{M} \upharpoonright a = \mathfrak{K} \upharpoonright a;$
2. $\mathfrak{M} \upharpoonright \exp(a^k) \subseteq \mathfrak{K}$, *for all standard* $k;$
3. $\mathfrak{K} \models \neg\mathsf{Con_T}(a^c);$
4. *for all standard* k *we have* $\mathfrak{K} \models \mathsf{Con_T}(a^k);$
5. $\mathfrak{K} \models \exp(a^c) \downarrow.$

Proof. Essentially, we just need to formalize the proof of [VV94, Theorem 5.1] in $\mathsf{ACA_0}$. The only difference between our formulation and the formulation by Visser and Verbrugge is that we have replaced the requirement that the small reflection principle is provable in $\mathsf{I\Delta_0} + \Omega_1$ with a stronger requirement that states that the small reflection principle is provable in R. First, we show how to formalize the proof itself and then explain why the results used in the proof are formalizable in $\mathsf{ACA_0}$.

The only non-trivial part of the formalization of the proof itself is the issue with the lack of truth definition for the cut

$$\mathfrak{N} = \{u \in \mathfrak{M} \mid u < \exp(a^k), \text{ for some standard } k\}$$

of $\mathfrak{M}$. However, for the purposes of the proof, it would be enough for $\mathfrak{N}$ to be a weak model (i.e. poses truth definition only for axioms, [Sim09, Definition II.8.9]). Moreover, unlike the original proof of Visser and Verbrugge, we just need $\mathfrak{N}$ to be a weak model of $\mathsf{R} + \mathsf{B\Sigma_1}$ rather than a model of $\mathsf{B\Sigma_1} + \Omega_1$. And since R is externally fixed finitely axiomatizable theory, we could create the required truth definition straightforward using arithmetical comprehension. Other parts of the proof could be formalized without any complications.

The proof of [VV94, Theorem 5.1] used Wilkie and Paris result [WP89, Theorem 1], Pudlák results from [Pud86], and the Omitting Types Theorem. We have already formalized the Omitting Types Theorem in Appendix A. The proof of [WP89, Theorem 1] is trivial and could be easily formalized in $\mathsf{ACA_0}$. The technique of [Pud86] is purely finitistic and thus could be easily formalized in $\mathsf{ACA_0}$.

Now we want to derive the formalization of Theorem 2 from Theorem 4. In order to do it, we first need to fix some finite fragment $\mathsf{R} \subset \mathsf{I\Delta_0} + \Omega_1$. And next we need to show in ACA_0 that all the extensions of PA by finitely many axioms are Σ_1^b-axiomatizable extensions of $\mathsf{I}\Delta_0 + \Omega_1$ for which R proves the small reflection principle. Obviously, extensions of PA by finitely many axioms are Σ_1^b-axiomatizable (and it could be checked in ACA_0).

In [VV94, Theorem 4.20] it were established that $\mathsf{I}\Delta_0 + \Omega_1$ proves small reflection principle for $\mathsf{I}\Delta_0 + \Omega_1$. By inspecting the proof, it is easy to see that it is possible to use only finitely many axioms of $\mathsf{I}\Delta_0 + \Omega_1$ in order to prove all the instances of the small reflection principle. Now we will indicate how to modify the proof of [VV94, Theorem 4.20] in order to prove in a finite fragment of $\mathsf{I}\Delta_0 + \Omega_1$ all the instances of the small reflection principle for all the extensions of PA by finitely many axioms. Actually, the only part of the proof that should be changed is [VV94, Lemma 4.16] that were needed to deal with the schema of Δ_0-induction schema in the case of $\mathsf{I}\Delta_0 + \Omega_1$-provability. For our adaptation we need to replace it with the analogous lemma that will deal with schema of full induction in the case of provability in PA. This analogous lemma could be proved essentially in the same way as [VV94, Lemma 4.16] itself with the only difference that the last part of the proof that were reducing an instance of induction schema to an instance of Δ_0-induction schema will not be needed any longer. This concludes the proof of Theorem 2 in ACA_0.

References

[Ber90] Berarducci, A.: The interpretability logic of Peano arithmetic. J. Symbolic Logic 55(3), 1059–1089 (1990)

[Boo95] Boolos, G.: The Logic of Provability. Cambridge University Press, Cambridge (1995)

[BV06] Beklemishev, L., Visser, A.: Problems in the logic of provability. In: Gabbay, D.M., Goncharov, S.S., Zakharyaschev, M. (eds.) Mathematical Problems from Applied Logic I. International Mathematical Series, pp. 77–136. Springer, New York (2006)

[CK90] Chang, C.C., Keisler, H.J.: Model Theory, vol. 73. Elsevier, Amsterdam (1990)

[dJJM91] de Jongh, D., Jumelet, M., Montagna, F.: On the proof of Solovay's theorem. Studia Logica: Int. J. Symbolic Logic 50(1), 51–69 (1991)

[Gö33] Gödel, K.: Ein Interpretation des intuitionistischen Aussagenkalküls. In Ergebnisse eines mathematischen Kolloquiums, vol. 4, pp. 39–40. Oxford (1933). Reprinted: An Interpretation of the Intuitionistic Propositional Calculus, Feferman, S, ed. Gödel Collected Works I publications, 1929–1936

[Há84] Hájek, P.: On a new notion of partial conservativity. In: Börger, E., Oberschelp, W., Richter, M.M., Schinzel, B., Thomas, W. (eds.) Computation and Proof Theory. LNM, vol. 1104, pp. 217–232. Springer, Heidelberg (1984). doi:10.1007/BFb0099487

[HSS09] Hirschfeldt, D., Shore, R., Slaman, T.: The atomic model theorem and type omitting. Trans. Am. Math. Soc. 361(11), 5805–5837 (2009)

[Jap86] Japaridze, G.K.: The modal logical means of investigation of provability. Thesis in Philosophy, in Russian, Moscow (1986)

[Kot96] Kotlarski, H.: An addition to Rosser's theorem. J. Symbolic Logic 61(1), 285–292 (1996)

[KP89] Krajíček, J., Pudlák, P.: On the structure of initial segments of models of arithmetic. Arch. Math. Logic **28**(2), 91–98 (1989)

[Lö55] Löb, M.H.: Solution of a problem of Leon Henkin. J. Symbolic Logic **20**(02), 115–118 (1955)

[Pak16] Pakhomov, F.: Semi-provability predicates and extensions of GL. In: 11th International Conference on Advances in Modal Logic, Short Presentations, pp. 110–115 (2016)

[Pud86] Pudlák, P.: On the length of proofs of finitistic consistency statements in first order theories. In: Logic Colloquium, vol. 84, pp. 165–196. Amsterdam, North-Holland (1986)

[Seg71] Segerberg, K.: An essay in classical modal logic. Ph.D. thesis, Uppsala:: Filosofiska Föreningen och Filosofiska Institutionen vid Uppsala Universitet (1971)

[Sha88] Shavrukov, V.Y.: The logic of relative interpretability over Peano arithmetic. Technical report, (5) (1988). Moscow: Steklov Mathematical Institute (in Russian)

[Shi09] Shipman, J.: Only one proof, 2009. FOM mailing list. http://cs.nyu.edu/pipermail/fom/2009-August/013994.html

[Sim09] Simpson, S.G.: Subsystems of Second Order Arithmetic, vol. 1. Cambridge University Press, Cambridge (2009)

[Smi63] Smiley, T.J.: The logical basis of ethics. Acta Philosophica Fennica **16**, 237–246 (1963)

[Sol76] Solovay, R.M.: Provability interpretations of modal logic. Israel J. Math. **25**, 287–304 (1976)

[Sol86] Solovay, R.M.: 12 May 1986. Letter by R. Solovay to E. Nelson

[Sol89] Solovay, R.M.: Injecting inconsistencies into models of PA. Ann. Pure Appl. Logic **44**(1–2), 101–132 (1989)

[VV94] Verbrugge, R., Visser, A.: A small reflection principle for bounded arithmetic. J. Symbolic Logic **59**(03), 785–812 (1994)

[WP89] Wilkie, A., Paris, J.: On the existence of end extensions of models of bounded induction. Stud. Logic Found. Math. **126**, 143–161 (1989)

An Epistemic Generalization of Rationalizability

Rohit Parikh[⊠]

City University of New York, New York, USA
rparikh@gc.cuny.edu

Abstract. Rationalizability, originally proposed by Bernheim and Pearce, generalizes the notion of Nash equilibrium. Nash equilibrium requires common knowledge of strategies. Rationalizability only requires common knowledge of rationality. However, their original notion assumes that the payoffs are common knowledge.

I.e. agents do know what world they are in, but may be ignorant of what other agents are playing.

We generalize the original notion of rationalizability to consider situations where agents do not know what world they are in, or where some know but others do not know. Agents who know something about the world can take advantage of their superior knowledge. It may also happen that both Ann and Bob know about the world but Ann does not know that Bob knows. How might they act?

We will show how a notion of rationalizability in the context of partial knowledge, represented by a Kripke structure, can be developed.

1 Introduction

Agenthood has become an important subject of study in psychology, cognitive science, artificial intelligence and of course in economics where the notion is fundamental.

But what are agents? Do we impose a metaphysical requirement that an agent have an *inside*, a self from which it looks at the world? Or do we go the route of Dennett's intentional stance where something is an agent if we can usefully *treat it* as one.

Clearly we do not regard a shopbot as something which has an internal state but something to which the BDI theory of Bratman and others applies. And if we do treat a shopbot as an agent then we have already dropped the metaphysical worries.

> Anything that is usefully and voluminously predictable from the intentional stance is, by definition, an intentional system. The intentional stance is the strategy of interpreting the behavior of an entity (person, animal, artifact, whatever) by treating it as if it were a rational agent who governed its choice of action by a consideration of its beliefs and desires [1].

Some of the ideas in this paper were included in [7], but the discussion on Dennett, and Theorem 4.1 are new.

© Springer-Verlag GmbH Germany 2017
J. Kennedy and R.J.G.B. de Queiroz (Eds.): WoLLIC 2017, LNCS 10388, pp. 295–303, 2017.
DOI: 10.1007/978-3-662-55386-2_21

However Premack [11] imposes another requirement, of second order agent-hood. A second order agent is one who is aware of other agents and has a theory of mind. We shall go along with Premack, not because we have to but because it is a much more interesting notion to think of. A second order agent is an agent which is aware of other agents and before asking "What shall I do?" first asks "What is she likely to do?".

Even though it is not clear that animals are second order agents, they often behave as if they were. Chimps and corvids both exhibit strategic behavior which makes sense if they had a theory of other chimps or corvids. A female chimp cheating on the *big boss* will refrain from uttering the noises of pleasure which she would make if he was not around[1]. Out in the wild, jays and other corvids will hide food in the ground. And if the birds were being watched when they hid their food, they rushed to move it to another hiding place as soon as the other watching birds were out of sight.

So it may be that the theory we propose has a wide field of application.

The actions of sophisticated agents take place in a world of desires, knowledge (or beliefs) and abilities. And quite often not only their own beliefs and desires are involved but also what they know about the desires and beliefs of others.

Savage [13] worked out a theory in which by observing an agent's willingness to accept or reject certain bets we can discover both his beliefs (subjective probability) and his desire (utility). This theory has been questioned and has some difficulties pointed out by Ellsberg, Allais, and Kahneman and Tversky. But the theory is much respected and still taught routinely.

But Savage did not have a theory of what we do when other agents are involved and we know something about their desires and beliefs. In this paper we will generalize this theory to consider agents who think about other agents and consider their probable actions.

1.1 Monk

Here is a story about the TV detective Adrian Monk. A woman has fallen off a fifteenth floor balcony and is lying on the pavement, dead of course. A policeman arrives and a little later Adrian Monk arrives. "We don't know yet if it was murder or suicide" says the policeman.

"It was murder," says Monk.
"But you just arrived! How can you know?"
"She was in the middle of painting her nails," says Monk.

Now we all agree that a woman does not paint half her nails and then commit suicide without painting the other half. Perhaps she was in despair because she ran out of paint but that is not very likely. Murder is the far more plausible explanation. Since Monk is unable to answer the question, "Why might she commit suicide at *that* moment?" he concludes that it was murder.

---

[1] She is separated from him by a big rock so he cannot see her but *could* hear her cries of pleasure.

This paper attempts connect beliefs and desires with actions. We offer some examples from literature, from real life and offer a formal framework. But we will be guided by the following intuition. *If, given an agent's desires and beliefs, action s is definitely worse than action s' then the agent will not do action s. If the agent does do action s then we can conclude that we were wrong about the beliefs or desires.*

Of course the agent might be irrational. All of us have met irrational people. But the scenarios we consider are so simple that irrationality is unlikely to be an explanation. And even an irrational person is predictable to some extent. Someone who claims to be Napoleon will most likely speak (bad) French but will not pretend to speak Mandarin.

2   Animal Cognition

2.1   Inducing False Beliefs in the Tigers of the Sundarbans

The Sundarbans are an area at the border of India and Bangladesh where lush forests grow and tiger attacks on humans have been common [14].

Fishermen and bushmen then created masks made to look like faces to wear on the back of their heads because tigers always attack from behind. The payoff matrix for the tiger is below.

	face	noface
attack	−20	100
not attack	0	0

If the tiger sees a face then his dominant strategy is not to attack since there might be resistance. Thus not attacking is dominant. If the tiger does not see a face then attacking is the dominant strategy.

Thus the fishermen changed the dominant strategy of the tiger by changing its beliefs.[2] In 1987 no one wearing a mask was killed by a tiger, but 29 people without masks were killed.

Unfortunately the tigers eventually realized it was a hoax, and the attacks resumed.[3]

2.2   The Tiger in the Bathroom

Suppose I know T, that there is a tiger in your bathroom. I also know that you need to go.

If $\sim K_y(T)$ then you will proceed to the bathroom (The y stands for you).

If $K_y(T)$ then you will go the neighbor's apartment and ask if you can use his bathroom. Or perhaps you will call your mother for advice.

So I can infer what you know from what you do.

---

[2] I am using the word belief in a weak sense in which we can use it for non-linguistic creatures.

[3] Something which puzzles me is how they passed the knowledge "it is a hoax" from one tiger to another. Tigers are solitary beasts and do not have cellphones.

There is one proposition, "a tiger is in the bathroom" which may be true or false, and two possible actions for you.

	tiger	no tiger
use own bathroom	-20000	10
neighbor bathroom	-5	-5

The -5 in the second row has to do with the fact that going to the neighbor's bathroom has a social cost.

If you know about the tiger then the bottom row dominates the top row. If you do not know about the tiger and "no tiger" is your default assumption, then the top row dominates.

What about me? If I see you heading to your own bathroom then I conclude you do not know about the tiger. Under most circumstances my own dominant strategy is to tell you about the tiger. But perhaps I want you to be eaten by the tiger. Then I will not tell you. So my own payoffs are also involved in what I do.

For people not familiar with epistemic logic or Kripke structures, there is a short appendix just before the bibliography.

3   A formal framework

We have n players and some propositions P about the world whose truth value they may or may not know. T is all truth assignments on P.

We define an epistemic game with n players to be a map F from T (truth assignments) and S (strategy profiles) to P (payoff profiles).

So $(F(t, S))_i$ is the payoff to player i when the truth values are according to t and the strategy profile is S.

We let $S_i^- = S''$ to mean the strategy profile of all players other than i. We will drop the subscript i when clear from the context.

Let s, s' be strategies for i. we let $s <_t^i s'$ to mean $(\forall S'')(F(t, (s, S''))_i < F(t, (s', S''))_i))$.

(We will usually assume that payoffs for i are never the same (the game is generic) so that we need not worry about $<$ and $\leq$.) In other words s' is better than s no matter what the other players do. We will also drop i when clear from the context.

If ϕ is a formula, we write $s <_\phi s'$ to mean that

for all $t \models \phi$, $s <_t^i s'$.

So if i knows ϕ and i is rational, i will not play s.

Theorem 3.1. If $s <_\phi s'$ and $\psi \models \phi$ then $s <_\psi s'$

If $s <_\phi s'$ and $s <_\psi s'$ then $s <_{\phi \vee \psi} s'$.

Corollary 3.2. The set $\{\phi | s <_\phi s'\}$ is a filter in the boolean algebra.

Note that if a rational player knows ϕ and $s <_\phi s'$ then the agent will not play s.

Moreover if j knows that i knows ϕ and $s <_\phi s'$ then j knows that the agent i will not play s, and j only needs to respond to strategies other than s.

Indeed what j knows about what other players know allows j to reduce the strategy profiles that he needs to respond to.

Definition 3.3. An *infon* α *for agent* i *is a pair* (t, S) *where* t *is a truth assignment and* S *is an* $(n-1)$-*tuple of the other players' strategies.*

In a generic game an infon α generates a unique strategy $b(\alpha)$ for agent i which yields the highest value given the infon.

A *state of knowledge* for agent i is a set X of infons.

A strategy s is *rational* for agent i relative to a state of knowledge X if $s \in \{b(\alpha) | \alpha \in X\}$.

If $X \subseteq Y$ and s is rational relative to X then it is rational relative to Y.

The less you know, the more rational you are!

A strategy profile $(s_1, \ldots, s_n)$ is rational for the agents relative to a tuple of knowledge $(X_1, \ldots, X_n)$ if each s_i is rational relative to X_i.

However, not all n-tuples $(X_1, \ldots, X_n)$ are possible. For instance if i knows that j knows P then j cannot be playing a strategy t which is dominated when P is true. And i herself cannot be playing a strategy which is dominated when j is not playing t.

So there are connections among the X_i which we have yet to fully investigate.

4   Payoffs

As an agent comes to know more about the world and about the other players, her rational strategies decrease. We show now that this has the effect of increasing her minimum payoffs.[4]

Given an agent i and an infon α we can define $p(s, \alpha)$ where $\alpha = (t, S)$. Let S^+ be the strategy obtained by combining s with S, i.e. combining i's strategy with the strategy profile of the other players.

Then $p(s, \alpha) = F(t, S^+)_i$, i.e. the payoff to i when the world is according to t and the total strategy profile is S^+.

We define $g(s, X)$ the guaranteed payoff to i given that i plays s, and the actual world is among the infons in X.

$$g(s, X) = \min[p(s, \alpha) : \alpha \in X]$$

And finally we define the guaranteed payoff to i, $G(i, x)$, if i acts rationally to be

$$\max[g(s, X) : s \in S_i] \text{ where } S_i \text{ is all possible strategies for } i$$

---

[4] We have not yet defined 'rational' so we will temporarily rely on an intuitive meaning of the word. A precise definition will be provided in the next section.

It does not matter here if we count all strategies for i or only the ones which are rational since the maximum will not be achieved by a strategy which is not rational (or if it is, it will be achieved also by one which is rational).

Theorem 4.1. If $X \subseteq Y$ then $G(i, Y) \leq G(i, X)$.

Proof: clearly $\min[p(s, \alpha) : \alpha \in Y] \leq \min[p(s, \alpha) : \alpha \in X]$.

And since $G(i, X) = \max[g(s, X) : s \in S_i]$ and $G(i, Y) = \max[g(s, Y) : s \in S_i]$, $G(i, Y) \leq G(i, X)$.

What i can guarantee if she knows X is more than what she can guarantee if she knows Y.   □

Alas, the possible maximum can come down when i knows more. But she should not worry about her dashed hopes because these were not real anyway. But the important point we are making is that exchange of information leads to a decrease in the various X_i, states of knowledge of different agents, and an increase in the minimum possible payoff for each agent.

5   Rationalizability

The notion of rationalizability and dominated strategy have been much discussed in the literature [12]. When there is common knowledge of rationality then player i knows that player j will not play a dominated strategy. Given this some of player i's own strategies can become dominated and i will eliminate them in turn. When this process of elimination of dominated strategies ends, the strategies which remain are the rationalizable ones.

Let us give an example. Suppose player 1 has three strategies a, b, e. Player 2 also has three strategies c, d, f.

a is the best reply to c, f. c is the best reply to b, e. b is the best reply to d. And d is the best reply to a.

So e and f are not best replies to anything and are not rationalizable.

The other four strategies a, b, c, d are rationalizable but there is no (pure) Nash equilibrium.

For instance (a, c) is not a Nash equilibrium because while a is the best reply to c, d is a better reply than c to a.

However, in this situation we have only one payoff matrix known to everyone. But if the payoff matrix depends on the world and different players have different knowledge about the world then the issue becomes complex. But we are confident that the goal described below can be achieved.

Goal: To define the notion of rationalizability relative to a given Kripke structure and an epistemic game.

Conjecture: Every strategy rationalizable relative to a Kripke structure is rationalizable in the usual sense. The reverse of course is not true.

Here is a rough argument. Every piece of knowledge you acquire in terms of the world or in terms of what another agent does reduces the possibilities of strategy profiles you face. That means that the relation of dominance becomes larger. So some strategies which might have been rational are so no longer. This happens not only when you yourself learn about the world but also when you learn that someone else has learned about the world, or even about the knowledge of a third agent.

For instance, the strategy of using your own bathroom was rationalizable for you before you knew about the tiger. But once you know about the tiger, that strategy is not rationalizable. There are *fewer* rationalizable strategies when we know more.

See [15] for some related work involving knowledge growth through asynchronous messages.

Conjecture: Two non-bisimilar Kripke structures yield different sets of rationalizable strategies in *at least one* epistemic game.

The rationale here is that in order to find out what an agent believes, about himself or about other agents, we set up an experiment. The conjecture is that *some* experiment will tell us what we want to know.

6   Conclusion

This paper is work in progress. We have described an approach towards understanding the behavior of groups of agents, by inferring beliefs by watching actions, and by affecting actions by affecting beliefs. Almost all we have said is common sense and the only new thing is to suggest that a formal framework is possible. We have in fact gone a long way towards defining such a framework but some more is to come.

Readers of this paper might now enjoy watching episodes of Adrian Monk, Columbo, or Sherlock Holmes with a new eye and see how the framework applies.

If I may be so daring, even the actions of Obama or Clinton or Trump, and the way they differ from their statements become explainable.

7   Appendix on Epistemic Logic

We create a language to talk about various knowledge properties in the following way.

- An atomic predicate P is a formula
- If A, B are formulas then so are $\neg A$ and $A \wedge B$
- If A is a formula and i is an agent then $K_i(A)$ is a formula
- there are no other formulas.

7.1   Intuition

Intuitively $K_i(A)$ means that the agent i knows the fact expressed by the formula A. $K_j K_i(A)$ means that j knows that i knows A.

7.2   Kripke Structures

Kripke structures are used to interpret the language above.

Kripke structure M for knowledge for n knowers consists of a space W of states and for each knower i a relation $R_i \subseteq W \times W$.[5]

There is a map π from $W \times A \longrightarrow \{0,1\}$ which decides the truth value of atomic formulas at each state.

We now define the truth values of formulas as follows:

1. $M, w \models P$ iff $\pi(w, P) = 1$
2. $M, w \models \neg A$ iff $M, w \not\models A$
3. $M, w \models A \wedge B$ iff $M, w \models A$ and $M, w \models B$
4. $M, w \models K_i(A)$ iff $(\forall t)(wR_i t \rightarrow M, t \models A)$

$K_i(A)$ holds at w, (i knows A at w) iff A holds at all states t which are R_i accessible from w.

7.3   Axiom System

1. All tautologies of the propositional calculus
2. $K_i(A \rightarrow B) \rightarrow (K_i(A) \rightarrow K_i(B))$
3. $K_i(A) \rightarrow A$
4. $K_i(A) \rightarrow K_i K_i(A)$
5. $\neg K_i(A) \rightarrow K_i(\neg K_i(A))$

There are also two rules of inference. Modus Ponens, to infer B from A and $A \rightarrow B$. And the other is generalization, to infer $K_i(A)$ from A.

The second rule *does not* say that if A is true than i knows it. Only that if A is a logical truth then i knows it.

These rules are complete. All valid formulas are provable using the axioms and rules. For a bit more detail, see [6].

7.4   Revising Kripke Structures When an Announcement is Made

Suppose we are given a Kripke structure $\mathcal{M}$. Then some formula ϕ is announced publicly.

The new Kripke structure is then obtained by deleting all states in $\mathcal{M}$ where ϕ did not hold. See for instance [9].

Acknowledgement. Dov Samet very kindly showed me some related work of his [2] which does talk about dominated strategies. This work was done independently of ours and has some elegant ideas. But he and his co-author do not make use of Kripke structures, or, for that matter, detectives and tigers! Thanks to David Makinson for comments. This research was supported by grants from the CUNY Faculty research assistance program.

---

[5] The R_i are often assumed to be equivalence relations and we shall follow this tradition.

References

1. Dennett, D.: Intentional systems theory. In: The Oxford Handbook of Philosophy of Mind, pp. 339–350 (2009)
2. Hillas, J., Samet, D.: Weak dominance, a mystery cracked (2015, unpubished)
3. Kahneman, D., Tversky, A.: Prospect theory: an analysis of decision under risk. Econometrica: J. Econom. Soc. 263–291 (1979)
4. Lurz, R.W.: Mindreading Animals: The Debate Over What Animals Know About Other Minds. MIT Press, Cambridge (2011)
5. Lurz, R.: If chimpanzees are mindreaders, could behavioral science tell? Toward a solution of the logical problem. Philos. Psychol. **22**(3), 305–328 (2009)
6. Parikh, R.: Recent issues in reasoning about knowledge. In: Proceedings of the 3rd Conference on Theoretical Aspects of Reasoning About Knowledge, pp. 3–10. Morgan Kaufmann Publishers Inc. (1990)
7. Parikh, R.: Knowledge and action in groups. Stud. Log. **8**(4), 108–123 (2015). (Sun Yat Sen University)
8. Parikh, R., Taşdemir, Ç., Witzel, A.: The power of knowledge in games. Int. Game Theory Rev. **15**(04), 1340030 (2013)
9. Plaza, J.: Logics of public communications. Synthese **158**(2), 165–179 (2007)
10. Premack, D., Woodruff, G.: Does the chimpanzee have a theory of mind? Behav. Brain Sci. **1**(04), 515–526 (1978)
11. Premack, D.: The codes of man and beasts. Behav. Brain Sci. **6**(01), 125–136 (1983)
12. Rubinstein, A.: Lecture Notes in Microeconomic Theory: The Economic Agent. Princeton University Press, Princeton (2012)
13. Savage, L.J.: The Foundations of Statistics. Courier Corporation, Chicago (1972)
14. Simons, M.: Face masks fool the Bengal tigers. The New York Times, 5 September 1989
15. Stambaugh, T.: Rationalizability in epistemic games with asynchronous messages. In: Talk Given at the Stony Brook Game Theory Conference, 18 July 2016
16. Tversky, A., Kahneman, D.: The framing of decisions and the psychology of choice. Science **211**(4481), 453–458 (1981)
17. Wimmer, H., Perner, J.: Beliefs about beliefs: representation and constraining function of wrong beliefs in young children's understanding of deception. Cognition **13**(1), 103–128 (1983)

Knowledge Is a Diamond

Vít Punčochář[(⊠)]

Institute of Philosophy, Czech Academy of Sciences,
Jilská 1, 11000 Prague, Czech Republic
vit.puncochar@centrum.cz

Abstract. In the standard epistemic logic, the knowledge operator is represented as a box operator, a universal quantifier over a set of possible worlds. There is an alternative approach to the semantics of knowledge, according to which an agent a knows α iff a has a reliable (e.g. sensory) evidence that supports α. In this interpretation, knowledge is viewed rather as an existential, i.e. a diamond modality. In this paper, we will propose a formal semantics for substructural logics that allows to model knowledge on the basis of this intuition. The framework is strongly motivated by a similar semantics introduced in [3]. However, as we will argue, our framework overcomes some unintuitive features of the semantics from [3]. Most importantly, knowledge does not distribute over disjunction in our logic.

Keywords: Epistemic logic · Substructural logic · Knowledge

1 Introduction

In the standard epistemic logic, as is described for example in [5], the knowledge operator is usually equipped with the following semantic clause: An agent a knows that α iff α is true in every possible world accessible to the agent a. The accessible worlds represent possibilities that are not excluded by the a's information state. So, in the terminology of modal logic, knowledge is standardly modeled as a box operator, a universal quantifier that quantifies over a set of possible worlds. However, there is also an alternative approach to the logic of knowledge: The agent a knows that α iff a has a reliable (e.g. sensory) evidence that supports α. In this interpretation, knowledge is viewed rather as an existential modality, similarly to the "diamond operator" of modal logic. In this paper, we will propose a formal semantics for substructural logics that allows to model knowledge on the basis of this intuition. The framework is strongly motivated by a similar semantics introduced in [3], which in turn builds on [2,6]. This kind of semantics was also discussed in [11]. We will provide a detailed comparison of our framework with the semantics from [3] and we will argue that the former overcomes some of the unintuitive features of the latter. Most importantly, our logic does not validate the following problematic law:

V. Punčochář—The work on this paper was supported by grant no. 16-07954J of the Czech Science Foundation.

J. Kennedy and R.J.G.B. de Queiroz (Eds.): WoLLIC 2017, LNCS 10388, pp. 304–320, 2017.
DOI: 10.1007/978-3-662-55386-2_22

if a knows that $\alpha \vee \beta$, then a knows that α or a knows that β.

Distributivity over disjunction is characteristic of the diamond operators. However, this principle is not intuitively acceptable if the operator is intended to represent knowledge.

The idea of regarding knowledge as an existential modality might be reminiscent of the recently developed evidence logic [12], in which $\Box\alpha$ is provided with an existential semantic condition relative to neighbourhood structures, and the formula is interpreted as "the agent has evidence for α". One of the differences between this and our approach is that evidence logic is based on classical logic, which, however, suffers from the well-known paradoxes of material implication and irrelevance.

We will adopt the strategy of [3,11] in that we will model knowledge over a basic substructural logic λ_0. There are several reasons for this strategy: The basic logic is very weak, so the class of its extensions is large and it encompasses many important non-classical logical systems including, for example, relevant logics, fuzzy logics, and superintuitionistic logics. These systems can be obtained syntactically by adding additional axioms and/or semantically by imposing further constraints on the semantic models. So, rather than a unique intended logical system, our general epistemic framework will provide a uniform basis from which one can obtain, by further specifications, various epistemic logics based on various non-classical systems.

The basic logic allows for avoiding the validity of the formulas that are usually regarded as paradigmatic examples of the "paradoxes of material implication and irrelevance": e.g. $p \rightarrow (q \rightarrow p)$, $\neg p \rightarrow (p \rightarrow q)$, $(p \wedge \neg p) \rightarrow q$, $p \rightarrow (q \vee \neg q)$.

We will work with two languages. The first one will be a propositional language that is used in substructural logics and is defined in the following way:

$$\alpha ::= p \mid t \mid \bot \mid \top \mid \neg\,\alpha \mid \alpha \rightarrow \alpha \mid \alpha \wedge \alpha \mid \alpha \& \alpha \mid \alpha \vee \alpha.$$

We will call this language $\mathcal{L}$ an the formulas of this language $\mathcal{L}$-formulas. Besides the standard vocabulary of propositional logic ($\bot$, $\top$, $\neg$, $\rightarrow$, $\wedge$, $\vee$) $\mathcal{L}$ has two extra symbols: one additional propositional constant t that represents logical truth (which will not be interpreted in the same way as the universal truth $\top$); and the so called intentional conjunction $\&$ that will differ from the standard "extensional" conjunction $\wedge$. The intentional conjunction $\&$ is a crucial part of the language $\mathcal{L}$. In substructural logics it is the residual of implication in the sense that $(\alpha \& \beta) \rightarrow \gamma$ is equivalent to $\alpha \rightarrow (\beta \rightarrow \gamma)$. Interestingly, this conjunction enables us to represent some form of non-monotonic reasoning. The conjunction is not monotone in the following sense: in the basic logic λ_0, the formula $p \rightarrow q$ does not entail $(p\&r) \rightarrow q$.

We will work also with a language denoted as $\mathcal{L}_K$, which is obtained from $\mathcal{L}$ by the addition of the knowledge operator K:

$$\alpha ::= p \mid t \mid \bot \mid \top \mid \neg\,\alpha \mid \alpha \rightarrow \alpha \mid \alpha \wedge \alpha \mid \alpha \& \alpha \mid \alpha \vee \alpha \mid K\alpha.$$

2   Two Semantics for Substructural Logics

In this section, we will provide a comparison of two alternative frameworks for substructural logics. The first one is taken from [10] and it was used in [3] as a basis for a non-standard epistemic logic. The second one was proposed and used in [8] as a basis for a logic of questions. It can be viewed as an extension of the framework that was originally introduced in [4]. This section will be concerned only with the language $\mathcal{L}$. We will present some results that show that with respect to this language the two semantics are closely related, and in some sense equivalent. In the next section, we will show that if we enrich the language with the knowledge operator, the second framework is significantly richer than the first one.

First, we will present the semantics for substructural logics that is used in [3]. The semantic structures will be called substructural models. These are structures of the following kind:

$$\mathcal{M} = \langle W, L, \leq, R, C, V \rangle,$$

satisfying the following: W is a nonempty set (of *states*), $\leq$ is a partial order on W, R is a ternary relation on W satisfying the following two conditions:

- if $Rxyz$ and $x' \leq x, y' \leq y$, and $z \leq z'$, then $Rx'y'z'$,
- if $Rxyz$, then $Ryxz$.

The set of *logical states* L is a nonempty subset of W that is upward closed w.r.t. $\leq$, and it holds:

- $x \leq y$ iff there is $z \in L$ such that $Rzxy$.

C is a binary *compatibility relation* on W satisfying:

- if xCy and $x' \leq x$, then $x'Cy$,
- if xCy, then yCx.

The *valuation* V assigns to every atomic formula an upward closed subset of W.

Given a substructural model, a relation $\Vdash$ is defined between the elements of W and $\mathcal{L}$-formulas in the following way. For the atomic formulas and constants, we define:

- $x \Vdash p$ iff $x \in V(p)$; $x \Vdash t$ iff $x \in L$; $x \nVdash \bot$; $x \Vdash \top$.

For the complex formulas, we adopt the following semantic clauses:

- $x \Vdash \neg\alpha$ iff for any y, if xCy, then $y \nVdash \alpha$,
- $x \Vdash \alpha \rightarrow \beta$ iff for any y, z, if $Rxyz$ and $y \Vdash \alpha$, then $z \Vdash \beta$,
- $x \Vdash \alpha \wedge \beta$ iff $x \Vdash \alpha$ and $x \Vdash \beta$,
- $x \Vdash \alpha \& \beta$ iff there are y, z such that $y \Vdash \alpha$, $z \Vdash \beta$, and $Ryzx$,
- $x \Vdash \alpha \vee \beta$ iff $x \Vdash \alpha$ or $x \Vdash \beta$.

The relation ⊩ is called a satisfaction relation in [3]. We will denote it as a *support relation*.

An $\mathcal{L}$-formula α is valid in a substructural model $\mathcal{M}$ iff α is supported by every logical state of $\mathcal{M}$, i.e. for every $x \in L$, $x \Vdash \alpha$. A consequence of this definition is that a formula of the form $\alpha \to \beta$ is valid in $\mathcal{M}$ iff for any state $x \in W$, if x supports α, then x supports β. We will denote the set of formulas that are valid in $\mathcal{M}$ as $Log_{\Vdash}(\mathcal{M})$.

An intuitive interpretation of this kind of semantics, and especially of the ternary relation, was worked out for example in [1,7,9]. We will compare this framework with an alternative semantics that can be interpreted in a similar fashion and that is a version of the semantics used in [8]. We will call the semantic structures of this alternative framework *information models*. These are structures of this kind:

$$\mathcal{N} = \langle\, A, +, \times, \cdot, 0, 1, C, V \,\rangle,$$

where A is an arbitrary nonempty set (of *states*); $+$, $\times$ and $\cdot$ are binary operations on A (*join, meet,* and *fusion* of states); 0 and 1 are two different elements of A (*trivially inconsistent* and *logical* state); C is again a binary relation among states that is interpreted as a *compatibility relation*; and V is a *valuation*. We assume that $\langle A, +, \times \rangle$ is a distributive lattice[1] that determines an ordering $\leq$ on A defined in this way: $a \leq b$ iff $a + b = b$. Moreover, we assume that the following hold generally:

- $0 + a = a$; $0 \cdot a = 0$; $1 \cdot a = a$;
- $a \cdot b = b \cdot a$; $a \cdot (b + c) = (a \cdot b) + (a \cdot c)$;
- there is no a such that $0Ca$; if aCb, then bCa; $(a + b)Cc$ iff aCc or bCc;
- $V(p)$ is an ideal, i.e. $0 \in V(p)$; $a + b \in V(p)$ iff $a \in V(p)$ and $b \in V(p)$.

Note that 0 is the least element of the structure but 1 does not have to be the top element. With respect to a given information model $\mathcal{N}$ a support relation $\vDash$ between its states and $\mathcal{L}$-formulas is defined in the following way. For atomic formulas and the constants we have:

- $a \vDash p$ iff $a \in V(p)$; $a \vDash t$ iff $a \leq 1$; $a \vDash \perp$ iff $a = 0$; $a \vDash \top$.

For complex formulas, we define:

- $a \vDash \neg\alpha$ iff for any b, if aCb, then $b \nvDash \alpha$,
- $a \vDash \alpha \to \beta$ iff for any b, if $b \vDash \alpha$, then $a \cdot b \vDash \beta$,
- $a \vDash \alpha \wedge \beta$ iff $a \vDash \alpha$ and $a \vDash \beta$,
- $a \vDash \alpha \& \beta$ iff there are b, c such that $b \vDash \alpha$, $c \vDash \beta$, and $a \leq b \cdot c$,
- $a \vDash \alpha \vee \beta$ iff there are b, c such that $b \vDash \alpha$, $c \vDash \beta$, and $a \leq b + c$.

---

[1] The operation $\times$ will not be directly used in the semantic conditions determining the support relation. However, its presence and the requirement that the lattice $\langle A, +, \times \rangle$ is distributive results in the validity of the axiom $(\alpha \wedge (\beta \vee \gamma)) \to ((\alpha \wedge \beta) \vee (\alpha \wedge \gamma))$ that is a part of the system from [3] (the system is formulated below).

An $\mathcal{L}$-formula α is valid in an information model $\mathcal{N}$ iff α is supported by the state 1 of $\mathcal{N}$, i.e. $1 \vDash \alpha$. Similarly to substructural models, a formula of the form $\alpha \to \beta$ is valid in $\mathcal{N}$ iff for any $a \in A$, if a supports α, then a supports β. We will denote the set of formulas that are valid in $\mathcal{N}$ as $Log_{\vDash}(\mathcal{N})$.

Let $\mathcal{M}$ be a substructural model and $\mathcal{N}$ an information model. The set of states of $\mathcal{M}$ that support α w.r.t. the relation $\Vdash$ will be called the proposition expressed by α in $\mathcal{M}$ and it will be denoted as $||\alpha||_{\mathcal{M}}$. Similarly, the set of states of $\mathcal{N}$ that support α w.r.t. the relation $\vDash$ will be called the proposition expressed by α in $\mathcal{N}$ and it will be denoted as $||\alpha||_{\mathcal{N}}$.

The following result shows an important difference between the two frameworks. It says that propositions (meanings of formulas) in substructural models are upward closed sets of states, while propositions in information models are ideals.

Theorem 1. *Let α be an $\mathcal{L}$-formula and let $\mathcal{M}$ be a substructural model and $\mathcal{N}$ an information model. Then it holds:*

(a) $||\alpha||_{\mathcal{M}}$ is an upward closed set in $\mathcal{M}$,
(b) $||\alpha||_{\mathcal{N}}$ is an ideal in $\mathcal{N}$.

An important connection between the two frameworks is revealed in the following construction. For any substructural model $\mathcal{M} = \langle W, L, \leq, R, C, V \rangle$ we can construct an "equivalent" information model $\mathcal{M}^i = \langle A, +, \times, \cdot, 0, 1, C^i, V^i \rangle$ in the following way:

- A is the set of upward closed sets of states in $\mathcal{M}$;
- $u + v = u \cup v$; $u \times v = u \cap v$;
- $u \cdot v = \{z \in W; \text{ there are } x \in u \text{ and } y \in v \text{ such that } Rxyz\}$;
- $0 = \emptyset$; $1 = L$;
- $uC^i v$ iff there are $x \in u$ and $y \in v$ such that xCy;
- $u \in V^i(p)$ iff $u \subseteq V(p)$.

Note that the ordering on $\mathcal{M}^i$ is inclusion $\subseteq$.

Theorem 2. *If $\mathcal{M}$ is a substructural model, then $\mathcal{M}^i$ is an information model.*

Theorem 3. *Let α be an $\mathcal{L}$-formula, $\mathcal{M}$ a substructural model, and u an upward closed set of its states. Then it holds:*

$$u \vDash \alpha \text{ in } \mathcal{M}^i \text{ iff } u \subseteq ||\alpha||_{\mathcal{M}}.$$

Corollary 1. *Let $\mathcal{M}$ be a substructural model. Then it holds: $Log_{\Vdash}(\mathcal{M}) = Log_{\vDash}(\mathcal{M}^i)$.*

The set of $\mathcal{L}$-formulas that are valid in all substructural models is axiomatized in [3] by a Hilbert axiomatic system consisting of the following axioms:

$$\alpha \to \alpha \qquad (\alpha \wedge (\beta \vee \gamma)) \to ((\alpha \wedge \beta) \vee (\alpha \wedge \gamma))$$
$$(\alpha \wedge \beta) \to \alpha \qquad (\alpha \wedge \beta) \to \beta$$
$$\alpha \to (\alpha \vee \beta) \qquad \beta \to (\alpha \vee \beta)$$
$$\alpha \to \top \qquad \bot \to \alpha$$

and the following one-sided (with /) and two-sided (with //) rules:[2]

---

[2] A two-sided rule of the form $\varphi // \psi$ means that φ and ψ are mutually inferable.

$$\alpha, \alpha \to \beta / \beta \qquad\qquad \alpha \to \beta, \beta \to \gamma / \alpha \to \gamma$$
$$\gamma \to \alpha, \gamma \to \beta / \gamma \to (\alpha \wedge \beta) \qquad \alpha \to \gamma, \beta \to \gamma / (\alpha \vee \beta) \to \gamma$$
$$\alpha \to (\beta \to \gamma) / / (\alpha \& \beta) \to \gamma \qquad \alpha \to (\beta \to \gamma) / / \beta \to (\alpha \to \gamma)$$
$$\alpha \to \neg\beta / / \beta \to \neg\alpha \qquad\qquad t \to \alpha / / \alpha$$

The logic is a distributive, commutative and non-associative full Lambek calculus with a paraconsistent negation.[3] We will denote the axiomatic system, as well as the logic determined by the system, as λ_0. Completeness can be proved via a canonical model construction. We will present the construction in a general form that can be applied to any logic extending λ_0.

Definition 1. *An $\mathcal{L}$-logic over λ_0 is any set of $\mathcal{L}$-formulas λ that satisfies the following three conditions: (a) λ contains all the axioms of λ_0, (b) λ is closed under the rules of λ_0, (c) λ is closed under uniform substitutions of $\mathcal{L}$-formulas.*

Definition 2. *Let λ be an $\mathcal{L}$-logic over λ_0. A nonempty set of $\mathcal{L}$-formulas Δ is an λ-theory if it satisfies the following two conditions:*

(a) if $\alpha \in \Delta$ and $\beta \in \Delta$, then $\alpha \wedge \beta \in \Delta$,
(b) if $\alpha \in \Delta$ and $\alpha \to \beta \in \lambda$, then $\beta \in \Delta$.

Let Δ be a λ-theory. We say that Δ is prime if $\Delta \neq \mathcal{L}$ and it holds:

(c) if $\alpha \vee \beta \in \Delta$, then $\alpha \in \Delta$ or $\beta \in \Delta$.

In accordance with [10], we will introduce a general construction of a canonical model for any $\mathcal{L}$-logic over λ_0.

Definition 3. *Let λ be an $\mathcal{L}$-logic over λ_0. The canonical substructural model of λ is the structure*

- $\mathcal{M}^\lambda = \langle W^\lambda, L^\lambda, \leq^\lambda, R^\lambda, C^\lambda, V^\lambda \rangle$, *where*
- W *is the set of all prime λ-theories,*
- $\Delta \in L^\lambda$ *iff* $t \in \Delta$,
- $\Delta \leq^\lambda \Gamma$ *iff* $\Delta \subseteq \Gamma$,
- $R^\lambda \Delta \Gamma \Omega$ *iff for all $\mathcal{L}$-formulas α, β, if $\alpha \to \beta \in \Delta$ and $\alpha \in \Gamma$, then $\beta \in \Omega$,*
- $\Delta C^\lambda \Gamma$ *iff for every $\mathcal{L}$-formula α, if $\neg\alpha \in \Delta$, then $\alpha \notin \Gamma$,*
- $\Delta \in V(p)$ *iff $p \in \Delta$.*

Now it can be verified that $\mathcal{M}^\lambda$ is indeed a substructural model. Moreover, it holds for every prime λ-theory Δ and every $\mathcal{L}$-formula α that

$$\Delta \Vdash \alpha \text{ in } \mathcal{M}^\lambda \text{ iff } \alpha \in \Delta.$$

As a consequence, λ is exactly the set of formulas valid in $\mathcal{M}^\lambda$. It follows that the system λ_0 is complete with respect to the class of all substructural models: if α is not provable in λ_0, then it is not valid in the substructural model $\mathcal{M}^{\lambda_0}$. Soundness can be verified in a mechanical way.

---

[3] The logic is paraconsistent in the sense that $(p \wedge \neg p) \to q$ is not a derivable formula in the system.

The completeness of λ_0 with respect to information models can be proved in a similar fashion.[4] The canonical model is not constructed from prime λ-theories but from all λ-theories.

Definition 4. *Let λ be an $\mathcal{L}$-logic over λ_0. The canonical information model of λ is the structure*

- *$\mathcal{N}_\lambda = \langle A_\lambda, +_\lambda, \times_\lambda, \cdot_\lambda, 0_\lambda, 1_\lambda, C_\lambda, V_\lambda \rangle$, where*
- *A_λ is the set of all λ-theories,*
- *$\Delta +_\lambda \Gamma = \Delta \cap \Gamma$,*
- *$\Delta \times_\lambda \Gamma = \{\alpha;$ for some $\delta \in \Delta$ and $\gamma \in \Gamma, (\gamma \wedge \delta) \to \alpha \in \lambda\}$,*
- *$\Delta \cdot_\lambda \Gamma = \{\alpha;$ for some $\delta \in \Delta$ and $\gamma \in \Gamma, (\gamma \& \delta) \to \alpha \in \lambda\}$,*
- *0_λ is the set of all $\mathcal{L}$-formulas,*
- *$1_\lambda = \lambda$,*
- *$\Delta C_\lambda \Gamma$ iff for every $\mathcal{L}$-formula α, if $\neg \alpha \in \Delta$, then $\alpha \notin \Gamma$,*
- *$\Delta \in V_\lambda(p)$ iff $p \in \Delta$.*

Note that the induced ordering is the superset relation $\supseteq$. Again, it can be verified that $\mathcal{N}_\lambda$ is indeed an information model and that for every λ-theory Δ and every $\mathcal{L}$-formula α it holds that

$$\Delta \Vdash \alpha \text{ in } \mathcal{N}_\lambda \text{ iff } \alpha \in \Delta.$$

As in the case of substructural models, completeness of λ_0 w.r.t. the class of all information models follows from this fact.

3   Knowledge as an Existential Modality

In this section, we will discuss extensions of the two frameworks that provide semantics to the knowledge operator. The main contribution of the paper [3] is an extension of substructural models by an additional binary relation S that is interpreted as a relation of "being a reliable source".

Definition 5. *An epistemic substructural model is a pair $\langle \mathcal{M}, S \rangle$, where $\mathcal{M} = \langle W, L, \leq, R, C, V \rangle$, is a substructural model and S is a binary relation on W satisfying:*

(a) if xSy, then $x \leq y$,
(b) if xSy, then xCy,
(c) if xSy, $x' \leq x$, and $y \leq y'$, then $x'Sy'$.

"xSy" is read as "the state x is a reliable source of the state y". The conditions (a)-(c) express basic restrictions that are required for such a relation. The condition (a) guarantees that every state incorporates all the information that is contained in all of its reliable sources. The condition (b) says that every state

---

[4] In [8] a slightly modified version of this result was proved. In that paper we do not assume commutativity of fusion and distributivity of the lattice. Completeness of a weaker system w.r.t. the resulting class of models was proved.

is compatible with its reliable sources. And the condition (c) says that if x is a reliable source of y, then every state that is informationally contained in x is a reliable source for every state that extends y.

The relation $\Vdash$ can now be defined for the language $\mathcal{L}_K$. The semantics for the logical symbols of the language $\mathcal{L}$ are determined by the same clauses as in the previous section. The semantics of the knowledge operator K is determined with the help of the relation S. For a state x of a given epistemic substructural model, let $S(x)$ denote the set of reliable sources of x, that is $S(x) = \{y \in W; ySx\}$.

$$x \Vdash K\alpha \text{ iff there is } y \in S(x) \text{ such that } y \Vdash \alpha.$$

We will assume that the knowledge operator is relative to an agent. The semantic clause says that the agent knows α in the state x iff α is supported by a reliable source of x.

A basic feature of the semantics is that the support of all $\mathcal{L}_K$-formulas is upward persistent, i.e. propositions expressed by $\mathcal{L}_K$-formulas are upward closed sets of states, which extends Theorem 1-(a).

Let x be a state of an epistemic substructural model. Let $|x|$ denote the set of $\mathcal{L}_K$-formulas α that are supported by x. The set $|x|$ can be very well inconsistent in the sense that it can contain a formula as well as its negation. It follows from the required conditions that the operator K selects a consistent part of x. That is, if we define $|x|_K = \{\alpha \in \mathcal{L}_K; K\alpha \in |x|\}$, then there is no $\mathcal{L}_K$-formula β such that both β and $\neg\beta$ would be in $|x|_K$.

The semantics based on the class of epistemic substructural models determines a logic for the language $\mathcal{L}_K$. It can be axiomatized as an extension of the system λ_0 by the following three axioms and one rule:

R1 $\alpha \to \beta / K\alpha \to K\beta$.
A1 $K\alpha \to \alpha$,
A2 $(\neg\alpha \wedge K\alpha) \to \bot$,
A3 $K(\alpha \vee \beta) \to (K\alpha \vee K\beta)$,

Let us denote the system consisting of the axioms and rules of λ_0 plus the rule $R1$ as λ_0^K. If one or more axioms from A1-A3 are added to λ_0^K, this will be indicated by the respective indexes. For example, the system consisting of the axioms and rules of λ_0^K plus the axioms A1 and A2 will be denoted as λ_{12}^K.

It was proved in [3] that the system λ_{123}^K is sound and complete with respect to the class of epistemic substructural models. Completeness was proved by an extension of the canonical model construction.

Definition 6. *An $\mathcal{L}_K$-logic over λ_0^K is any set of $\mathcal{L}_K$-formulas λ that satisfies the following three conditions: (a) λ contains all the axioms of λ_0^K, (b) λ is closed under the rules of λ_0^K, (c) λ is closed under uniform substitutions of $\mathcal{L}_K$-formulas.*

For any $\mathcal{L}_K$-logic over λ_0^K, denoted as λ, we define the notion of a λ-theory and a prime λ-theory in the same way as in Definition 2.

Definition 7. *Let λ be an $\mathcal{L}_K$-logic over λ_0^K that contains all instances of the axioms A1-A3. The canonical epistemic substructural model of λ is the structure $\mathcal{M}_K^\lambda = \langle \mathcal{M}^\lambda, S^\lambda \rangle$, where $\mathcal{M}^\lambda$ is constructed from the prime λ-theories in the same way as in Definition 3, and S^λ is defined in the following way:*

$$\Delta S^\lambda \Gamma \text{ iff for every } \mathcal{L}_K - formula \ \alpha, if \ \alpha \ \in \ \Delta, then \ K \ \alpha \ \in \ \Gamma.$$

Under the assumption that λ contains all instances of the axioms A1-A3, the resulting structure is an epistemic substructural model and it holds for any prime λ-theory Δ and any $\mathcal{L}_K$-formula α that

$$\Delta \Vdash \alpha \text{ in } \mathcal{M}_K^\lambda \text{ iff } \alpha \in \Delta.$$

This gives us the completeness of the system λ_{123}^K w.r.t. the class of all epistemic substructural models.

Let us now informally discuss the principles A1–A3 and R1 as candidates for the principles characterizing a logic of knowledge based on substructural logics. The rule R1 can be criticized on the basis that it leads to a weak version of the problem of omniscience. However, we will not concentrate on this problem in the present paper and we will treat this rule as unproblematic. In other words, we assume that some form of implicit knowledge counts also as knowledge, which has to be the case if the validity of R1 is assumed (i.e. the claim that α logically implies β entails the claim that knowing α logically implies knowing β).

A1 is usually regarded as a characteristic principle of the knowledge operator which distinguishes it from belief. We do not regard this principle as controversial. However, the axiom A1 is denoted as "truth principle" in [3], which is slightly misleading in the context of the whole framework of (epistemic) substructural models. It is reasonable to use the term "truth" if one uses the notion of a possible world. Truth can be understood as a relation between sentences (formulas) and possible worlds. However, the (epistemic) substructural models consist of information states, bodies of information that can be incomplete and/or inconsistent. Formulas are evaluated with respect to these bodies and there is no need of a "reality" behind these bodies that would enter into the semantic picture. Then the principle A1 should be read as "the information supporting the claim that the agent knows α supports also α" rather than as "if it is true that the agent knows α, then α has to be true as well".

Before we discuss the plausibility of A2 let us reflect more carefully on the role of the knowledge operator in the framework. What is the intended meaning of the metaclaim "$x \Vdash K\alpha$". This claim has two alternative readings. According to the first reading "$x \Vdash K\alpha$" means that

r-a: if x is the agent's information state, then the agent knows α.

The second reading is that

r-b: according to the information state x, the agent knows α.

It seems that r-a is the intended reading in [3]. However, we will prefer r-b (which resembles the interpretation from [11]) for the following two reasons: first, if we adopt r-a, it is rather difficult to provide a coherent interpretation of iterated knowledge operator (as in KKp); second, if r-b is adopted, support of $K\alpha$ by a state and support of the other types of complex formulas are interpreted along the same lines. We will see that the two readings have different impacts on the acceptability of some logical principles.

A2 would be equivalent to A1 if the background logic for the language $\mathcal{L}$ was classical. However, the two formulas have "different meaning" over substructural logics. In the framework of epistemic substructural models validity of the axiom A2 amounts to saying that there cannot be a state that supports both $\neg\alpha$ and $K\alpha$. However, this principle seems to be problematic in both readings r-a and r-b. Consider r-a first. For example, let p represent the sentence "Peter is in Prague". Suppose that the agent has a strong evidence supporting this claim. For example, she herself is in Prague and just has seen clearly Peter walking on a street. This evidence serves as a reliable source of the agent's state, so her state supports Kp. But at the same time, someone told the agent yesterday that Peter would not be in Prague. So, the agent's state supports also $\neg p$. These two pieces of information are incompatible but that is fine, since we do not exclude states that support conflicting information.

A principle that would be a more plausible alternative to A2 with respect to the reading r-a, is the following modification of A2:

A4 $(K\neg\alpha \wedge K\alpha) \rightarrow \bot$.

In the presence of A1 (but not without A1), A4 is weaker than A2. It amounts to saying that there cannot be a state that would support both $K\neg\alpha$ and $K\alpha$. It is an acceptable idealizing assumption that reliable sources cannot give us conflicting information. We obtain an adequate semantics for λ_{134}^K if we define epistemic substructural models in such a way that we replace the condition (b) in Definition 5 by

(b') if xSz and ySz, then xCy.

The system λ_{134}^K is sound and complete with respect to this modified semantics.

However, if we adopt the reading r-b, even the weaker axiom A4 seems to be problematic. If we do not exclude information states that support incompatible information, it is also reasonable not to exclude states that support incompatible information about the knowledge of the agent. As there can be some weak evidence that α and at the same time some weak evidence that $\neg\alpha$, there can also be some weak evidence that the agent knows α and at the same time some weak evidence that the agent knows $\neg\alpha$.

But it is easy to avoid the validity of A2, or its modification A4, completely. It suffices simply to require in Definition 5 neither the condition (b), nor its modification (b'). λ_{13}^K is sound and complete w.r.t. the resulting class of models.

Far the most problematic and the most controversial from the four principles, if we interpret K as knowledge, is the axiom A3. It should be possible to have information states according to which the agent knows a disjunction without knowing any of its disjuncts. However, the axiom holds throughout the class of all substructural models. If a state x in a substructural model supports $K(\alpha \vee \beta)$ this means that there is a source that supports $\alpha \vee \beta$. But this means that the source supports α or β, so x supports $K\alpha \vee K\beta$. The standard condition for disjunction, in the interaction with the existential modality, leads straightforwardly to the validity of the axiom. So, it is not possible to avoid the validity of the axiom A3 as easily as the validity of A2.

If one wants to define knowledge by an existential clause and at the same time not to validate A3, one needs an alternative semantic clause for disjunction, as in the semantics based on information models. We would like to stress that such a non-standard condition is natural, if the points in the semantic structures are interpreted as bodies of information. A body of information may naturally support a disjunction without supporting any of its disjuncts. The standard condition for disjunction is natural in frameworks that are based on truth conditions relative to possible worlds (a disjunction is true iff at least one of the disjuncts is true) but not in frameworks that are based on support conditions relative to bodies of information.

Definition 8. *Let $\mathcal{E}$ be a pair $\langle \mathcal{N}, S \rangle$, where $\mathcal{N} = \langle A, +, \times, \cdot, 0, 1, C, V \rangle$ is an information model, and S is a binary relation on A. Consider the following conditions:*

(a) $0S0$,
(b) if aSb, $a \leq a'$, and $b' \leq b$, then $a'Sb'$,
(c) if aSb and aSc, then $aS(b+c)$,
(d) if aSb, then $b \leq a$,
(e) if aSb and $b \neq 0$, then aCb,
(f) if aSb, cSb, and $b \neq 0$, then aCc.

We say that $\mathcal{E}$ is an epistemic information model if it satisfies the conditions (a)–(c). $\mathcal{E}$ is an intended epistemic information model if it satisfies (a)–(d). $\mathcal{E}$ is a strong epistemic information model if it satisfies (a)–(e). $\mathcal{E}$ is a full epistemic information model if it satisfies (a)–(d) and (f).

Again, let us define for any state a of any epistemic information model the set of its sources $S(a) = \{b \in A; bSa\}$ and add to the semantics the following clause:

$$a \vDash K\alpha \text{ iff there is } b \in S(a) \text{ such that } b \vDash \alpha.$$

It holds for any epistemic information model that propositions expressed by $\mathcal{L}_K$-formulas are ideals, which extends Theorem 1-(b). We will show that there are (even strong and full) epistemic information models where A3 fails. Consider the following lattice:

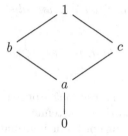

We define an epistemic information model in the following way: A is the set $\{0, a, b, c, 1\}$; $+$ is the join of the lattice; $\times$ is identical with $\cdot$ and it is the meet of the lattice; xCy iff $x \neq 0$ and $y \neq 0$; $S = \{\langle d, 0 \rangle; d \in A\} \cup \{\langle 1, a \rangle\}$; $V(p)$ is the ideal generated by b, i.e. $V(p) = \{0, a, b\}$, and $V(q)$ is the ideal generated by q, i.e. $V(q) = \{0, a, c\}$. The resulted structure is a strong and full epistemic information model and it holds: a supports $K(p \vee q)$ but a does not support $Kp \vee Kq$.

Definition 9. *Let λ be an $\mathcal{L}_K$-logic over λ_0^K. The canonical epistemic information model of λ is the structure $\mathcal{N}_\lambda^K = \langle \mathcal{N}_\lambda, S_\lambda \rangle$, where $\mathcal{N}_\lambda$ is constructed from the λ-theories in the same way as in Definition 4, and S_λ is defined in the following way:*

$\Delta S_\lambda \Gamma$ iff for every $\mathcal{L}_K$-formula α, if $\alpha \in \Delta$, then $K\alpha \in \Gamma$.

Theorem 4. *Let λ be an $\mathcal{L}_K$-logic over λ_0^K. The structure $\mathcal{N}_\lambda^K$ is an epistemic information model. Moreover, it holds for any λ-theory Δ and any $\mathcal{L}_K$-formula α that*

$$\Delta \vDash \alpha \ in \ \mathcal{N}_\lambda^K \ iff \ \alpha \in \Delta.$$

Theorem 5. *Let λ be an $\mathcal{L}_K$-logic over λ_0^K. Then (a) λ_0^K is sound and complete w.r.t. the class of epistemic information models; (b) λ_1^K is sound and complete w.r.t. the class of intended epistemic information models; (c) λ_{12}^K is sound and complete w.r.t. the class of strong epistemic information models; (d) λ_{14}^K is sound and complete w.r.t. the class of full epistemic information models.*

We have succeeded in avoiding the unwanted law A3. Now, we will characterize the class of frames that validate the law.

Definition 10. *An epistemic information frame is an epistemic information model without the valuation. If an epistemic information frame $\mathcal{F}$ is obtained from an epistemic information model $\mathcal{E}$ in this way, we say that $\mathcal{E}$ is a model on $\mathcal{F}$. An $\mathcal{L}_K$-formula α is valid in an epistemic information frame $\mathcal{F}$ if α is valid in every model on $\mathcal{F}$. A schema characterizes a class of frames $\mathcal{C}$ if it holds that every instance of the schema is valid in $\mathcal{F}$ iff $\mathcal{F} \in \mathcal{C}$.*

Definition 11. *An epistemic information frame is called source-distributive if its source relation S has the following property:*

if $(a + b)Sc$ then there are d, e such that aSd and bSe and $c \leq d + e$.

An epistemic information model on a source-distributive frame is also called source-distributive.

Theorem 6. *The schema A3 characterizes the class of source-distributive epistemic information frames.*

Now we can extend the constructions from the previous section revealing a relation between the framework of epistemic substructural models and framework of source-distributive strong epistemic information models. Let $\mathcal{D} = \langle \mathcal{M}, S \rangle$ be an epistemic substructural model. We define $\mathcal{D}^i$ as the pair $\langle \mathcal{M}^i, S^i \rangle$, where $\mathcal{M}^i$ is defined as in the previous section from the upward closed sets in $\mathcal{M}$ and S^i is defined as follows. For any upward closed sets u, v in $\mathcal{M}$:

$$uS^i v \text{ iff for every } y \in v \text{ there is } x \in u \text{ such that } xSy.$$

Theorem 7. *If $\mathcal{D}$ is an epistemic substructural model, then $\mathcal{D}^i$ is a source-distributive strong epistemic information model.*

Theorem 8. *Let α be an $\mathcal{L}_K$-formula, $\mathcal{D}$ an epistemic substructural model, and u an upward closed set of its states. Then it holds:*

$$u \vDash \alpha \text{ in } \mathcal{D}^i \text{ iff } u \subseteq ||\alpha||_{\mathcal{D}}.$$

Corollary 2. *Let $\mathcal{D}$ be an epistemic substructural model. Then it holds: $Log_{\Vdash}(\mathcal{D}) = Log_{\vDash}(\mathcal{D}^i)$.*

4 Conclusion

In this paper, we have compared two semantics for substructural logics, one of them, based on substructural models, used in [3] as a basis for a non-standard epistemic logic, and the second one, based on information models, used in [8] as a basis for a logic of questions. We have presented results showing that the two frameworks are intimately related and determine the same logic as regards the language without modalities. However, if we add an existential modality into these two frameworks, the framework of information models is richer and especially suitable if we interpret the modality as knowledge, since in it, unlike in the framework of substructural models, we are not forced to accept the principle that the modality distributes over disjunction.

Appendix

In the Appendix, we will provide proofs of the results of the paper. For the proof of Theorem 1-(a), see [10]. The proof of 1-(b) is contained in [8].

Proof of Theorem 2: We are proving that $\mathcal{M}^i = \langle A, +, \times, \cdot, 0, 1, C^i, V^i \rangle$ is an information model, so we have to verify that all conditions from the definition of information models are satisfied. Since $+$ is union and $\times$ intersection, it is immediate that $\langle A, +, \times \rangle$ is a distributive lattice. Since 0 is the empty set, we also immediately get $0 + a = 0$ and $0 \cdot a = 0$. The definition of $\cdot$ leads also directly to its distributivity over union. Moreover, $\cdot$ is commutative, since $Rxyz$ implies $Ryxz$.

We will verify that $1 \cdot a = a$. We have to show that for any upward closed set u it holds that $L \cdot u = u$. First, assume that $z \in L \cdot u$. So, there is $x \in L$ and $y \in u$ such that $Rxyz$. Then it holds that $y \leq z$. Since u is upward closed, it follows that $z \in u$. Second, assume $z \in u$. Since $z \leq z$, there is some $x \in L$ such that $Rxzz$. It follows that $z \in L \cdot u$.

The conditions related to the compatibility relation can be easily verified. Moreover, since the ordering in $\mathcal{M}^i$ is inclusion, it holds for any atomic formula p that $V^i(p)$ is the principal ideal generated by $V(p)$.

Proof of Theorem 3: We are proving that for every upward closed set of states u it holds that $u \vDash \alpha$ in $\mathcal{M}^i$ iff for all $x \in u$, $x \Vdash \alpha$ in $\mathcal{M}$. We can proceed by induction on the complexity of α. The case of atomic formulas as well as the constants $\bot, \top, t$ is straightforward. As the induction hypothesis we assume that the statement holds for some $\mathcal{L}$-formulas α and β.

Negation: First, assume that $u \nvDash \neg\beta$. Then for some v, uC^iv and $v \vDash \beta$, i.e. for all $x \in v$, $x \Vdash \beta$. Then for some v there are $y \in u$ and $z \in v$ such that yCz and $z \Vdash \beta$. So for some $y \in u$, $y \nVdash \neg\beta$.

Second, assume that for some $y \in u$, $y \nVdash \neg\beta$, i.e. for some z, yCz and $z \Vdash \beta$. Take $v = \{w \in W; z \leq w\}$. Then uC^iv and for all $x \in v$, $x \Vdash \beta$, i.e. $v \vDash \beta$. So $u \nvDash \neg\beta$.

Implication: First, assume that $u \nvDash \alpha \to \beta$. This means that there is an upward closed set v such that $v \vDash \alpha$ but $u \cdot v \nvDash \beta$. Then it follows from the induction hypothesis that every state of v supports α in $\mathcal{M}$ and there is a state $z \in u \cdot v$ such that $z \nVdash \beta$. So, there are $x \in u$ and $y \in v$ such that $Rxyz$. Since $y \Vdash \alpha$ and $z \nVdash \beta$, it follows that $x \nVdash \alpha \to \beta$.

Second, assume that for some $x \in u$, $x \nVdash \alpha \to \beta$. That means that there are y, z such that $Rxyz$, and $y \Vdash \alpha$ and $z \nVdash \beta$. Let v be the set $\{w \in W; y \leq w\}$. Then $z \in u \cdot v$ and it follows that $u \cdot v \nvDash \beta$. Moreover, $v \vDash \alpha$, so $u \nvDash \alpha \to \beta$.

Conjunctions: The step for the conjunction $\wedge$ is straightforward. We will show the step for the intentional conjunction $\&$. First, assume that $u \vDash \alpha\&\beta$. So, there are v_1, v_2 such that $v_1 \vDash \alpha$, $v_2 \vDash \beta$, and $u \subseteq v_1 \cdot v_2$. Take an arbitrary state $x \in u$. Then $x \in v_1 \cdot v_2$, which means that there are $y \in v_1$ (so $y \Vdash \alpha$) and $z \in v_2$ (so $z \Vdash \beta$) such that $Ryzx$. Therefore, $x \Vdash \alpha\&\beta$.

Second, assume that for every $x \in u$, $x \Vdash \alpha\&\beta$. So, for every $x \in u$, there are $y, z \in W$ such that $y \Vdash \alpha$, $z \Vdash \beta$, and $Ryzx$. For every $x \in u$, let us select some y, z with this property and denote them y_x and z_x. Now we define

$$v_1 = \bigcup_{x \in u} \{w \in W; y_x \leq w\},$$

$$v_2 = \bigcup_{x \in u} \{w \in W; z_x \leq w\}.$$

Then $v_1 \vDash \alpha$ and $v_2 \vDash \beta$. Moreover, $u \subseteq v_1 \cdot v_2$. It follows that $u \vDash \alpha\&\beta$.

Disjunction: First, assume that $u \vDash \alpha \vee \beta$. So, there are v_1, v_2 such that $v_1 \vDash \alpha$, $v_2 \vDash \beta$, and $u \subseteq v_1 \cup v_2$. Take an arbitrary state $x \in u$. Then $x \in v_1$ or $x \in v_2$, so $x \Vdash \alpha$ or $x \Vdash \beta$. Therefore, $x \Vdash \alpha \vee \beta$.

Second, assume that for every $x \in u$, $x \Vdash \alpha \vee \beta$. So, for every $x \in u$, $x \Vdash \alpha$ or $x \Vdash \beta$. Take $v_1 = ||\alpha||_{\mathcal{M}}$ and $v_2 = ||\beta||_{\mathcal{M}}$. Then $v_1 \vDash \alpha$ and $v_2 \vDash \beta$. Moreover, $u \subseteq v_1 \cup v_2$. It follows that $u \vDash \alpha \vee \beta$.

Proof of Theorem 4: To show that $\mathcal{N}_\lambda^K$ is an epistemic information model we have to verify the conditions (a)–(c) from Definition 8 which is straightforward. We will show the inductive step for K in the proof of the claim that $\Delta \vDash \alpha$ iff $\alpha \in \Delta$. Assume that the claim holds for some $\mathcal{L}_K$-formula β. We have to prove that $K\beta \in \Delta$ iff there is a λ-theory Γ such that $\Gamma S_\lambda \Delta$ and $\beta \in \Gamma$.

First, assume that there is a λ-theory Γ such that $\Gamma S_\lambda \Delta$ and $\beta \in \Gamma$. Then we have directly from the definition of S_λ that $K\beta \in \Delta$.

Second, assume $K\beta \in \Delta$. We define $\Gamma = \{\gamma; \beta \to \gamma \in \lambda\}$. Γ is a λ-theory, due to the application of the rules: $\beta \to \gamma, \gamma \to \delta / \beta \to \delta$ and $\beta \to \delta, \beta \to \epsilon / \beta \to (\delta \wedge \epsilon)$. Moreover, due to the axiom $\beta \to \beta$, $\beta \in \Gamma$. Now, it will suffice to prove that $\Gamma S_\lambda \Delta$. Assume that $\gamma \in \Gamma$, i.e. $\beta \to \gamma \in \lambda$. Then $K\beta \to K\gamma \in \lambda$ due to $R1$. Since $K\beta \in \Delta$, then also $K\gamma \in \Delta$, since Δ is a λ-theory.

Proof of Theorem 5: Proving soundness is a matter of a mechanical verification of the respective axioms in the respective class of models. We will discuss only completeness. (a) is a consequence of Theorem 4. To prove (b), we have to show that the canonical model of $\lambda = \lambda_1^K$ is intended, i.e. it holds: if $\Delta S_\lambda \Gamma$, then $\Delta \subseteq \Gamma$. Assume that $\Delta S_\lambda \Gamma$ and $\alpha \in \Delta$. Then $K\alpha \in \Gamma$ and since we assume that $K\alpha \to \alpha \in \lambda$, we have $\alpha \in \Gamma$.

To prove (c), we have to show that the canonical model of $\lambda = \lambda_{12}^K$ is strong. So we have to verify that in that case if $\Delta S_\lambda \Gamma$ and $\Gamma \neq \mathcal{L}_K$, then $\Delta C \Gamma$. Assume $\Delta S_\lambda \Gamma$ and $\Gamma \neq \mathcal{L}_K$. For the contradiction, assume that it does not hold that $\Delta C \Gamma$. Then it also does not hold that $\Gamma C \Delta$, so there is an $\mathcal{L}_K$-formula α such that $\neg \alpha \in \Gamma$ and $\alpha \in \Delta$. Then $K\alpha \in \Gamma$, and since we assume $A2$, it follows $\bot \in \Gamma$. Then $\Gamma = \mathcal{L}_K$, which is a contradiction. The proof of (d) is similar.

Proof of Theorem 6: Let a be a state of a model on a source-distributive epistemic information frame and let α be any $\mathcal{L}_K$-formula. Assume that $a \vDash K(\alpha \vee \beta)$. Then there is some $b \in S(a)$ such that $b \vDash \alpha \vee \beta$. So, there are c, d such that $c \vDash \alpha$, $d \vDash \beta$, and $b \leq c + d$. It follows that $(c + d)Sa$. Since the model is on a source-distributive frame, there are c', d' such that cSc', dSd', and $a \leq c' + d'$. Then $c' \vDash K\alpha$ and $d' \vDash K\beta$, and, as a consequence, $a \vDash K\alpha \vee K\beta$.

We have proved that A3 is valid in every model on any source-distributive epistemic information frame. Now consider an arbitrary epistemic information model on a frame that is not source-distributive. So, there are some states a, b, c such that $(a + b)Sc$ and there are no d, e such that aSd, bSe, and $c \leq d + e$. Define a valuation V on the model in such a way that $V(p) = \{d; d \leq a\}$ and $V(q) = \{e; e \leq b\}$. It holds that $a + b \vDash p \vee q$, so $c \vDash K(p \vee q)$. We want to show that $c \nvDash Kp \vee Kq$. For the sake of contradiction, assume that $c \vDash Kp \vee Kq$. Then there are d, e such that $d \vDash Kp$, $e \vDash Kq$, and $c \leq d + e$. But then there are $d' \in S(d)$ and $e' \in S(e)$ such that $d' \vDash p$ and $e' \vDash q$. It follows that aSd and bSe, which is in contradiction with our assumption.

Proof of Theorem 7: This result is an extension of Theorem 2. We have to verify that $\mathcal{D}^i$ satisfies the conditions (a)–(e) from Definition 8 and the condition from Definition 11. The conditions (a)–(c) are quite straightforward. To prove (e), we have to show that if $uS^i v$, then $v \subseteq u$. So, assume that $uS^i v$ and $x \in v$. Then there is $y \in u$ such that ySx. But then $y \leq x$, and so $x \in u$, which is what we wanted to show. Now we will show that $\mathcal{D}^i$ is source distributive. Assume $(u \cup v)S^i w$. Let us define

$$u' = \{y; \text{ there is } x \in u \text{ such that } xSy\},$$
$$v' = \{y; \text{ there is } x \in v \text{ such that } xSy\}.$$

Both u' and v' are upward closed. Moreover, $uS^i u'$, $vS^i v'$, and $w \subseteq u' \cup v'$.

Proof of Theorem 8: This result is an extension of Theorem 3. So, we are proving that for every upward closed set of states u it holds that $u \vDash \alpha$ in $\mathcal{D}^i$ iff for all $x \in u$, $x \Vdash \alpha$ in $\mathcal{D}$, and we have to show the inductive step for K. Assume that the statement holds for some $\mathcal{L}_K$-formula β. First, assume that $u \vDash K\beta$ in $\mathcal{D}^i$. So, there is $v \in S^i(u)$ such that $v \vDash \beta$. Then for every $x \in u$ there is some $y \in v$ such that ySx, and for every $y \in v$, $y \Vdash \beta$. Therefore, for every $x \in u$ there is y such that ySx, and $y \Vdash \beta$, that is, for every $x \in u$, $x \Vdash K\beta$.

Second, suppose for every $x \in u$, $x \Vdash K\beta$, that is, for every $x \in u$ there is x' such that $x'Sx$, and $x' \Vdash \beta$. Define $v = \{z; \text{ there is } x \in u \text{ such that } x' \leq z\}$. Obviously, v is upward closed, and for all $z \in v$, $z \Vdash \beta$. Moreover, for every $x \in u$ there is $z \in v$ (namely x') such that zSx. It follows that there is v such that $vS^i u$ and $v \vDash \beta$. Therefore, $u \vDash K\beta$ in $\mathcal{D}^i$.

References

1. Beall, J., et al.: On the ternary relation and conditionality. J. Philos. Logic **41**, 595–612 (2012)
2. Bílková, M., Majer, O., Peliš, M., Restall, G.: Relevant agents. In: Beklemishev, L., Goranko, V., Shehtman, V. (eds.) Advances in Modal Logic, vol. 8, pp. 22–38. College Publications, London (2010)
3. Bílková, M., Majer, O., Peliš, M.: Epistemic logics for skeptical agents. J. Logic Comput. **26**, 1815–1841 (2016)
4. Došen, K.: Sequent systems and groupoid models, part 2. Stud. Logica. **48**, 41–65 (1989)
5. Fagin, R., Halpern, J.Y., Moses, Y., Vardi, M.Y.: Reasoning About Knowledge. MIT Press, Cambridge (1995)
6. Majer, O., Peliš, M.: Epistemic logic with relevant agents. In: Peliš, M. (ed.) The Logica Yearbook 2008, pp. 123–135. College Publications, London (2009)
7. Mares, E.D.: Relevant Logics. A Philosophical Interpretation. Cambridge University Press, Cambridge (2004)

8. Punčochář, V.: Substructural inquisitive logics. Rev. Symbol. Logic (Under Consideration)
9. Restall, G.: Information Flow and Relevant Logic. In: Seligman, J., Westerståhl, D. (eds.) Logic, Language, and Computation: The 1994 Moraga Proceedings, pp. 463–477. CSLI Publications, Stanford (1995)
10. Restall, G.: An Introduction to Substructural Logics. Routledge, London (2000)
11. Sedlár, I.: Substructural epistemic logics. J. Appl. Non-Class. Logics **25**, 256–285 (2015)
12. van Benthem, J., Fernández-Duque, D., Pacuit, E.: Evidence logic: a new look at neighborhood structures. In: Bolander, T., Bräunder, T., Ghilardi, S., Moss, L. (eds.) Advances in Modal Logic, vol. 9, pp. 97–118. College Publications, London (2012)

Cut-Elimination for the Modal Grzegorczyk Logic via Non-well-founded Proofs

Yury Savateev[1(✉)] and Daniyar Shamkanov[1,2]

[1] National Research University Higher School of Economics, Moskva, Russia
yury.savateev@gmail.com
[2] Steklov Mathematical Institute of the Russian Academy of Sciences,
Moskva, Russia

Abstract. We present a sequent calculus for the modal Grzegorczyk logic Grz allowing non-well-founded proofs and obtain the cut-elimination theorem for it by constructing a continuous cut-elimination mapping acting on these proofs.

Keywords: Non-well-founded proofs · Grzegorczyk logic · Cut elimination

1 Introduction

The Grzegorczyk logic Grz is a well-known modal logic [3], which can be characterized by reflexive partially ordered Kripke frames without infinite ascending chains. This logic is complete w.r.t. the arithmetical semantics, where the modal connective $\square$ corresponds to the strong provability operator *"... is true and provable"* in Peano arithmetic.

Recently a new proof-theoretic description for the Gödel-Löb provability logic GL in the form of a sequent calculus allowing so-called cyclic, or circular, proofs was given in [6]. A feature of cyclic proofs is that the graph underlying a proof is not a finite tree but is allowed to contain cycles. Since GL and Grz are closely connected, we wonder whether cyclic and, more generally, non-well-founded proofs can be fruitfully considered in the case of Grz.

In this paper, we present a sequent calculus for the modal Grzegorczyk logic allowing non-well-founded proofs and obtain the cut-elimination theorem for it by constructing a continuous cut-elimination mapping acting on these proofs.

In Sect. 2, we recall an ordinary sequent calculus for Grz. In Sect. 3 we introduce the infinitary proof system Grz_∞. In Sect. 4 we establish the cut elimination result for Grz_∞ syntactically. Then, in Sect. 5 we prove the equivalence of the two systems. In Sect. 6 we discuss possible applications of the new system.

D. Shamkanov—The article was prepared within the framework of the Basic Research Program at the National Research University Higher School of Economics (HSE) and supported within the framework of a subsidy by the Russian Academic Excellence Project '5-100'. Both authors also acknowledge support from the Russian Foundation for Basic Research (grant no. 15-01-09218a).

J. Kennedy and R.J.G.B. de Queiroz (Eds.): WoLLIC 2017, LNCS 10388, pp. 321–335, 2017.
DOI: 10.1007/978-3-662-55386-2_23

2   Preliminaries

In this section we recall the modal Grzegorczyk logic Grz and define an ordinary sequent calculus for it.

Formulas of Grz, denoted by A, B, C, are built up as follows:

$$A ::= \bot \mid p \mid (A \to A) \mid \Box A,$$

where p stands for atomic propositions. We treat other boolean connectives and the modal operator $\Diamond$ as abbreviations:

$$\neg A := A \to \bot, \qquad \top := \neg\bot, \qquad A \wedge B := \neg(A \to \neg B),$$
$$A \vee B := (\neg A \to B), \qquad \Diamond A := \neg\Box\neg A.$$

The Hilbert-style axiomatization of Grz is given by the following axioms and inference rules:

Axioms:
  (i) Boolean tautologies;
  (ii) $\Box(A \to B) \to (\Box A \to \Box B)$;
  (iii) $\Box A \to \Box\Box A$;
  (iv) $\Box A \to A$;
  (v) $\Box(\Box(A \to \Box A) \to A) \to \Box A$.

Rules: modus ponens, $A/\Box A$.

Now we define an ordinary sequent calculus for Grz. A *sequent* is an expression of the form $\Gamma \Rightarrow \Delta$, where Γ and Δ are finite multisets of formulas. For a multiset of formulas $\Gamma = A_1, \ldots, A_n$, we set $\Box\Gamma := \Box A_1, \ldots, \Box A_n$.

The system $\mathsf{Grz_{Seq}}$, which is a variant of the sequent calculus from [2], is defined by the following initial sequents and inference rules:

$$\mathsf{Id} : \Gamma, A \Rightarrow A, \Delta \,, \qquad \bot : \Gamma, \bot \Rightarrow \Delta \,,$$

$$\to_\mathsf{L}: \frac{\Gamma, B \Rightarrow \Delta \quad \Gamma \Rightarrow A, \Delta}{\Gamma, A \to B \Rightarrow \Delta} \,, \qquad \to_\mathsf{R}: \frac{\Gamma, A \Rightarrow B, \Delta}{\Gamma \Rightarrow A \to B, \Delta} \,,$$

$$\mathsf{refl} : \frac{\Gamma, B, \Box B \Rightarrow \Delta}{\Gamma, \Box B \Rightarrow \Delta} \,, \qquad \Box_\mathsf{Grz} : \frac{\Box\Pi, \Box(A \to \Box A) \Rightarrow A}{\Gamma, \Box\Pi \Rightarrow \Box A, \Delta} \,.$$

Fig. 1. The system $\mathsf{Grz_{Seq}}$

Lemma 2.1. *The rule*

$$\text{weak}: \frac{\Gamma \Rightarrow \Delta}{\Pi, \Gamma \Rightarrow \Delta, \Sigma}$$

is admissible in $\mathsf{Grz_{Seq}}$.

Proof. Standard induction on the structure of a proof of the sequent $\Gamma \Rightarrow \Delta$. $\square$

The cut rule has the form

$$\text{cut}: \frac{\Gamma \Rightarrow A, \Delta \qquad \Gamma, A \Rightarrow \Delta}{\Gamma \Rightarrow \Delta},$$

where A is called the *cut formula* of the given inference.

Lemma 2.2. $\mathsf{Grz_{Seq}} + \mathsf{cut} \vdash \Gamma \Rightarrow \Delta$ *if and only if* $\mathsf{Grz} \vdash \bigwedge \Gamma \to \bigvee \Delta$.

Proof. Standard transformations of proofs. $\square$

Theorem 2.1. *If* $\mathsf{Grz_{Seq}} + \mathsf{cut} \vdash \Gamma \Rightarrow \Delta$, *then* $\mathsf{Grz_{Seq}} \vdash \Gamma \Rightarrow \Delta$.

A syntactic cut-elimination proof for the logic Grz was obtained by Borga and Gentilini in [2]. In this paper, we will give another proof of this cut-elimination theorem.

3   Non-well-founded Proofs

Now we define a sequent calculus for the logic Grz allowing non-well-founded proofs. The cut-elimination theorem for it will be proved in the next section.

Inference rules and initial sequents of the sequent calculus $\mathsf{Grz_\infty}$ have the following form:

$$\mathsf{Id_p}: \Gamma, p \Rightarrow p, \Delta, \qquad \bot: \Gamma, \bot \Rightarrow \Delta,$$

$$\to_\mathsf{L}: \frac{\Gamma, B \Rightarrow \Delta \qquad \Gamma \Rightarrow A, \Delta}{\Gamma, A \to B \Rightarrow \Delta}, \qquad \to_\mathsf{R}: \frac{\Gamma, A \Rightarrow B, \Delta}{\Gamma \Rightarrow A \to B, \Delta},$$

$$\mathsf{refl}: \frac{\Gamma, A, \Box A \Rightarrow \Delta}{\Gamma, \Box A \Rightarrow \Delta}, \qquad \Box: \frac{\Gamma, \Box \Pi \Rightarrow A, \Delta \qquad \Box \Pi \Rightarrow A}{\Gamma, \Box \Pi \Rightarrow \Box A, \Delta}.$$

Fig. 2. The system $\mathsf{Grz_\infty}$

The system $\mathsf{Grz_\infty} + \mathsf{cut}$ is defined by adding the rule (cut) to the system $\mathsf{Grz_\infty}$. An ∞-*proof* in $\mathsf{Grz_\infty}$ ($\mathsf{Grz_\infty} + \mathsf{cut}$) is a (possibly infinite) tree whose nodes are marked by sequents and whose leaves are marked by initial sequents and that is constructed according to the rules of the sequent calculus. In addition, every infinite branch in an ∞-proof must pass through a right premise of the rule $\Box$

infinitely many times. A sequent $\Gamma \Rightarrow \Delta$ is *provable* in Grz_∞ ($\mathsf{Grz}_\infty + \mathsf{cut}$) if there is an ∞–proof in Grz_∞ ($\mathsf{Grz}_\infty + \mathsf{cut}$) with the root marked by $\Gamma \Rightarrow \Delta$.

The *main fragment* of an ∞–proof is a finite tree obtained from the ∞–proof by cutting every infinite branch at the nearest to the root right premise of the rule $(\Box)$. The *local height* $|\pi|$ of an ∞–proof π is the length of the longest branch in its main fragment. An ∞–proof only consisting of an initial sequent has height 0.

For instance, consider an ∞–proof of the sequent $\Box(\Box(p \to \Box p) \to p) \Rightarrow p$:

$$
\begin{array}{c}
\cfrac{
  \cfrac{
    \mathsf{Id_p}\ \cfrac{F, p \Rightarrow p}{\ }
  }{\ }
  \quad
  \cfrac{
    \mathsf{Id_p}\ \cfrac{F, p \Rightarrow \Box p, p}{F \Rightarrow p \to \Box p, p}\ {\scriptstyle \to R}
    \quad\quad
    \Box\ \cfrac{
      \cfrac{p, F \Rightarrow p \quad\quad F \Rightarrow p \ \vdots}{p, F \Rightarrow \Box p}
      {F \Rightarrow p \to \Box p}\ {\scriptstyle \to R}
    }{F \Rightarrow \Box(p \to \Box p), p}\ \Box
  }{\ }
}{\ }
\end{array}
$$

$$
\mathsf{refl}\ \cfrac{\Box(p \to \Box p) \to p, F \Rightarrow p}{\Box(\Box(p \to \Box p) \to p) \Rightarrow p}\ {\scriptstyle \to L} \quad ,
$$

where $F = \Box(\Box(p \to \Box p) \to p)$. The local height of this ∞–proof equals to 4 and its main fragment has the form

$$
\mathsf{refl}\ \cfrac{
  \mathsf{Id_p}\ \cfrac{F, p \Rightarrow p}{\ } \quad
  \Box\ \cfrac{
    {\scriptstyle \to R}\ \cfrac{\mathsf{Id_p}\ F, p \Rightarrow \Box p, p}{F \Rightarrow p \to \Box p, p}
  }{F \Rightarrow \Box(p \to \Box p), p}
}{\dfrac{\Box(p \to \Box p) \to p, F \Rightarrow p}{F \Rightarrow p}}\ {\scriptstyle \to L} \quad .
$$

By $\mathcal{P}$ we denote the set of all ∞–proofs in $\mathsf{Grz}_\infty + \mathsf{cut}$. For $n \in \mathbb{N}$, we define binary relations $\sim_n$ on $\mathcal{P}$ by simultaneous induction:

1. $\pi \sim_0 \tau$ for any π, τ;
2. if $|\pi| = 0$, then $\pi \sim_n \pi$;
3. if π and τ are obtained by the same instance of inference rules $(\to L)$, (cut) from π', π'' and τ', τ'', where $\pi' \sim_n \tau'$ and $\pi'' \sim_n \tau''$, then $\pi \sim_n \tau$;
4. if π and τ are obtained by the same instance of inference rules $(\to R)$, (refl) from π' and τ', where $\pi' \sim_n \tau'$, then $\pi \sim_n \tau$;
5. if π and τ are obtained by the same instance of an inference rule $(\Box)$ from π', π'' and τ', τ'', where π', τ' are ∞–proofs for the left premises of $(\Box)$, and $\pi' \sim_{n+1} \tau'$, $\pi'' \sim_n \tau''$, then $\pi \sim_{n+1} \tau$.

Notice that $\pi \sim_1 \tau$ if and only if π and τ have the same main fragment.

Lemma 3.1. *For any $n \in \mathbb{N}$, we have that*

1. *the relation $\sim_n$ is an equivalence relation;*

2. *the relation $\sim_{n+1}$ is finer than the relation $\sim_n$ (i.e. $\pi \sim_{n+1} \tau$ implies $\pi \sim_n \tau$).*

In addition, the intersection of all relations $\sim_n$ is exactly the equality relation over $\mathcal{P}$.

Now we define a sequence $\mathcal{P}_n$ of subsets of $\mathcal{P}$ by simultaneous induction:

1. $\pi \in \mathcal{P}_0$ for any π;
2. if $|\pi| = 0$, then $\pi \in \mathcal{P}_n$;
3. if π is obtained by an instance of an inference rule ($\rightarrow_L$) from π' and π'', where $\pi', \pi'' \in \mathcal{P}_n$, then $\pi \in \mathcal{P}_n$;
4. if π is obtained by an instance of inference rules ($\rightarrow_R$), (refl) from π', where $\pi' \in \mathcal{P}_n$, then $\pi \in \mathcal{P}_n$;
5. if π is obtained by an instance of an inference rule ($\square$) from π' and π'', where π' is an ∞-proof for the left premise of ($\square$), and $\pi' \in \mathcal{P}_{n+1}$, $\pi'' \in \mathcal{P}_n$, then $\pi \in \mathcal{P}_{n+1}$.

Notice that $\mathcal{P}_0 = \mathcal{P}$ and $\mathcal{P}_1$ consists of the ∞-proofs that do not contain the cut rule in their main fragment.

Lemma 3.2. *We have that $\mathcal{P}_{n+1} \subset \mathcal{P}_n$ for any $n \in \mathbb{N}$. In addition, the intersection of all sets $\mathcal{P}_n$ consists exactly of the ∞-proofs in Grz_∞.*

For $\pi, \tau \in \mathcal{P}$, we define $d(\pi, \tau) = 2^{-\sup\{n\in\mathbb{N}\,|\,\pi\sim_n\tau\}}$, where by convention $2^{-\infty} = 0$. We see that an equivalence $\pi \sim_n \tau$ holds if and only if $d(\pi, \tau) \leqslant 2^{-n}$.

Proposition 3.3. *$(\mathcal{P}, d)$ is a complete metric space.*

A mapping $\mathcal{U} \colon \mathcal{P}^k \to \mathcal{P}$ is *nonexpansive* if for any $n \in \mathbb{N}$

$$\pi_1 \sim_n \tau_1, \ldots, \pi_k \sim_n \tau_k \Rightarrow \mathcal{U}(\pi_1, \ldots, \pi_k) \sim_n \mathcal{U}(\tau_1, \ldots, \tau_k),$$

which is equivalent to the standard condition

$$d(\mathcal{U}(\pi_1, \ldots, \pi_k), \mathcal{U}(\tau_1, \ldots, \tau_k)) \leqslant \max\{d(\pi_1, \tau_1), \ldots, d(\pi_k, \tau_k)\}.$$

Trivially, any nonexpansive mapping is continuous.

A nonexpansive mapping $\mathcal{U} \colon \mathcal{P} \to \mathcal{P}$ is called *adequate* if $\mathcal{U}(\mathcal{P}_1) \subset \mathcal{P}_1$ and $|\mathcal{U}(\pi)| \leqslant |\pi|$ for any $\pi \in \mathcal{P}$.

Recall that an inference rule is called *admissible* (in a given proof system) if, for any instance of the rule, the conclusion is provable whenever all premises are provable. In $\mathsf{Grz}_\infty + \mathsf{cut}$, we call a single-premise inference rule *strongly admissible* if there is an adequate mapping $\mathcal{U} \colon \mathcal{P} \to \mathcal{P}$ that maps any ∞-proof of the premise of the rule to an ∞-proof of the conclusion.

Lemma 3.4. *For any finite multisets of formulas Π and Σ, the inference rule*

$$\mathsf{wk}_{\Pi,\Sigma} : \frac{\Gamma \Rightarrow \Delta}{\Pi, \Gamma \Rightarrow \Delta, \Sigma}$$

is strongly admissible in $\mathsf{Grz}_\infty + \mathsf{cut}$.

Lemma 3.5. *For any formulas A and B, the rules*

$$\mathsf{li}_{A\to B} : \frac{\Gamma, A\to B \Rightarrow \Delta}{\Gamma, B \Rightarrow \Delta} \qquad \mathsf{ri}_{A\to B} : \frac{\Gamma, A\to B \Rightarrow \Delta}{\Gamma \Rightarrow A, \Delta}$$

$$\mathsf{i}_{A\to B} : \frac{\Gamma \Rightarrow A\to B, \Delta}{\Gamma, A \Rightarrow B, \Delta} \qquad \mathsf{i}_\perp : \frac{\Gamma \Rightarrow \perp, \Delta}{\Gamma \Rightarrow \Delta} \qquad \mathsf{li}_{\Box A} : \frac{\Gamma \Rightarrow \Box A, \Delta}{\Gamma \Rightarrow A, \Delta}$$

are strongly admissible in $\mathsf{Grz}_\infty + \mathsf{cut}$.

Lemma 3.6. *For any atomic proposition p, the rules*

$$\mathsf{acl}_p : \frac{\Gamma, p, p \Rightarrow \Delta}{\Gamma, p \Rightarrow \Delta} \qquad \mathsf{acr}_p : \frac{\Gamma \Rightarrow p, p, \Delta}{\Gamma \Rightarrow p, \Delta}$$

are strongly admissible in $\mathsf{Grz}_\infty + \mathsf{cut}$.

These lemmas can be obtained in a standard way, so we omit the proofs.

4   Cut Elimination

In this section we construct a continuous cut elimination mapping from $\mathcal{P}$ to $\mathcal{P}$, which eliminates all applications of the cut rule from any ∞-proof in $\mathsf{Grz}_\infty + \mathsf{cut}$. In what follows, we use nonexpansive mappings $wk_{\Pi,\Sigma}$, $\mathsf{li}_{A\to B}$, $\mathsf{ri}_{A\to B}$, $\mathsf{i}_{A\to B}$, $\mathsf{i}_\perp$, $\mathsf{li}_{\Box A}$, acl_p, acr_p from Lemmas 3.4, 3.5 and 3.6.

For a modal formula A, a nonexpansive mapping $\mathcal{R}$ from $\mathcal{P}_1 \times \mathcal{P}_1$ to $\mathcal{P}_1$ is called A-*reducing* if $\mathcal{R}(\pi', \pi'')$ is an ∞-proof of $\Gamma \Rightarrow \Delta$ whenever π' is an ∞-proof of $\Gamma \Rightarrow \Delta, A$ and π'' is an ∞-proof of $A, \Gamma \Rightarrow \Delta$.

Lemma 4.1. *For any atomic proposition p there is a p-reducing mapping* $\mathcal{R}_p$.

Lemma 4.2. *Given a B-reducing mapping* $\mathcal{R}_B$, *there is a* $\Box B$-*reducing mapping* $\mathcal{R}_{\Box B}$.

The proof of these two lemmas can be found in the Appendix.

Lemma 4.3. *For any formula A, there is an A-reducing mapping* $\mathcal{R}_A$.

Proof. We define $\mathcal{R}_A$ by induction on the structure of the formula A.

Case 1. The formula A has the form p. In this case, $\mathcal{R}_p$ is defined in Lemma 4.1.

Case 2. The formula A has the form $\perp$. Then we put $\mathcal{R}_\perp(\pi', \pi'') := \mathsf{i}_\perp(\pi')$, where $\mathsf{i}_\perp$ is the nonexpansive mapping from Lemma 3.5.

Case 3. The formula A has the form $B \to C$. Then we put

$$\mathcal{R}_{B\to C}(\pi', \pi'') := \mathcal{R}_C(\mathcal{R}_B(wk_{\emptyset,C}(\mathsf{ri}_{B\to C}(\pi'')), \mathsf{i}_{B\to C}(\pi')), \mathsf{li}_{B\to C}(\pi'')),$$

where $ri_{B \to C}$, $i_{B \to C}$, $li_{B \to C}$ are nonexpansive mappings from Lemma 3.5 and $wk_{\emptyset,C}$ is a nonexpansive mapping from Lemma 3.4.

Case 4. The formula A has the form $\Box B$. By the induction hypothesis, there is a B-reducing mapping $\mathcal{R}_B$. By Lemma 4.2 there is a $\Box B$-reducing mapping $\mathcal{R}_{\Box B}$. $\Box$

A mapping $\mathcal{U} \colon \mathcal{P} \to \mathcal{P}$ is called *root-preserving* if it maps ∞-proofs to ∞-proofs of the same sequents. The set of all root-preserving nonexpansive mappings from $\mathcal{P}$ to $\mathcal{P}$ is denoted by $\mathcal{N}$. We consider $\mathcal{N}$ as a metric space with the uniform metric:

$$\rho(\mathcal{U}, \mathcal{V}) = \sup_{\pi \in \mathcal{P}} d(\mathcal{U}(\pi), \mathcal{V}(\pi)).$$

Lemma 4.4. $(\mathcal{N}, \rho)$ *is a non-empty complete metric space.*

Proof. By Proposition 3.3, $\mathcal{P}$ is a complete metric space. Consequently the set $C(\mathcal{P}, \mathcal{P})$ of all continuous mappings from $\mathcal{P}$ to $\mathcal{P}$ with the uniform metric forms a complete metric space. It can be easily shown that $\mathcal{N}$ is a closed subset of $C(\mathcal{P}, \mathcal{P})$. In addition, the set $\mathcal{N}$ is non-empty, because the identity mapping belongs to $\mathcal{N}$. Thus $(\mathcal{N}, \rho)$ is a non-empty complete metric space. $\Box$

We define $\mathcal{N}_n := \{\mathcal{U} \in \mathcal{N} \mid \mathcal{U}(\mathcal{P}) \subset \mathcal{P}_n\}$.

Lemma 4.5. *There exists a mapping $\mathcal{E}^* \in \mathcal{N}_1$.*

Proof. Assume we have an ∞-proof π. We define $\mathcal{E}^*(\pi)$ by induction on $|\pi|$.

If $|\pi| = 0$, then we put $\mathcal{E}^*(\pi) = \pi$. Otherwise, consider the last application of an inference rule in π and define $\mathcal{E}^*$ as follows:

$$\to_L \frac{\overset{\displaystyle \pi_1}{\Gamma, B \Rightarrow \Delta} \qquad \overset{\displaystyle \pi_2}{\Gamma \Rightarrow A, \Delta}}{\Gamma, A \to B \Rightarrow \Delta} \;\longmapsto\; \to_L \frac{\overset{\displaystyle \mathcal{E}^*(\pi_1)}{\Gamma, B \Rightarrow \Delta} \qquad \overset{\displaystyle \mathcal{E}^*(\pi_2)}{\Gamma \Rightarrow A, \Delta}}{\Gamma, A \to B \Rightarrow \Delta} \;,$$

$$\to_R \frac{\overset{\displaystyle \pi_0}{\Gamma, A \Rightarrow B, \Delta}}{\Gamma \Rightarrow A \to B, \Delta} \;\longmapsto\; \to_R \frac{\overset{\displaystyle \mathcal{E}^*(\pi_0)}{\Gamma, A \Rightarrow B, \Delta}}{\Gamma \Rightarrow A \to B, \Delta} \;,$$

$$\text{refl} \frac{\overset{\displaystyle \pi_0}{\Gamma, A, \Box A \Rightarrow \Delta}}{\Gamma, \Box A \Rightarrow \Delta} \;\longmapsto\; \text{refl} \frac{\overset{\displaystyle \mathcal{E}^*(\pi_0)}{\Gamma, A, \Box A \Rightarrow \Delta}}{\Gamma, \Box A \Rightarrow \Delta} \;,$$

$$\Box \frac{\overset{\displaystyle \pi_1}{\Gamma, \Box \Pi \Rightarrow A, \Delta} \qquad \overset{\displaystyle \pi_2}{\Box \Pi \Rightarrow A}}{\Gamma, \Box \Pi \Rightarrow \Box A, \Delta} \;\longmapsto\; \Box \frac{\overset{\displaystyle \mathcal{E}^*(\pi_1)}{\Gamma, \Box \Pi \Rightarrow A, \Delta} \qquad \overset{\displaystyle \pi_2}{\Box \Pi \Rightarrow A}}{\Gamma, \Box \Pi \Rightarrow \Box A, \Delta} \;,$$

$$\text{cut} \; \frac{\overset{\pi_1}{\Gamma \Rightarrow \Delta, A} \quad \overset{\pi_2}{A, \Gamma \Rightarrow \Delta}}{\Gamma \Rightarrow \Delta} \; \longmapsto \; \mathcal{R}_A(\mathcal{E}^*(\pi_1), \mathcal{E}^*(\pi_2)).$$

Clearly, the mapping $\mathcal{E}^*$ is root-preserving, and $\mathcal{E}^*(\mathcal{P}) \subset \mathcal{P}_1$. We also see that $\mathcal{E}^*$ is nonexpansive, i.e. for any $n \in \mathbb{N}$ and any $\pi, \tau \in \mathcal{P}$

$$\pi \sim_n \tau \Rightarrow \mathcal{E}^*(\pi) \sim_n \mathcal{E}^*(\tau).$$

$\square$

Recall that a contraction mapping, or an operator, on a metric space (M, d) is a function f from M to itself, with the property that there is some real number $0 \leqslant k < 1$ such that for all x and y in M holds $d(f(x), f(y)) \leqslant k d(x, y)$.

Now we define a contraction operator $\mathcal{F} \colon \mathcal{N} \to \mathcal{N}$. The required cut-elimination mapping will be obtained as the fixed-point of $\mathcal{F}$.

For a root-preserving nonexpansive mapping $\mathcal{U}$ and an ∞-proof π of a sequent $\Gamma \Rightarrow \Delta$, we define $\mathcal{F}(\mathcal{U})(\pi)$. In the case $\pi \in \mathcal{P}_1$, $\mathcal{F}(\mathcal{U})(\pi)$ is introduced by induction on $|\pi|$. If $|\pi| = 0$, then we put $\mathcal{F}(\mathcal{U})(\pi) = \pi$. Otherwise, consider the last application of an inference rule in π and define $\mathcal{F}(\mathcal{U})$ as follows:

$$\rightarrow\text{L} \; \frac{\overset{\pi_1}{\Gamma, B \Rightarrow \Delta} \quad \overset{\pi_2}{\Gamma \Rightarrow A, \Delta}}{\Gamma, A \to B \Rightarrow \Delta} \; \longmapsto \; \rightarrow\text{L} \; \frac{\overset{\mathcal{F}(\mathcal{U})(\pi_1)}{\Gamma, B \Rightarrow \Delta} \quad \overset{\mathcal{F}(\mathcal{U})(\pi_2)}{\Gamma \Rightarrow A, \Delta}}{\Gamma, A \to B \Rightarrow \Delta},$$

$$\rightarrow\text{R} \; \frac{\overset{\pi_0}{\Gamma, A \Rightarrow B, \Delta}}{\Gamma \Rightarrow A \to B, \Delta} \; \longmapsto \; \rightarrow\text{R} \; \frac{\overset{\mathcal{F}(\mathcal{U})(\pi_0)}{\Gamma, A \Rightarrow B, \Delta}}{\Gamma \Rightarrow A \to B, \Delta},$$

$$\text{refl} \; \frac{\overset{\pi_0}{\Gamma, A, \Box A \Rightarrow \Delta}}{\Gamma, \Box A \Rightarrow \Delta} \; \longmapsto \; \text{refl} \; \frac{\overset{\mathcal{F}(\mathcal{U})(\pi_0)}{\Gamma, A, \Box A \Rightarrow \Delta}}{\Gamma, \Box A \Rightarrow \Delta},$$

$$\Box \; \frac{\overset{\pi_1}{\Gamma, \Box\Pi \Rightarrow A, \Delta} \quad \overset{\pi_2}{\Box\Pi \Rightarrow A}}{\Gamma, \Box\Pi \Rightarrow \Box A, \Delta} \; \longmapsto \; \Box \; \frac{\overset{\mathcal{F}(\mathcal{U})(\pi_1)}{\Gamma, \Box\Pi \Rightarrow A, \Delta} \quad \overset{\mathcal{U}(\pi_2)}{\Box\Pi \Rightarrow A}}{\Gamma, \Box\Pi \Rightarrow \Box A, \Delta}.$$

The mapping $\mathcal{F}(\mathcal{U})$ is well defined on the set $\mathcal{P}_1$. If $\pi \notin \mathcal{P}_1$, then we put $\mathcal{F}(\mathcal{U})(\pi) := \mathcal{F}(\mathcal{U})(\mathcal{E}^*(\pi))$.

It can easily be checked that $\mathcal{F}(\mathcal{U})$ is a root-preserving nonexpansive mapping.

Lemma 4.6. *For any mappings $\mathcal{U}, \mathcal{V} \in \mathcal{N}$ we have that*

$$\rho(\mathcal{F}(\mathcal{U}), \mathcal{F}(\mathcal{V})) \leqslant \frac{1}{2} \cdot \rho(\mathcal{U}, \mathcal{V}).$$

Proof. Let us write $\mathcal{U} \sim_n \mathcal{V}$ if $\mathcal{U}(\pi) \sim_n \mathcal{V}(\pi)$ for any $\pi \in \mathcal{P}$. We claim that for any $n \in \mathbb{N}$ the following holds:

$$\mathcal{U} \sim_n \mathcal{V} \Rightarrow \mathcal{F}(\mathcal{U}) \sim_{n+1} \mathcal{F}(\mathcal{V}).$$

Assume we have an ∞-proof π and $\mathcal{U} \sim_n \mathcal{V}$. It can be easily proved by induction on $|\pi|$ that $\mathcal{F}(\mathcal{U})(\pi) \sim_{n+1} \mathcal{F}(\mathcal{V})(\pi)$.

Further, we see that $\mathcal{U} \sim_n \mathcal{V}$ if and only if $\rho(\mathcal{U}, \mathcal{V}) \leqslant 2^{-n}$. Thus, the condition

$$\forall n \, (\mathcal{U} \sim_n \mathcal{V} \Rightarrow \mathcal{F}(\mathcal{U}) \sim_{n+1} \mathcal{F}(\mathcal{V}))$$

is equivalent to $\rho(\mathcal{F}(\mathcal{U}), \mathcal{F}(\mathcal{V})) \leqslant \frac{1}{2} \cdot \rho(\mathcal{U}, \mathcal{V})$. $\qquad \square$

Lemma 4.7. *If $\mathcal{U} \in \mathcal{N}_n$, then $\mathcal{F}(\mathcal{U}) \in \mathcal{N}_{n+1}$.*

Proof. Assume we have an ∞-proof π and $\mathcal{U} \in \mathcal{N}_n$. We claim $\mathcal{F}(\mathcal{U})(\pi) \in \mathcal{P}_{n+1}$.

If $\pi \in \mathcal{P}_1$, then it is not hard to prove by induction on $|\pi|$ that $\mathcal{F}(\mathcal{U})(\pi) \in \mathcal{P}_{n+1}$. If $\pi \notin \mathcal{P}_1$, then $\mathcal{E}^*(\pi) \in \mathcal{P}_1$ by Lemma 4.5. Thus $\mathcal{F}(\mathcal{U})(\pi) = \mathcal{F}(\mathcal{U})(\mathcal{E}^*(\pi)) \in \mathcal{P}_{n+1}$ by the previous case. $\qquad \square$

Lemma 4.8. *There exists a mapping $\mathcal{E}$ such that $\mathcal{E} \in \mathcal{N}_n$ for any $n \in \mathbb{N}$.*

Proof. The operator $\mathcal{F} \colon \mathcal{N} \to \mathcal{N}$ is a contraction operator (by Lemma 4.6) on a complete metric space (Lemma 4.4). By the Banach fixed-point theorem, there exists a root-preserving nonexpansive mapping $\mathcal{E}$ such that $\mathcal{F}(\mathcal{E}) = \mathcal{E}$. Trivially, $\mathcal{E} \in \mathcal{N}_0 = \mathcal{N}$. Hence $\mathcal{E}$ belongs to the intersection of all $\mathcal{N}_n$ for $n \in \mathbb{N}$ by Lemma 4.7. $\qquad \square$

Theorem 4.9. (cut-elimination). *If $\mathsf{Grz}_\infty + \mathsf{cut} \vdash \Gamma \Rightarrow \Delta$, then $\mathsf{Grz}_\infty \vdash \Gamma \Rightarrow \Delta$.*

Proof. Take an ∞-proof of the sequent $\Gamma \Rightarrow \Delta$ in the system $\mathsf{Grz}_\infty + \mathsf{cut}$ and apply the mapping $\mathcal{E}$ from Lemma 4.8 to it. You will get an ∞-proof of the same sequent in the system Grz_∞. $\qquad \square$

5 Ordinary and Non-well-founded Proofs

In this section we define two translations that connect ordinary and non-well-founded sequent calculi for Grz.

Lemma 5.1. *We have $\mathsf{Grz}_\infty \vdash \Gamma, A \Rightarrow A, \Delta$ for any multisets Γ, Δ and any formula A.*

Proof. Standard induction on the structure of A. $\qquad \square$

Lemma 5.2. *We have $\mathsf{Grz}_\infty \vdash \Box(\Box(A \to \Box A) \to A) \Rightarrow A$ for any formula A.*

Proof. Consider an example of ∞–proof for the sequent $\Box(\Box(p \to \Box p) \to p) \Rightarrow p$ from Sect. 3. We transform this example into an ∞–proof for $\Box(\Box(A \to \Box A) \to A) \Rightarrow A$ by replacing p with A and adding required ∞–proofs instead of initial sequents using Lemma 5.1. $\qquad \square$

Theorem 5.3. *If* $\mathsf{Grz_{Seq}} + \mathsf{cut} \vdash \Gamma \Rightarrow \Delta$, *then* $\mathsf{Grz_{\infty}} + \mathsf{cut} \vdash \Gamma \Rightarrow \Delta$.

Proof. Assume π is a proof of $\Gamma \Rightarrow \Delta$ in $\mathsf{Grz_{Seq}} + \mathsf{cut}$. By induction on the size of π we prove $\mathsf{Grz_{\infty}} + \mathsf{cut} \vdash \Gamma \Rightarrow \Delta$.

If $\Gamma \Rightarrow \Delta$ is an initial sequent of $\mathsf{Grz_{Seq}} + \mathsf{cut}$, then it is provable in $\mathsf{Grz_{\infty}} + \mathsf{cut}$ by Lemma 5.1. Otherwise, consider the last application of an inference rule in π.

The only non-trivial case is when the proof π has the form

$$\square_{\mathsf{Grz}} \frac{\overset{\pi'}{\square \Pi, \square(A \to \square A) \Rightarrow A}}{\Sigma, \square \Pi \Rightarrow \square A, \Lambda},$$

where $\Sigma, \square \Pi = \Gamma$ and $\square A, \Lambda = \Delta$. By the induction hypothesis there is an ∞–proof ξ of $\square \Pi, \square(A \to \square A) \Rightarrow A$ in $\mathsf{Grz_{\infty}} + \mathsf{cut}$.

We have the following ∞–proof λ of $\square \Pi \Rightarrow A$ in $\mathsf{Grz_{\infty}} + \mathsf{cut}$:

$$\mathsf{cut} \frac{\square \frac{\to_{\mathsf{R}} \dfrac{wk_{\emptyset,A}(\xi)}{\dfrac{\square \Pi, \square(A \to \square A) \Rightarrow A, A}{\square \Pi \Rightarrow G, A}}}{\square \Pi \Rightarrow \square G, A} \qquad \to_{\mathsf{R}} \dfrac{\overset{\xi}{\square \Pi, \square(A \to \square A) \Rightarrow A}}{\square \Pi \Rightarrow G} \qquad wk_{\square \Pi, \emptyset}(\theta) \dfrac{}{\square \Pi, \square G \Rightarrow A}}{\square \Pi \Rightarrow A},$$

where $G = \square(A \to \square A) \to A$, $wk_{-,-}$ is a nonexpansive mapping from Lemma 3.4, and θ is an ∞–proof of $\square G \Rightarrow A$, which exists by Lemma 5.2. The required ∞–proof for $\Sigma, \square \Pi \Rightarrow \square A, \Delta$ has the form

$$\square \frac{\overset{wk_{\Sigma,\Lambda}(\lambda)}{\Sigma, \square \Pi \Rightarrow A, \Lambda} \qquad \overset{\lambda}{\square \Pi \Rightarrow A}}{\Sigma, \square \Pi \Rightarrow \square A, \Lambda},$$

The cases of other inference rules being last in π are straightforward, so we omit them. $\square$

For a sequent $\Gamma \Rightarrow \Delta$, let $Sub(\Gamma \Rightarrow \Delta)$ be the set of all subformulas of the formulas from $\Gamma \cup \Delta$. For a finite set of formulas Λ, set $\Lambda^* := \{\square(A \to \square A) \mid A \in \Lambda\}$.

Lemma 5.4. *If* $\mathsf{Grz_{\infty}} \vdash \Gamma \Rightarrow \Delta$, *then* $\mathsf{Grz_{Seq}} \vdash \Lambda^*, \Gamma \Rightarrow \Delta$ *for any finite set of formulas* Λ.

Proof. Assume π is an ∞–proof of the sequent $\Gamma \Rightarrow \Delta$ in $\mathsf{Grz_{\infty}}$ and Λ is a finite set of formulas. By induction on the number of elements in the finite set $Sub(\Gamma \Rightarrow \Delta) \backslash \Lambda$ with a subinduction on $|\pi|$, we prove $\mathsf{Grz_{Seq}} \vdash \Lambda^*, \Gamma \Rightarrow \Delta$.

If $|\pi| = 0$, then $\Gamma \Rightarrow \Delta$ is an initial sequent. We see that the sequent $\Lambda^*, \Gamma \Rightarrow \Delta$ is an initial sequent and it is provable in $\mathsf{Grz_{Seq}}$. Otherwise, consider the last application of an inference rule in π. **Case 1.** The ∞-proof π has the form

$$\to_{\mathsf{R}} \frac{\overset{\pi'}{\Gamma, A \Rightarrow B, \Sigma}}{\Gamma \Rightarrow A \to B, \Sigma},$$

where $A \to B, \Sigma = \Delta$. Notice that $|\pi'| < |\pi|$. By the induction hypothesis for π' and Λ, the sequent $\Lambda^*, \Gamma, A \Rightarrow B, \Sigma$ is provable in $\mathsf{Grz}_{\mathsf{Seq}}$. Applying the rule $(\to_\mathsf{R})$ to it, we obtain the sequent $\Lambda^*, \Gamma \Rightarrow \Delta$.**Case 2.** The ∞-proof π has the form

$$\to_\mathsf{L} \frac{\overset{\pi'}{\Sigma, B \Rightarrow \Delta} \quad \overset{\pi''}{\Sigma \Rightarrow A, \Delta}}{\Sigma, A \to B \Rightarrow \Delta} ,$$

where $\Sigma, A \to B = \Gamma$. We see that $|\pi'| < |\pi|$. By the induction hypothesis for π' and Λ, the sequent $\Lambda^*, \Sigma, B \Rightarrow \Delta$ is provable in $\mathsf{Grz}_{\mathsf{Seq}}$. Also we have $\mathsf{Grz}_{\mathsf{Seq}} \vdash \Lambda^*, \Sigma \Rightarrow A, \Delta$. Applying the rule $(\to_\mathsf{L})$ to these sequents, we obtain the sequent $\Lambda^*, \Sigma, A \to B \Rightarrow \Delta$.**Case 3.** The ∞-proof π has the form

$$\mathsf{refl} \frac{\overset{\pi'}{\Sigma, A, \Box A \Rightarrow \Delta}}{\Sigma, \Box A \Rightarrow \Delta} ,$$

where $\Sigma, \Box A = \Gamma$. We see that $|\pi'| < |\pi|$. By the induction hypothesis for π' and Λ, the sequent $\Lambda^*, \Sigma, A, \Box A \Rightarrow \Delta$ is provable in $\mathsf{Grz}_{\mathsf{Seq}}$. Applying the rule (refl), we obtain the sequent $\Lambda^*, \Sigma, \Box A \Rightarrow \Delta$. **Case 4.** The ∞-proof π has the form

$$\Box \frac{\overset{\pi'}{\Phi, \Box\Pi \Rightarrow A, \Sigma} \quad \overset{\pi''}{\Box\Pi \Rightarrow A}}{\Phi, \Box\Pi \Rightarrow \Box A, \Sigma} ,$$

where $\Phi, \Box\Pi = \Gamma$ and $\Box A, \Sigma = \Delta$.

Subcase 4.1. The formula A belongs to the set Λ. We see that $|\pi'| < |\pi|$. By the induction hypothesis for π' and Λ, the sequent $\Lambda^*, \Phi, \Box\Pi \Rightarrow A, \Sigma$ is provable in $\mathsf{Grz}_{\mathsf{Seq}}$. Then we see

$$\mathsf{refl} \frac{\to_\mathsf{L} \dfrac{\mathsf{Id} \dfrac{}{\Lambda^*, \Box A, \Phi, \Box\Pi \Rightarrow \Box A, \Sigma} \quad \mathsf{weak} \dfrac{\Lambda^*, \Phi, \Box\Pi \Rightarrow A, \Sigma}{\Lambda^*, \Phi, \Box\Pi \Rightarrow A, \Box A, \Sigma}}{(\Lambda\backslash\{A\})^*, A \to \Box A, \Box(A \to \Box A), \Phi, \Box\Pi \Rightarrow \Box A, \Sigma}}{(\Lambda\backslash\{A\})^*, \Box(A \to \Box A), \Phi, \Box\Pi \Rightarrow \Box A, \Sigma} ,$$

where the rule (weak) is admissible by Lemma 2.1.

Subcase 4.2. The formula A doesn't belong to the set Λ. We have that the number of elements in $Sub(\Box\Pi \Rightarrow A)\backslash(\Lambda \cup \{A\})$ is strictly less than the number of elements in $Sub(\Phi, \Box\Pi \Rightarrow \Box A, \Sigma)\backslash\Lambda$. Therefore, by the induction hypothesis for π'' and $\Lambda \cup \{A\}$, the sequent $\Lambda^*, \Box(A \to \Box A), \Box\Pi \Rightarrow A$ is provable in $\mathsf{Grz}_{\mathsf{Seq}}$. Then we have

$$\Box_{\mathsf{Grz}} \frac{\Lambda^*, \Box(A \to \Box A), \Box\Pi \Rightarrow A}{\Lambda^*, \Phi, \Box\Pi \Rightarrow \Box A, \Sigma} .$$

$\square$

From Lemma 5.4 we immediately obtain the following theorem.

Theorem 5.5. *If* $\mathsf{Grz}_\infty \vdash \Gamma \Rightarrow \Delta$*, then* $\mathsf{Grz}_{\mathsf{Seq}} \vdash \Gamma \Rightarrow \Delta$*.*

Theorem 2.1 is now established as a direct consequence of Theorems 5.3, 4.9, and 5.5.

6   Conclusion and Future Work

Recall that the Craig interpolation property for a logic L says that if A implies B, then there is an interpolant, that is, a formula I containing only common variables of A and B such that A implies I and I implies B. The Lyndon interpolation property is a strengthening of the Craig one that also takes into consideration negative and positive occurrences of the shared propositional variables; that is, the variables occurring in I positively (negatively) must also occur both in A and B positively (negatively).

Though the Grzegorczyk logic has the Lyndon interpolation property [4], there were seemingly no syntactic proofs of this result. It is unclear how Lyndon interpolation can be obtained from previously introduced sequent systems for Grz [1,2,5] by direct proof-theoretic arguments because these systems contain inference rules in which a polarity change occurs under the passage from the principal formula in the conclusion to its immediate ancestors in the premise. Using our system Grz_∞ we believe that we can obtain a syntactic proof of Lyndon interpolation for the modal Grzegorczyk logic as an application of our cut-elimination theorem.

We also believe that every provable Grz_∞ sequent has a proof that is a regular tree (has only finite amount of distinct subtrees). This gives a possibility of a proof system for the logic Grz with cyclical proofs, like the system introduced in [6].

Appendix

Proof of Lemma 4.1

Let π' be an ∞-proof of $\Gamma \Rightarrow \Delta, p$ and π'' be an ∞-proof of $p, \Gamma \Rightarrow \Delta$.

We define $\mathcal{R}_p(\pi', \pi'')$ by induction on $|\pi'|$.

If $|\pi'| = 0$, then $\Gamma \Rightarrow \Delta, p$ is an initial sequent. Suppose that $\Gamma \Rightarrow \Delta$ is also an initial sequent. Then $\mathcal{R}_p(\pi', \pi'')$ is defined as the ∞-proof consisting only of this initial sequent. Otherwise, Γ has the form p, Φ, and π'' is an ∞-proof of $p, p, \Phi \Rightarrow \Delta$. Applying the nonexpansive mapping acl_p from Lemma 3.6, we put $\mathcal{R}_p(\pi', \pi'') := acl_p(\pi'')$.

Now suppose that $|\pi'| > 0$. We consider the last application of an inference rule in π'.

Case 1. The ∞-proof π' has the form

$$\to_R \frac{\pi_0' \quad \Gamma, A \Rightarrow B, \Sigma, p}{\Gamma \Rightarrow A \to B, \Sigma, p} \; ,$$

where $A \to B, \Sigma = \Delta$. Notice that $|\pi_0'| < |\pi'|$. In addition, π'' is an ∞-proof of $p, \Gamma \Rightarrow A \to B, \Sigma$. We define $\mathcal{R}_p(\pi', \pi'')$ as

$$\to_R \frac{\mathcal{R}_p(\pi_0', i_{A \to B}(\pi''))}{\Gamma, A \Rightarrow B, \Sigma} \; ,$$

where $i_{A \to B}$ is a nonexpansive mapping from Lemma 3.5.

Case 2. The ∞-proof π' has the form

$$\to_L \frac{\pi_0' \qquad\qquad \pi_1''}{\frac{\Sigma, B \Rightarrow \Delta, p \quad\quad \Sigma \Rightarrow A, \Delta, p}{\Sigma, A \to B \Rightarrow \Delta, p}} \; ,$$

where $\Sigma, A \to B = \Gamma$. We see that $|\pi_0'| < |\pi'|$ and $|\pi_1'| < |\pi'|$. Also, π'' is an ∞-proof of $p, \Sigma, A \to B \Rightarrow \Delta$. We define $\mathcal{R}_p(\pi', \pi'')$ as

$$\to_L \frac{\mathcal{R}_p(\pi_0', li_{A \to B}(\pi'')) \qquad \mathcal{R}_p(\pi_1', ri_{A \to B}(\pi''))}{\frac{\Sigma, B \Rightarrow \Delta, p \qquad\qquad \Sigma \Rightarrow A, \Delta, p}{\Sigma, A \to B \Rightarrow \Delta, p}} \; ,$$

where $li_{A \to B}$ and $ri_{A \to B}$ are nonexpansive mappings from Lemma 3.5.

Case 3. The ∞-proof π' has the form

$$\text{refl} \; \frac{\pi_0' \quad \Sigma, A, \square A \Rightarrow \Delta, p}{\Sigma, \square A \Rightarrow \Delta, p} \; ,$$

where $\Sigma, \square A = \Gamma$. We have that $|\pi'| < |\pi|$. Define $\mathcal{R}_p(\pi', \pi'')$ as

$$\text{refl} \; \frac{\mathcal{R}_p(\pi_0', wk_{A, \emptyset}(\pi''))}{\frac{\Sigma, A, \square A \Rightarrow \Delta}{\Sigma, \square A \Rightarrow \Delta}} \; ,$$

where $wk_{A, \emptyset}$ is the nonexpansive mapping from Lemma 3.4.

Case 4. Now consider the final case when π' has the form

$$\square \; \frac{\pi_0' \qquad\qquad \pi_1'}{\frac{\Phi, \square \Pi \Rightarrow A, \Sigma, p \quad\quad \square \Pi \Rightarrow A}{\Phi, \square \Pi \Rightarrow \square A, \Sigma, p}} \; ,$$

where $\Phi, \Box\Pi = \Gamma$ and $\Box A, \Sigma = \Delta$. Notice that $|\pi_0'| < |\pi'|$. In addition, π'' is an ∞-proof of $p, \Phi, \Box\Pi \Rightarrow \Box A, \Sigma$. We define $\mathcal{R}_p(\pi', \pi'')$ as

$$\Box \frac{\mathcal{R}_p(\pi_0', li_{\Box A}(\pi''))\qquad\qquad \pi_1'}{\dfrac{\Phi, \Box\Pi \Rightarrow A, \Sigma\qquad\quad \Box\Pi \Rightarrow A}{\Phi, \Box\Pi \Rightarrow \Box A, \Sigma}},$$

where $li_{\Box A}$ is a nonexpansive mapping from Lemma 3.5.

The mapping $\mathcal{R}_p$ is well defined. It remains to check that $\mathcal{R}_p$ is nonexpansive, i.e. for any $n \in \mathbb{N}$ and any $\pi', \pi'', \tau', \tau''$ from $\mathcal{P}_0$

$$(\pi' \sim_n \tau' \wedge \pi'' \sim_n \tau'') \Rightarrow \mathcal{R}_p(\pi', \pi'') \sim_n \mathcal{R}_p(\tau', \tau'') .$$

This condition is checked by structural induction on the inductively defined relation $\pi' \sim_n \tau'$ in a straightforward way. So we omit further details. $\qquad\Box$

Proof of Lemma 4.2

Let π' be an ∞-proof of $\Gamma \Rightarrow \Delta, \Box B$ and π'' be an ∞-proof of $\Box B, \Gamma \Rightarrow \Delta$.

We define $\mathcal{R}_{\Box B}(\pi', \pi'')$ by induction on $|\pi'| + |\pi''|$.

If $|\pi'| = 0$ or $|\pi''| = 0$, then $\Gamma \Rightarrow \Delta$ is an initial sequent. Then $\mathcal{R}_{\Box B}(\pi', \pi'')$ is defined as the ∞-proof consisting only of this initial sequent.

The only interesting cases are when the formula $\Box B$ is the principal formula in both π' and π''.

So the ∞-proof π' has the form

$$\Box \frac{\pi_0'\qquad\qquad\qquad \pi_1'}{\dfrac{\Phi, \Box\Pi \Rightarrow B, \Sigma\qquad\quad \Box\Pi \Rightarrow B}{\Phi, \Box\Pi \Rightarrow \Box B, \Sigma}},$$

The cases for the ∞-proof π'' are the following:

Case 1. The ∞-proof π'' has the form

$$\text{refl } \frac{\pi_0''}{\dfrac{\Gamma, B, \Box B \Rightarrow \Delta}{\Gamma, \Box B \Rightarrow \Delta}} .$$

Since that $|\pi_0''| < |\pi''|$, we can define $\mathcal{R}_{\Box B}(\pi', \pi'')$ as

$$\mathcal{R}_B(\pi_0', \mathcal{R}_{\Box B}(wk_{B,\emptyset}(\pi'), \pi_0'')).$$

where $wk_{-,-}$ is a nonexpansive mapping from Lemma 3.4.

Case 2. The ∞-proof π'' has the form

$$\Box \frac{\pi_0''\qquad\qquad\qquad \pi_1''}{\dfrac{\Phi', \Box B, \Box\Pi' \Rightarrow C, \Sigma'\qquad\quad \Box B, \Box\Pi' \Rightarrow C}{\Phi', \Box B, \Box\Pi' \Rightarrow \Box C, \Sigma'}},$$

Since $|\pi_0''| < |\pi''|$ and the sequent $\Phi', \Box\Pi' \Rightarrow \Box C, \Sigma'$, the sequent $\Phi, \Box\Pi \Rightarrow \Sigma$, and the sequent $\Gamma \Rightarrow \Delta$ are equal, we can define $\mathcal{R}_{\Box B}(\pi', \pi'')$ as

$$
\Box\ \cfrac{\mathcal{R}_{\Box B}(li_{\Box C}(\pi'), \pi_0'')\quad \Box\ \cfrac{wk_{\Box\Pi'\backslash\Box\Pi,C}(\pi_1')\quad wk_{\Box\Pi'\backslash\Box\Pi,\varnothing}(\pi_1')}{\text{cut}\ \cfrac{\Box\Pi\cup\Box\Pi'\Rightarrow B,C\qquad \Box\Pi\cup\Box\Pi'\Rightarrow B}{\Box\Pi\cup\Box\Pi'\Rightarrow\Box B,C}\qquad wk_{\Box\Pi\backslash\Box\Pi',\varnothing}(\pi_1'')\ \cfrac{}{\Box\Pi\cup\Box\Pi',\Box B\Rightarrow C}}{\cfrac{\Phi',\Box\Pi'\Rightarrow C,\Sigma'\qquad\qquad \Box\Pi\cup\Box\Pi'\Rightarrow C}{\Phi',\Box\Pi'\Rightarrow\Box C,\Sigma'}}}{}
$$

where $wk_{-,-}$ is a nonexpansive mapping from Lemma 3.4 and $li_{\Box A}$ is a nonexpansive mapping from Lemma 3.5. Since the instance of the rule cut is not in the main fragment, this proof is in $\mathcal{P}_1$. $\qquad\Box$

References

1. Avron, A.: On modal systems having arithmetical interpretations. J. Symb. Log. **49**(3), 935–942 (1984)
2. Borga, M., Gentilini, P.: On the proof theory of the modal logic Grz. Math. Log. Q. **32**(10–12), 145–148 (1986)
3. Maksimova, L.L.: On modal Grzegorczyk logic. Fundamenta Informaticae **81**(1–3), 203–210 (2008). Topics in Logic, Philosophy and Foundations of Mathematics and Computer Science, In: Recognition of Professor Andrzej Grzegorczyk
4. Maksimova, L.L.: The Lyndon property and uniform interpolation over the Grzegorczyk logic. Sib. Math. J. **55**(1), 118–124 (2014)
5. Dyckhoff, R., Negri, S.: A cut-free sequent system for Grzegorczyk logic, with an application to the Gödel-McKinsey-Tarski embedding. J. Log. Comput. **26**(1), 169–187 (2016)
6. Shamkanov, D.S.: Circular proofs for the Gödel-Löb provability logic. Math. Notes **96**(3), 575–585 (2014)

On Two Concepts of Ultrafilter Extensions of First-Order Models and Their Generalizations

Nikolai L. Poliakov[1] and Denis I. Saveliev[2(✉)]

[1] Financial University, Moscow, Russia
[2] Institute for Information Transmission Problems of the Russian Academy of Sciences, Steklov Mathematical Institute of the Russian Academy of Sciences, Moscow, Russia
d.i.saveliev@gmail.com

Abstract. There exist two known concepts of ultrafilter extensions of first-order models, both in a certain sense canonical. One of them [1] comes from modal logic and universal algebra, and in fact goes back to [2]. Another one [3,4] comes from model theory and algebra of ultrafilters, with ultrafilter extensions of semigroups [5] as its main precursor. By a classical fact, the space of ultrafilters over a discrete space is its largest compactification. The main result of [3,4], which confirms a canonicity of this extension, generalizes this fact to discrete spaces endowed with a first-order structure. An analogous result for the former type of ultrafilter extensions was obtained in [6].

Here we offer a uniform approach to both types of extensions. It is based on the idea to extend the extension procedure itself. We propose a generalization of the standard concept of first-order models in which functional and relational symbols are interpreted rather by ultrafilters over sets of functions and relations than by functions and relations themselves. We provide two specific operations which turn generalized models into ordinary ones, and establish necessary and sufficient conditions under which the latter are the two canonical ultrafilter extensions of some models.

1. Fix a first-order language and consider an arbitrary model

$$\mathfrak{A} = (X, F, \ldots, R, \ldots)$$

with the universe X, operations $F, \ldots$, and relations $R, \ldots$. Let us define an abstract *ultrafilter extension* of $\mathfrak{A}$ as a model $\mathfrak{A}'$ (in the same language) of form

$$\mathfrak{A}' = (\beta X, F', \ldots, R', \ldots)$$

where βX is the set of ultrafilters over X (one lets $X \subseteq \beta X$ by identifying each $x \in X$ with the principal ultrafilter given by x), and operations $F', \ldots$ and relations $R', \ldots$ on βX extend $F, \ldots$ and $R, \ldots$ resp. There are essentially *two*

The second author was supported by Grant 16-01-00615 of the Russian Foundation for Basic Research.

J. Kennedy and R.J.G.B. de Queiroz (Eds.): WoLLIC 2017, LNCS 10388, pp. 336–348, 2017.
DOI: 10.1007/978-3-662-55386-2_24

known ways to extend relations by ultrafilters, and *one* to extend maps. Partial cases of these extensions were discovered by various authors in different time and different areas, typically, without a knowledge of parallel studies in adjacent areas.

Recall that βX carries a natural topology generated by basic open sets

$$\widetilde{A} = \{\mathfrak{u} \in \beta X : A \in \mathfrak{u}\}$$

for all $A \subseteq X$. Easily, the sets are also closed, so the space βX is zero-dimensional. In fact, βX is compact, Hausdorff, extremally disconnected (the closure of any open set is open), and the largest compactification of the discrete space X. This means that X is dense in βX and every (trivially continuous) map h of X into any compact Hausdorff space Y uniquely extends to a continuous map $\widetilde{h}$ of βX into Y:

$$\begin{array}{ccc} \beta X & & \\ \big\uparrow & \diagdown \ \widetilde{h} & \\ \big\uparrow & & \diagdown \\ X & \xrightarrow{\ h\ } & Y \end{array}$$

The largest compactification of Tychonoff spaces was discovered independently by Čech [7] and Stone [8]; then Wallman [9] did the same for T_1 spaces (by using ultrafilters on lattices of closed sets); see [5, 10, 11] for more information.

The ultrafilter extensions of *unary* maps F and relations R are exactly $\widetilde{F}$ and $\widetilde{R}$ (for $F : X \to X$ let $Y = \beta X$); thus in the unary case the procedure gives classical objects known in 30s. As for mappings and relations of greater arities, several instances of their ultrafilter extensions were discovered only in 60s.

Studying ultraproducts, Kochen [12] and Frayne et al. [13] considered a "multiplication" of ultrafilters, which actually is the ultrafilter extension of the n-ary operation of taking n-tuples. They shown that the successive iteration of ultrapowers by ultrafilters $\mathfrak{u}_1, \ldots, \mathfrak{u}_n$ is isomorphic to a single ultrapower by their "product". This has leaded to the general construction of iterated ultrapowers, invented by Gaifman and elaborated by Kunen, which has become common in model theory and set theory (see [14, 15]).

Ultrafilter extensions of semigroups appeared in 60s as subspaces of function spaces; the first explicit construction of the ultrafilter extension of a group is due to Ellis [16]. In 70s Galvin and Glazer applied them to give an easy proof of what now known as Hindman's Finite Sums Theorem; the key idea was to use idempotent ultrafilters. The method was developed then by Blass, van Douwen, Hindman, Protasov, Strauss, and many others, and gave numerous Ramsey-theoretic applications in number theory, algebra, topological dynamics, and ergodic theory. The book [5] is a comprehensive treatise of this area, with an historical information. This technique was applied also for obtaining analogous results for certain non-associative algebras (see [17, 18]).

Ultrafilter extensions of arbitrary n-ary maps have been introduced independently by Goranko [1] and Saveliev [3, 4]. For $F : X_1 \times \ldots \times X_n \to Y$, the extended map $\widetilde{F} : \beta X_1 \times \ldots \times \beta X_n \to \beta Y$ is defined by letting

$$\widetilde{F}(\mathfrak{u}_1, \ldots, \mathfrak{u}_n) =$$
$$\{A \subseteq Y : \{x_1 \in X_1 : \ldots \{x_n \in X_n : F(x_1, \ldots, x_n) \in A\} \in \mathfrak{u}_n \ldots\} \in \mathfrak{u}_1\}.$$

One can simplify this cumbersome notation by introducing *ultrafilter quanti-fiers*: let $(\forall^{\mathfrak{u}} x)\,\varphi(x, \ldots)$ means $\{x : \varphi(x, \ldots)\} \in \mathfrak{u}$. In fact, this is a second-order quantifier: $(\forall^{\mathfrak{u}} x)$ is equivalent to $(\forall A \in \mathfrak{u})(\exists x \in A)$, and also (since $\mathfrak{u}$ is ultra) to $(\exists A \in \mathfrak{u})(\forall x \in A)$. Such quantifiers are self-dual, i.e. $\forall^{\mathfrak{u}}$ and $\exists^{\mathfrak{u}}$ coincide, and generally do not commute with each other, i.e. $(\forall^{\mathfrak{u}} x)(\forall^{\mathfrak{v}} y)$ and $(\forall^{\mathfrak{v}} y)(\forall^{\mathfrak{u}} x)$ are not equivalent. Then the definition above is rewritten as follows:

$$\widetilde{F}(\mathfrak{u}_1, \ldots, \mathfrak{u}_n) = \{A \subseteq Y : (\forall^{\mathfrak{u}_1} x_1) \ldots (\forall^{\mathfrak{u}_n} x_n)\, F(x_1, \ldots, x_n) \in A\}.$$

The map $\widetilde{F}$ can be also described as the composition of the ultrafilter exten-sion of taking n-tuples, which maps $\beta X_1 \times \ldots \times \beta X_n$ into $\beta(X_1 \times \ldots \times X_n)$, and the continuous extension of F considered as a unary map, which maps $\beta(X_1 \times \ldots \times X_n)$ into βY.

One type of ultrafilter extensions of relations goes back to a seminal paper by Jónsson and Tarski [2] where they have been appeared implicitly, in terms of representations of Boolean algebras with operators. For binary relations, their representation theory was rediscovered in modal logic by Lemmon [19] who cred-ited much of this work to Scott, see footnote 6 on p. 204 (see also [20]). Goldblatt and Thomason [21] used this to characterize modal definability (where Sect. 2 was entirely due to Goldblatt); the term "ultrafilter extension" has been intro-duced probably in the subsequent work by van Benthem [22] (for modal defin-ability see also [23,24]). Later Goldblatt [25] generalized the extension to n-ary relations.

Let us give an equivalent formulation: for $R \subseteq X_1 \times \ldots \times X_n$, the extended relation $R^* \subseteq \beta X_1 \times \ldots \times \beta X_n$ is defined by letting

$$R^*(\mathfrak{u}_1, \ldots, \mathfrak{u}_n) \quad \text{iff}$$
$$(\forall A_1 \in \mathfrak{u}_1) \ldots (\forall A_n \in \mathfrak{u}_n)(\exists x_1 \in A_1) \ldots (\exists x_n \in A_n)\, R(x_1, \ldots, x_n).$$

Another type of ultrafilter extensions of n-ary relations has been recently discovered in [3,4]:

$$\widetilde{R}(\mathfrak{u}_1, \ldots, \mathfrak{u}_n) \quad \text{iff}$$
$$\{x_1 \in X_1 : \ldots \{x_n \in X_n : R(x_1, \ldots, x_n)\} \in \mathfrak{u}_n \ldots\} \in \mathfrak{u}_1,$$

or rewritting this via ultrafilter quantifiers,

$$\widetilde{R}(\mathfrak{u}_1, \ldots, \mathfrak{u}_n) \quad \text{iff} \quad (\forall^{\mathfrak{u}_1} x_1) \ldots (\forall^{\mathfrak{u}_n} x_n)\, R(x_1, \ldots, x_n).$$

Or else, by decoding ultrafilter quantifiers, this can be rewritten by

$$\widetilde{R}(\mathfrak{u}_1, \ldots, \mathfrak{u}_n) \quad \text{iff}$$
$$(\forall A_1 \in \mathfrak{u}_1)(\exists x_1 \in A_1) \ldots (\forall A_n \in \mathfrak{u}_n)(\exists x_n \in A_n)\, R(x_1, \ldots, x_n),$$

whence it is clear that $\widetilde{R} \subseteq R^*$. For unary R both extensions coincide with the basic open set given by R. If R is functional then R^* (but not $\widetilde{R}$) coincides with the above-defined extension of R as a map. An easy instance of $^{\sim}$-extensions (where R are linear orders) is studied in [26].

A systematic comparative study of both extensions (for binary R) is undertaken in [6]. In particular, there is shown that the *- and the $^{\sim}$-extensions have a dual character w.r.t. relation-algebraic operations: the former commutes with composition and inversion but not Boolean operations except for union, while the latter commutes with all Boolean operations but neither composition nor inversion. Also [6] contains topological characterizations of $\widetilde{R}$ and R^* in terms of appropriate closure operations and in terms of Vietoris-type topologies (regarding R as multi-valued maps).

Ultrafilter extensions of arbitrary first-order models were considered for the first time independently in [1] with *-extensions of relations, and in [3] with their $^{\sim}$-extensions. We shall denote them by $\mathfrak{A}^*$ and $\widetilde{\mathfrak{A}}$ resp. Thus for a model $\mathfrak{A} = (X, F, \ldots, R, \ldots)$ we let

$$\mathfrak{A}^* = (\beta X, \widetilde{F}, \ldots, R^*, \ldots) \quad \text{and} \quad \widetilde{\mathfrak{A}} = (\beta X, \widetilde{F}, \ldots, \widetilde{R}, \ldots).$$

The following is the main result of [1]:

Theorem 1. *If h is a homomorphism between models $\mathfrak{A}$ and $\mathfrak{B}$, then the continuous extension $\widetilde{h}$ is a homomorphism between $\mathfrak{A}^*$ and $\mathfrak{B}^*$:*

$$
\begin{array}{ccc}
\mathfrak{A}^* & \xrightarrow{\widetilde{h}} & \mathfrak{B}^* \\
\uparrow & & \uparrow \\
\mathfrak{A} & \xrightarrow{h} & \mathfrak{B}
\end{array}
$$

A full analog of Theorem 1 for the $^{\sim}$-extensions has been appeared in [3] (called the First Extension Theorem in [4]):

Theorem 2. *If h is a homomorphism between models $\mathfrak{A}$ and $\mathfrak{B}$, then the continuous extension $\widetilde{h}$ is a homomorphism between $\widetilde{\mathfrak{A}}$ and $\widetilde{\mathfrak{B}}$:*

$$
\begin{array}{ccc}
\widetilde{\mathfrak{A}} & \xrightarrow{\widetilde{h}} & \widetilde{\mathfrak{B}} \\
\uparrow & & \uparrow \\
\mathfrak{A} & \xrightarrow{h} & \mathfrak{B}
\end{array}
$$

Moreover, both theorems remain true for embeddings and some other model-theoretic interrelations (see [1,3,4]).

Theorem 2 is actually is a partial case of a much stronger result of [3] (called the Second Extension Theorem in [4]). To formulate this, we need the following concepts (introduced in [3]).

Let $X_1, \ldots, X_n, Y$ be topological spaces, and let $A_1 \subseteq X_1, \ldots, A_{n-1} \subseteq X_{n-1}$. An n-ary function $F : X_1 \times \ldots \times X_n \to Y$ is *right continuous w.r.t.* $A_1, \ldots, A_{n-1}$

iff for each i, $1 \leqslant i \leqslant n$, and every $a_1 \in A_1, \ldots, a_{i-1} \in A_{i-1}$ and $x_{i+1} \in X_{i+1}, \ldots, x_n \in X_n$, the map

$$x \mapsto F(a_1, \ldots, a_{i-1}, x, x_{i+1}, \ldots, x_n)$$

of X_i into Y is continuous. An n-ary relation $R \subseteq X_1 \times \ldots \times X_n$ is *right open* (*right closed*, etc.) w.r.t. $A_1, \ldots, A_{n-1}$ iff for each i, $1 \leqslant i \leqslant n$, and every $a_1 \in A_1, \ldots, a_{i-1} \in A_{i-1}$ and $x_{i+1} \in X_{i+1}, \ldots, x_n \in X_n$, the set

$$\{x \in X_i : R(a_1, \ldots, a_{i-1}, x, x_{i+1}, \ldots, x_n)\}$$

is open (closed, etc.) in X_i.

Theorem 3 [3,4] characterizes topological properties of $\sim$-extensions, it is a base of Theorem 4 (the Second Extension Theorem of [4]).

Theorem 3. *Let $\mathfrak{A}$ be a model. In the extension $\widetilde{\mathfrak{A}}$, all operations are right continuous and all relations right clopen w.r.t. the universe of $\mathfrak{A}$.*

Theorem 4. *Let $\mathfrak{A}$ and $\mathfrak{C}$ be two models, h a homomorphism of $\mathfrak{A}$ into $\mathfrak{C}$, and let $\mathfrak{C}$ carry a compact Hausdorff topology in which all operations are right continuous and all relations are right closed w.r.t. the image of the universe of $\mathfrak{A}$ under h. Then $\widetilde{h}$ is a homomorphism of $\widetilde{\mathfrak{A}}$ into $\mathfrak{C}$:*

Theorem 2 (for homomorphisms) easily follows: take $\widetilde{\mathfrak{B}}$ as $\mathfrak{C}$. The meaning of Theorem 4 is that it generalizes the classical Čech–Stone result to the case when the underlying discrete space X carries an arbitrary first-order structure.

A natural question is whether $*$-extensions also canonical in a similar sense. The answer is positive; two following theorems are counterparts of Theorems 3 and 4 resp. (essentially both have been proved in [6]).

Theorem 5. *Let $\mathfrak{A}$ be a model. In the extension $\mathfrak{A}^*$, all relations are closed (and all operations are right continuous w.r.t. the universe of $\mathfrak{A}$).*

Theorem 6. *Let $\mathfrak{A}$ and $\mathfrak{C}$ be two models, h a homomorphism of $\mathfrak{A}$ into $\mathfrak{C}$, and let $\mathfrak{C}$ carry a compact Hausdorff topology in which all operations are right continuous w.r.t. the image of the universe of $\mathfrak{A}$ under h, and all relations are closed. Then $\widetilde{h}$ is a homomorphism of $\mathfrak{A}^*$ into $\mathfrak{C}$.*

Similarly, Theorem 1 (for homomorphisms) follows from Theorem 6. The latter also generalizes the Čech–Stone result for discrete spaces to discrete models but with a narrow class of target models $\mathfrak{C}$: having relations rather closed than right closed in Theorem 4.

2. The immediate purpose of this section is to provide a uniform approach to both types of extensions. This approach will lead us to certain structures, called here generalized models, which generalize ultrafilter extensions of each of the two types.

First we shall show that the $*$-extension can be described in terms of the basic (cl)open sets and the continuous extension of maps. For this, let us consider the continuous extension of the continuous extension operation *itself*. To make notation easier, denote by ext the operation of continuous extension of maps; i.e. $\mathrm{ext}(f)$ is another notation for $\widetilde{f}$:

$$\mathrm{ext}(f) = \widetilde{f}.$$

So if we consider maps of X into Y, then ext is a map of Y^X into $C(\beta X, \beta Y)$. Since $C(\beta X, \beta Y)$ with the standard (i.e. pointwise convergence) topology is a compact Hausdorff space, ext continuously extends to the map $\widetilde{\mathrm{ext}}$ of $\beta(Y^X)$ into this space:

$$\beta(Y^X)$$

The extended map $\widetilde{\mathrm{ext}}$ is surjective and non-injective.

Lemma 1. *Let $R \subseteq Y^X$. Then $\widetilde{\mathrm{ext}}$ maps the closure of R in the space $\beta(Y^X)$ onto the closure of R in the space $C(\beta X, \beta Y)$:*

$$\left\{\widetilde{\mathrm{ext}}(\mathfrak{f}) : \mathfrak{f} \in \mathrm{cl}_{\beta(Y^X)} R\right\} = \mathrm{cl}_{C(\beta X, \beta Y)} R.$$

For our purpose, let $X = n$. Then $\beta X = n$ and $C(\beta X, \beta Y)$ is $(\beta Y)^n$, which can be identified with $\beta Y \times \ldots \times \beta Y$ (n times). Now the required description of the $*$-extension follows from Theorem 5:

Theorem 7. *Let $R \subseteq X \times \ldots \times X$. Then $R^* \subseteq \beta X \times \ldots \times \beta X$ is (identified with) the image of $\mathrm{cl}_{\beta(X^n)} R$ under $\widetilde{\mathrm{ext}}$.*

Using ultrafilters over maps leads to the following concept. Given a language, we define a *generalized* (or *ultrafilter*) *interpretation* (the term is ad hoc) as a map $\imath$ that takes each n-ary functional symbol F to an ultrafilter over the set of n-ary operations on X, and each n-ary predicate symbol R to an ultrafilter over the set of n-ary relations on X; let also v be an *ultrafilter valuation* of variables, i.e. a valuation which takes each variable x to an ultrafilter over a given set X:

$$v(x) \in \beta X, \quad \imath(F) \in \beta(X^{X \times \ldots \times X}), \quad \imath(R) \in \beta \mathcal{P}(X \times \ldots \times X).$$

The set $(\beta X, \imath(F), \ldots, \imath(R), \ldots)$ is a *generalized model*. Now we are going to define the satisfiability relation in generalized models, which will be denoted by the symbol $\Vdash$.

First, given an interpretation $\imath$ of non-logical symbols, we expand any valuation v of variables to the map $v_\imath$ defined on all terms as follows. Let $\mathrm{app} : X_1 \times \ldots \times X_n \times Y^{X_1 \times \ldots \times X_n} \to Y$ be the *application* operation:

$$\mathrm{app}(a_1, \ldots, a_n, f) = f(a_1, \ldots, a_n).$$

Extend it to the map $\widetilde{\mathrm{app}} : \beta X_1 \times \ldots \times \beta X_n \times \beta(Y^{X_1 \times \ldots \times X_n}) \to \beta Y$ right continuous w.r.t. the principal ultrafilters, in the usual way:

$$
\begin{array}{ccc}
\beta X_1 \times \ldots \times \beta X_n \times \beta(Y^{X_1 \times \ldots \times X_n}) & \xrightarrow{\ \widetilde{\mathrm{app}}\ } & \beta Y \\
\uparrow & & \uparrow \\
X_1 \times \ldots \times X_n \times Y^{X_1 \times \ldots \times X_n} & \xrightarrow{\ \mathrm{app}\ } & Y
\end{array}
$$

Let $v_\imath$ coincide with v on variables, and if $v_\imath$ has been already defined on terms $t_1, \ldots, t_n$, we let

$$v_\imath(F(t_1, \ldots, t_n)) = \widetilde{\mathrm{app}}(v_\imath(t_1), \ldots, v_\imath(t_n), \imath(F)).$$

Further, given a generalized model $\mathfrak{A} = (\beta X, \imath(F), \ldots, \imath(R), \ldots)$, define the satisfiability in $\mathfrak{A}$ as follows. Let $\mathrm{in} \subseteq X_1 \times \ldots \times X_n \times \mathcal{P}(X_1 \times \ldots \times X_n)$ be the *membership* predicate:

$$\mathrm{in}\,(a_1, \ldots, a_n, R) \quad \text{iff} \quad (a_1, \ldots, a_n) \in R.$$

Extend it to the relation $\widetilde{\mathrm{in}} \subseteq \beta X_1 \times \ldots \times \beta X_n \times \beta \mathcal{P}(X_1 \times \ldots \times X_n)$ right clopen w.r.t. principal ultrafilters. Let

$$\mathfrak{A} \Vdash t_1 = t_2 \, [v] \quad \text{iff} \quad v_\imath(t_1) = v_\imath(t_2).$$

If $R(t_1, \ldots, t_n)$ is an atomic formula in which R is not the equality predicate, we let

$$\mathfrak{A} \Vdash R(t_1, \ldots, t_n) \, [v] \quad \text{iff} \quad \widetilde{\mathrm{in}}\,(v_\imath(t_1), \ldots, v_\imath(t_n), \imath(P)).$$

(Equivalently, we could define the satisfiability of atomic formulas by identifying predicates with their characteristic functions and using the satisfiability of equalities of the resulting terms.) Finally, if $\varphi(t_1, \ldots, t_n)$ is obtained by negation, conjunction, or quantification from formulas for which $\Vdash$ has been already defined, we define $\mathfrak{A} \Vdash \varphi \, [v]$ in the standard way.

When needed, we shall use variants of notation commonly used for ordinary models and satisfiability, for the generalized ones. E.g. for a generalized model $\mathfrak{A}$ with the universe βX, a formula $\varphi(x_1, \ldots, x_n)$, and elements $\mathfrak{u}_1, \ldots, \mathfrak{u}_n$ of βX, the notation $\mathfrak{A} \Vdash \varphi \, [\mathfrak{u}_1, \ldots, \mathfrak{u}_n]$ means that φ is satisfied in $\mathfrak{A}$ under a valuation taking the variables $x_1, \ldots, x_n$ to the ultrafilters $\mathfrak{u}_1, \ldots, \mathfrak{u}_n$.

Generalized models actually generalize not all ordinary models but those that are ultrafilter extensions of some models. It is worth also pointing out that whenever a generalized interpretation is *principal,* i.e. all non-logical symbols are interpreted by principal ultrafilters, we naturally identify it with the obvious ordinary interpretation with the same universe βX; however, not every ordinary interpretation with the universe βX is of this form. Precise relationships between generalized models, ordinary models, and ultrafilter extensions will be described in Theorems 9 and 10.

An ultrafilter valuation v is *principal* iff it takes any variable to a principal ultrafilter.

Lemma 2. *Let two generalized models* $\mathfrak{A} = (\beta X, \imath(F), \ldots, \imath(R), \ldots)$ *and* $\mathfrak{B} = (\beta X, \jmath(F), \ldots, \jmath(R), \ldots)$ *have the same universe* βX. *If for all functional symbols* F, *predicate symbols* R, *variables* $x_1, \ldots, x_n$, *and principal valuations* v,

$$\widetilde{\mathrm{app}}(v(x_1), \ldots, v(x_n), \imath(F)) = \widetilde{\mathrm{app}}(v(x_1), \ldots, v(x_n), \jmath(F)),$$
$$\widetilde{\mathrm{in}}\,(v(x_1), \ldots, v(x_n), \imath(R)) \ \textit{iff} \ \widetilde{\mathrm{in}}\,(v(x_1), \ldots, v(x_n), \jmath(R)),$$

then for all formulas φ, *terms* $t_1, \ldots, t_n$, *and valuations* v,

$$\mathfrak{A} \Vdash \varphi(t_1, \ldots, t_n)\,[v] \ \textit{iff} \ \mathfrak{B} \Vdash \varphi(t_1, \ldots, t_n)\,[v].$$

Corollary 1. *Let* $\mathfrak{A} = (\beta X, \imath(F), \ldots, \imath(R), \ldots)$ *be a generalized model and* $\mathfrak{B} = (\beta X, \jmath(F), \ldots, \jmath(R), \ldots)$ *the generalized model having the same universe* βX *and such that* $\jmath$ *coincides with* $\imath$ *on functional symbols and for each predicate symbol* R, $\jmath(R)$ *is the principal ultrafilter given by*

$$\big\{(a_1, \ldots, a_n) \in X^n : \widetilde{\mathrm{in}}\,(a_1, \ldots, a_n, \imath(R))\big\}.$$

Then for all valuations v, *formulas* φ, *and terms* $t_1, \ldots, t_n$,

$$\mathfrak{A} \Vdash \varphi(t_1, \ldots, t_n)\,[v] \ \textit{iff} \ \mathfrak{B} \Vdash \varphi(t_1, \ldots, t_n)\,[v].$$

Let us say that an ultrafilter $\mathfrak{f}$ over functions is *pseudo-principal* iff $\widetilde{\mathrm{app}}$ takes any tuple consisting of principal ultrafilters together with $\mathfrak{f}$ to a principal ultrafilter, i.e. for $\mathfrak{f} \in \beta(Y^{X_1 \times \cdots \times X_n})$,

$$a_1 \in X_1, \ldots, a_n \in X_n \text{ implies } \widetilde{\mathrm{app}}(a_1, \ldots, a_n, \mathfrak{f}) \in Y.$$

Every principal $\mathfrak{f}$ is pseudo-principal, and there exist pseudo-principal ultrafilters that are not principal as well as ultrafilters that are not pseudo-principal. A generalized interpretation $\imath$ is *pseudo-principal on functional symbols* iff $\imath(F)$ is a pseudo-principal ultrafilter for each functional symbol F (and then, for each term t).

Corollary 2. *Let* $\mathfrak{A} = (\beta X, \imath(F), \ldots, \imath(R), \ldots)$ *be a generalized model with* $\imath$ *pseudo-principal on functional symbols. Let* $\mathfrak{B} = (\beta X, \jmath(F), \ldots, \jmath(R), \ldots)$ *be the generalized model having the same universe* βX *and such that* $\jmath$ *coincides with* $\imath$ *on predicate symbols and for each functional symbol* F, $\jmath(F)$ *is the principal ultrafilter given by* $f : X^n \to X$ *defined by letting*

$$f(a_1, \ldots, a_n) = \widetilde{\mathrm{app}}(a_1, \ldots, a_n, \imath(F)).$$

Then for all valuations v, *formulas* φ, *and terms* $t_1, \ldots, t_n$,

$$\mathfrak{A} \Vdash \varphi(t_1, \ldots, t_n)\,[v] \ \textit{iff} \ \mathfrak{B} \Vdash \varphi(t_1, \ldots, t_n)\,[v].$$

It follows that for any generalized model $\mathfrak{A}$ whose interpretation is pseudo-principal on functional symbols, by replacing its relations as in Corollary 1 and its operations as in Corollary 2, one obtains an ordinary model $\mathfrak{B}$ with the same universe such that for all formulas φ and elements $\mathfrak{u}_1, \ldots, \mathfrak{u}_n$ of the universe, $\mathfrak{A} \Vdash \varphi [\mathfrak{u}_1, \ldots, \mathfrak{u}_n]$ iff $\mathfrak{B} \vDash \varphi [\mathfrak{u}_1, \ldots, \mathfrak{u}_n]$.

We do not formulate this fact as a separate theorem since we shall be able to establish stronger facts soon. In Theorem 8, we shall establish that for any generalized model $\mathfrak{A}$, not only one with a pseudo-principal interpretation, one can construct a certain ordinary model $e(\mathfrak{A})$ satisfying the same formulas; and then, in Theorem 9, that whenever $\mathfrak{A}$ has a pseudo-principal interpretation, $e(\mathfrak{A})$ is nothing but the $\tilde{\ }$-extension of some model. In fact, in the latter case, $e(\mathfrak{A})$ coincides with $\mathfrak{B}$ from the previous paragraph.

Now we provide two operations, e and E, which turn generalized models into certain ordinary models that generalize $*$- and $\tilde{\ }$-extensions. Both operations are surjective and non-injective.

Define a map e on ultrafilters over functions to functions over ultrafilters,

$$e : \beta(Y^{X_1 \times \ldots \times X_n}) \to \beta Y^{\beta X_1 \times \ldots \times \beta X_n},$$

by induction on n. For $n = 1$, let e coincide with $\widetilde{\mathrm{ext}}$. Assume that e has been already defined for n. First we identify $Y^{X_1 \times X_2 \times \ldots \times X_{n+1}}$ with $(Y^{X_2 \times \ldots \times X_{n+1}})^{X_1}$ (by the so-called evaluation map, or carrying). Under this identification, each $\mathfrak{f} \in \beta(Y^{X_1 \times X_2 \times \ldots \times X_{n+1}})$ corresponds to a certain $\mathfrak{f}' \in \beta((Y^{X_2 \times \ldots \times X_{n+1}})^{X_1})$. Now we define $e(\mathfrak{f})$ by letting

$$e(\mathfrak{f})(\mathfrak{u}_1, \mathfrak{u}_2, \ldots, \mathfrak{u}_{n+1}) = e(e(\mathfrak{f}')(\mathfrak{u}_1))(\mathfrak{u}_2, \ldots, \mathfrak{u}_{n+1})$$

(since e has been already defined on $\mathfrak{f}'$ and $e(\mathfrak{f}')(\mathfrak{u}_1)$ by induction hypothesis).

Alternatively, we can define e as follows. Expand the domain of ext by letting

$$\mathrm{ext}(f) = \tilde{f}$$

for n-ary functions f with any n, not only unary ones. Thus, if we consider functions of $X_1 \times \ldots \times X_n$ into Y, then ext maps $Y^{X_1 \times \ldots \times X_n}$ into $RC_{X_1, \ldots, X_{n-1}}(\beta X_1 \times \ldots \times \beta X_n, \beta Y)$, the set of all functions of $\beta X_1 \times \ldots \times \beta X_n$ into βY that are right continuous w.r.t. $X_1, \ldots, X_{n-1}$. It can be shown that the latter set forms a closed subspace in the compact Hausdorff space $\beta Y^{\beta X_1 \times \ldots \times \beta X_n}$ of all functions of $\beta X_1 \times \ldots \times \beta X_n$ into βY with the standard (i.e. pointwise convergence) topology, and hence, is compact Hausdorff too. Therefore, ext continuously extends to the map $\widetilde{\mathrm{ext}}$ of $\beta(Y^{X_1 \times \ldots \times X_n})$ into it:

$$\beta(Y^{X_1 \times \ldots \times X_n})$$

$$Y^{X_1 \times \ldots \times X_n} \xrightarrow{\ \mathrm{ext}\ } RC_{X_1, \ldots, X_{n-1}}(\beta X_1 \times \ldots \times \beta X_n, \beta Y)$$

Now we can identify e with $\widetilde{\mathrm{ext}}$ in this expanded meaning.

By identifying relations with their characteristic functions, we can also let that e takes ultrafilters over relations to relations over ultrafilters:

$$e : \beta \mathcal{P}(X_1 \times \ldots \times X_n) \to \mathcal{P}(\beta X_1 \times \ldots \times \beta X_n).$$

In fact, e and $\widetilde{app}$ (or $\widetilde{in}$) are expressed via each other:

Lemma 3. *For all* $\mathfrak{f} \in \beta(Y^{X_1 \times \ldots \times X_n})$, $\mathfrak{r} \in \beta \mathcal{P}(X_1 \times \ldots \times X_n)$, *and* $\mathfrak{u}_1 \in \beta X_1, \ldots, \mathfrak{u}_n \in \beta X_n$,

$$e(\mathfrak{f})(\mathfrak{u}_1, \ldots, \mathfrak{u}_n) = \widetilde{app}(\mathfrak{u}_1, \ldots, \mathfrak{u}_n, \mathfrak{f}),$$

$$e(\mathfrak{r})(\mathfrak{u}_1, \ldots, \mathfrak{u}_n) \text{ iff } \widetilde{in}(\mathfrak{u}_1, \ldots, \mathfrak{u}_n, \mathfrak{r}).$$

In other words,

$$e(\mathfrak{f}) = \{(\mathfrak{u}_1, \ldots, \mathfrak{u}_n, \mathfrak{v}) \in \beta X_1 \times \ldots \times \beta X_n \times \beta Y : \widetilde{app}(\mathfrak{u}_1, \ldots, \mathfrak{u}_n, \mathfrak{f}) = \mathfrak{v}\},$$

$$e(\mathfrak{r}) = \{(\mathfrak{u}_1, \ldots, \mathfrak{u}_n) \in \beta X_1 \times \ldots \times \beta X_n : \widetilde{in}(\mathfrak{u}_1, \ldots, \mathfrak{u}_n, \mathfrak{r})\}.$$

Corollary 3. *For all generalized models* $\mathfrak{A} = (\beta X, \imath(F), \ldots, \imath(R), \ldots)$ *and valuations* v,

$$v_\imath(F(t_1, \ldots, t_n)) = e(\imath(F))(v_\imath(t_1), \ldots, v_\imath(t_n)),$$

$$\mathfrak{A} \Vdash R(t_1, \ldots, t_n) \ [v] \text{ iff } e(\imath(R))(v_\imath(t_1), \ldots, v_\imath(t_n)).$$

For a generalized model $\mathfrak{B} = (\beta X, \mathfrak{f}, \ldots, \mathfrak{r}, \ldots)$, let

$$e(\mathfrak{B}) = (\beta X, e(\mathfrak{f}), \ldots, e(\mathfrak{r}), \ldots).$$

Note that $e(\mathfrak{B})$ is an ordinary model.

Theorem 8. *If* $\mathfrak{A}$ *is a generalized model, then for all formulas* φ *and elements* $\mathfrak{u}_1, \ldots, \mathfrak{u}_n$ *of the universe of* $\mathfrak{A}$,

$$\mathfrak{A} \Vdash \varphi \ [\mathfrak{u}_1, \ldots, \mathfrak{u}_n] \text{ iff } e(\mathfrak{A}) \vDash \varphi \ [\mathfrak{u}_1, \ldots, \mathfrak{u}_n].$$

Define a map E, with the same domain and range that the map e has, as follows: E and e coincide on $\beta(Y^{X_1 \times \ldots \times X_n})$, and if $\mathfrak{r} \in \beta \mathcal{P}(X_1 \times \ldots \times X_n)$ then

$$E(\mathfrak{r}) = \{\widetilde{ext}(\mathfrak{q}) : \mathfrak{q} \in \widetilde{ext}(\mathfrak{r})\}.$$

Here $\widetilde{ext}(\mathfrak{r})$ is a clopen subset of $\beta(X_1 \times \ldots \times X_n)$, if $\mathfrak{q} \in \widetilde{ext}(\mathfrak{r})$ then $\widetilde{ext}(\mathfrak{q})$ is identified with an element of the space $\beta X_1 \times \ldots \times \beta X_n$ (as in Theorem 7), and the resulting $E(\mathfrak{r})$ is closed in the space.

Lemma 4. *Let* $\mathfrak{r} \in \beta \mathcal{P}(X_1 \times \ldots \times X_n)$. *Then*

$$e(\mathfrak{r}) = \tilde{R} \text{ and } E(\mathfrak{r}) = R^*$$

for $R = e(\mathfrak{r}) \cap (X_1 \times \ldots \times X_n) = E(\mathfrak{r}) \cap (X_1 \times \ldots \times X_n) = \bigcap_{S \in \mathfrak{r}} \bigcup S.$

One may write up this R more explicitly:

$$R = \big\{ (a_1, \ldots, a_n) \in X_1 \times \ldots \times X_n : (\forall S \in \mathfrak{r})\,(\exists Q \in S)\, Q(a_1, \ldots, a_n) \big\}.$$

For a generalized model $\mathfrak{B} = (\beta X, \mathfrak{f}, \ldots, \mathfrak{r}, \ldots)$, let

$$E(\mathfrak{B}) = (\beta X, E(\mathfrak{f}), \ldots, E(\mathfrak{r}), \ldots).$$

Then $E(\mathfrak{B})$, like $e(\mathfrak{B})$, is an ordinary model.

By Lemma 4, relations of the model $e(\mathfrak{B})$ are $\sim$-extensions of some relations on X, while relations of the model $E(\mathfrak{B})$ are $*$-extensions of the same relations. Whether the whole models $e(\mathfrak{B})$ and $E(\mathfrak{B})$ are ultrafilter extensions of some models depends only on the (generalized) interpretation of functional symbols in $\mathfrak{B}$:

Theorem 9. *Let $\mathfrak{B}$ be a generalized model with the universe βX. The following are equivalent:*

(i) $e(\mathfrak{B}) = \widetilde{\mathfrak{A}}$ for a model $\mathfrak{A}$ with the universe X,
(ii) $E(\mathfrak{B}) = \mathfrak{A}^*$ for a model $\mathfrak{A}$ with the universe X,
(iii) The interpretation in $\mathfrak{B}$ is pseudo-principal on functional symbols.

Moreover, the model $\mathfrak{A}$ in (i) and (ii) is the same.

Finally, we point out that the fact whether an ordinary model with the universe βX is of form $e(\mathfrak{B})$, and whether it is of form $E(\mathfrak{B})$, for some generalized model $\mathfrak{B}$ (clearly, with the same universe βX) depends only on its topological properties:

Theorem 10. *Let $\mathfrak{A}$ be a model with the universe βX. Then:*

(i) $\mathfrak{A} = e(\mathfrak{B})$ for a generalized model $\mathfrak{B}$ iff in $\mathfrak{A}$ all operations are right continuous and all relations right clopen w.r.t. X,
(ii) $\mathfrak{A} = E(\mathfrak{B})$ for a generalized model $\mathfrak{B}$ iff in $\mathfrak{A}$ all operations are right continuous w.r.t. X and all relations closed.

Since by Theorem 9, e and E applied to generalized models with pseudo-principal interpretations give the $\sim$- and $*$-extensions of ordinary models, Theorem 10 can be considered as a generalization of Theorems 3 and 5.

In conclusion, let us mention that various characterizations of both types of ultrafilter extensions lead to a spectrum of similar extensions as proposed at the end of [6]; so natural tasks are to study all of the spectrum as well as to isolate special features of the two canonical extensions among others.

Acknowledgement. We are indebted to Professor Robert I. Goldblatt who provided some useful historical information concerning the $*$-extension of relations.

References

1. Goranko, V.: Filter and ultrafilter extensions of structures: universal-algebraic aspects (2007, preprint)
2. Jónsson, B., Tarski, A.: Boolean algebras with operators. Part I: Amer. J. Math. **73**(4), 891–939 (1951). Part II: ibid. 74(1), 127–162 (1952)
3. Saveliev, D.I.: Ultrafilter extensions of models. In: Banerjee, M., Seth, A. (eds.) ICLA 2011. LNCS, vol. 6521, pp. 162–177. Springer, Heidelberg (2011). doi:10.1007/978-3-642-18026-2_14
4. Saveliev, D.I.: On ultrafilter extensions of models. In: Friedman, S.-D., et al. (eds.) The Infinity Project Proceedings CRM Documents 11, Barcelona, pp. 599–616 (2012)
5. Hindman, N., Strauss, D.: Algebra in the Stone-Čech compactification, 2nd edn. (2012). de Gruyter, W.: Revised and expanded, Berlin-New York
6. Saveliev, D.I.: On two concepts of ultrafilter extensions of binary relations (2014, preprint)
7. Čech, E.: On bicompact spaces. Ann. Math. **38**(2), 823–844 (1937)
8. Stone, M.H.: Applications of the theory of Boolean rings to general topology. Trans. Amer. Math. Soc. **41**, 375–481 (1937)
9. Wallman, H.: Lattices and topological spaces. Ann. Math. **39**, 112–126 (1938)
10. Comfort, W.W., Negrepontis, S.: The Theory of Ultrafilters. Springer, Berlin (1974)
11. Engelking, R.: General topology. Monogr. Matem. 60, Warszawa (1977)
12. Kochen, S.: Ultraproducts in the theory of models. Ann. Math. **74**(2), 221–261 (1961)
13. Frayne, T., Morel, A.C., Scott, D.S.: Reduced direct products. Fund. Math. **51**, 195–228 (1962). **53**, 117 (1963)
14. Chang, C.C., Keisler, H.J.: Model Theory. North-Holland, Amsterdam-London-New York (1973)
15. Kanamori, A.: The Higher Infinite: Large Cardinals in Set Theory from Their Beginnings, 2nd edn. Springer, Berlin (2005)
16. Ellis, R.: Lectures on Topological Dynamics. Benjamin, New York (1969)
17. Saveliev, D.I.: On idempotents in compact left topological universal algebras. Topol. Proc. **43**, 37–46 (2014)
18. Saveliev, D.I.: On Hindman sets (2008, preprint)
19. Lemmon, E.J.: Algebraic semantics for modal logic. Part II: J. Symb. Logic **31**(2), 191–218 (1966)
20. Lemmon, E.J., Scott, D.S.: An Introduction to Modal Logic. Blackwell, Oxford (1977)
21. Goldblatt, R.I., Thomason, S.K.: Axiomatic classes in propositional modal logic. In: Crossley, J.N. (ed.) Algebra and Logic. LNM, vol. 450, pp. 163–173. Springer, Heidelberg (1975). doi:10.1007/BFb0062855
22. van Benthem, J.F.A.K.: Canonical modal logics and ultrafilter extensions. J. Symb. Logic **44**(1), 1–8 (1979)
23. van Benthem, J.F.A.K.: Notes on modal definability. Notre Dame J. Formal Logic **30**(1), 20–35 (1988)
24. Venema, Y.: Model definability, purely modal. In: Gerbrandy, J., et al. (eds.) JFAK. Essays Dedicated to Johan van Benthem on the Occasion on his 50th Birthday, Amsterdam (1999)

25. Goldblatt, R.I.: Varieties of complex algebras. Ann. Pure Appl. Logic **44**, 173–242 (1989)
26. Saveliev, D.I.: Ultrafilter extensions of linearly ordered sets. Order **32**(1), 29–41 (2015)

Substructural Logics with a Reflexive Transitive Closure Modality

Igor Sedlár[(✉)] [iD]

Institute of Philosophy, Czech Academy of Sciences, Prague, Czech Republic
sedlar@flu.cas.cz

Abstract. Reflexive transitive closure modalities represent a number of important notions, such as common knowledge in a group of agents or non-deterministic iteration of actions. Normal modal logics with such modalities are well-explored but weaker logics are not. We add a reflexive transitive closure box modality to the modal non-associative commutative full Lambek calculus with a simple negation. Decidability and weak completeness of the resulting system are established and extensions of the results to stronger substructural logics are discussed. As a special case, we obtain decidability and weak completeness for intuitionistic modal logic with the reflexive transitive closure box.

Keywords: Substructural logics · Modal logic · Reflexive transitive closure · Intuitionistic modal logic

1 Introduction

Modalities semantically interpreted using a reflexive transitive closure of a modal accessibility relation model a number of important notions. For instance, they represent common knowledge in epistemic logic [4] and program iteration in dynamic logic [6].

Normal modal logics with such modalities are well-explored but weaker logics are not. This paper adds a reflexive transitive closure modality to a weak modal substructural logic, namely, the modal non-associative commutative full Lambek calculus with a simple negation. Decidability and weak completeness of the resulting system are established utilising the notion of filtration used by Bílková et al. [1]. Extensions of our theorems to stronger substructural logics are also discussed. It is shown that completeness and decidability proofs for intuitionistic logic with common knowlegde [7] follow as a corollary. Our results are

This work has been supported by the joint project of the German Science Foundation (DFG) and the Czech Science Foundation (GA ČR) number 16-07954J (*SEGA: From shared evidence to group attitudes*). The author would like to thank the anonymous reviewers for a number of suggestions, and to Adam Přenosil for reading a draft of the paper. A preliminary version of the paper was presented at the 8th International Workshop on Logic and Cognition in Guangzhou, China; the author is indebted to the audience for valuable feedback.

© Springer-Verlag GmbH Germany 2017
J. Kennedy and R.J.G.B. de Queiroz (Eds.): WoLLIC 2017, LNCS 10388, pp. 349–357, 2017.
DOI: 10.1007/978-3-662-55386-2_25

expected to find applications in substructural epistemic logics [1,10] extended with a common knowledge operator and non-classical versions of propositional dynamic logic PDL [11].

The paper is organised as follows. Section 2 introduces our basic modal substructural Lambek calculus and Sect. 3 adds to it a reflexive transitive closure modality. Decidability and completeness of the resulting system are established in Sect. 4. Extensions of the results to stronger substructural logics are briefly discussed in Sect. 5.

2   A Modal Lambek Calculus

Our basic logic is a modal version of the non-associative commutative full Lambek calculus **DFNLe** [2] extended with a simple negation.[1] This logic is chosen for the sake of syntactic simplicity (one implication and one negation), but also because it is often taken as basic in the literature on substructural epistemic logics [1,10,12].

Our results can be established for a non-commutative background logic with a pair of negations as well. As noted in Sect. 5, however, non-associativity is an important prerequisite of the applicability of the present technique.

The language $\mathcal{L}$ contains a countable set of atomic formulas Var and the set of 0-ary connectives $\{t, \top, \bot\}$, unary connectives $\{\neg, \Box\}$ and binary connectives $\{\wedge, \vee, \to, \otimes\}$. The set of formulas $Frm(\mathcal{L})$ is defined in the usual manner. The variable p ranges over Var; α, β and φ, ψ, χ etc. range over $Frm(\mathcal{L})$.

Definition 1. *A $\mathcal{L}$-model is a tuple $M = \langle P, \leq, L, C, S, R, [\![\cdot]\!]_M \rangle$ such that P is a non-empty set, $\leq$ is a partial order on P, L is a (upwardly) $\leq$-closed subset of P (let the set of such subsets be denoted $\mathrm{Up}(P)$), C and S are binary relations on P, R is a ternary relation on P and $[\![\cdot]\!]_M$ is a mapping from $Frm(\mathcal{L})$ to 2^P such that $[\![p]\!] \in \mathrm{Up}(P)$, for all $p \in Var$. It is required that every model satisfies the following conditions:*

$$x' \leq x \implies (Cxy \implies Cx'y) \tag{1}$$

$$x' \leq x \implies (Sxy \implies Sx'y) \tag{2}$$

$$x' \leq x \implies (Rxyz \implies Rx'yz) \tag{3}$$

$$x \leq y \iff (\exists z)(z \in L \ \& \ Rzxy) \tag{4}$$

$$Rxyz \implies Ryxz \tag{5}$$

$$Cxy \implies Cyx \tag{6}$$

Moreover, the truth-set mapping $[\![\cdot]\!]_M$ (mapping each formula to the set of states in which the formula is true) is required to satisfy the following conditions:

$$[\![\top]\!]_M = P, \ [\![\bot]\!]_M = \emptyset \ and \ [\![t]\!]_M = L \tag{7}$$

$$[\![\varphi \wedge \psi]\!]_M = [\![\varphi]\!]_M \cap [\![\psi]\!]_M \ and \ [\![\varphi \vee \psi]\!]_M = [\![\varphi]\!]_M \cup [\![\psi]\!]_M \tag{8}$$

---

[1] Due to space limitations, we do not provide an introduction to substructural logics and their relational semantics. See [9], for example.

$$[\![\neg\varphi]\!]_M = \{x \mid (\forall y)(Cxy \implies y \notin [\![\varphi]\!]_M)\} \tag{9}$$

$$[\![\Box\varphi]\!]_M = \{x \mid (\forall y)(Sxy \implies y \in [\![\varphi]\!]_M)\} \tag{10}$$

$$[\![\varphi \to \psi]\!]_M = \{x \mid (\forall yz)(Rxyz \;\&\; y \in [\![\varphi]\!]_M \implies z \in [\![\psi]\!]_M)\} \tag{11}$$

$$[\![\varphi \otimes \psi]\!]_M = \{x \mid (\exists yz)(Ryzx \;\&\; y \in [\![\varphi]\!]_M \;\&\; z \in [\![\psi]\!]_M)\} \tag{12}$$

A formula φ is valid in M ($M \Vdash \varphi$) iff $L \subseteq [\![\varphi]\!]_M$; φ is $\mathcal{L}$-valid ($\mathcal{L} \Vdash \varphi$) iff $M \Vdash \varphi$ for all $\mathcal{L}$-models M. A formula φ entails ψ in M ($\varphi \Vdash_M \psi$) iff $[\![\varphi]\!]_M \subseteq [\![\psi]\!]_M$.

The frame conditions (1)–(3) entail that $[\![\cdot]\!]_M$ is a mapping from *Frm* to $\mathrm{Up}(P)$. This, together with condition (4) implies that $M \Vdash \varphi \to \psi$ iff $\varphi \Vdash_M \psi$.

We do not have space to provide a full informal interpretation of the semantics,[2] but it will perhaps be helpful to think of $x \in P$ as "bodies of information" in some general sense and $\leq$ as "informational containment" ($x \leq y$ means that every piece of information supported by x is supported by y). We can then think of L a set of "logical" bodies of information (i.e. those that support logically valid formulas), C as a relation of compatibility and R as a relation associated with combining bodies of information ($Rxyz$ means, roughly, that the result of combining x and y is at least as strong as z).

Theorem 1. $\mathcal{L} \Vdash \varphi$ iff φ is a theorem of the axiom system H, consisting of axioms:

- $\varphi \to \varphi$
- $\varphi \land \psi \to \varphi$ and $\varphi \land \psi \to \psi$
- $\varphi \to \varphi \lor \psi$ and $\psi \to \varphi \lor \psi$
- $\varphi \to \top$ and $\bot \to \varphi$

- $\varphi \land (\psi \lor \chi) \to (\varphi \land \psi) \lor (\varphi \land \chi)$
- $\Box\varphi \land \Box\psi \to \Box(\varphi \land \psi)$
- $\top \to \Box\top$

and inference rules ('$//$' indicates a two-way rule):

- $\varphi, \varphi \to \psi \,/\, \psi$
- $\varphi \to \psi, \psi \to \chi \,/\, \varphi \to \chi$
- $\chi \to \varphi, \chi \to \psi \,/\, \chi \to (\varphi \land \psi)$
- $\varphi \to \chi, \psi \to \chi \,/\, (\varphi \lor \psi) \to \chi$
- $\varphi \to \psi \,/\, \Box\varphi \to \Box\psi$

- $\varphi \to (\psi \to \chi) \,//\, (\psi \otimes \varphi) \to \chi$
- $\varphi \to (\psi \to \chi) \,//\, \psi \to (\varphi \to \chi)$
- $t \to \varphi \,//\, \varphi$
- $\varphi \to \neg\psi \,//\, \psi \to \neg\varphi$

Proof. This is a standard result [1,5].

Example 1. It is easily seen that the modal $\Box$ distributes over $\land, \lor$ in the expected way; both

$$\Box(\varphi \land \psi) \leftrightarrow (\Box\varphi \land \Box\psi) \text{ and } \Box\varphi \lor \Box\psi \to \Box(\varphi \lor \psi)$$

are valid in every M. However, the "K-axiom"

$$\Box(\varphi \to \psi) \to (\Box\varphi \to \Box\psi)$$

---

[2] See [8], for example.

is not valid in every M (observe that the set $\{\varphi \mid x \in [\![\varphi]\!]_M\}$ is closed under Modus Ponens only if $Rxxx$). For a similar reason, $\Box$ does not distribute over $\otimes$, i.e.

$$\Box(\varphi \otimes \psi) \rightarrow (\Box\varphi \otimes \Box\psi)$$

is not valid in every M.[3]

3   Adding a Reflexive Transitive Closure Modality

The language $\mathcal{L}^*$ extends $\mathcal{L}$ with a unary connective $\Box^*$. The set of formulas $Frm(\mathcal{L}^*)$ of $\mathcal{L}^*$ is defined in the usual manner and all the syntactic metavariables are now taken to range over $Frm(\mathcal{L}^*)$. Γ, Δ etc. range over subsets of $Frm(\mathcal{L}^*)$. Let $\Gamma/U = \{\varphi \mid U\varphi \in \Gamma\}$ for all $U \in \{\neg, \Box, \Box^*\}$.

Definition 2. *A $\mathcal{L}^*$-model is $M = \langle P, \leq, L, C, S, S^*, R, [\![\cdot]\!]_M \rangle$ where everything is as in Definition 1 and, in addition,*

$$S^* \text{ is the reflexive transitive closure of } S \tag{13}$$

$$[\![\Box^*\varphi]\!]_M = \{x \mid (\forall y)(S^*xy \implies y \in [\![\varphi]\!]_M)\} \tag{14}$$

In general, if $\Gamma^{\downarrow}$ is closed under subformulas, then a $\Gamma^{\downarrow}$-model is a structure that satisfies all the conditions required for $\mathcal{L}^*$-models, but the truth-set conditions (7)–(12) and (14) are required to hold only for $\Gamma^{\downarrow}$.

Lemma 1. *Let $\varphi \in \Gamma^{\downarrow}$. If there is a $\Gamma^{\downarrow}$-model M such that $M \not\Vdash \varphi$, then there is a $\mathcal{L}^*$-model M' such that $M' \not\Vdash \varphi$. If M is finite then so is M'.*

We note that $[\![\Box^*\varphi]\!]_M \in \mathrm{Up}(P)$, so $\varphi \Vdash_M \psi$ iff $M \Vdash \varphi \rightarrow \psi$ for all $\Gamma^{\downarrow}$-models M where $\varphi \rightarrow \psi \in \Gamma^{\downarrow}$.

One immediate consequence of the truth condition for $\Box^*\varphi$ is that our logic is *not compact*. To see this, observe that every finite subset of $\{\neg\Box^*p\} \cup \{p\} \cup \{\Box^n p \mid n \in \omega\}$ is satisfiable, but the set itself is not.

4   Axiomatization and Decidability

Our main result is a weakly complete axiomatization of the set of $\mathcal{L}^*$-valid formulas and a proof that this set is decidable. We use a generalisation of the standard filtration technique [4,6]. In particular, we build on the notion of filtration used in [1].

Definition 3. *Let H^* be the axiom system obtained from H by adding the axiom and rule schemas shown in Fig. 1 (called stars).*

---

[3] Stated more precisely, counterexamples to the K-axiom can be constructed if the frame property $Syx \implies Rxxx$ fails. Similarly, counterexamples to distributivity of $\Box$ over $\otimes$ can be found if we have Swx and $Ryzx$ but also $Ry'z'w$ and $Sy'u$ with $y \not\leq u$ for some u. Counterexamples to the converse implication can be found if a similar frame condition holds.

$$(*1)\ \Box^*\varphi \wedge \Box^*\psi \to \Box^*(\varphi \wedge \psi)$$
$$(*2)\ \top \to \Box^*\top$$
$$(*3)\ \Box^*\varphi \leftrightarrow (\varphi \wedge \Box\Box^*\varphi)$$
$$(*4)\ \varphi \to \psi\ /\ \Box^*\varphi \to \Box^*\psi$$
$$(*5)\ \varphi \to \Box\varphi\ /\ \varphi \to \Box^*\varphi$$

Fig. 1. The stars.

We note that axiomatizations of normal modal logics with a reflexive transitive closure modality usually contain *the induction axiom*

$$(\varphi \wedge \Box^*(\varphi \to \Box\varphi)) \to \Box^*\varphi$$

inhstead of the *loop invariance rule* ($*5$) (see [6], for example). These two are equivalent in the classical setting (in the sense that the rule preserves validity iff the axiom is valid), but not so in our framework. In fact, it can be shown that the induction axiom is not valid in every model. The reason is closely related to the failure of the K-axiom pointed out in Example 1.

We write $\vdash \varphi$ if φ is a theorem of H^*, $\varphi \vdash \psi$ if $\vdash \varphi \to \psi$ and $\Gamma \vdash \Delta$ if there are finite $\Gamma' \subseteq \Gamma$ and $\Delta' \subseteq \Delta$ such that $\bigwedge \Gamma' \vdash \bigvee \Delta'$.

A set of formulas Γ is a *proper prime theory* iff $\Gamma \neq Frm(\mathcal{L}^*)$ and $\Gamma \vdash \varphi \vee \psi$ only if $\varphi \in \Gamma$ or $\psi \in \Gamma$. Note that, for all proper prime theories Γ, $\Gamma \vdash \varphi$ only if $\varphi \in \Gamma$ since $\Gamma \vdash \varphi$ implies $\Gamma \vdash \varphi \vee \varphi$. Note also that if $\Delta \vdash \varphi$, then $\Delta \subseteq \Gamma$ only if $\varphi \in \Gamma$.

Theorem 2. *If $\Gamma \nvdash \Delta$, then there is a proper prime theory $\Gamma' \supseteq \Gamma$ disjoint from Δ.*

Proof. This is established by using a variant of the Pair Extension Theorem [9, p. 94].

Definition 4. *The* canonical structure

$$M_c = \langle P_c, \leq_c, L_c, S_c, S_c^*, C_c, R_c, [\![\cdot]\!]_c \rangle$$

is defined as follows:

- P_c *is the set of all proper prime theories;*
- $\leq_c$ *is set inclusion;*
- $L_c = \{\Gamma \mid t \in \Gamma\};$
- $S_c = \{\langle \Gamma, \Delta \rangle \mid \Gamma/\Box \subseteq \Delta\};$
- $S_c^* = \{\langle \Gamma, \Delta \rangle \mid \Gamma/\Box^* \subseteq \Delta\};$
- $C_c = \{\langle \Gamma, \Delta \rangle \mid \Gamma/\neg \cap \Delta = \emptyset\};$
- $R_c = \{\langle \Gamma, \Delta, \Theta \rangle \mid (\forall \varphi \psi)(\varphi \to \psi \in \Gamma\ \&\ \varphi \in \Delta \implies \psi \in \Theta)\};$
- $[\![\varphi]\!]_c = \{\Gamma \mid \varphi \in \Gamma\}.$

It is a standard observation that the canonical structure is not a $\mathcal{L}^*$-model, for it fails to meet condition (13). In general, S_c^* contains the reflexive transitive closure $(S_c)^*$ of S_c, but it is not identical to it.[4] Nevertheless, the conditions (1)–(12) and (14) are met [1,9]. (For instance, let us check condition (14). The left-to-right inclusion is trivial. The right-to-left inclusion is established by a variant of the Witness Lemma [9, p. 255]. If $\square^*\varphi \notin \Gamma$, then $\Gamma/\square^* \nvdash \varphi$. Hence, by the Pair Extension Theorem, there is $\Delta \in P_c$ such that $S_c^*\Gamma\Delta$ and $\Delta \notin [\![\varphi]\!]_c$.)

Definition 5. *The closure of φ, $cl(\varphi)$, is the smallest set of formulas such that*

- $\{\varphi, t\} \subseteq cl(\varphi)$;
- $\psi \in cl(\varphi)$ *for all subformulas ψ of φ;*
- *if $\square^*\psi \in cl(\varphi)$, then $\square\square^*\psi \in cl(\varphi)$.*

For every φ, let $\Gamma \preceq_\varphi \Delta$ iff $(\Gamma \cap cl(\varphi)) \subseteq \Delta$ and $\Gamma \sim_\varphi \Delta$ iff $\Gamma \preceq_\varphi \Delta$ and $\Delta \preceq_\varphi \Gamma$. Moreover, let $[\Gamma]_\varphi = \{\Delta \mid \Gamma \sim_\varphi \Delta\}$.

It is plain that $cl(\varphi)$ is finite for all φ.

Definition 6. *Fix a formula φ. The φ-filtration of the canonical structure is a structure $M_\varphi = \langle P_\varphi, \leq_\varphi, L_\varphi, S_\varphi, S_\varphi^*, C_\varphi, R_\varphi, [\![\cdot]\!]_\varphi \rangle$ defined as follows:*

- $P_\varphi = \{[\Gamma]_\varphi \mid \Gamma \in P_c\}$;
- $[\Gamma]_\varphi \leq_\varphi [\Delta]_\varphi$ *iff $\Gamma \preceq_\varphi \Delta$;*
- $L_\varphi = \{[\Gamma]_\varphi \mid (\exists\Gamma' \preceq_\varphi \Gamma)(\Gamma' \in L_c)\}$;
- $S_\varphi = \{\langle [\Gamma]_\varphi, [\Delta]_\varphi \rangle \mid (\exists\Gamma' \succeq_\varphi \Gamma)(S_c\Gamma'\Delta)\}$;
- $S_\varphi^* = (S_\varphi)^*$;
- $C_\varphi = \{\langle [\Gamma]_\varphi, [\Delta]_\varphi \rangle \mid (\exists\Gamma' \succeq_\varphi \Gamma, \exists\Delta' \succeq_\varphi \Delta)(C_c\Gamma'\Delta')\}$;
- $R_\varphi = \{\langle [\Gamma]_\varphi, [\Delta]_\varphi, [\Theta]_\varphi \rangle \mid (\exists\Gamma' \succeq_\varphi \Gamma, \exists\Delta' \succeq_\varphi \Delta)(R_c\Gamma'\Delta'\Theta)\}$;
- *for all $\alpha \in cl(\varphi)$, $[\![\alpha]\!]_\varphi = \{[\Gamma]_\varphi \mid \alpha \in \Gamma\}$; for $\alpha \notin cl(\varphi)$, $[\![\alpha]\!]_\varphi = \emptyset$.*

The crucial difference between the canonical structure and its filtration (in addition to the fact that the latter is finite) is the fact that, in a φ-filtration, S_φ^* is *defined* to be the reflexive transitive closure of S_φ. However, one needs to check that the φ-filtration of the canonical structure is a $cl(\varphi)$-model. In what follows, we drop the subscript 'φ' whenever possible.

Theorem 3. *For all φ, the φ-filtration of the canonical structure is a $cl(\varphi)$-model.*

Proof. The relation $\leq_\varphi$ is obviously a partial order on P_φ. The fact that C_φ, R_φ and L_φ satisfy the conditions (1), (6), (3), (5) and (4), respectively, and that L_φ is closed under $\leq_\varphi$ are established similarly as in [1]. (13) holds by definition and (2) is established as follows. If $S[\Gamma][\Delta]$ then $S_c\Gamma'\Delta$ for some $\Gamma' \succeq \Gamma$. But if $[\Theta] \leq [\Gamma]$ then $\Theta \preceq \Gamma$ and, consequently, $\Theta \preceq \Gamma'$. Hence, $S[\Theta][\Delta]$.

---

[4] The reason is that if $\square^*\varphi \in \Gamma$, then $\varphi, \square^n\varphi \in \Gamma$ for all $n \in \omega$ by (*3), but the converse implication cannot be established (our axiomatization is finitary).

It remains to be shown that $[\![\cdot]\!]_\varphi$ satisfies the conditions (7)–(12) and (14) when applied to $\psi \in cl(\varphi)$. The cases where the main connective of ψ is in $\{\top, \bot, t, \neg, \wedge, \vee, \rightarrow, \otimes\}$ are established as in [1].

Next, assume that $\psi = \Box\alpha$. We have to show that

$$\Box\alpha \in \Gamma \iff (\forall[\Delta])(S[\Gamma][\Delta] \implies \alpha \in \Delta)$$

($\alpha \in \Delta$ means $[\Delta] \in [\![\alpha]\!]$) Assume first that $\Box\alpha \in \Gamma$ and $S[\Gamma][\Delta]$. It follows that $S_c\Gamma'\Delta$ for some $\Gamma' \succeq \Gamma$. But then $\Box\alpha \in \Gamma'$ and, by the definition of S_c, $\alpha \in \Delta$. Conversely, if $\Box\alpha \notin \Gamma$, then the Witness Lemma [9, p. 255] entails that there is Δ such that $S_c\Gamma\Delta$ and $\alpha \notin \Delta$. But it is plain that $S_c\Gamma\Delta$ only if $S[\Gamma][\Delta]$.

Finally, assume that $\psi = \Box^*\alpha$. We have to show that

$$\Box^*\alpha \in \Gamma \iff (\forall[\Delta])(S^*[\Gamma][\Delta] \implies \alpha \in \Delta)$$

If $\Box^*\alpha \notin \Gamma$ then, by a variation of the Witness Lemma, there is Δ such that $\alpha \notin \Delta$ and $S_c^*\Gamma\Delta$. It is sufficient to show that there is Θ such that $S^*[\Gamma][\Theta]$ and $\Theta \preceq \Delta$.

Let

$$E = \{\Phi \mid (\exists\Theta)(S^*[\Gamma][\Theta] \,\&\, \Theta \preceq \Phi)\}$$

(E is closed under $\leq_c$, but $E' = \{\Phi \mid S^*[\Gamma][\Phi]\}$ is not. Recall that $(S_c)^* \subseteq S_c^*$, but not necessarily vice versa.) We have to show that $\Delta \in E$. For all $[\Phi] \in P$, define

$$\psi_{[\Phi]} = \bigwedge\{\alpha \in cl(\varphi) \mid \alpha \in \Phi\}$$

and

$$\psi_E = \bigvee\{\psi_{[\Phi]} \mid S^*[\Gamma][\Phi]\}.$$

Note that ψ_E is well-defined since P_φ is finite. We establish two claims.

Claim 1. E is closed under S_c, i.e., if $\Phi \in E$ and $S_c\Phi\Psi$, then $\Psi \in E$. If $\Phi \in E$, then there is Θ such that $S^*[\Gamma][\Theta]$ and $\Theta \preceq \Phi$. It follows from $S_c\Phi\Psi$ and $\Theta \preceq \Phi$ that $S[\Theta][\Psi]$. But S^* is the reflexive transitive closure of S, so it follows that $S^*[\Gamma][\Psi]$. Hence, $\Psi \in E$.

Claim 2. $E = [\![\psi_E]\!]_c$. First, assume that $\Phi \in E$, i.e., there is Ψ such that $S^*[\Gamma][\Psi]$ and $\Psi \preceq \Phi$. Now $\Psi \preceq \Phi$ implies $\psi_{[\Psi]} \in \Phi$ (proper prime theories are closed under forming conjunctions). But $\Psi \in E$, so $\psi_{[\Psi]} \vdash \psi_E$. Consequently, $\psi_E \in \Phi$, i.e., $\Phi \in [\![\psi_E]\!]_c$. Conversely, assume that $\psi_E \in \Phi$. Φ is a prime theory, so $\psi_{[\Theta]} \in \Phi$ for some Θ such that $S^*[\Gamma][\Theta]$. It follows that $\Theta \preceq \Phi$. Hence, $\Phi \in E$.

The two claims imply that $[\![\psi_E]\!]_c \subseteq [\![\Box\psi_E]\!]_c$. (If $\psi_E \in \Delta$, then $\Delta \in E$ by Claim 2. But then, if $S_c\Delta\Theta$ for some Θ, then $\Theta \in E$ by Claim 1. By Claim 2, if $S_c\Delta\Theta$, then $\psi_E \in \Theta$. Hence, $\Box\psi_E \in \Delta$.) Consequently, $\psi_E \rightarrow \Box\psi_E \in \bigcap\{\Phi \mid \Phi \in L_c\}$. Now since $\varphi \rightarrow \Box\varphi / \varphi \rightarrow \Box^*\varphi$ is an inference rule of H^*, $\psi_E \rightarrow \Box^*\psi_E \in \bigcap\{\Phi \mid \Phi \in L_c\}$ and $[\![\psi_E]\!]_c \subseteq [\![\Box^*\psi_E]\!]_c$. Now we show that $S_c^*\Gamma\Delta$ implies that there is Θ such that $S^*[\Gamma][\Theta]$ and $\Theta \preceq \Delta$. It is plain that $\Gamma \in E$. By Claim

2, $\Gamma \in [\![\psi_E]\!]_c$ and, consequently, $\Gamma \in [\![\square^*\psi_E]\!]_c$. Now $\Delta \in [\![\psi_E]\!]_c$ since $S_c^*\Gamma\Delta$. In other words, $\Delta \in E$. But this means that there is Θ such that $S^*[\Gamma][\Theta]$ and $\Theta \preceq \Delta$.

The final thing to show is that if $\square^*\alpha \in \Gamma$ and $S^*[\Gamma][\Delta]$, then $\alpha \in \Delta$. Our assumption $S^*[\Gamma][\Delta]$ entails that either $[\Gamma] = [\Delta]$ or there is $n \geq 1$ such that $S[\Gamma][\Delta_1]\cdots[\Delta_n] = [\Delta]$. If $[\Gamma] = [\Delta]$, then $\Gamma \sim \Delta$ and $\square^*\alpha \in \Delta$. Since $\vdash \square^*\alpha \rightarrow \alpha$, $\alpha \in \Delta$ and we are done. Assume that there is $n \geq 1$ such that $S[\Gamma][\Delta_1]\cdots[\Delta_n] = [\Delta]$. We show by induction that for all $n \geq 1$, Δ_n contains α and $\square^*\alpha$. $\square^*\alpha \in \Gamma$ entails that $\square\square^*\alpha \in \Gamma$ (as $\vdash \square^*\alpha \rightarrow \square\square^*\alpha$). By the clause $\psi = \square\beta$ established above, $S[\Gamma][\Delta_1]$ entails that $\square^*\alpha \in \Delta_1$, so $\alpha \in \Delta_1$ as well. Let us now assume that the claim holds for $k < n$. We show that it holds for $k + 1$ as well. Assume that $\alpha, \square^*\alpha \in \Delta_k$ and $S[\Delta_k][\Delta_{k+1}]$. Again, $\square\square^*\alpha \in \Delta_k$ and $\square^*\alpha \in \Delta_{k+1}$ by the clause $\psi = \square\beta$ and, consequently, $\alpha \in \Delta_{k+1}$.

Theorem 4. *If $\nvdash \varphi$, then there is a finite $\mathcal{L}^*$-model M such that $M \nVdash \varphi$.*

Proof. If $\nvdash \varphi$, then $t \nVdash \varphi$. By the Pair Extension Theorem [9], there is a proper prime theory $\Gamma \in L_c$ such that $\varphi \notin \Gamma$. The φ-filtration M_φ of the canonical structure is a finite $cl(\varphi)$-model by Theorem 3. Moreover, $[\Gamma] \in L_\varphi$ and $[\Gamma] \notin [\![\varphi]\!]_\varphi$. So, $M_\varphi \nVdash \varphi$. By Lemma 1, there is a finite $\mathcal{L}^*$-model M such that $M \nVdash \varphi$.

Theorem 5. *H^* is a sound and weakly complete axiomatisation of the set of formulas valid in every $\mathcal{L}^*$-model. This set is decidable.*

Proof. Soundness of H^* is left to the reader. Weak completeness follows from Theorem 4. Decidability follows from the fact that any φ-filtration of the canonical structure is finite (and, in fact, bounded by the size of φ).

5   Extensions

It is easily seen that our results can be extended to stronger substructural logics.

Theorem 6. *Let $\mathbf{L}$ be a substructural logic (in $\mathcal{L}$) axiomatised by $H(\mathbf{L})$ and characterised by a class of models $Mod(\mathbf{L})$. Assume that $M \in Mod(\mathbf{L})$ iff M satisfies a set of frame conditions $Con(\mathbf{L})$ such that the $H(\mathbf{L})$-canonical structure and the φ-filtration (for arbitrary φ) of the canonical structure both satisfy $Con(\mathbf{L})$. Then the extension of $\mathbf{L}$ by a reflexive transitive closure modality is decidable and axiomatised by $H(\mathbf{L})$ plus the stars.*

Proof. If it is assumed that the φ-filtration of the canonical structure satisfies all the relevant frame conditions, then the fact that $[\![\cdot]\!]_\varphi$ satisfies the conditions (7)–(12) and (14) when applied to $\psi \in cl(\varphi)$ is established exactly as in the proof of Theorem 3 above. But this means that the φ-filtration of the canonical structure is a finite $cl(\varphi)$-model. The rest of the argument is as before.

This observation also hints at potential limitations of the present technique. In general, if a frame condition is not preserved under forming filtrations (i.e. if the canonical structure satisfies the condition, then its φ-filtration for arbitrary φ does so as well) then the present technique cannot be applied to logics complete with respect to models satisfying the frame condition. For instance, the frame condition corresponding to associativity[5]

$$Rxyv \ \& \ Rvzw \implies (\exists u)(Rxuw \ \& \ Ryzu)$$

is not preserved by standard notions of filtration such as the one we have used in the present paper.

If one is interested in adding a reflexive transitive closure modality to modal intuitionistic logic (as in [7]), however, the problem with associativity can be avoided by interpreting $\rightarrow$ directly in terms of $\leq$:

$$[\![\varphi \rightarrow \psi]\!] = \{x \mid (\forall y)(x \leq y \implies (y \in [\![\varphi]\!] \implies y \in [\![\psi]\!]))\}$$

An inspection of our proof of Theorem 3 reveals that, after adding the stars to any axiomatization of intuitionistic logic with $\square$ [3], our argument can be repeated without modification.

References

1. Bílková, M., Majer, O., Peliš, M.: Epistemic logics for sceptical agents. J. Logic Comput. **26**(6), 1815–1841 (2016)
2. Buszkowski, W., Farulewski, M.: Nonassociative lambek calculus with additives and context-free languages. In: Grumberg, O., Kaminski, M., Katz, S., Wintner, S. (eds.) Languages: From Formal to Natural. LNCS, vol. 5533, pp. 45–58. Springer, Heidelberg (2009). doi:10.1007/978-3-642-01748-3_4
3. Božić, M., Došen, K.: Models for normal intuitionistic modal logics. Stud. Logica. **43**(3), 217–245 (1984)
4. Fagin, R., Halpern, J.Y., Moses, Y., Vardi, M.Y.: Reasoning About Knowledge. MIT Press, Cambridge (1995)
5. Fuhrmann, A.: Models for relevant modal logics. Stud. Logica. **49**(4), 501–514 (1990)
6. Harel, D., Kozen, D., Tiuryn, J.: Dynamic Logic. MIT Press, Cambridge (2000)
7. Jäger, G., Marti, M.: Intuitionistic common knowledge or belief. J. Appl. Logic **18**(C), 150–163 (2016)
8. Mares, E.: Relevant Logic: A Philosophical Interpretation. Cambridge University Press, Cambridge (2004)
9. Restall, G.: An Introduction to Substructural Logics. Routledge, London (2000)
10. Sedlár, I.: Substructural epistemic logics. J. Appl. Non-Class. Logics **25**(3), 256–285 (2015)
11. Sedlár, I.: Propositional dynamic logic with Belnapian truth values. In: Beklemishev, L., Demri, S., Máté, A. (eds.) Advances in Modal Logic 2016, pp. 503–519. College Publications, London (2016)
12. Sequoiah-Grayson, S.: Epistemic closure and commutative, nonassociative residuated structures. Synthese **190**(1), 113–128 (2013)

---

[5] Note that $(\varphi \otimes \psi) \otimes \chi \rightarrow \varphi \otimes (\psi \otimes \chi)$ is valid in M if the model satisfies this frame condition. On the other hand, (5) entails the validity of $\varphi \otimes \psi \rightarrow \psi \otimes \varphi$.

Global Neighbourhood Completeness of the Gödel-Löb Provability Logic

Daniyar Shamkanov[✉]

Steklov Mathematical Institute of Russian Academy of Sciences,
Gubkina str. 8, 119991 Moscow, Russia
daniyar.shamkanov@gmail.com

Abstract. The Gödel-Löb provability logic GL is strongly neighbourhood complete in the case of the so-called local semantic consequence relation. In the given paper, we consider Hilbert-style non-well-founded derivations in GL and establish that GL with the obtained derivability relation is strongly neighbourhood complete in the case of the global semantic consequence relation.

Keywords: Provability logic · Neighbourhood semantics · Global consequence relations · Non-well-founded proofs

1 Introduction

The Gödel-Löb provability logic GL is a modal logic describing all universally valid principles of the formal provability in Peano arithmetic [13]. This logic is complete with respect to its Kripke semantics [9], yet it is not strongly complete. In this paper, we study so-called neighbourhood semantics of the given logic.

Neighbourhood semantics is a generalization of Kripke semantics independently developed by Montague and Scott in [7,8]. A neighbourhood frame can be defined as a pair $(X, \Box)$, where X is a set and $\Box$ is an unary operation in $\mathcal{P}(X)$. The logic GL is not compact with respect to Kripke semantics, but it is neighbourhood compact, which immediately implies that GL is strongly neighbourhood complete (see [1,11]). We stress that this strong completeness result was obtained for the case of the so-called local semantic consequence relation. Recall that, over neighbourhood GL-models, a formula A is a local semantic consequent of Γ if for any neighbourhood GL-model $\mathcal{M}$ and any world x of $\mathcal{M}$

$$(\forall B \in \Gamma \ \mathcal{M}, x \vDash B) \Rightarrow \mathcal{M}, x \vDash A.$$

A formula A is a global semantic consequent of Γ if for any neighbourhood GL-model $\mathcal{M}$

$$(\forall B \in \Gamma \ \mathcal{M} \vDash B) \Rightarrow \mathcal{M} \vDash A.$$

This paper studies the case of the global semantic consequence relation.

This work was supported by the Russian Science Foundation (grant no. 14-50-00005).

J. Kennedy and R.J.G.B. de Queiroz (Eds.): WoLLIC 2017, LNCS 10388, pp. 358–370, 2017.
DOI: 10.1007/978-3-662-55386-2_26

Recently a proof-theoretic presentation of GL in the form of a sequent calculus allowing non-well-founded proofs was given in [5,10]. In this paper, we consider Hilbert-style non-well-founded derivations in GL and establish that GL with the obtained derivability relation is strongly neighbourhood complete with respect to the global semantic consequence relation.

At the same time we should emphasize that, in the case of GL, the ordinary global syntactic consequence relation, which is a derivability relation standardly defined without non-well-founded derivations, is not neighbourhood complete (see Corollary 7.6 in [6]).

The plan of the paper is as follows. In Sect. 2, we recall the Gödel-Löb provability logic GL and its neighbourhood semantics. Then we recall a connection between scattered topological spaces and GL-frames in Sect. 3. In the next section, we define global semantic consequence relations over GL-frames and introduce corresponding derivability relations using non-well-founded derivation trees. In Sect. 5, we obtain a form of neighbourhood compactness using the ultrabouquet construction from [11]. In Sect. 6, we present a sequent calculus for GL allowing non-well-founded proof trees. The final section is devoted to establishing global neighbourhood completeness of GL with non-well-founded proofs.

2   Preliminaries

In this section we recall the Gödel-Löb provability logic GL and its neighbourhood semantics.

Formulas of GL (also called *modal formulas*) are built from the countable set of variables $PV = \{p, q, \dots\}$ and the constant $\bot$ using propositional connectives $\rightarrow$ and $\Box$. We treat other boolean connectives and the modal connective $\Diamond$ as abbreviations:

$$\neg A := A \rightarrow \bot, \qquad \top := \neg\bot, \qquad A \wedge B := \neg(A \rightarrow \neg B),$$
$$A \vee B := (\neg A \rightarrow B), \qquad \Diamond A := \neg\Box\neg A.$$

In the sequel, the set of modal formulas is denoted by Fm.

The Gödel-Löb provability logic GL is defined via its Hilbert-style axiomatization.

Axiom Schemes:

(i) the tautologies of classical propositional logic;
(ii) $\Box(A \rightarrow B) \rightarrow (\Box A \rightarrow \Box B)$;
(iii) $\Box A \rightarrow \Box\Box A$;
(iv) $\Box(\Box A \rightarrow A) \rightarrow \Box A$.

Inference Rules:

$$\text{mp}\frac{A \quad A \rightarrow B}{B}, \qquad \text{nec}\frac{A}{\Box A}.$$

A relation of derivability from assumptions in GL is inductively defined in the following way. A formula A *is derivable from the set of assumptions* Γ (cf. [4]), if A is in Γ, or A is one of the axioms of GL, or follows from derivable formulas through applications of the inference rules (mp) and (nec) so that the rule (nec) can be applied only to derivations without assumptions. We denote this derivability relation by $\vdash_l$, where l stands for 'local'.

Following [11], we define a *neighbourhood frame* $\mathcal{X} = (X, \Box)$ as a set X together with an operator on its subsets $\Box : \mathcal{P}(X) \to \mathcal{P}(X)$. Elements of X are called *worlds* of the frame $\mathcal{X}$. A *valuation (over $\mathcal{X}$)* is a function $\theta : Fm \to \mathcal{P}(X)$ such that $\theta(\bot) = \varnothing$, $\theta(V_1 \to V_2) = (X \setminus \theta(V_1)) \cup \theta(V_2)$ and $\theta(\Box V) = \Box\theta(V)$. A *neighbourhood model* is defined as a pair $\mathcal{M} = (\mathcal{X}, \theta)$, where $\mathcal{X}$ is a neighbourhood frame and θ is a valuation over it.

A formula A is *true at a world* x of a model $\mathcal{M}$, written as $\mathcal{M}, x \vDash A$, if $x \in \theta(A)$. A formula A is called *true in* $\mathcal{M}$, written as $\mathcal{M} \vDash A$, if A is true at all worlds of $\mathcal{M}$. In addition, A is *valid in a frame* $\mathcal{X}$ if A is true at all worlds of $\mathcal{X}$ under all valuations. A GL-*frame* is a frame in which all formulas provable in GL are valid. A GL-*model* is a neighbourhood model over a GL-frame.

Now we define a local (pointwise) semantic consequence relation over GL-frames. Given a set of modal formulas Γ and a formula A, we set $\Gamma \vDash_l A$ if for any GL-model $\mathcal{M}$ and any its world x

$$(\forall B \in \Gamma \; \mathcal{M}, x \vDash B) \Rightarrow \mathcal{M}, x \vDash A.$$

The following strong completeness result was obtained by Shehtman using the so-called ultrabouquet construction (see [11]).

Proposition 1. $\Gamma \vdash_l A \Longleftrightarrow \Gamma \vDash_l A.$

3   Scattered Topological Spaces

In this section we briefly recall a connection between scattered topological spaces and GL-frames (cf. [2]).

In a topological space, an open set U containing a point x is called a *neighbourhood* of x. A set U is a *punctured neighbourhood* of x if $x \notin U$ and $U \cup \{x\}$ is open. For a topological space (X, τ) and its subset V the *derivative set* $d_\tau(V)$ *of* V is the set of limit points of V:

$$x \in d_\tau(V) \Longleftrightarrow \forall U \in \tau \; (x \in U \Rightarrow \exists y \neq x \; (y \in U \cap V)).$$

The *co-derivative set* $cd_\tau(V)$ *of* V is defined as $X \setminus d_\tau(X \setminus V)$. By definition, $x \in cd_\tau(V)$ if and only if there is a punctured neighbourhood of x entirely contained in V.

In a topological space, a point having an empty punctured neighbourhood is called *isolated*. A topological space is *scattered* if each non-empty subset of X (as a topological space with the inherited topology) has an isolated point.

Proposition 2 (Esakia [3]). *If* $(X, \Box)$ *is a* GL-*frame, then* X *bears a unique topology* τ *for which* $\Box = cd_\tau$. *Moreover, the space* (X, τ) *is scattered.*

Proposition 3 (Simmons [12], Esakia [3]). *If (X, τ) is a scattered topological space, then (X, cd_τ) is a GL-frame.*

In the sequel, we don't distinguish GL-frames and corresponding topological spaces so that we use the topological terminology referring to (X, τ) for the frame (X, cd_τ). For example, we say that a subset U is *open* in $(X, \square)$ if it is open in the corresponding topological space (which is equivalent to $U \subset \square U$).

For a topological space (X, τ), we define transfinite iterations of the co-derivative-set operator by

- $cd_\tau^0(V) = V$, $cd_\tau^{\alpha+1}(V) = cd_\tau(cd_\tau^\alpha(V))$,
- $cd_\tau^\alpha(V) = \bigcup_{\beta < \alpha}(cd_\tau^\beta(V))$ if α is a limit ordinal.

Proposition 4 (Cantor). *A topological space (X, τ) is scattered if and only if $cd_\tau^\alpha(\varnothing) = X$ for some α.*

For a scattered topological space (X, τ) and a point $x \in X$, the *rank* $\rho_\tau(x)$ of x is the least ordinal α such that $x \in cd_\tau^{\alpha+1}(\varnothing)$.

Lemma 1 (cf. Lemma 3.11 from [1]). *In a scattered topological space (X, τ), we have*

- $cd_\tau^\alpha(\varnothing)$ *is an open set for any* α,
- $cd_\tau^\alpha(\varnothing) \subset cd_\tau^\beta(\varnothing)$ *if* $\alpha \leqslant \beta$.

Lemma 2. *For any scattered topological space (X, τ) and any $x \in X$, the set $\{y \in X \mid \rho_\tau(y) < \rho_\tau(x)\}$ is a punctured neighbourhood of x.*

Proof. Let us check that $\{x\} \cup \{y \in X \mid \rho_\tau(y) < \rho_\tau(x)\}$ is an open set. We have

$$\{y \in X \mid \rho_\tau(y) < \rho_\tau(x)\} = cd_\tau^{\rho_\tau(x)}(\varnothing), \quad \{x\} \subset cd_\tau^{\rho_\tau(x)+1}(\varnothing),$$
$$cd_\tau^{\rho_\tau(x)}(\varnothing) \subset cd_\tau^{\rho_\tau(x)+1}(\varnothing).$$

Hence

$$\{x\} \cup \{y \in X \mid \rho_\tau(y) < \rho_\tau(x)\} \subset cd_\tau(\{y \in X \mid \rho_\tau(y) < \rho_\tau(x)\}) \subset$$
$$\subset cd_\tau(\{x\} \cup \{y \in X \mid \rho_\tau(y) < \rho_\tau(x)\}).$$

Notice that, in any topological space, a set U is open if and only if $U \subset cd_\tau(U)$. Thus $\{x\} \cup \{y \in X \mid \rho_\tau(y) < \rho_\tau(x)\}$ is an open set.

4 Global Consequence Relations

In this section we define global semantic consequence relations over GL-frames and introduce corresponding derivability relations using non-well-founded derivation trees. We obtain completeness results for these derivability relations in Sect. 7.

Given a set of modal formulas Γ and a formula A, we set $\Gamma \vDash_g A$ if for any GL-model $\mathcal{M}$

$$(\forall B \in \Gamma \; \mathcal{M} \vDash B) \Rightarrow \mathcal{M} \vDash A \,.$$

We also set $\Sigma; \Gamma \vDash A$ if for any GL-model $\mathcal{M}$ and any its world x

$$((\forall y \neq x \; \forall B \in \Sigma \; \mathcal{M}, y \vDash B) \wedge (\forall C \in \Gamma \; \mathcal{M}, x \vDash C)) \Rightarrow \mathcal{M}, x \vDash A.$$

Notice that the relation $\vDash$ is a generalization of $\vDash_l$ and $\vDash_g$ that is $\varnothing; \Gamma \vDash A \Leftrightarrow \Gamma \vDash_l A$ and $\Gamma; \Gamma \vDash A \Leftrightarrow \Gamma \vDash_g A$.

An ∞-*derivation* in GL is a (possibly infinite) tree whose nodes are marked by modal formulas and that is constructed according to the rules (mp) and (nec). In addition, any infinite branch in an ∞-derivation must contain infinitely many applications of the rule (nec). An *assumption leaf* of an ∞-derivation is a leaf that is not marked by an axiom of GL. An assumption leaf is *boxed* if there is an application of the rule (nec) on the path from this leaf to the root of the tree.

The *main fragment* of an ∞-derivation is a finite tree obtained from the ∞-derivation by cutting every infinite branch at the nearest to the root application of the rule (nec). The *local height* $|\pi|$ of an ∞-*derivation* π is the length of the longest branch in its main fragment. An ∞-derivation only consisting of a single formula has height 0.

For example, consider the following ∞-derivation

$$\begin{array}{c} \vdots \\ \mathsf{mp} \; \dfrac{\Box p_3 \qquad \Box p_3 \to p_2}{} \\ \mathsf{nec} \; \dfrac{p_2}{\Box p_2} \qquad \Box p_2 \to p_1 \\ \mathsf{mp} \; \dfrac{}{} \\ \mathsf{nec} \; \dfrac{p_1}{\Box p_1} \qquad \Box p_1 \to p_0 \\ \mathsf{mp} \; \dfrac{}{p_0} \end{array} \,,$$

where assumption leafs are marked by formulas of the form $\Box p_{n+1} \to p_n$. The local height of this ∞-derivation equals to 1 and its main fragment has the form

$$\mathsf{mp} \; \dfrac{\Box p_1 \qquad\qquad \Box p_1 \to p_0}{p_0} \,.$$

We set $\Gamma \vdash_g A$ if there is an ∞-derivation with the root marked by A in which all assumption leafs are marked by some elements of Γ. We also set $\Sigma; \Gamma \vdash A$ if there is an ∞-derivation with the root marked by A in which all boxed assumption leafs are marked by some elements of Σ and all non-boxed assumption leafs are marked by some elements of Γ. Note that $\Gamma; \Gamma \vdash A \Leftrightarrow \Gamma \vdash_g A$.

Now we introduce an auxiliary semantic consequence relation $\vDash^*$, which will be proved to be equivalent to $\vDash$. We set $\Sigma; \Gamma \vDash^* A$ if for any GL-model $\mathcal{M}$, any its world x and any punctured neighbourhood O of x

$$((\forall y \in O \; \forall B \in \Sigma \; \mathcal{M}, y \vDash B) \wedge (\forall C \in \Gamma \; \mathcal{M}, x \vDash C)) \Rightarrow \mathcal{M}, x \vDash A.$$

Lemma 3. $\Sigma; \Gamma \vdash A \Rightarrow \Sigma; \Gamma \vDash^* A$.

Proof. Assume π is an ∞-derivation with the root marked by A in which all boxed assumption leafs are marked by some elements of Σ and all non-boxed assumption leafs are marked by some elements of Γ. In addition, assume $\mathcal{M} = (\mathcal{X}, \theta)$ is a GL-model, x is a world of $\mathcal{M}$ and O is a punctured neighbourhood of x such that

$$\forall y \in O \; \forall C \in \Sigma \; \mathcal{M}, y \vDash C \quad \text{and} \quad \forall D \in \Gamma \; \mathcal{M}, x \vDash D.$$

Let (X, τ) be the scattered topological space corresponding to the GL-frame $\mathcal{X}$. We prove $\mathcal{M}, x \vDash A$ by transfinite induction on $\rho_\tau(x)$ and a subinduction on $|\pi|$.

If $|\pi| = 0$, then A is an axiom of GL or an element of Γ. We obtain $\mathcal{M}, x \vDash A$ immediately. Otherwise, consider the lowermost application of an inference rule in π.

Case 1. Suppose that π has the form

$$
\text{mp} \; \frac{\overset{\pi'}{\overset{\vdots}{B}} \quad \overset{\pi''}{\overset{\vdots}{B \to A}}}{A}.
$$

By the induction hypotheses for π' and π'', we have $\mathcal{M}, x \vDash B$ and $\mathcal{M}, x \vDash B \to A$. Consequently $\mathcal{M}, x \vDash A$.

Case 2. Suppose that π has the form

$$
\text{nec} \; \frac{\overset{\pi'}{\overset{\vdots}{B}}}{\Box B},
$$

where $\Box B = A$. We see that $\Sigma; \Sigma \vdash B$. We also have

$$\forall y \in O \; (\rho_\tau(y) < \rho_\tau(x) \Rightarrow \forall C \in \Sigma \; \mathcal{M}, y \vDash C).$$

Notice that

$$\{y \in O \mid \rho_\tau(y) < \rho_\tau(x)\} = O \cap \{y \in X \mid \rho_\tau(y) < \rho_\tau(x)\} = O \cap cd_\tau^{\rho_\tau(x)}(\varnothing).$$

Hence, by Lemma 1, the set $\{y \in O \mid \rho_\tau(y) < \rho_\tau(x)\}$ is open as the intersection of two open sets. By the induction hypothesis for any point of $\{y \in O \mid \rho_\tau(y) < \rho_\tau(x)\}$ we obtain

$$\forall y \in O \; (\rho_\tau(y) < \rho_\tau(x) \Rightarrow \mathcal{M}, y \vDash B).$$

Thus we have

$$\{y \in O \mid \rho_\tau(y) < \rho_\tau(x)\} \subset \theta(B).$$

Now recall that for any world z of $\mathcal{M}$

$$\mathcal{M}, z \vDash \Box B \Leftrightarrow z \in \theta(\Box B) \Leftrightarrow z \in cd_\tau(\theta(B)) \, .$$

By definition, the co-derivative set $cd_\tau(\theta(B))$ contains a world z if and only if there is a punctured neighbourhood of z entirely contained in $\theta(B)$. We have that $\{y \in O \mid \rho_\tau(y) < \rho_\tau(x)\} \subset \theta(B)$ and $\{y \in O \mid \rho_\tau(y) < \rho_\tau(x)\}$ is a punctured neighbourhood of x by Lemma 2. Thus $x \in cd_\tau(\theta(B))$ and $\mathcal{M}, x \vDash \Box B$.

Lemma 4. $\Sigma; \Gamma \vDash^* A \Rightarrow \Sigma; \Gamma \vDash A$.

Proof. Assume $\Sigma; \Gamma \vDash^* A$. In addition, assume $\mathcal{M} = (\mathcal{X}, \theta)$ is a GL-model and x is a world of $\mathcal{M}$ such that

$$\forall y \neq x \, \forall C \in \Sigma \, \mathcal{M}, y \vDash C \quad \text{and} \quad \forall D \in \Gamma \, \mathcal{M}, x \vDash D.$$

Let (X, τ) be the scattered topological space corresponding to the GL-frame $\mathcal{X}$. We see that $X \setminus \{x\}$ is a punctured neighbourhood of x. Hence $\mathcal{M}, x \vDash A$.

5   Neighbourhood Compactness

In this section we prove that if $\Sigma; \Gamma \vDash A$, then there is a finite subset Γ_0 of Γ such that $\Sigma; \Gamma_0 \vDash A$. This neighbourhood compactness result is obtained applaying the ultrabouquet construction from [11].

Notice that an open set in a scattered topological space is scattered (as a topological space with the inherited topology). Hence an open set in a GL-frame $\mathcal{X}$ defines the GL-frame, which is called an *open subframe of $\mathcal{X}$*.

Lemma 5. *If $(X_0, \Box_0)$ is an open subframe of a GL-frame $(X_1, \Box_1)$, then $\Box_0 V = X_0 \cap \Box_1 V$ for any $V \subset X_0$.*

Lemma 6 (see Lemma 6 from [11]). *For GL-models $(\mathcal{X}_0, \theta_0)$ and $(\mathcal{X}_1, \theta_1)$, where $\mathcal{X}_0$ is an open subframe of $\mathcal{X}_1$ and $\theta_0(p) = X_0 \cap \theta_1(p)$ for $p \in PV$, we have that $\theta_0(A) = X_0 \cap \theta_1(A)$ for any formula A.*

For any $n \in \mathbb{N}$, let $\mathcal{X}_n = (X_n, \tau_n)$ be a topological space and x_n be a closed point in it. Let $\mathcal{U}$ be a non-principal ultrafilter in $\mathbb{N}$. The *ultrabouquet* $\bigvee_\mathcal{U}(\mathcal{X}_n, x_n)$ is a topological space obtained as a set from the disjoint union $\bigsqcup_{n \in \mathbb{N}} X_n$ by iden- tifying all points x_n. A set U is *open in* $\bigvee_\mathcal{U}(\mathcal{X}_n, x_n)$ if and only if

- for any $n \in \mathbb{N}$ the set $U \cap (X_n \setminus \{x_n\})$ is open in $\mathcal{X}_n$,
- $\{n \in \mathbb{N} \mid U \cap X_n \text{ is open in } \mathcal{X}_n\} \in \mathcal{U}$ whenever $x_* \in U$,

where x_* is the point of $\bigvee_\mathcal{U}(\mathcal{X}_n, x_n)$ obtained by identifying points x_n.

Clearly, an ultrabouquet of scattered topological spaces is a scattered topo- logical space. Hence we can construct a GL-frame as an ultrabouquet of a count- able family of GL-frames.

For $n \in \mathbb{N}$, let θ_n be a valuation over a GL-frame $\mathcal{X}_n = (X_n, \Box_n)$. Let θ be a valuation over $\bigvee_\mathcal{U}(\mathcal{X}_n, x_n)$ defined on the set of propositional variables as follows:

- the restriction of $\theta(p)$ to $X_n \setminus \{x_n\}$ is equal to $\theta_n(p)$;
- $x_* \in \theta(p)$ if and only if $\{n \in \mathbb{N} \mid x_n \in \theta_n(p)\} \in \mathcal{U}$.

We denote this valuation θ by $\bigvee_{\mathcal{U}}(\theta_n, x_n)$.

Lemma 7 (see Lemmas 22 and 27 from [11]). *For any $n \in \mathbb{N}$, let $(\mathcal{X}_n, \theta_n)$ be a GL-model and x_n be a closet point in it. Let $\mathcal{U}$ be a non-principal ultrafilter in $\mathbb{N}$ and $\theta = \bigvee_{\mathcal{U}}(\theta_n, x_n)$. Then for any formula A we have*

- $\theta(A) \cap (X_n \setminus \{x_n\}) = \theta_n(A)$ *for any $n \in \mathbb{N}$;*
- $x_* \in \theta(A)$ *if and only if $\{n \in \mathbb{N} \mid x_n \in \theta_n(A)\} \in \mathcal{U}$.*

A topological space is T_d *(local T_1)* if any point in it is closed in some of its neighbourhoods.

Lemma 8 (see Corollary 1 from [2] or Lemma 61 from [11]). *Any scattered space is T_d.*

Theorem 1. *If $\Sigma; \Gamma \vDash A$, then there is a finite subset Γ_0 of Γ such that $\Sigma; \Gamma_0 \vDash A$.*

Proof. Assume $\Sigma; \Gamma \vDash A$. We prove that there exists the required finite subset Γ_0 of Γ by *reductio ad absurdum*.

Suppose that for any finite subset Γ_0 of Γ we have $\Sigma; \Gamma_0 \nvDash A$. Let $\Gamma = \{B_n \mid n \in \mathbb{N}\}$ and $C_n = \bigwedge_{i=0}^{i=n} B_i$. Then, for any $n \in \mathbb{N}$, there exist a GL-frame $\mathcal{X}_n = (X_n, \square_n)$, a valuation θ_n over it and a world x_n such that

$$\forall y \neq x_n \ \forall D \in \Sigma \ (\mathcal{X}_n, \theta_n), y \vDash D, \quad (\mathcal{X}_n, \theta_n), x_n \vDash C_n \quad \text{and} \quad (\mathcal{X}_n, \theta_n), x_n \nvDash A.$$

Let $\mathcal{Y}_n$ be an open subframe of $\mathcal{X}_n$, in which x_n is closed. We define a valuation ψ_n over $\mathcal{Y}_n$ obtained by restricting θ_n to $\mathcal{Y}_n$. By Lemma 6, we have

$$\forall y \neq x_n \ \forall D \in \Sigma \ (\mathcal{Y}_n, \psi_n), y \vDash D, \quad (\mathcal{Y}_n, \psi_n), x_n \vDash C_n \quad \text{and} \quad (\mathcal{Y}_n, \psi_n), x_n \nvDash A.$$

We take an non-principal ultrafilter $\mathcal{U}$ in $\mathbb{N}$ and consider the ultrabouquet $\mathcal{Y} = \bigvee_{\mathcal{U}}(\mathcal{Y}_n, x_n)$ together with the valuation $\psi = \bigvee_{\mathcal{U}}(\psi_n, x_n)$ over $\mathcal{Y}$. From Lemma 7, we have

$$\forall y \neq x_n \ \forall D \in \Sigma \ (\mathcal{Y}, \psi), y \vDash D, \quad \forall B \in \Gamma \ (\mathcal{Y}, \psi), x_n \vDash B \quad \text{and} \quad (\mathcal{Y}, \psi), x_n \nvDash A.$$

We obtain a contradiction with the assumption $\Sigma; \Gamma \vDash A$. Therefore there exists a finite subset Γ_0 of Γ such that $\Sigma; \Gamma_0 \vDash A$.

6   Sequent Calculus

In this section we define a calculus corresponding to the global consequence relation $\vDash$.

A *sequent* is an expression of the form $\Sigma; \Gamma \Rightarrow \Delta$, where Γ and Δ are finite multisets of formulas, and Σ is an arbitrary set of formulas. For a multiset of formulas $\Gamma = A_1, \ldots, A_n$, we set $\Box\Gamma := \Box A_1, \ldots, \Box A_n$.

Initial sequents and inference rules of the sequent calculus S have the following form:

$$\Sigma; \Gamma, p \Rightarrow p, \Delta, \qquad \Sigma; \Gamma, \bot \Rightarrow \Delta,$$

$$\to_{\mathsf{L}} \frac{\Sigma; \Gamma, B \Rightarrow \Delta \qquad \Sigma; \Gamma \Rightarrow A, \Delta}{\Sigma; \Gamma, A \to B \Rightarrow \Delta}, \qquad \to_{\mathsf{R}} \frac{\Sigma; \Gamma, A \Rightarrow B, \Delta}{\Sigma; \Gamma \Rightarrow A \to B, \Delta},$$

$$\Box \frac{\Sigma; \Sigma_0, \Gamma, \Box\Gamma \Rightarrow A}{\Sigma; \Pi, \Box\Gamma \Rightarrow \Box A, \Delta} \quad (\Sigma_0 \text{ is a finite subset of } \Sigma).$$

An ∞-*proof in* S is a (possibly infinite) tree whose nodes are marked by sequents and whose leaves are marked by initial sequents and that is constructed according to the rules of the sequent calculus. A sequent $\Sigma; \Gamma \Rightarrow \Delta$ is *provable in* S if there is an ∞-proof π with the root marked by $\Sigma; \Gamma \Rightarrow \Delta$. In this case π is called an ∞-*proof of* $\Sigma; \Gamma \Rightarrow \Delta$.

A sequent $\Sigma; \Gamma \Rightarrow \Delta$ is called *valid* if $\Sigma; \{\bigwedge \Gamma\} \vDash \bigvee \Delta$.

Lemma 9. *If a sequent* $\Sigma; \Gamma, A \to B \Rightarrow \Delta$ *is valid, then sequents* $\Sigma; \Gamma, B \Rightarrow \Delta$ *and* $\Sigma; \Gamma \Rightarrow A, \Delta$ *are valid. If* $\Sigma; \Gamma \Rightarrow A \to B, \Delta$ *is valid, then* $\Sigma; \Gamma, A \Rightarrow B, \Delta$ *is also valid.*

A sequent $\Sigma; \Gamma \Rightarrow \Delta$ is called *saturated* if Γ and Δ do not contain formulas of the form $A \to B$.

Lemma 10. *If* $\Sigma; \Pi, \Box\Gamma \Rightarrow \Delta$ *is a valid non-initial saturated sequent, where* Π *consists only of propositional variables, then there are a finite subset* Σ_0 *of* Σ *and a formula* $\Box A$ *from* Δ *such that* $\Sigma; \Sigma_0, \Gamma, \Box\Gamma \Rightarrow A$ *is a valid sequent.*

Proof. Assume $\Sigma; \Pi, \Box\Gamma \Rightarrow \Delta$ is a valid non-initial saturated sequent, where Π consists only of propositional variables. We claim that there is a formula $\Box A$ from Δ such that $\Sigma; \Sigma \cup \{\bigwedge \Gamma \wedge \bigwedge \Box\Gamma\} \vDash A$. We prove this claim by *reductio ad absurdum*.

Let $\Box A_1, \ldots, \Box A_n$ be all elements from Δ of the form $\Box A$. Suppose that for any $i \in \{1, \ldots, n\}$ there exist a GL-frame $\mathcal{X}_i = (X_i, \Box_i)$, a valuation θ_i over it and a world x_i such that

$$\forall y \in X_i \,\forall B \in \Sigma \,(\mathcal{X}_i, \theta_i), y \vDash B, \quad (\mathcal{X}_i, \theta_i), x_i \vDash \bigwedge \Gamma \wedge \bigwedge \Box\Gamma \quad \text{and} \quad (\mathcal{X}_i, \theta_i), x_i \nvDash A_i.$$

We consider a topological space $\mathcal{X}$ obtained from the disjoint union of $\mathcal{X}_i$ by adding a new world x. A subset U of $\mathcal{X}$ is open if and only if the following conditions hold:

– the restriction of U to $\mathcal{X}_i$ is open for any $i \in \{1, \ldots, n\}$;
– if $x \in U$, then $x_i \in U$ for any $i \in \{1, \ldots, n\}$.

Clearly, the topological space $\mathcal{X}$ is scattered. Hence we can consider it as a GL-frame. Let θ be a valuation over $\mathcal{X}$ defined on the set of propositional variables as follows:

– the restriction of $\theta(p)$ to $\mathcal{X}_i$ is equal to $\theta_i(p)$ for any $i \in \{1, \ldots, n\}$;
– $x \in \theta(p)$ if and only if $p \in \Pi$.

We shall show that

$$\forall y \neq x \; \forall B \in \Sigma \; (\mathcal{X}, \theta), y \vDash B, \quad (\mathcal{X}, \theta), x \vDash \bigwedge \Pi \wedge \bigwedge \Box \Gamma \quad \text{and} \quad (\mathcal{X}, \theta), x \nvDash \bigvee \Delta.$$

Every GL-frame $\mathcal{X}_i$ is an open subframe of $\mathcal{X}$. Thus, by Lemma 6, the condition $\forall y \neq x \; \forall B \in \Sigma \; (\mathcal{X}, \theta), y \vDash B$ follows from $\forall i \in \{1, \ldots, n\} \; \forall y \in X_i \; \forall B \in \Sigma \; (\mathcal{X}_i, \theta_i), y \vDash B$. Further, we have $(\mathcal{X}, \theta), x \vDash p$ for $p \in \Pi$ and $(\mathcal{X}, \theta), x \nvDash p$ for $p \in \Delta$ by definition of θ.

Let us check that $(\mathcal{X}, \theta), x \vDash \bigwedge \Box \Gamma$. For any formula C from Γ and any $i \in \{1, \ldots, n\}$ we have $(\mathcal{X}, \theta), x_i \vDash C \wedge \Box C$ by Lemma 6. This yields that for any $i \in \{1, \ldots, n\}$ there is a neighbourhood U_i of x_i such that $U_i \subset \theta(C)$. We have that $\bigcup_{1 \leqslant i \leqslant n} U_i \subset \theta(C)$, where $\bigcup_{1 \leqslant i \leqslant n} U_i$ is a punctured neighbourhood of x. Hence $(\mathcal{X}, \theta), x \vDash \Box C$ for any $C \in \Gamma$.

It remains to check that $(\mathcal{X}, \theta), x \nvDash \Box A_i$ for $i \in \{1, \ldots, n\}$. For any punctured neighbourhood U of x, there is a world $x_i \in U$ such that $(\mathcal{X}, \theta), x_i \nvDash A_i$. Hence $(\mathcal{X}, \theta), x \nvDash \Box A_i$.

We obtain that the sequent $\Sigma; \Pi, \Box \Gamma \Rightarrow \Delta$ is not valid, which is a contradiction. Therefore there is a formula $\Box A$ from Δ such that $\Sigma; \Sigma \cup \{\bigwedge \Gamma \wedge \bigwedge \Box \Gamma\} \vDash A$. In addition, by Theorem 1, there is a finite subset Σ_0 of Σ such that $\Sigma; \Sigma_0 \cup \{\bigwedge \Gamma \wedge \bigwedge \Box \Gamma\} \vDash A$. Hence we find the required valid sequent $\Sigma; \Sigma_0, \Gamma, \Box \Gamma \Rightarrow A$. $\qquad \square$

Theorem 2. *Any valid sequent is provable in* S.

Proof. Let us consider a valid sequent $\Sigma; \Gamma \Rightarrow \Delta$. If this sequent is not saturated, then it can be obtained by an application of the rule $(\rightarrow_R)$ or $(\rightarrow_L)$ from other valid sequents using Lemma 9. If this sequent is saturated, then it is initial or can be obtained by an application of the rule $(\Box)$ from another valid sequent using Lemma 10. Therefore any valid sequent is initial sequent of the sequent calculus S or can be obtained by an application of an inference rule from other valid sequents. Thus, for any valid sequent, its ∞-proof in S is immediately defined travelling upwards from conclusions to premises by co-recursion. $\qquad \square$

Corollary 1. *If* $\Sigma; \Gamma \vDash A$, *then the sequent* $\Sigma; \Gamma \Rightarrow A$ *is provable in* S.

Proof. If $\Sigma; \Gamma \vDash A$, then the sequent $\Sigma; \Gamma \Rightarrow A$ is valid by definition. Hence this sequent is provable in S by Theorem 2. $\qquad \square$

7 Global Completeness

Theorem 3. *If a sequent $\Sigma; \Gamma \Rightarrow \Delta$ is provable in S, then $\Sigma; \varnothing \vdash \bigwedge \Gamma \to \bigvee \Delta$.*

Proof. Assume π is an ∞-proof of $\Sigma; \Gamma \Rightarrow \Delta$ in S. We define the required ∞-derivation $f(\pi)$ in GL travelling upwards from conclusions to premises by co-recursion.

If $\Sigma; \Gamma \Rightarrow \Delta$ is an initial sequent of the sequent calculus S, then the formula $\bigwedge \Gamma \to \bigvee \Delta$ is provable in GL by a finite proof. Let $f(\pi)$ be such a proof.

Otherwise, consider the final application of an inference rule in π.

Case 1. If π has the form

$$
\to_L \frac{\begin{array}{c}\pi' \\ \vdots \\ \Sigma; \Gamma, B \Rightarrow \Delta\end{array} \qquad \begin{array}{c}\pi'' \\ \vdots \\ \Sigma; \Gamma, \Rightarrow A, \Delta\end{array}}{\Sigma; \Gamma, A \to B \Rightarrow \Delta} ,
$$

then we define $f(\pi)$ as

$$
\mathrm{mp}\ \frac{\begin{array}{c}f(\pi') \\ \vdots \\ G\end{array} \qquad \mathrm{mp}\ \dfrac{\begin{array}{c}f(\pi'') \\ \vdots \\ F\end{array} \qquad \begin{array}{c}\xi \\ \vdots \\ F \to (G \to H)\end{array}}{G \to H}}{H} ,
$$

where $F = \bigwedge \Gamma \to \bigvee(\{A\} \cup \Delta)$, $G = \bigwedge(\Gamma \cup \{B\}) \to \bigvee \Delta$, $H = \bigwedge(\Gamma \cup \{A \to B\}) \to \bigvee \Delta$ and ξ is a finite proof of the formula $F \to (G \to H)$ in GL.

Case 2. If π has the form

$$
\to_R \frac{\begin{array}{c}\pi' \\ \vdots \\ \Sigma; \Gamma, A \Rightarrow B, \Delta\end{array}}{\Sigma; \Gamma \Rightarrow A \to B, \Delta} ,
$$

then we define $f(\pi)$ as

$$
\mathrm{mp}\ \frac{\begin{array}{c}f(\pi') \\ \vdots \\ F\end{array} \qquad \begin{array}{c}\xi \\ \vdots \\ F \to G\end{array}}{G} ,
$$

where $F = \bigwedge(\Gamma \cup \{A\}) \to \bigvee(\{B\} \cup \Delta)$, $G = \bigwedge \Gamma \to \bigvee(\{A \to B\} \cup \Delta)$ and ξ is a finite proof of the formula $F \to G$ in GL.

Case 3. Now consider the final case when π has the form

$$
\Box\ \frac{\begin{array}{c}\pi' \\ \vdots \\ \Sigma; \Sigma_0, \Gamma, \Box\Gamma \Rightarrow A\end{array}}{\Sigma; \Pi, \Box\Gamma \Rightarrow \Box A, \Delta}\ (\Sigma_0 \text{ is a finite subset of } \Sigma).
$$

We define $f(\pi)$ as

$$
\begin{array}{ccc}
 & f(\pi') & \zeta \\
\eta & \vdots & \vdots \\
\vdots & \mathrm{mp}\ \dfrac{F \qquad F \to (\bigwedge \Sigma_0 \to G)}{\bigwedge \Sigma_0 \to G} & \xi \\
\mathrm{mp}\ \dfrac{\bigwedge \Sigma_0}{} & & \vdots \\
 & \mathrm{nec}\ \dfrac{G}{\Box G} & \\
 & \mathrm{mp}\ \dfrac{\Box G \qquad\qquad\qquad \Box G \to H}{H} & ,
\end{array}
$$

where $F = \bigwedge(\Sigma_0 \cup \Gamma \cup \Box\Gamma) \to A$, $G = \bigwedge(\Gamma \cup \Box\Gamma) \to A$, $H = \bigwedge(\Pi \cup \Box\Gamma) \to \bigvee(\{\Box A\} \cup \Delta)$ and η is a finite derivation of the formula $\bigwedge \Sigma_0$ in GL from the set of assumptions Σ_0. In addition, ζ and ξ are finite proofs in GL of the corresponding formulas $F \to (\bigwedge \Sigma_0 \to G)$ and $\Box G \to H$.

It is not hard to prove that every infinite branch in $f(\pi)$ contains infinitely many applications of the rule (nec) and, in addition, any assumption leaf of $f(\pi)$ is boxed and is marked by an element of Σ_0. Hence $f(\pi)$ is the required ∞-derivation.

Corollary 2. $\Sigma; \Gamma \vDash A \Leftrightarrow \Sigma; \Gamma \vDash^* A \Leftrightarrow \Sigma; \Gamma \vdash A$.

Proof. The right-to-left implications follow from Lemmas 3 and 4. Assume $\Sigma; \Gamma \vDash A$. By Theorem 1, there is a finite subset Γ_0 of Γ such that $\Sigma; \Gamma_0 \vDash A$. By Corollary 1, the sequent $\Sigma; \Gamma_0 \Rightarrow A$ is provable in S. From Theorem 3, we have $\Sigma; \varnothing \vdash \bigwedge \Gamma_0 \to A$. Now it easily follows that $\Sigma; \Gamma \vdash A$.

Corollary 3. $\Gamma \vDash_g A \Leftrightarrow \Gamma \vdash_g A$.

Acknowledgements. The development of main ideas of this paper took place during my stay in Tash-Bulak village, the Kyrgyz Republic, in 2015. I heartily thank my uncle Kengebek Shamkanov and his family for their hospitality. In addition, I am grateful to Tadeusz Litak, whose constructive comments have helped me to improve the manuscript.

References

1. Aguilera, J.P., Fernández-Duque, D.: Strong completeness of provability logic for ordinal spaces (2015). arXiv:1511.05882v1
2. Beklemishev, L., Gabelaia, D.: Topological interpretations of provability logic. In: Bezhanishvili, G. (ed.) Leo Esakia on Duality in Modal and Intuitionistic Logics. OCL, vol. 4, pp. 257–290. Springer, Dordrecht (2014). doi:10.1007/978-94-017-8860-1_10
3. Esakia, L.: Diagonal constructions, Löb's formula and Cantor's scattered space. Stud. Logic Semant. **132**(3), 128–143 (1981). (in Russian)
4. Hakli, R., Negri, S.: Does the deduction theorem fail for modal logic? Synthese **187**(3), 849–867 (2011)

5. Iemhoff, R.: Reasoning in circles. In: van Eijck, J., et al. (eds.) Liber Amicorum Alberti. A Tribute to Albert Visser, pp. 165–178. College Publications, London (2016)
6. Litak, T.: An algebraic approach to incompleteness in modal logic. Ph.D. thesis. Japan Advanced Institute of Science and Technology (2005)
7. Montague, R.: Universal grammar. Theoria **36**(3), 373–398 (1970)
8. Scott, D.: Advice in modal logic. In: Lambert, K. (ed.) Philosophical Problems in Logic. Reidel, Kufstein (1970)
9. Segerberg, K.: An Essay in Classical Modal Logic. Filosofiska Studier, vol. 13. Uppsala University (1971)
10. Shamkanov, D.: Circular proofs for the Gödel-Löb provability logic. Math. Not. **96**(3), 575–585 (2014)
11. Shehtman, V.: On neighbourhood semantics thirty years later. In: Artemov, S., et al. (eds.) We Will Show Them! Essays in Honour of Dov Gabbay, vol. 2, pp. 663–692. College Publications, London (2005)
12. Simmons, H.: Topological aspects of suitable theories. Proc. Edinb. Math. Soc. **19**(4), 383–391 (1975)
13. Solovay, R.: Provability interpretations of modal logic. Isr. J. Math. **25**, 287–304 (1976)

Coherent Diagrammatic Reasoning in Compositional Distributional Semantics

Gijs Jasper Wijnholds[(✉)]

Queen Mary University of London, London, UK
g.j.wijnholds@qmul.ac.uk

Abstract. The framework of Categorical Compositional Distributional models of meaning [3], inspired by category theory, allows one to compute the meaning of natural language phrases, given basic meaning entities assigned to words. Composing word meanings is the result of a functorial passage from syntax to semantics. To keep one from drowning in technical details, diagrammatic reasoning is used to represent the information flow of sentences that exists independently of the concrete instantiation of the model. Not only does this serve the purpose of clarification, it moreover offers computational benefits as complex diagrams can be transformed into simpler ones, which under coherence can simplify computation on the semantic side. Until now, diagrams for compact closed categories and monoidal closed categories have been used (see [2,3]). These correspond to the use of pregroup grammar [12] and the Lambek calculus [9] for syntactic structure, respectively. Unfortunately, the diagrammatic language of Baez and Stay [1] has not been proven coherent. In this paper, we develop a graphical language for the (categorical formulation of) the nonassociative Lambek calculus [10]. This has the benefit of modularity where extension of the system are easily incorporated in the graphical language. Moreover, we show the language is coherent with monoidal closed categories without associativity, in the style of Selinger's survey paper [17].

Keywords: Diagrammatic reasoning · Coherence theorem · Proof nets · Compositional distributional semantics

1 Background, Motivation

Having a form of visual representation of information flow is pervasive in the natural sciences: in physics, graphical languages have been developed coming from the ideas of Penrose [16], to formalise reasoning about matrix multiplication, for instance. Computer science and electronic engineering makes extensive use of diagrammatic representation of systems, circuits etc. In logic, there has been a great interest in graphical notation since the development of natural deduction and sequent calculi, but mostly after the introduction of proof nets [5]. Most of these languages can be greatly generalised; for instance, within physics, graphical

© Springer-Verlag GmbH Germany 2017
J. Kennedy and R.J.G.B. de Queiroz (Eds.): WoLLIC 2017, LNCS 10388, pp. 371–386, 2017.
DOI: 10.1007/978-3-662-55386-2_27

notation usually is describing the structure of a monoidal category; a similar situation occurs with electronic circuits, which have a notion of serial and parallel execution, categorically speaking a notion of composition and tensor, the basic principles of a monoidal category. In logic, different deductive systems capture different types of category, e.g. intuitionistic logic is captured by cartesian closed categories [13], a special instance of monoidal categories. Hence, proof nets for intuitionistic logic should bear a relation to a graphical language for cartesian closed categories (though the latter, as far as the author knows, does not exist).

While graphical languages are visually appealing, they will not convince the practicing scientist to be useful unless they are also precise. In other words, we want a graphical language to be coherent (i.e. sound and complete) with respect to the category it is describing. In recent work on categorical compositional distributional semantics [2], the clasp language of [1] was assumed, in order to describe the morphisms of a monoidal biclosed category. The reason is that the authors relied on a functorial passage from the Lambek calculus (the logic of monoidal biclosed categories) to finite dimensional vector spaces (an instance of a compact closed category). Here, the diagrammatic notation does not only describe the combination of those two systems, it has an additional functional role in simplifying calculations: by rewriting a diagram as much as possible, one will obtain the same results with smaller computational effort. Sadly, no attempt at proving the coherence of the clasp language is known to the author; moreover, one of the inventors of the language has stated to have no interest in doing so[1]. Hence, there is a need to obtain either a coherence result for this language, or to introduce another graphical language and show its coherence. In this paper, we will take the latter option and motivate it by pointing out some potential problems with the clasp language.

Originally out of interest in proof nets and their relation to categorical diagrammatic reasoning, we will introduce a graphical notation for morphisms in a biclosed monoidal category without associativity/units (!) which will give a fairly easy way of showing coherence. This comes from considerations on structural rules in the Lambek calculus, and the need for modularity in said systems. We will then argue how the addition of associativity and units can be easily incorporated graphically, so that we obtain a modular way of describing categories, moreover giving a coherent language for monoidal biclosed categories, which would be first successful attempt.

The rest of this paper is organised as follows: in Sect. 2 we describe some related work on graphical languages for monoidal categories and the clasp language for monoidal closed categories. Then Sect. 3 discusses definitions and notation for a graphical language for non-associative non-unital monoidal closed categories. Section 4 contains our main result, and we give some extensions in Sect. 5, after which we conclude in Sect. 6 with some avenues for future work.

---

[1] John Baez, personal communication, 2014.

2   Related Work: Visualising Monoidal Categories

There has been a fair body of research devoted to displaying several variants of monoidal categories using a graphical language. For instance, diagrammatic reasoning for compact closed categories was introduced and shown to be coherent by Kelly and Laplaza [8], and the research of Joyal and Street has led to graphical languages for, amongst others, planar monoidal and braided monoidal categories (see [6,7]). For some cases of autonomous categories (monoidal categories with dual objects) there are some results as well [4]. In this paper, however, we follow the presentation style of the survey of Selinger [17], as it is uniform, and in our opinion a gentle introduction to graphical languages. We will introduce the basic constructs of a graphical language for monoidal categories and then review the clasp language for the general case of monoidal closure, proposed by Baez and Stay [1]. As we will consider monoidal categories without associativity too, which for lack of a better name we will refer to as magmatic categories[2], we split the definition of a monoidal category:

Definition 1. *We say that a category* **C** *with objects A, B and morphisms $f : A \to B$ between objects has a **magma structure** if it has*

1. *An object $A \otimes B$ for all objects A, B and a morphims $f \otimes g : A \otimes B \to C \otimes D$ for any pair of morphisms $f : A \to C, g : B \to D$*

*And has a **monoidal structure** if for this magma structure it has*

1. *A natural isomorphism $\alpha : (A \otimes B) \otimes C \cong A \otimes (B \otimes C)$,*
2. *A unit object I with natural isomorphisms $\lambda : I \otimes A \cong A$ and $\rho : A \otimes I \cong A$.*

In addition, a monoidal category also satisfies the so-called pentagon and triangle identities, expressing the behavior of associativity and the interaction of λ, ρ with associativity.

Representing any category visually is fairly straightforward, as is shown in the following diagram:

Object	Morphism	Identity	Composition
A	$f : A \to B$	$id_A : A \to A$	$g \circ f$

---

[2] This terminology comes from the algebraic concept of a magma, a monoid with no associativity or unit properties. We refer the reader to a blog post that advocates the name *magmatic*: https://bartoszmilewski.com/2014/09/29/how-to-get-enriched-over-magmas-and-monoids/.

To obtain a graphical language for a monoidal category, one represents the tensor of two objects by juxtaposing their arrows, and similar for the tensor product of two morphisms. The unit is represented as an empty arrow, and in general a morphism with a tensor of objects in its domain and codomain is represented by allowing several wires to be input/output to the box of that morphism:

Tensor	Unit	Morphism	Tensor
$A \otimes B$	I	$f : A_1 \otimes \ldots \otimes A_n \to B_1 \otimes \ldots \otimes B_m$	$f \otimes g$

Because we do not draw the unit explicitly, and there is no bracketing around the wires, the associativity and unit equations get automatically satisfied. Bifunctoriality is satisfied by the fact that horizontal and vertical composition in any order will result in the same diagram. A full coherence theorem for this language can be found in [6].

We recall that a closed monoidal category is obtained by considering bifunctors that form an adjunction with the tensor when one of the arguments of the bifunctor is fixed. In this way we may obtain two right adjoints for the tensor. Note that so far we have not considered symmetric monoidal categories (ones in which $A \otimes B \cong B \otimes A$), so that there is a difference between a left closed and a right closed monoidal category.

Definition 2. *If a monoidal category* $\mathbf{C}$ *has a contravariant-covariant bifunctor* $\backslash$*, that is, for* $f : A \to C, g : B \to D$ *we get* $f\backslash g : B\backslash C \to A\backslash D$*, and when there is additionally a natural isomorphism* $\beta : Hom_{\mathbf{C}}(A \otimes B, C) \cong Hom_{\mathbf{C}}(B, A\backslash C)$ *for a fixed* A*, we say that* $\mathbf{C}$ *is* **left closed**. *In a similar fashion we can define a* **right closed** *monoidal category by requiring a covariant-contravariant bifunctor* $/$ *that gives* $f/g : A/D \to C/B$ *for* $f : A \to C, g : B \to D$ *and a natural isomorphism* $\gamma : Hom_{\mathbf{C}}(A \otimes B, C) \cong Hom_{\mathbf{C}}(A, C/B)$ *for a fixed* B*. Objects of the form* $A\backslash B$ *and* B/A *are often referred to as the* **internal hom** *of the category.*

To represent closure on monoidal categories, a language was introduced by Baez and Stay in [1]. Their proposal amounts to representing internal homs with upward pointing arrows that are attached with a "clasp" as in the table below (we only include left closure here, as right closure is symmetric):

Closure	Closure	Currying	Uncurrying
$A\backslash B$	$f\backslash g$	$\beta(f)$	$\beta^{-1}(g)$

The reason to draw clasps and bend arrows around is that biclosed monoidal categories in general do not allow "dual behavior". Hence, arrows pointing upwards need to be attached to a downward pointing arrow, and arrow bending must be containing within a box. Unfortunately, the clasp language has not been proved to be coherent with respect to biclosed monoidal categories. We will point out why we doubt whether the clasp language can be shown to be coherent.

Problem: Yanking is required. In order to satisfy the categorical equations $id_A\backslash id_B = id_{A\backslash B}$ one has to allow yanking inside a box. However, one draws boxes every time an arrow is bent around, which implies that yanking inside a box means the same as having yanking everywhere, in turn rendering the clasp language a graphical language for compact closed categories instead of monoidal closed categories! The situation is similar when one wants to show isomorphicity of β, i.e. $\beta(\beta^{-1}(g)) = g$ and $\beta^{-1}(\beta(f)) = f$.

In the next sections, we define an alternative graphical language that does not suffer from the issue described above. It is a language defined for magmatic closed categories, following the type logical grammar philosophy. Starting out with such a restricted system has the benefit of modularity; different categorical concepts can be added by simply extending the diagrams. We will shed a light on this in Sect. 5, after establishing the base language and showing its coherence.

3   A Graphical Language for Closed Magmatic Categories

As noted near the end of Selingers survey, once one goes beyond a single tensor product in a category, simply juxtaposing arrows presents an ambiguity in a graphical language. So it is the case when one takes out associativity from monoidal categories. Within research on *proof nets* (see [5,15]) every bifunctor will get its own representation by means of *links*. Usually proof nets are defined restrictively: once the links are defined, correctness criteria decide whether a graph built by said links is a proof net or not. We will follow the philosophy of proof nets, but instead of defining proof nets by correctness criteria, we give an inductive definition since it is equivalent to the restrictive definition ([18], p. 42) and easier to work with.

For every bifunctor we define two labelled links that make the merging and unmerging of objects explicit:

Definition 3 (Links). *For every bifunctor we define constructor and destructor links that respect the variance of the bifunctor:*

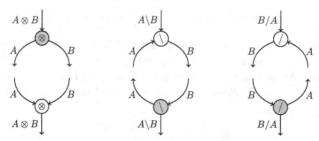

Because the list of correction criteria on these proof nets is quite extensive, we will skip to the inductive definition of proof nets for reasons of space.

Definition 4 (Proof Nets). *Any diagram from Fig. 1 (see appendix) is a proof net, given that the boxes for N_1, N_2 are proof nets. N^* refers to a proof net that has been drawn upside down.*

Besides having a definition for proof nets themselves, we also define equations on proof nets to establish the categorical equations we want to capture. Interestingly enough, these equations correspond to *cut elimination* in sequent calculi in logic!

Definition 5 (Proof Net Equations). *All the equations from Fig. 2 hold on proof nets. Note that N^* is a proof net drawn upside down, which explains why we have equations for "sliding" a net upside down.*

4   Main Result

Following along the lines of [17], we prove coherence by means of a freeness theorem. It works as follows: first, we define for any biclosed magmatic Σ the associated proof net language $\mathbf{PN}(\Sigma)$ and show that it is in turn biclosed. Then, we show coherence by proving that $\mathbf{PN}(\Sigma)$ is the free biclosed magmatic category over Σ.

Definition 6. *A **biclosed magmatic signature** $\Sigma = (\Sigma_0, \Sigma_1, dom, cod)$ has:*

1. *a set Σ_0 of object variables,*
2. *a set Σ_1 of morphism variables,*
3. *two maps $dom, cod : \Sigma_1 \to CT(\Sigma_0)$.*

where $CT(\Sigma_0)$ is the free $(\otimes, \backslash, /)$-algebra generated by Σ_0.

Definition 7. *Given a biclosed magmatic signature Σ and a biclosed magmatic category $\mathbf{C}$, an **interpretation** $i : \Sigma \to \mathbf{C}$ consists of:*

1. *an object map $i_0 : \Sigma_0 \to Ob(\mathbf{C})$ such that*

$$i_0(A \otimes B) = i_0(A) \otimes i_0(B)$$
$$i_0(A\backslash B) = i_0(A)\backslash i_0(B)$$
$$i_0(B/A) = i_0(B)/i_0(A),$$

2. *for every $f \in \Sigma_1$ a morphism $i_1(f) : i_0(dom(f)) \to i_0(cod(f))$.*

Definition 8. *A biclosed magmatic category $\mathbf{C}$ is a **free biclosed magmatic category** over a biclosed magmatic signature Σ if there is an interpretation $i : \Sigma \to \mathbf{C}$ such that for any biclosed magmatic category $\mathbf{D}$ and biclosed magmatic interpretation $j : \Sigma \to \mathbf{D}$, there is a unique biclosed magmatic functor $F : \mathbf{C} \to \mathbf{D}$ such that $j = F \circ i$.*

Given the preceding definitions, we can define the category of proof nets over a signature, and show that it is in fact the *free* category over that signature.

Definition 9. *Given* $\Sigma = (\Sigma_0, \Sigma_1, dom, cod)$ *a biclosed magmatic signature, the* **proof net category** $\mathbf{PN}(\Sigma)$ *over* Σ *is defined as follows:*

1. *The objects of* $\mathbf{PN}(\Sigma)$ *are the elements of* Σ_0,
2. *For every object* A *of* $\mathbf{PN}(\Sigma)$, *the identity net is defined as in Definition 4,*
3. *For every element* f *in* Σ_1 *with* $dom(f) = A$, $cod(f) = B$, *we stipulate a proof net*

4. *Composition of morphisms and applying bifunctors* $\otimes, \backslash, /$ *to morphisms are given by composition and monotonicity in Definition 4,*
5. *Left closure* $\beta(N_1) : B \to A\backslash C$ *for a morphism* $N_1 : A \otimes B \to C$ *and its inverse* $\beta^{-1}(N_2) : A \otimes B \to C$ *for* $N_2 : B \to A\backslash C$ *are given by*

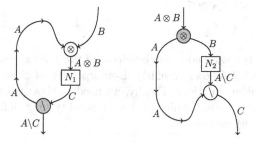

Right closure is treated similarly, where $\gamma(N_1) : A \to C/B$ *for a morphism* $N_1 : A \otimes B \to C$ *and its inverse* $\gamma^{-1}(N_2) : A \otimes B \to C$ *for* $N_2 : A \to C/B$ *are given by*

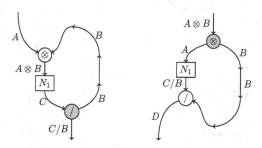

6. *All the proof net equations from Definition 5 hold.*

Proposition 1. *For any biclosed magmatic signature* Σ, $\mathbf{PN}(\Sigma)$ *is a biclosed magmatic category.*

Proof. The basic categorical axioms are trivially satisfied: the associativity of gluing gives associativity of composition, and for any morphism $N_1 : A \to B$ we have that $id_B \circ N_1$ is the result of gluing an extra piece of wire on the bottom and $N_1 \circ id_A$ is the result of gluing extra piece of wire on the top, which is just the same as the original morphism. $id_A \otimes id_B = id_{A \otimes B}$ is also trivially satisfied by the identy cut equation for $\otimes$ (similarly for $\backslash$ and $/$. Bifunctioriality of $\otimes$, is also satisfied because $(k \otimes h) \circ (g \otimes f) = (k \circ g) \otimes (h \circ f)$ translates to

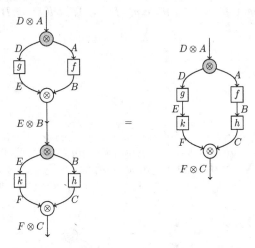

which definitely is a valid equation by virtue of the general cut equation. Bifunctoriality of $\backslash$ and $/$ follows similarly. Isomorphicity of β and γ is easily verified using the snake equations. Finally, we need to show naturality of β and γ. We show (half of) the naturality of β as naturality for γ follows similarly. For $(g \backslash k) \circ ((\beta(f)) \circ h)$ we have

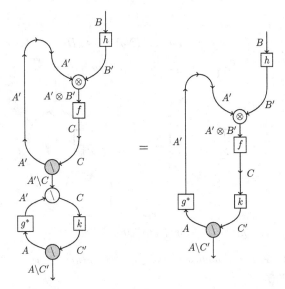

and these nets are obviously equal, given that we can bend around g to g^* and vice versa.

In order to show freeness of the proof net category, we need a way to obtain a categorical morphism from a diagram. In previous work we achieved such a translation as a two step process: diagrams are translated to sequent proofs, which then get translated to morphisms, which is the reason this process is called *sequentialisation*. In this paper we simply give the composed translation to save space. This translation is given in Fig. 3, where we use two extra notations, defined below:

Definition 10. *We write, in sequent calculus style, $\Gamma[B]$ ($\Gamma[\Delta]$) for a tensor of objects Γ that "contains" the object B (Δ). Formally, we can define $\Gamma[]$ as any object of a biclosed category with a "hole" in it, that, is:*

$$\Gamma[] := [] \mid \Gamma'[] \otimes \Delta \mid \Delta \otimes \Gamma'[]$$

Then, we say that $\Gamma[B]$ ($\Gamma[\Delta]$) is the object with its "hole" replace by B (Δ). Given a morphism $f : A \to B$, we construct $op(f) : \Gamma[A] \to \Gamma[B]$ for the morphism that acts as the identity on $\Gamma[]$ but applies f to A. Formally, we have

1. *$op(f) = f$ for $\Gamma[] = []$,*
2. *$op(f) = op'(f) \otimes id_\Delta$ for $\Gamma[] = \Gamma'[] \otimes \Delta$,*
3. *$op(f) = id_\Delta \otimes op'(f)$ for $\Gamma[] = \Delta \otimes \Gamma'[]$.*

Now that we have fully defined the translation from diagrams to categorical morphisms, we can state the sequentialisation property:

Proposition 2. *Every proof net sequentialises using the translation in Fig. 3 and all the equalities between diagrams are preserved under this sequentialisation.*

Proof. This is shown for the two step translation by Wijnholds ([18], pp. 50–61).

Given that any diagram can be transformed back into a categorical morphism, we are ready to prove freeness and thus coherence of the proof net category.

Theorem 1. *For any biclosed magmatic signature Σ, the proof net category $\mathbf{PN}(\Sigma)$ is the free biclosed category over Σ.*

Proof. We need to give an interpretation $i : \Sigma \to \mathbf{PN}(\Sigma)$ and for any biclosed magmatic category $\mathbf{D}$ and biclosed magmatic interpretation $j : \Sigma \to \mathbf{D}$ give a unique biclosed magmatic functor $F : \mathbf{PN}(\Sigma) \to \mathbf{D}$ such that $j = F \circ i$.

First, we define $i = \langle i_0, i_1 \rangle$ with i_0 the identity on Σ_0 and i_1 the map that sends morphism variables f to

$$dom(f) \downarrow \boxed{f} \atop cod(f) \downarrow$$

Now let $\mathbf{D}$ be any biclosed magmatic category and let $j : \Sigma \to \mathbf{D}$ be an interpretation. We define $F : \mathbf{PN}(\Sigma) \to \mathbf{D}$ as follows:

1. On objects we define $F(A) = j_0(A)$,
2. On morphisms/nets we define $F(N : A \to B) = \hat{j}_1 \circ S$ where
   - S is the sequentialization that turns a proof net into a categorical morphism module morphism variables,
   - $\hat{j}_1$ sends all objects A to $j_0(A)$, and on morphisms $\hat{j}_1$ sends morphism variables $f : A \to B$ to $j_1(f) : j_0(A) \to j_0(B)$ but otherwise preserves the bifunctors $\otimes, \backslash, /$ and composition.

We need to show that F is well defined as a biclosed magmatic functor, that $j = F \circ i$ and that F is unique.

1. F preserves the closed structure of $\mathbf{PN}(\Sigma)$ in $\mathbf{D}$. This is easy to see since F acts as j_0 on objects and therefore strictly preserves the object structure of $\mathbf{PN}(\Sigma)$. On morphisms, we first note that $\hat{j}_1$ preserving morphism structure and only maps morphism variables to their corresponding variables in $\mathbf{D}$. We then note that for β and β^{-1} in $\mathbf{PN}(\Sigma)$, the proof net $\beta^{-1}(\beta(N)) : A \otimes B \to C$ is equal to N by the proof net equations, hence F will send it to the morphism $F(N) : F(A) \to F(B)$. The case for the converse composition and for γ with γ^{-1} is similar.
2. $j = F \circ i$. We note that i_0 is the identity and F on objects is j_0, hence $j_0 = F \circ i_0$. On morphism variables $f : A \to B$, we note that these simply are encoded in $\mathbf{PN}(\Sigma)$ as a box labelled with f with one ingoing arrow labelled with A and an outgoing arrow labelled with B by i. Then, F will send this net to the morphism $j_1(f) : j_0(A) \to j_0(B)$, and so we have that $j_1 = F \circ i_1$.
3. F is unique. Let $G : \mathbf{PN}(\Sigma) \to \mathbf{D}$ be a biclosed magmatic functor such that $j = G \circ i$. As i_0 is the identity, we must have that (on objects) $G = j_0 = F$. On morphism variables, note that S and i_1 are inverse as i_1 turns a morphism variable $f : A \to B$ into its graphical version, whereas S recovers the morphism itself. Hence, we can state that

$$F = F \circ i_1 \circ S = G \circ i_1 \circ S = G$$

By showing the freeness of the proof net category, we have shown that any equation of a biclosed magmatic category will hold if and only if it holds in the proof net category. In other words, this proof net category allows us to reason about biclosed magmatic categories graphically in a coherent way. One may consider the relevance of biclosed magmatic categories in particular and wonder whether the proof net category is more general. In the next section we highlight two extensions of the graphical language, one of which is shown to easily lead to a coherent graphical language for biclosed monoidal categories, the kind of categories that the above-mentioned clasp language was supposed to capture.

5   Extensions

Associativity. The graphical language we have considered so far is inspired by proof nets for the nonassociative Lambek calculus. Of course, a logical step is to consider associativity as well, in order to coherently capture monoidal closed categories, as the language of Baez and Stay tries to do. A simple solution is to add the associators as two hardcoded diagrams:

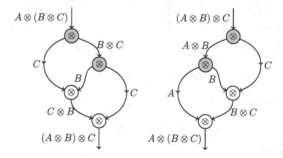

If we define the proof nets using correctness criteria, the so-called *operator balance* criterium would fail. However, when we stick to the inductive proof net definition, we can easily verify that the resulting proof net category is in fact monoidal biclosed (soundness), which we leave as an exercise to the reader. The completeness part requires a bit more thought, but basically amounts to adapting the sequentialisation such that every time a destructor link for $\otimes$ is rewritten, one composes with the associator α when it happens that the domain of f is not compatible with the domain that is obtained when A and B are conjoined into $A \otimes B$. Once we do this, one can show freeness and thus coherence of the graphical language.

Modalities. So far we have argued for modularity in defining diagrammatic calculi. Much in the spirit of Lambek's deductive systems view [11], and the development of proof nets for linguistic analysis [15], the language we developed so far enjoys a modular approach by adding structural properties as diagrams that satisfy coherence under the equations imposed on those diagrams. We wish to go a bit further in this approach by arguing that modalities as used for linguistic purposes [14] can also be incorporated in our proof net language. In a nutshell, one adds two unary connectives to the nonassociative Lambek calculus that exhibit residuating behavior. Categorically speaking, this corresponds to covariant adjunction. Then, one adds structural rules governing special behavior of the unary connectives. Again, these are added in the form of extra diagrams, much like the case of associativity. In pictures, we define the links for $\Diamond, \Box$ as follows:

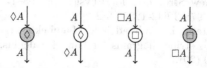

Then, structural rules take the form

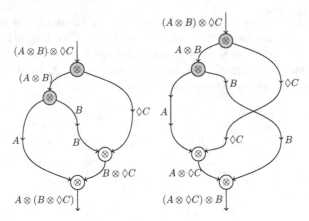

6   Conclusion

In this paper, we argued for a modular approach to diagrammatic reasoning in the context of categorical compositional distributional semantics. We developed a graphical language for closed magmatic categories, inspired by proof nets for Lambek calculi. We then showed coherence for this *proof net category* and argued that associativity can be recovered by adding two diagrams to the language, effectively obtaining a graphical language for monoidal closed categories. Finally, we offered some thoughts on adding diagrams that in logical terminology are called unary residuated connectives but categorically can be thought of as pairs of adjoint functors. Working out coherence for these unary modalities, as well as incorporating bialgebras and Frobenius algebras graphically, constitutes future work.

Acknowledgements. The author is greatly indebted for many fruitful discussions with Michael Moortgat during the writing of the MSc thesis on which this paper is largely based. Also, a thanks goes out to Mehrnoosh Sadrzadeh for discussions culminating in the existence of this paper. A thanks as well to John Baez and Peter Selinger for giving some advice a long time ago on the topic of diagrammatic reasoning. Finally, the author would like to thank the two anonymous referees of this paper. The author was supported by a Queen Mary Principal's Research Studentship during the writing of this paper.

References

1. Baez, J., Stay, M.: Physics, topology, logic and computation: a rosetta stone. In: Coecke, B. (ed.) New Structures for Physics, pp. 95–172. Springer, Heidelberg (2010). doi:10.1007/978-3-642-12821-9_2
2. Coecke, B., Grefenstette, E., Sadrzadeh, M.: Lambek vs. Lambek: functorial vector space semantics and string diagrams for lambek calculus. Ann. Pure Appl. log. **164**(11), 1079–1100 (2013)

3. Coecke, B., Sadrzadeh, M., Clark, S.: Mathematical foundations for a compositional distributional model of meaning. arXiv preprint arXiv:1003.4394 (2010)
4. Freyd, P., Yetter, D.N.: Coherence theorems via knot theory. J. Pure Appl. Algebr. **78**(1), 49–76 (1992)
5. Girard, J.Y.: Linear logic. Theor. Comput. Sci. **50**(1), 1–101 (1987)
6. Joyal, A., Street, R.: The geometry of tensor calculus, I. Adv. Math. **88**(1), 55–112 (1991)
7. Joyal, A., Street, R.: Braided tensor categories. Adv. Math. **102**(1), 20–78 (1993)
8. Kelly, G.M., Laplaza, M.L.: Coherence for compact closed categories. J. Pure Appl. Algebr. **19**, 193–213 (1980)
9. Lambek, J.: The mathematics of sentence structure. Am. Math. Mon. **65**(3), 154–170 (1958)
10. Lambek, J.: On the calculus of syntactic types. Struct. Lang. Math. Asp. **166**, C178 (1961)
11. Lambek, J.: Deductive systems and categories. Theory Comput. Syst. **2**(4), 287–318 (1968)
12. Lambek, J.: Type grammar revisited. In: Lecomte, A., Lamarche, F., Perrier, G. (eds.) LACL 1997. LNCS, vol. 1582, pp. 1–27. Springer, Heidelberg (1999). doi:10.1007/3-540-48975-4_1
13. Lambek, J., Scott, P.J.: Introduction to Higher-Order Categorical Logic, vol. 7. Cambridge University Press, Cambridge (1988)
14. Moortgat, M.: Multimodal linguistic inference. J. Log. Lang. Inf. **5**(3–4), 349–385 (1996)
15. Moot, R.: Proof Nets for Linguistic Analysis. Ph.D. thesis, Utrecht University (2002)
16. Penrose, R.: Applications of negative dimensional tensors. In: Combinatorial Mathematics and its Applications, vol. 1, pp. 221–244 (1971)
17. Selinger, P.: A survey of graphical languages for monoidal categories. In: Coecke, B. (ed.) New Structures for Physics, pp. 289–355. Springer, Heidelberg (2010)
18. Wijnholds, G.J.: Categorical foundations for extended compositional distributional models of meaning. MSc. thesis, University of Amsterdam (2014)

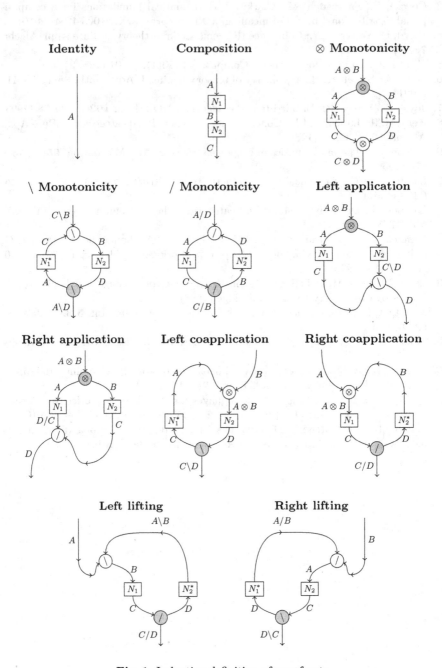

Fig. 1. Inductive definition of proof nets

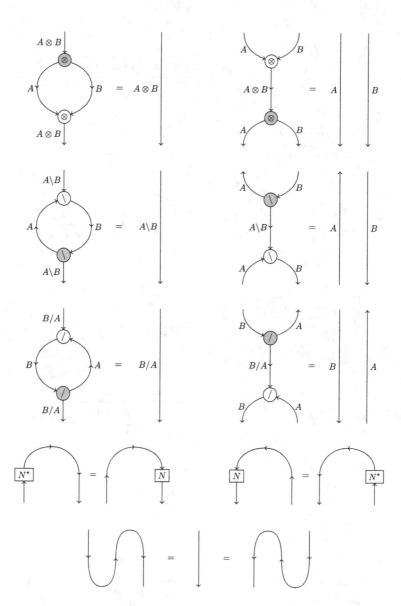

Fig. 2. Equations on proof nets

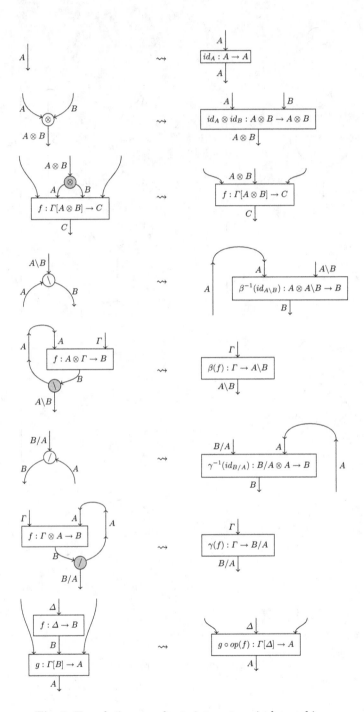

Fig. 3. Translating proof nets into categorical morphisms

Algorithmic Sahlqvist Preservation for Modal Compact Hausdorff Spaces

Zhiguang Zhao[✉]

Delft University of Technology, Delft, The Netherlands
zhaozhiguang23@gmail.com

Abstract. In this paper, we use the algorithm ALBA to reformulate the proof in [1,2] that over modal compact Hausdorff spaces, the validity of Sahlqvist sequents are preserved from open assignments to arbitrary assignments. In particular, we prove an adapted version of the topological Ackermann lemma based on the Esakia-type lemmas proved in [1,2].

1 Introduction

Canonicity, i.e. the preservation of validity of formulas from descriptive general frames to their underlying Kripke frames, is an important notion in modal logic, since it provides a uniform strategy for proving the strong completeness of axiomatic extensions of a basic (normal modal) logic. Thanks to its importance, the notion of canonicity has been explored also for other non-classical logics. In [26], Jónsson gave a purely algebraic reformulation of the frame-theoretic notion of canonicity, which he defined as the preservation of validity under taking canonical extensions, and proved the canonicity of Sahlqvist identities in a purely algebraic way. The construction of canonical extension was introduced by Jónsson and Tarski [27] as a purely algebraic encoding of the Stone spaces dual to Boolean algebras. In particular, the denseness requirement in the definition of canonical extension directly relates to the zero-dimensionality of Stone spaces. A natural question is then for which classes of formulas do canonicity-type preservation results hold in topological settings in which compactness is maintained and zero-dimensionality is generalized to the Hausdorff separation condition. This question has been addressed in [1,2]. Specifically, in [1], Bezhanishvili et al. gave a canonicity-type preservation result for Sahlqvist formulas from modal compact Hausdorff spaces to their underlying Kripke frames, and in [2], Bezhanishvili and Sourabh generalized this result to modal fixed point formulas.

In the present paper, some preliminary results are collected which reformulate the canonicity-type preservation results in [1,2] in an algebraic and algorithmic way. These results link the preservation results of [1,2] with the theory of *unified correspondence*, which aims at identifying the underlying principles of Sahlqvist-type canonicity and correspondence for non-classical logics, which is useful in the completeness proofs. As explained in [10,14], this theory is grounded on the Stone-type dualities between the algebraic and the relational semantics of non-classical logics, and explains the "Sahlqvist phenomenon" in terms of the

© Springer-Verlag GmbH Germany 2017
J. Kennedy and R.J.G.B. de Queiroz (Eds.): WoLLIC 2017, LNCS 10388, pp. 387–400, 2017.
DOI: 10.1007/978-3-662-55386-2_28

order-theoretic properties of the algebraic interpretations of the connectives of a non-classical logic. The focus on these properties has been crucial to the possibility of generalizing the Sahlqvist-type results from modal logic to a wide array of non-classical logics, including intuitionistic and distributive and general (non-distributive) lattice-based (modal) logics [9,11,13], non-normal (regular) modal logics [31], monotone modal logic [19], hybrid logics [17], many valued logics [5] and bi-intuitionistic and lattice-based modal mu-calculus [6–8]. In addition, unified correspondence has effectively provided overarching techniques unifying different methods for proving both canonicity and correspondence: in [30], the methodology pioneered by Jónsson [26] and the one pioneered by Sambin-Vaccaro [32] were unified; in [12,15], constructive canonicity proposed by Ghilardi and Meloni [21] was unified with the Sambin-Vaccaro methodology; in [16], the Sambin-Vaccaro correspondence has been unified with the methodology of correspondence via translation introduced by Gehrke et al. in [20]. Recently, a very surprising connection has been established between the notions and techniques developed in unified correspondence and structural proof theory, which made it possible to solve a problem, opened by Kracht [28], concerning the characterization of the axioms which can be transformed into analytic structural rules [23,29]. The main tools of unified correspondence are a purely order-theoretic definition of *inductive formulas/inequalities*, and the algorithm ALBA (Ackermann Lemma Based Algorithm, cf. [11]), which computes the first-order correspondent of input formulas/inequalities (i.e. the first-order formula which is valid on the same class of Kripke frames as the input formula/inequality) and is guaranteed to succeed on the inductive class, which is a strictly larger class of formulas/inequalities than the Sahlqvist class.

In the present paper, we adapt the algorithm ALBA to the setting of modal compact Hausdorff spaces, show the soundness of the algorithm with respect to the interpretation over modal compact Hausdorff spaces. In particular, an adapted version of the topological Ackermann lemma (cf. [11, Lemma 9.3]) is proved using the Esakia-type lemma for the modal language over modal compact Hausdorff spaces. The results of the present paper pave the way to extend the tools of unified correspondence to canonicity-type preservation results based on different dualities than Stone duality.

This paper is organized as follows: Sect. 2 collects preliminaries on modal compact Hausdorff spaces and the semantic interpretation for the modal language. Section 3 discusses the main ideas for the preservation results. Section 4 provides the expanded modal language of the algorithm as well as its interpretations, together with the syntactic definition of Sahlqvist sequents. The Ackermann Lemma Based Algorithm (ALBA) for modal compact Hausdorff space is given in Sect. 5. Sections 6 and 8 respectively shows the soundness of the algorithm with respect to modal compact Hausdorff spaces and the success of the algorithm on Sahlqvist sequents.

2   Preliminaries

2.1   Modal Compact Hausdorff Spaces

In the present subsection, we collect the preliminaries for modal compact Hausdorff spaces. For more details, the readers are referred to [1, 2, 25].

We will use the following notations: given a binary relation R on W, we denote $R[X] = \{w \in W \mid (\exists x \in X)Rxw\}$ and $R^{-1}[X] = \{w \in W \mid (\exists x \in X)Rwx\}$, $R[w] := R[\{w\}]$ and $R^{-1}[w] := R^{-1}[\{w\}]$, respectively.

A topological space $\mathcal{T} = (W, \tau)$ is

1. *compact* if for any collection $\{X_i\}_{i \in I}$ of open sets, if $W = \bigcup_{i \in I} X_i$, then there is a finite subset $I_0 \subseteq I$ such that $W = \bigcup_{i \in I_0} X_i$;
2. *Hausdorff* if for any two distinct points $x, y \in W$, there exist $X, Y \in \tau$ such that $x \in X$, $y \in Y$ and $X \cap Y = \varnothing$.

It is well-known that singletons are closed in Hausdorff spaces.

Definition 1 *(cf. e.g. [1, Definition 2.14]). A modal compact Hausdorff space is a triple $\mathcal{T} = (W, R, \tau)$ such that (W, τ) is a compact Hausdorff space and R is continuous, i.e.*

1. *$R[w]$ is closed for any $w \in W$;*
2. *$R^{-1}[X]$ is closed for any closed set X;*
3. *$R^{-1}[X]$ is open for any open set X.*

We let $\mathbb{F}_{\mathcal{T}} = (W, R)$ denote the underlying Kripke frame[1] of $\mathcal{T}$.

As is well known, open sets of topological spaces are captured algebraically by the notion of *frame*. A frame $\mathbb{L}$ is a complete lattice validating the following identity: $a \wedge \bigvee X = \bigvee \{a \wedge x \mid x \in X\}$ for any $X \subseteq \mathbb{L}$. A frame is *compact* if for any $X \subseteq \mathbb{L}$ such that $\bigvee X = \top$, there is a finite subset $Y \subseteq X$ such that $\bigvee Y = \top$. For any frame $\mathbb{L}$ and any $a \in \mathbb{L}$, the *pseudocomplement* of a is $\neg a := \bigvee \{b \mid b \wedge a = \bot\}$. For $a, b \in \mathbb{L}$, a is *well inside* b (notation: $a \prec b$) if $\neg a \vee b = \top$. A frame $\mathbb{L}$ is *regular* if $a = \bigvee \{b \mid b \prec a\}$ for all $a \in \mathbb{L}$. For any topological space $\mathcal{T} = (W, \tau)$, its associated frame is defined as $\mathbb{L}_{\mathcal{T}} := (\tau, \cap, \bigcup)$. If $\mathcal{T}$ is compact Hausdorff, then $\mathbb{L}_{\mathcal{T}}$ is compact regular.

Definition 2 *(cf. e.g. [1, Definition 3.5]). A modal compact regular frame is a triple $\mathbb{L} = (L, \Box, \Diamond)$ where L is a compact regular frame, and $\Box, \Diamond$ are unary operations on L such that:[2]*

1. *$\Box\top = \top$ and $\Box(a \wedge b) = \Box a \wedge \Box b$;*
2. *$\Diamond\bot = \bot$ and $\Diamond(a \vee b) = \Diamond a \vee \Diamond b$;*

---

[1] Notice that the name "frame" occurs in two different ways in the present paper, one is in point-free topology, the other is in modal logic. Here we use the name "Kripke frame" to refer to the notion in modal logic and "frame" to refer to the notion in point-free topology.

[2] The condition 3 is well-known in [18].

3. $\square(a \vee b) \leq \square a \vee \Diamond b$ and $\square a \wedge \Diamond b \leq \Diamond(a \wedge b)$;
4. for any upward directed $X \subseteq L$, $\Diamond \bigvee X = \bigvee \{\Diamond x \mid x \in X\}$ and $\square \bigvee X = \bigvee \{\square x \mid x \in X\}$.

One can readily show (cf. [1, Proposition 3.10]) that if $\mathcal{T} = (W, R, \tau)$ is a modal compact Hausdorff space, then $\mathbb{L}_{\mathcal{T}} := (\tau, \square_{\mathcal{T}}, \Diamond_{\mathcal{T}})$ is a modal compact regular frame where $\square_{\mathcal{T}} X = (R^{-1}[X^c])^c$ and $\Diamond_{\mathcal{T}} X = R^{-1}[X]$.

2.2 Language and Interpretation

Given a set Prop of propositional variables, the positive modal language $\mathcal{L}$ is recursively defined as follows:

$$\varphi ::= p \mid \bot \mid \top \mid \varphi \wedge \varphi \mid \varphi \vee \varphi \mid \square \varphi \mid \Diamond \varphi,$$

where $p \in$ Prop. We let $\mathsf{Prop}(\alpha)$ denote the set of propositional variables occuring in α.

In [18], the positive fragment of basic normal modal logic is completely axiomatized as follows.

$$\varphi \vdash \varphi \qquad \varphi \vdash \varphi \vee \psi \qquad \psi \vdash \varphi \vee \psi \qquad \varphi \wedge \psi \vdash \varphi \qquad \varphi \wedge \psi \vdash \psi$$

$$\varphi \wedge (\psi \vee \chi) \vdash (\varphi \wedge \psi) \vee (\varphi \wedge \chi)$$

$$\Diamond(\varphi \vee \psi) \dashv\vdash \Diamond \varphi \vee \Diamond \psi \qquad \square(\varphi \wedge \psi) \dashv\vdash \square \varphi \wedge \square \psi$$

$$\Diamond \varphi \wedge \square \psi \vdash \Diamond(\varphi \wedge \psi) \qquad \square(\varphi \vee \psi) \vdash \square \varphi \vee \Diamond \psi$$

$$\frac{\varphi \vdash \psi \quad \psi \vdash \chi}{\varphi \vdash \chi} \qquad \frac{\varphi \vdash \chi \quad \psi \vdash \chi}{\varphi \vee \psi \vdash \chi} \qquad \frac{\chi \vdash \varphi \quad \chi \vdash \psi}{\chi \vdash \varphi \wedge \psi} \quad \frac{\varphi \vdash \psi}{\Diamond \varphi \vdash \Diamond \psi} \qquad \frac{\varphi \vdash \psi}{\square \varphi \vdash \square \psi}$$

In the following sections we will typically work with *inequalities* $\varphi \leq \psi$, and *quasi-inequalities* $\varphi_1 \leq \psi_1 \ \& \ \ldots \ \& \ \varphi_n \leq \psi_n \Rightarrow \varphi \leq \psi$ (cf. [11]), where & is the meta-conjunction and $\Rightarrow$ is the meta-implication.

Interpretation on Modal Compact Hausdorff Spaces. Modal compact Hausdorff spaces play the role played by descriptive general frames in the Stone-based setting. Accordingly, the counterparts of admissible valuations (i.e. valuations such that propositional variables are interpreted in a restricted class of subsets of the domain W, instead of arbitrary subsets) are the open valuations defined below.

A *modal compact Hausdorff model* is a pair $\mathbb{M} = (\mathcal{T}, V)$ where $\mathcal{T} = (W, R, \tau)$ is a modal compact Hausdorff space, and $V : \mathsf{Prop} \to \tau$ is an *open valuation* on $\mathcal{T}$. The satisfaction relation on modal compact Hausdorff models is defined as standard in modal logic. We let $[\![\varphi]\!]^{\mathbb{M}} = \{w \in W \mid \mathbb{M}, w \Vdash \varphi\}$ denote the *truth set* of φ in $\mathbb{M}$.

An inequality $\varphi \leq \psi$ is valid on a modal compact Hausdorff space $\mathcal{T}$ if $[\![\varphi]\!]^{\mathbb{M}} \subseteq [\![\psi]\!]^{\mathbb{M}}$ for every model $\mathbb{M}$ based on $\mathcal{T}$ (i.e. for every open valuation into τ). A quasi-inequality $\varphi_1 \leq \psi_1 \ \& \ \ldots \ \& \ \varphi_n \leq \psi_n \Rightarrow \varphi \leq \psi$ is valid on $\mathcal{T}$ if, for every model $\mathbb{M}$ based on $\mathcal{T}$, if $[\![\varphi_i]\!]^{\mathbb{M}} \subseteq [\![\psi_i]\!]^{\mathbb{M}}$ for all i then $[\![\varphi]\!]^{\mathbb{M}} \subseteq [\![\psi]\!]^{\mathbb{M}}$.

Interpretation on Algebras. In what follows, we let $\mathbb{B}$ denote a Boolean algebra with operator (BAO) (cf. [3, Definition 5.19]). We let $\theta : \mathsf{Prop} \to \mathbb{B}$ denote an *assignment* on $\mathbb{B}$, and let $\varphi^{(\mathbb{B},\theta)}$ or $\theta(\varphi)$ denote the value of φ in $\mathbb{B}$ under θ. We write $(\mathbb{B},\theta) \vDash \varphi \leq \psi$ to indicate that $\varphi \leq \psi$ is *true* on $\mathbb{B}$ under θ, and $\mathbb{B} \vDash \varphi \leq \psi$ to indicate that $\varphi \leq \psi$ is *valid* on $\mathbb{B}$. Notations for truth and validity for quasi-inequalities are similar.

Another useful way to look at a formula $\varphi(p_1, \ldots, p_n)$ is to interpret it as an n-ary function $\varphi^{\mathbb{B}} : \mathbb{B}^n \to \mathbb{B}$ such that $\varphi^{\mathbb{B}}(a_1, \ldots, a_n) = \theta(\varphi)$ where $\theta : \mathsf{Prop} \to \mathbb{B}$ satisfies $\theta(p_i) = a_i$, $i = 1, \ldots, n$.

For any Kripke frame $\mathbb{F} = (W, R)$, we let $\mathbb{B}_{\mathbb{F}} = (P(W), \varnothing, W, \cap, \cup, (\cdot)^c, \square_{\mathbb{B}_{\mathbb{F}}})$ denote the *dual BAO* of $\mathbb{F}$ (i.e. the complex algebra of $\mathbb{F}$), where $\square_{\mathbb{B}_{\mathbb{F}}} X = (R^{-1}[X^c])^c$ for any $X \in P(W)$. It it folklore that a Kripke frame $\mathbb{F}$ and its dual BAO validate the same (quasi-)inequalities. In what follows, we let $\mathsf{At}(\mathbb{B}_{\mathbb{F}}) = \{\{w\} \mid w \in W\}$ and $\mathsf{CoAt}(\mathbb{B}_{\mathbb{F}}) = \{W - \{w\} \mid w \in W\}$ denote the set of atoms and coatoms of $\mathbb{B}_{\mathbb{F}}$ respectively.

Analogous notions and notations also apply to modal compact regular frames (cf. Definition 2). In particular, the dual algebra of the modal compact Hausdorff space $\mathcal{T} = (W, R, \tau)$ is the modal compact regular frame $\mathbb{L}_{\mathcal{T}} = (\tau, \square_{\mathcal{T}}, \Diamond_{\mathcal{T}})$, which provides a natural interpretation for the positive modal language. In addition, $\mathbb{L}_{\mathcal{T}}$ can be naturally embedded as a modal subframe[3] into the complex algebra $\mathbb{B}_{\mathbb{F}_{\mathcal{T}}}$ of the underlying Kripke frame $\mathbb{F}_{\mathcal{T}}$ of $\mathcal{T}$. Hence, all connectives in the positive modal language are interpreted in the same way when restricting the valuation of propositional variables to open subsets. Therefore, validity in $\mathbb{L}_{\mathcal{T}}$ (denoted $\mathbb{L}_{\mathcal{T}} \vDash \varphi \leq \psi$) coincides with validity in $\mathbb{B}_{\mathbb{F}_{\mathcal{T}}}$ relative to *open assignments* (i.e. assignments into open subsets), denoted $\mathbb{B}_{\mathbb{F}_{\mathcal{T}}} \vDash_{\mathbb{L}_{\mathcal{T}}} \varphi \leq \psi$.

3 Main Ideas

3.1 From Stone to Modal Compact Hausdorff

As is well-known [22,27], every Boolean algebra $\mathbb{B}$ is dually equivalent to a descriptive general frame $\mathbb{G}$ (cf. [3, Definition 5.65]), and the underlying Kripke frame $\mathbb{F}_{\mathbb{G}}$ of $\mathbb{G}$ is dually equivalent to the canonical extension $\mathbb{B}^{\delta}$ (cf. [4, Chapter 6, Definition 104]) of $\mathbb{B}$, as illustrated in the left diagram below. The canonicity of an inequality (i.e., the preservation of its validity from any $\mathbb{G}$ to its $\mathbb{F}_{\mathbb{G}}$) can be equivalently rephrased as the preservation of its validity from any $\mathbb{B}$ to $\mathbb{B}^{\delta}$. This picture can be analogously generalized to the setting of modal compact Hausdorff spaces. In the right diagram, in the bottom line, every modal compact Hausdorff space $\mathcal{T}$ is dually equivalent (inducated by the symbol $\cong^{\partial}$) to its dual modal compact regular frame $\mathbb{L}_{\mathcal{T}}$ (cf. [1, Theorem 3.14]), and the forgetful functor U maps any $\mathcal{T}$ to its underlying Kripke frame $\mathbb{F}_{\mathcal{T}}$. On the dual algebraic side, $\mathbb{L}_{\mathcal{T}}$ is embedded into the dual BAO $\mathbb{B}_{\mathbb{F}_{\mathcal{T}}}$ of $\mathbb{F}_{\mathcal{T}}$. The canonical embedding $\mathbb{B} \hookrightarrow \mathbb{B}^{\delta}$ encodes the Stone-type duality between BAOs and descriptive general frames in

---

[3] That is, not only finite meets and complete joins are preserved, but also the modal operators, i.e. $\square_{\mathbb{B}_{\mathbb{F}}} X = \square_{\mathcal{T}} X$ and $(\square_{\mathbb{B}_{\mathbb{F}}} X^c)^c = \Diamond_{\mathcal{T}} X$ for all $X \in \tau$.

a purely algebraic way. Likewise, the Isbell duality (cf. [24]) can be encoded in the purely algebraic properties of the embedding $e : \mathbb{L} \hookrightarrow \mathbb{B}$ of a compact regular frame $\mathbb{L}$ into a complete atomic Boolean algebra $\mathbb{B}$, namely, that e be a frame homomorphism such that the following conditions hold (where we suppress the embedding):

(compactness) For any $S \subseteq \mathbb{L}$, if $\bigvee S = \top$, then $\bigvee S' = \top$ for some finite $S' \subseteq S$;

(Hausdorff) For any $x, y \in \mathsf{At}(\mathbb{B})$, if $x \neq y$, then $x \leq a, y \leq b$ for some $a, b \in \mathbb{L}$ such that $a \wedge b = \bot$.

In particular, $\mathbb{L}_\mathcal{T}$ can be identified with the collection $\mathsf{O}(\mathbb{B}_{\mathbb{F}_\mathcal{T}})$ of open subsets in $\mathbb{B}_{\mathbb{F}_\mathcal{T}}$, i.e. the open subsets in $\mathcal{T}$. The collection $\mathsf{K}(\mathbb{B}_{\mathbb{F}_\mathcal{T}})$ of closed subsets in $\mathbb{B}_{\mathbb{F}_\mathcal{T}}$ can be then identified as the relative complements of elements in $\mathbb{L}_\mathcal{T}$. We omit the subscripts when they are clear from the context.

3.2 Basic Proof Strategy for Preservation

In the present section, we explain the basic proof structure we will implement in Sect. 5. We will treat the preservation results as a generalized canonicity result which, using the algorithmic canonicity strategy, are typically proved by a "U-shaped" argument described in the figure below (see [10] for a more detailed discussion): In the present setting, the U-shaped argument can be sketched as follows:

$$
\begin{array}{ccc}
\mathbb{L}_\mathcal{T} \vDash \varphi \leq \psi & & \mathbb{B}_{\mathbb{F}_\mathcal{T}} \vDash \varphi \leq \psi \\
\Updownarrow & & \\
\mathbb{B}_{\mathbb{F}_\mathcal{T}} \vDash_{\mathbb{L}_\mathcal{T}} \varphi \leq \psi & & \Updownarrow \\
\Updownarrow & & \\
\mathbb{B}_{\mathbb{F}_\mathcal{T}} \vDash_{\mathbb{L}_\mathcal{T}} \mathrm{Pure}(\varphi \leq \psi) & \Leftrightarrow & \mathbb{B}_{\mathbb{F}_\mathcal{T}} \vDash \mathrm{Pure}(\varphi \leq \psi)
\end{array}
$$

Assume that the inequality $\varphi \leq \psi$ is valid on the modal compact regular frame $\mathbb{L}_\mathcal{T}$. This is equivalent to the validity on the BAO $\mathbb{B}_{\mathbb{F}_\mathcal{T}}$ over all open assignments. Then the algorithm ALBA can equivalently transform the input inequality into a set of pure quasi-inequalities $\mathrm{Pure}(\varphi \leq \psi)$, i.e. quasi-inequalities that contain no propositional variables, therefore their validity is invariant under replacing open assignments of propositional variables by arbitrary assignments of propositional variables. Then by the soundness of ALBA on perfect BAOs, the validity of $\mathrm{Pure}(\varphi \leq \psi)$ is equivalent to the validity of $\varphi \leq \psi$.

4 Language and Interpretation for ALBA

4.1 The Expanded Language for the Algorithm

In the present subsection, we will define the expanded modal language for the algorithm. Our treatment is similar to [11].

The expanded positive modal language $\mathcal{L}^+$ contains, in addition to the symbols in the positive modal language, two sets of special variables Nom of *nominals* and CoNom of *conominals*, and *connective* $\blacklozenge$. The nominals and conominals are interpreted as atoms and coatoms in $\mathbb{B}_{\mathbb{F}_\mathcal{T}}$ respectively, and $\blacklozenge$ is interpreted as the left adjoint of the operation interpreting $\Box$.

The formulas in the expanded modal language $\mathcal{L}^+$ is given as follows:

$$\varphi ::= p \mid \mathbf{i} \mid \mathbf{m} \mid \bot \mid \top \mid \varphi \wedge \varphi \mid \varphi \vee \varphi \mid \Box \varphi \mid \Diamond \varphi \mid \blacklozenge \varphi,$$

where $p \in$ Prop, $\mathbf{i} \in$ Nom and $\mathbf{m} \in$ CoNom.

For the expanded positive modal language $\mathcal{L}^+$, the valuation V and assignment θ extend to the nominals and conominals, such that $V(\mathbf{i}), \theta(\mathbf{i}) \in$ At$(\mathbb{B})$ and $V(\mathbf{m}), \theta(\mathbf{m}) \in$ CoAt$(\mathbb{B})$. The satisfaction relation for the additional symbols is given as follows:

Definition 3 *In any Kripke model* $\mathbb{M} = (W, R, V)$ *or any modal compact Hausdorff models* $\mathbb{M} = (W, R, \tau, V)$,

$$\mathbb{M}, w \Vdash \mathbf{i} \quad \textit{iff } w \in V(\mathbf{i}) \quad \textit{iff } V(\mathbf{i}) = \{w\};$$
$$\mathbb{M}, w \Vdash \mathbf{m} \quad \textit{iff } w \in V(\mathbf{m}) \quad \textit{iff } V(\mathbf{m}) \neq W - \{w\};$$
$$\mathbb{M}, w \Vdash \blacklozenge\varphi \textit{ iff } \exists v (Rvw \quad \textit{and } \mathbb{M}, v \Vdash \varphi)$$

Algebraically, $\blacklozenge^{\mathbb{B}} X = R[X]$.

4.2   1-Sahlqvist Inequalities

In the present section, we define the class of inequalities for which we prove the preservation result in Sect. 6.

Definition 4 (1-Sahlqvist inequalities). *The $\mathcal{L}$-inequality $\varphi \leq \psi$ is 1-Sahlqvist if $\varphi = \varphi'(\chi_1/z_1, \ldots \chi_n/z_n)$ such that*

1. *$\varphi'(z_1, \ldots, z_n)$ is built out of $\wedge, \vee, \Diamond$;*
2. *every χ is of the form $\Box^n p$, $\Box^n \top$, $\Box^n \bot$ for $n \geq 0$;*
3. *ψ is a formula in the positive modal language.*

Remark 1. As its name suggests, the definition above is the restriction of the general definition of ε-Sahlqvist inequalities of [11] to the order type ε which assigns every variable to 1. In the general notation of unified correspondence, the formula φ' corresponds to the Skeleton of φ, and the χ-formulas correspond to its PIA parts[4]. This definition is slightly more general than [1, Definition 7.12] since $\vee$ is allowed to occur in φ'. The inequalities captured by [1, Definition 7.12] correspond to those referred to as *definite* 1-Sahlqvist inequalities in [11].

---

[4] For these terminologies, see [10].

5   The Algorithm ALBA

In the present section, we will give the algorithm ALBA for modal compact Hausdorff spaces, which is similar to the version in [11].

ALBA receives an inequality $\varphi \leq \psi$ as input. Then the algorithm proceeds in three stages:

The first stage is the preprocessing stage, which eliminates all uniformly occurring propositional variables (i.e. propositional variables occurring uniformly positive or uniformly negative), and exhaustively applies the distribution and splitting rules. This stage produces a finite number of inequalities, $\varphi_i \leq \psi_i$, $1 \leq i \leq n$. Then for each inequality, the first approximation rule is applied, which produces a set of inequalities $\{\mathbf{i}_0 \leq \varphi_i, \psi_i \leq \mathbf{m}_0\}$.

The second stage is the reduction and elimination stage, which aims at rewriting the set $\{\mathbf{i}_0 \leq \varphi_i, \psi_i \leq \mathbf{m}_0\}$ into a set of inequalities which has no occurrence of propositional variables. In particular, the step which eliminates all propositional variables is called the Ackermann rule. After this stage, the algorithm produces a set S_i of inequalities.

The third stage is the output stage. If for some set $\{\mathbf{i}_0 \leq \varphi_i, \psi_i \leq \mathbf{m}_0\}$, the propositional variables cannot be eliminated, then the algorithm stops and outputs failure. Otherwise, the algorithm outputs the conjunction of the pure quasi-inequalities $\forall \mathbf{i} \forall \mathbf{m} (\& S_i \Rightarrow \mathbf{i}_0 \leq \mathbf{m}_0)$.

1. **Preprocessing and first approximation**:
   In the generation tree of φ,
   (a) Apply the distribution rules: Push down $\Diamond$ and $\wedge$, by distributing them over nodes labelled with $\vee$;
   (b) Apply the splitting rule 1:

$$\frac{\alpha \vee \beta \leq \gamma}{\alpha \leq \gamma \quad \beta \leq \gamma}$$

   (c) Apply the variable-elimination rules:

$$\frac{\alpha \leq \beta(p)}{\alpha \leq \beta(\bot)} \qquad \frac{\beta(p) \leq \alpha}{\beta(\top) \leq \alpha}$$

   for $\beta(p)$ containing p and α not containing p.
   We denote by Preprocess $(\varphi \leq \psi)$ the finite set $\{\varphi_i \leq \psi_i\}_{i \in I}$ of inequalities obtained after the exhaustive application of the previous rules. Then we apply the first approximation rule to each inequality in Preprocess $(\varphi \leq \psi)$:

$$\frac{\varphi_i \leq \psi_i}{\mathbf{i}_0 \leq \varphi_i \quad \psi_i \leq \mathbf{m}_0}$$

   Here, $\mathbf{i}_0$ and $\mathbf{m}_0$ are special nominals and co-nominals. Now we get a set of inequalities $\{\mathbf{i}_0 \leq \varphi_i, \psi_i \leq \mathbf{m}_0\}_{i \in I}$.

2. **Reduction and elimination:**

In this stage, for each $\{i_0 \leq \varphi_i, \psi_i \leq m_0\}$, we apply the following rules to eliminate all the proposition variables in $\{i_0 \leq \varphi_i, \psi_i \leq m_0\}$:

Residuation rule	Approximation rule	Splitting rule 2
$\dfrac{\alpha \leq \Box\beta}{\blacklozenge\alpha \leq \beta}$	$\dfrac{i \leq \Diamond\alpha}{j \leq \alpha \quad i \leq \Diamond j}$	$\dfrac{\alpha \leq \beta \wedge \gamma}{\alpha \leq \beta \quad \alpha \leq \gamma}$

The nominals introduced by the approximation rule must not occur in the system before applying the rule.

The right-handed Ackermann rule. This is the core rule of ALBA, which eliminates propositional variables. This rule operates on all inequalities in the system, instead of on a single inequality.

$$\frac{S_1 \cup \ldots \cup S_k \cup P \cup \{\psi_i(p_1, \ldots, p_k) \leq m_0\}}{P \cup \{\psi_i(\alpha_{(1,1)} \vee \ldots \vee \alpha_{(1,n_1)}, \ldots, \alpha_{(k,1)} \vee \ldots \vee \alpha_{(k,n_k)}) \leq m_0\}}$$

where $S_l = \{\alpha_{(l,j)} \leq p_l \mid 1 \leq j \leq n_l\}$, $P = \{\beta_l \leq \gamma_l \mid 1 \leq l \leq m\}$, and $\alpha_{(1,1)}, \ldots, \alpha_{(k,n_k)}, \beta_1, \ldots, \beta_m, \gamma_1, \ldots, \gamma_m$ do not contain propositional variables.

3. **Output:** If in the previous stage, some proposition variables cannot be eliminated by the application of the reduction rules, then the algorithm halts and outputs "failure". Otherwise, each initial tuple $\{i_0 \leq \varphi_i, \psi_i \leq m_0\}$ of inequalities after the first approximation has been reduced to a set $\text{Reduce}(\varphi_i \leq \psi_i)$ of pure inequalities, and then the output is a set of quasi-inequalities $\{\&\text{Reduce}(\varphi_i \leq \psi_i) \Rightarrow i_0 \leq m_0 : \varphi_i \leq \psi_i \in \text{Preprocess}(\varphi \leq \psi)\}$, which we denote as $\text{Pure}(\varphi \leq \psi)$.

6   Main Result

In the present section, we prove the preservation of the validity of 1-Sahlqvist inequalities we are after. This result follows from the soundness and success of ALBA. Specifically, we prove the soundness of ALBA with respect to the dual BAOs of the Kripke frames, both for open assignments and for arbitrary assignments. For the soundness with respect to arbitrary assignments and most of the rules with respect to open assignments, the argument is similar to existing settings [11], and hence omitted. We will focus on the right-handed Ackermann rule with respect to open assignments.

Theorem 1 (Soundness with respect to arbitrary assignments). *If ALBA succeeds on an input inequality $\varphi \leq \psi$ and outputs $\text{Pure}(\varphi \leq \psi)$, then for any modal compact Hausdorff space $\mathcal{T}$,*

$$\mathbb{B}_{\mathbb{F}_{\mathcal{T}}} \vDash \varphi \leq \psi \text{ iff } \mathbb{B}_{\mathbb{F}_{\mathcal{T}}} \vDash \text{Pure}(\varphi \leq \psi).$$

Proof The proof goes similarly to [11, Theorem 8.1]. Let $\varphi_i \leq \psi_i$, $1 \leq i \leq n$ denote the inequalities produced by preprocessing $\varphi \leq \psi$ after Stage 1, and

(S_i, Ineq_i), $1 \leq i \leq n$ denote the corresponding quasi-inequalities produced by ALBA after Stage 2. It suffices to show the equivalence from (1) to (4) given below:

$$\mathbb{B}_{\mathbb{F}_{\mathcal{T}}} \vDash \varphi \leq \psi \tag{1}$$

$$\mathbb{B}_{\mathbb{F}_{\mathcal{T}}} \vDash \varphi_i \leq \psi_i, \text{ for all } 1 \leq i \leq n \tag{2}$$

$$\mathbb{B}_{\mathbb{F}_{\mathcal{T}}} \vDash \mathbf{i}_0 \leq \varphi_i \ \& \ \psi_i \leq \mathbf{m}_0 \Rightarrow \mathbf{i}_0 \leq \mathbf{m}_0, \text{ for all } 1 \leq i \leq n \tag{3}$$

$$\mathbb{B}_{\mathbb{F}_{\mathcal{T}}} \vDash \ \& \operatorname{Reduce}(\varphi_i \leq \psi_i) \Rightarrow \mathbf{i}_0 \leq \mathbf{m}_0, \text{ for all } 1 \leq i \leq n \tag{4}$$

- for the equivalence of (1) and (2), it suffices to show the soundness of the rules in Stage 1, which can be proved in the same way as in [11, Lemma 8.3];
- the equivalence between (2) and (3) follows from the soundness of the first-approximation rule, which is similar to [11, Theorem 8.1];
- the equivalence between (3) and (4) follows from the soundness of rules in Stage 2, i.e. the soundness of the approximation rule, the residuation rule, the right-handed Ackermann rule and the splitting rule, which is similar to [11, Lemma 8.4].

For the soundness with respect to open assignments, most of the arguments are the same as the case for arbitrary assignments except for the right-handed Ackermann rule. The soundness of the right-handed Ackermann rule with respect to arbitrary assignments is justified by the following lemma:

Lemma 1 (Right-handed Ackermann lemma). *Let* $\varphi_1, \ldots, \varphi_n$ *be pure formulas,* $\psi(p_1, \ldots, p_n)$ *be an* $\mathcal{L}$*-formula,* $a \in \mathbb{B}$*. Then for any arbitrary assignment* θ*, the following are equivalent:*

1. $\psi^{\mathbb{B}}(\alpha_1^{\mathbb{B},h}, \ldots, \alpha_n^{\mathbb{B},h}) \leq a$;
2. *There exist* $b_1, \ldots, b_n \in \mathbb{B}$ *s.t.* $\alpha_i^{\mathbb{B},h} \leq b_i$ *for* $1 \leq i \leq n$ *and* $\psi^{\mathbb{B}}(b_1, \ldots, b_n) \leq a$.

As is discussed in e.g. [11, Section 9], the lemma above cannot be applied immediately to the setting of open assignments, since formulas in the expanded modal language $\mathcal{L}^+$ might be interpreted as non-open elements, thus the elements $b_1, \ldots, b_n$ might not be in $\mathsf{O}(\mathbb{B})$. We are going to apply similar adaptation strategies as in [11] in the current setting, namely adapt the Ackermann lemma based on syntactic restrictions of the formulas.

Definition 5 (Syntactically closed and open formulas)

1. A formula in $\mathcal{L}^+$ is syntactically closed *if it does not contain occurrences of* conominals;
2. A formula in $\mathcal{L}^+$ is syntactically open *if it does not contain occurrences of* nominals or $\blacklozenge$.

Lemma 2 *(cf. e.g. [1, Lemma 7.10]). For any modal compact Hausdorff space* $\mathcal{T} = (W, R, \tau)$*, if* X *is closed, then* $R[X]$ *is also closed.*

Lemma 3. *If* $\varphi(\mathbf{i}, \boldsymbol{p})$ *and* $\psi(\mathbf{m}, \boldsymbol{p})$ *are syntactically closed and open respectively,* then

1. $\varphi^{\mathbb{B}}(\boldsymbol{i}, \boldsymbol{c}) \in \mathsf{K}(\mathbb{B})$ *for any* $\boldsymbol{i} \in \mathsf{At}(\mathbb{B})$, $\boldsymbol{c} \in \mathsf{K}(\mathbb{B})$.
2. $\psi^{\mathbb{B}}(\boldsymbol{m}, \boldsymbol{o}) \in \mathsf{O}(\mathbb{B})$ *for any* $\boldsymbol{m} \in \mathsf{CoAt}(\mathbb{B})$, $\boldsymbol{o} \in \mathsf{O}(\mathbb{B})$.

Proof. By induction on the structure of the formulas. The basic case follows from the fact that singletons are closed in Hausdorff spaces, and their complements are open. The cases of $\wedge$ and $\vee$ are easy. The cases of $\Diamond$ and $\Box$ follow from Definition 1. The case of $\blacklozenge$ follows from Lemma 2.

Lemma 4 *(cf. e.g.* [1, *Lemma 7.8]). For any modal compact Hausdorff space* $\mathcal{T} = (W, R, \tau)$, *any* $\mathcal{L}$-*formula* $\varphi(p_1, \ldots, p_n)$, *any* $c_1, \ldots, c_n \in \mathsf{K}(\mathbb{B})$,

1. $\varphi(c_1, \ldots, c_n) = \bigwedge \{\varphi(o_1, \ldots, o_n) \mid c_i \leq o_i \text{ for } 1 \leq i \leq n \text{ and } o_i \in \mathsf{O}(\mathbb{B})\}$;
2. $\varphi(c_1, \ldots, c_n) = \bigwedge \{\varphi(\mathsf{cl}(o_1), \ldots, \mathsf{cl}(o_n)) \mid c_i \leq o_i \text{ for } 1 \leq i \leq n \text{ and } o_i \in \mathsf{O}(\mathbb{B})\}$, *where* $\mathsf{cl}(a)$ *denotes the least closed element* $\geq a$.

The lemma below justifies the soundness of right-handed Ackermann rule with respect to open assignments:

Lemma 5 (Right-handed topological Ackermann lemma). *Let* $\varphi_1, \ldots,$ φ_n *be pure and syntactically closed formulas,* $\psi(p_1, \ldots, p_n)$ *be an* $\mathcal{L}$-*formula,* $o \in \mathsf{O}(\mathbb{B})$. *Then for any open assignment* θ, *the following are equivalent:*

1. $\psi^{\mathbb{B}}(\alpha_1^{\mathbb{B}, \theta}, \ldots, \alpha_n^{\mathbb{B}, \theta}) \leq o$;
2. *There exist* $b_1, \ldots, b_n \in \mathsf{O}(\mathbb{B})$ *such that* $\alpha_i^{\mathbb{B}, \theta} \leq b_i$ *for* $1 \leq i \leq n$ *and* $\psi^{\mathbb{B}}(b_1, \ldots, b_n) \leq o$.

Proof $1 \Leftarrow 2$: By the monotonicity of $\psi^{\mathbb{B}}(p_1, \ldots, p_n)$ together with $\alpha_i^{\mathbb{B}, \theta} \leq b_i$ for $1 \leq i \leq n$, we have that $\psi^{\mathbb{B}}(\alpha_1^{\mathbb{B}, \theta}, \ldots, \alpha_n^{\mathbb{B}, \theta}) \leq \psi^{\mathbb{B}}(b_1, \ldots, b_n) \leq \boldsymbol{m}^{\mathbb{B}}$.
$2 \Rightarrow 1$: Suppose that $\psi^{\mathbb{B}}(\alpha_1^{\mathbb{B}, \theta}, \ldots, \alpha_n^{\mathbb{B}, \theta}) \leq o$. By Lemma 3, $\alpha_1^{\mathbb{B}, \theta}, \ldots, \alpha_n^{\mathbb{B}, \theta} \in \mathsf{K}(\mathbb{B})$. By Lemma 4, $o \geq \psi^{\mathbb{B}}(\alpha_1^{\mathbb{B}, \theta}, \ldots, \alpha_n^{\mathbb{B}, \theta}) = \bigwedge \{\psi^{\mathbb{B}}(\mathsf{cl}(o_1), \ldots, \mathsf{cl}(o_n)) \mid \alpha_i^{\mathbb{B}, \theta} \leq o_i \text{ and } o_i \in \mathsf{O}(\mathbb{B}) \text{ for } 1 \leq i \leq n\}$. Since $\mathsf{cl}(o_1), \ldots, \mathsf{cl}(o_n) \in \mathsf{K}(\mathbb{B})$, by Lemma 3, $\psi^{\mathbb{B}}(\mathsf{cl}(o_1), \ldots, \mathsf{cl}(o_n)) \in \mathsf{K}(\mathbb{B})$. By compactness, there exist $o_{1,j} \ldots, o_{n,j}$, $1 \leq j \leq m$ such that $o \geq \bigwedge_j \{\psi^{\mathbb{B}}(\mathsf{cl}(o_{1,j}), \ldots, \mathsf{cl}(o_{n,j})) \mid \alpha_i^{\mathbb{B}, \theta} \leq o_{i,j} \text{ and } o_{i,j} \in \mathsf{O}(\mathbb{B}) \text{ for } 1 \leq i \leq n\}$. Then

$$o \geq \bigwedge_j \psi^{\mathbb{B}}(\mathsf{cl}(o_{1,j}), \ldots, \mathsf{cl}(o_{n,j}))$$
$$\geq \psi^{\mathbb{B}}(\bigwedge_j \mathsf{cl}(o_{1,j}), \ldots, \bigwedge_j \mathsf{cl}(o_{n,j})) \text{ (Monotonicity of } \psi^{\mathbb{B}})$$
$$\geq \psi^{\mathbb{B}}(\bigwedge_j o_{1,j}, \ldots, \bigwedge_j o_{n,j}), \qquad \text{(Monotonicity of } \psi^{\mathbb{B}} \text{ and cl)}$$

Take $b_i := \bigwedge_j o_{i,j}$, then b_i is a finite meet of open elements, therefore $b_i \in \mathsf{O}(\mathbb{B})$. Since $\alpha_i^{\mathbb{B}, \theta} \leq o_{i,j}$ for $1 \leq i \leq n$ and $1 \leq j \leq m$, it follows that $\alpha_i^{\mathbb{B}, \theta} \leq b_i$.

The lemma above is formulated independently of the specific language. As to the language treated by this paper, in any concrete application of the Ackermann rule (see descriptions in Lemma 8), the inequalities $\alpha \leq p$ have the shape $\blacklozenge^n \mathbf{j} \leq p$, with $\blacklozenge^n \mathbf{j}$ syntactically closed by definition.

Main Result. As is shown in Sect. 8, we have:

Theorem 2 *(Success). ALBA succeeds on 1-Sahlqvist inequalities.*

As discussed in Sect. 3.2, the preservation result follows from Theorem 2 above and the soundness of ALBA with respect to both open assignments and arbitrary assignments:

Theorem 3. *For any 1-Sahlqvist inequality $\varphi \leq \psi$, if $\mathbb{L}_\mathcal{T} \vDash \varphi \leq \psi$, then $\mathbb{B}_{\mathbb{F}_\mathcal{T}} \vDash \varphi \leq \psi$.*

7   Conclusion

In this paper, we give an algorithmic account of the preservation results proved in [1]. The preservation result in the present paper concerns a slight generalization (cf. Remark 1) of the class of inequalities treated in [1] over the same language of positive modal logic. The algorithmic approach adopted here emphasizes the algebraic side of this preservation result and makes it more similar to the way in which Sahlqvist canonicity has been presented in an algebraic way in [26]. In particular, just in the same way in which the embedding map of algebras into their canonical extensions encodes Stone-type dualities, the canonical embedding of modal compact regular frames into the complex algebras of the underlying Kripke frames of their dual spaces encodes Isbell-type dualities. How to optimally characterize this embedding in a way which is aligned with the definition of Jónsson and Tarski [27] is ongoing work. Building on this algebraic perspective, the ALBA approach has unified many different strategies for canonicity, e.g. those of Jónsson [26], Sambin–Vaccaro [32], Ghilardi–Meloni [21], and Venema's pseudo-correspondence [33]. Having extended the algorithmic approach to the Isbell-type dualities paves the way to several different generalizations and extensions: to correspondence results over modal compact Hausdorff spaces, to richer languages such as arbitrary distributive lattice expansions, fixed point expansions of positive modal logics [2], but also to a non-distributive setting, to a constructive meta-theory, to more general syntactic shapes than 1-Sahlqvist (e.g. inductive formulas), and so on.

8   ALBA Succeeds on 1-Sahlqvist Inequalities

In the present section, we sketch the proof of Theorem 2. In the following lemmas, we will track the shape of term inequalities in each stage of execution of ALBA. The proofs of the lemmas are similar to those given in [11, Section 10], therefore we only report on the main line of argument and omit proofs.

Lemma 6. *Let $\varphi \leq \psi$ be a 1-Sahlqvist inequality. After stage 1, it becomes sets of inequalities $\{\mathbf{i}_0 \leq \varphi_i, \psi_i \leq \mathbf{m}_0\}_{i \in I}$ where ψ_i is a formula in the positive language $\mathcal{L}$, and every φ_i is built from $\square^n p$, $\square^n \top$, $\square^n \bot$ by applying $\wedge$ and $\Diamond$.*

Lemma 7. *Let $\{i_0 \leq \varphi_i, \psi_i \leq m_0\}$ as described in Lemma 6. By applying the approximation rule and the splitting rule 2 exhaustively, the system is transformed into one which contains the following types of inequalities:*

- $\psi_i \leq m_0$,
- $j \leq \Diamond k$ *where* j, k *are nominals,*
- $j \leq \Box^n p$,
- $j \leq \beta$ *where* β *is pure, i.e.* β *contains no propositional variables.*

Lemma 8. *Given a system as described in Lemma 7, by applying the residuation rule exhaustively, the system is transformed into one which contains the following types of inequalities:*

- $\psi_i \leq m_0$,
- $j \leq \Diamond k$ *where* j, k *are nominals,*
- $\blacklozenge^n j \leq p$,
- $\beta \leq \gamma$ *where* β, γ *are pure.*

The system described in Lemma 8 is in a shape in which the right-handed Ackermann rule can be applied and all propositional variables can be eliminated. Therefore the algorithm succeeds and we have proven Theorem 2.

References

1. Bezhanishvili, G., Bezhanishvili, N., Harding, J.: Modal compact hausdorff spaces. J. Logic Comput. **25**(1), 1–35 (2015)
2. Bezhanishvili, N., Sourabh, S.: Sahlqvist preservation for topological fixed-point logic. J. Logic Comput. **27**(3), 679–703 (2017)
3. Blackburn, P., de Rijke, M., Venema, Y.: Modal Logic. Cambridge Tracts in Theoretical Computer Science, vol. 53. Cambridge University Press, Cambridge (2001)
4. Blackburn, P., van Benthem, J.F., Wolter, F.: Handbook of Modal Logic, vol. 3. Elsevier, Amsterdam (2006)
5. Britz, C.: Correspondence theory in many-valued modal logics. Master's thesis, University of Johannesburg, South Africa (2016)
6. Conradie, W., Craig, A.: Canonicity results for mu-calculi: an algorithmic approach. J. Logic Comput. **27**(3), 705–748 (2017)
7. Conradie, W., Craig, A., Palmigiano, A., Zhao, Z.: Constructive canonicity for lattice-based fixed point logics. ArXiv preprint arXiv:1603.06547
8. Conradie, W., Fomatati, Y., Palmigiano, A., Sourabh, S.: Algorithmic correspondence for intuitionistic modal mu-calculus. Theoret. Comput. Sci. **564**, 30–62 (2015)
9. Conradie, W., Frittella, S., Palmigiano, A., Piazzai, M., Tzimoulis, A., Wijnberg, N.M.: Categories: how i learned to stop worrying and love two sorts. In: Proceedings of 23rd International Workshop on Logic, Language, Information, and Computation, WoLLIC 2016, Puebla, Mexico, 16–19th August 2016, pp. 145–164 (2016)
10. Conradie, W., Ghilardi, S., Palmigiano, A.: Unified correspondence. In: Baltag, A., Smets, S. (eds.) Johan van Benthem on Logic and Information Dynamics. OCL, vol. 5, pp. 933–975. Springer, Cham (2014). doi:10.1007/978-3-319-06025-5_36

11. Conradie, W., Palmigiano, A.: Algorithmic correspondence and canonicity for distributive modal logic. Ann. Pure Appl. Logic **163**(3), 338–376 (2012)
12. Conradie, W., Palmigiano, A.: Constructive canonicity of inductive inequalities (Submitted). ArXiv preprint arXiv:1603.08341
13. Conradie, W., Palmigiano, A.: Algorithmic correspondence and canonicity for non-distributive logics (Submitted). ArXiv preprint arXiv:1603.08515
14. Conradie, W., Palmigiano, A., Sourabh, S.: Algebraic modal correspondence: Sahlqvist and beyond. J. Logical Algebraic Methods Program. (2016)
15. Conradie, W., Palmigiano, A., Sourabh, S., Zhao, Z.: Canonicity and relativized canonicity via pseudo-correspondence: an application of ALBA (Submitted). ArXiv preprint arXiv:1511.04271
16. Conradie, W., Palmigiano, A., Zhao, Z.: Sahlqvist via translation (Submitted). ArXiv preprint arXiv:1603.08220
17. Conradie, W., Robinson, C.: On Sahlqvist theory for hybrid logic. J. Logic Comput. **27**(3), 867–900 (2017)
18. Dunn, J.M.: Positive modal logic. Stud. Logica. **55**(2), 301–317 (1995)
19. Frittella, S., Palmigiano, A., Santocanale, L.: Dual characterizations for finite lattices via correspondence theory for monotone modal logic. J. Logic Comput. **27**(3), 639–678 (2017)
20. Gehrke, M., Nagahashi, H., Venema, Y.: A Sahlqvist theorem for distributive modal logic. Ann. Pure Appl. Logic **131**(1–3), 65–102 (2005)
21. Ghilardi, S., Meloni, G.: Constructive canonicity in non-classical logics. Ann. Pure Appl. Logic **86**(1), 1–32 (1997)
22. Goldblatt, R.I.: Metamathematics of modal logic. Bull. Aust. Math. Soc. **10**(03), 479–480 (1974)
23. Greco, G., Ma, M., Palmigiano, A., Tzimoulis, A., Zhao, Z.: Unified correspondence as a proof-theoretic tool. J. Logic Comput. (2016) ArXiv preprint arXiv:1603.08204. doi:10.1093/logcom/exw022
24. Isbell, J.R.: Atomless parts of spaces. Math. Scand. **31**, 5–32 (1972)
25. Johnstone, P.T.: Stone Spaces, vol. 3. Cambridge University Press, Cambridge (1986)
26. Jónsson, B.: On the canonicity of Sahlqvist identities. Stud. Logica. **53**, 473–491 (1994)
27. Jónsson, B., Tarski, A.: Boolean algebras with operators. Am. J. Math. **74**, 127–162 (1952)
28. Kracht, M.: Power and weakness of the modal display calculus. In: Wansing, H. (ed.) Proof Theory of Modal Logic, pp. 93–121. Springer, Heidelebrg (1996). doi:10.1007/978-94-017-2798-3_7
29. Ma, M., Zhao, Z.: Unified correspondence and proof theory for strict implication. J. Logic Comput. **27**(3), 921–960 (2017)
30. Palmigiano, A., Sourabh, S., Zhao, Z.: Jónsson-style canonicity for ALBA-inequalities. J. Logic Comput. **27**(3), 817–865 (2017)
31. Palmigiano, A., Sourabh, S., Zhao, Z.: Sahlqvist theory for impossible worlds. J. Logic Comput. **27**(3), 775–816 (2017)
32. Sambin, G., Vaccaro, V.: A new proof of Sahlqvist's theorem on modal definability and completeness. J. Symbolic Logic **54**(3), 992–999 (1989)
33. Venema, Y.: Canonical pseudo-correspondence. Adv. Modal Logic **2**, 421–430 (2001)

Author Index

Printed in the United States
By Bookmasters